Data-Centric Systems and Applications

Intelligent data management is the backbone of all information processing and has hence been one of the core topics in computer science from its very start. This series is intended to offer an international platform for the timely publication of all topics relevant to the development of data-centric systems and applications. All books show a strong practical or application relevance as well as a thorough scientific basis. They are therefore of particular interest to both researchers and professionals wishing to acquire detailed knowledge about concepts of which they need to make intelligent use when designing advanced solutions for their own problems.

Special emphasis is laid upon:

- Scientifically solid and detailed explanations of practically relevant concepts and techniques
  (what does it do)
- Detailed explanations of the practical relevance and importance of concepts and techniques
  (why do we need it)
- Detailed explanation of gaps between theory and practice
  (why it does not work)

According to this focus of the series, submissions of advanced textbooks or books for advanced professional use are encouraged; these should preferably be authored books or monographs, but coherently edited, multi-author books are also envisaged (e.g. for emerging topics). On the other hand, overly technical topics (like physical data access, data compression etc.), latest research results that still need validation through the research community, or mostly product-related information for practitioners ("how to use Oracle 9i efficiently") are not encouraged.

More information about this series at http://www.springer.com/series/5258

David Taniar • Wenny Rahayu

Data Warehousing and Analytics

Fueling the Data Engine

 Springer

David Taniar
Faculty of Information Technology
Monash University
Clayton, VIC, Australia

Wenny Rahayu
School of Engineering and Mathematical
Sciences
La Trobe University
Bundoora, VIC, Australia

ISSN 2197-9723 ISSN 2197-974X (electronic)
Data-Centric Systems and Applications
ISBN 978-3-030-81978-1 ISBN 978-3-030-81979-8 (eBook)
https://doi.org/10.1007/978-3-030-81979-8

This Springer imprint is published by the registered company Springer Nature Switzerland AG.
The registered company address is: Gewerbestrasse 11, 6330 Cham, Switzerland

Foreword

I have known Professors David Taniar and Wenny Rahayu for 15 years as top-class educators and scholars in the field of data management, as well as great friends. In the fall of 2020, they told me that they were nearing completion of a comprehensive book on data warehousing and sought my advice on the title of the book. I suggested that data has become fuel for big data and machine learning and that they may view data warehousing as a key type of fuel for the data engine that drives big data and artificial intelligence (AI). I am honored that they reflected that spur-of-the-moment suggestion in the title of their book "*Data Warehousing and Analytics – Fueling the Data Engine*".

Before coming to Korea and starting what has turned out to be my second career as a university professor, I spent some 25 years engaged in research and development in the field of data management in industry. I worked at IBM Almaden Research Center and MCC (Microelectronics and Computer Technology Corporation) on relational databases and object-oriented databases. Then I launched a startup that created the first object-relational database system and a federated database system. At the same time, I served as Chair of ACM SIGMOD, Chair of ACM SIGKDD, and Editor-in-Chief of ACM Transactions on Database Systems and ACM Transactions on Internet Technology.

Through all these years, I have come to regard data as a "many-splendored" thing for two primary reasons. One is that data has to be looked at from many different perspectives in order to understand its meaning. In this regard, statistics, databases, and AI have pursued the same problem. Another is that managing data entails many difficult practical issues. They include data modeling, the ever-expanding variety of data sources, many data types, data quality, big data, security and privacy, migration, visualization, etc.

From these perspectives, there are three reasons I believe this book *Data Warehousing and Analytics – Fueling the Data Engine* is a noteworthy resource for students and professionals working in the field of data management. First, this is a comprehensive book on data warehousing. It takes the readers on an "end-to-end" journey on all data warehouse modeling concepts and techniques to create and use (i.e., query) data warehouses. In other words, the book deals with many of the difficult issues in understanding the meaning of data and using the data by modeling complex data warehouses, creating (i.e., populating the data warehouses), and extracting the desired information (by issuing SQL queries against the data warehouses).

To be sure, these days people gravitate towards topics such as driverless cars, humanoids, deep learning, chatbots, machine learning analytics, and the like rather than data warehousing, which has a long history. However, the criticality of data warehouses as a business resource has only increased over the years. Furthermore, the new-fangled technologies are, in a sense, data engines fueled by data, including data systematically cleaned and stored and updated in data warehouses.

Second, although this is a comprehensive book, I think it should be easy to read, requiring only an ordinary effort from readers with a reasonable knowledge of relational databases. In other words, the "end-to-end" journey should be a pleasant one for the readers. The authors start with a simple star schema, then bring in hierarchies and bridge tables, then delve into determinant dimensions and multi-fact schemas. They build up the most complex data warehouse incrementally in a methodical and disciplined way.

Third, although in each chapter, the authors introduce each data warehouse modeling concept and its implementation in an easy-to-understand way, they include case studies and exercises at the end of each chapter. I believe this approach helps the readers really understand the materials and prepare them to apply what they have learned. It is often the case that even when people think they understand something, they really do not, and so they cannot apply what they think they have learned. The authors are very experienced, student-oriented educators.

I would also like to invite the readers to embark on a great "end-to-end" journey through the modeling, creation, and use of powerful data warehouses.

Seoul, South Korea   Won Kim
March 2021

(Present)
Distinguished Professor and Vice President (Artificial Intelligence),
Gachon University, Seongnam, South Korea

(Past)
Recipient of "Medal, Order of Service Merit"
from the Government of the Republic of Korea;
Chair, ACM SIGMOD;
Chair, ACM SIGKDD;
Editor-in-Chief, ACM Transactions on Database Systems;
Editor-in-Chief, ACM Transactions on Internet Technology;
Founder and CEO (in the United States), UniSQL and CyberDatabase Solutions.

Preface

Learning by Doing – that's the motto of this book. Each concept is discussed and presented as a case study. The aim is not only to learn the concept as abstract knowledge but to put the knowledge into practise. Each concept is comprehensively applied to one or more case studies, highlighting not only the complexity of the concept but also the depth and breadth of the concept. Many data warehousing concepts are often thought to be simple but turn out to be challenging and complex at both the schema level and implementation level.

For readers (students, practitioners, and instructors) who would like to implement the case studies discussed in this book, the data, database scripts, data warehousing scripts, and SQL codes can be downloaded for free from the following website: http://www.davidtaniar.com/data-warehousing-book/

Instructors, lecturers, and professors who adopt this book as a textbook in their courses can download the complete slides (in PowerPoint format) and figures (in PNG format), so that the slides can be personalised more easily. Additionally, the solutions to all exercises can also be downloaded from the aforementioned website.

Clayton, VIC, Australia   David Taniar
Bundoora, VIC, Australia   Wenny Rahayu
March 2021

Acknowledgements

First and foremost, we would like to thank our respective employers, Monash University and La Trobe University, for giving us the opportunity to undertake this book project. We started writing this book during our sabbatical leave at Professor Takahiro Hara's lab, Osaka University, Japan, in the first half of 2019. If there is one positive thing about the COVID-19 pandemic, it is that the lockdown and work-from-home situation throughout 2020 allowed us to complete this book.

As the book contains a lot of codes, scripts and data, it is very important to make sure that they are free from error. We would like to particularly thank Shuyi Yolanda Sun, a research assistant at La Trobe University, who painstakingly checked and tested the codes, scripts and data, so that they run perfectly. She worked for more than a year on this project. Her tasks were enormous, including creating the data and scripts for some of the case studies, drawing some illustrations and graphs, fixing all the corrections suggested by the proofreader, checking the entire manuscript for the consistency and uniformity of terminologies used throughout the book, making the slides of all the chapters, which can be downloaded by instructors, to name a few.

We owe a debt of gratitude to Dr Kiki Adhinugraha, a lecturer in the Department of Computer Science and Information Technology, La Trobe University, who helped us in many ways, including troubleshooting for all sorts of problems, including LaTeX problems, figures and tables. He helped write and test the complex coding and created the slides for some of the chapters. He also created the website for the book so that supporting materials, such as slides, exercise solutions, data and codes, may be downloaded by qualified readers.

We would also like to thank Michele Mooney, a professional English language proofreader, who proofread the entire manuscript, correcting the grammatical mistakes, questioning phrases that were difficult to understand and suggesting alternatives. Our students who have studied the topic of data warehousing with us in the last 10 years or so have reshaped the contents of our teaching materials, most of which have found their place in this book. Their comments, feedback, questions and discussions over the years have improved the teaching materials which we have presented in this book.

Our special thanks to Professor Won Kim of Gachon University, South Korea, an ACM Fellow and one of the pioneers in database technology, who kindly agreed to write the foreword for this book. He has provided thoughtful comments and feedback on this book, not only the title, but most importantly his vision on the relationship between data sources (e.g. data warehousing) and machine learning, in which we have reflected these in some parts of the first and last chapters of the book.

We would also like to thank the editors-in-chief of the *Data-Centric Systems and Applications* book series, Professors Stefano Ceri and Michael Carey, who believed that this book would be a good addition to the series. They also provided useful comments and feedback to improve the draft manuscript.

Finally, we would like to thank Ralf Gerstner from Springer for his strong support and encouragement. His responses and comments have been comprehensive and deep, showing his extensive experience in book publishing, including data-related books.

Thank you all!

Clayton, VIC, Australia   David Taniar
Bundoora, VIC, Australia   Wenny Rahayu
March 2021

Contents

Chapter 1
Introduction

Data warehouse design and implementation requires a systematic and structured approach. This is the primary focus of this book—a systematic approach to data warehouse design and implementation. There are many perspectives on data warehouse design, such as business process models, semantic approaches, etc., but this book focuses on a different perspective, namely, a database management perspective. Consequently, the book contains not only star schemas but also a lot of SQL codes to explain the steps in data warehouse design and implementation.

There are many ways to learn data warehouse design. Many books discuss the concepts, including definitions and explanations, sometimes followed by several examples. These can be abstract and high level. This book approaches the topic from a different angle, that is, from the case study angle. Each concept in data warehouse design is explained using case studies. Each chapter presents one or more case studies to thoroughly explain the concepts. Hence, it is very practical in both design and implementation.

Each chapter has different levels of difficulty; hence, learning is incremental. Consequently, in the early chapters, it may seem trivial and may be incomplete. In the later chapters, some concepts will be enhanced and revised, and only then will the bigger picture emerge. Hence, it is suggested that you follow the chapters sequentially until the end.

1.1 Operational Databases

In the era of digital information, every company has a database system. This database system is used to operate the daily business activities of the company. This can be sales, finance, booking systems or any transaction events which are pivotal to the business. This database is then called an *operational database* (or also known as a *transactional database*), because the database is used to support the operation

D. Taniar, W. Rahayu, *Data Warehousing and Analytics*, Data-Centric Systems and Applications, https://doi.org/10.1007/978-3-030-81979-8_1

Fig. 1.1 A sample
supermarket transaction
receipt

Coles Supermarkets Australia Pty Ltd
Tax Invoice ABN: 45 004 189 708

coles

```
Store: 522 - CS GLENFERRIE
Store Manager: Alen
Phone:      03 9818 1234
Served By: Assisted Checkout
Register:  115                 Receipt: 3399
Date:        21/08/2019      Time:   14:12

Description                              $

% COLES BETTER BAG 1EACH            0.15
  SUNNY QUEEN 6 PACK E 350GRAM      3.90
  COLES WHITE SUGAR 1KG             1.00
  TRUSS TOMATOES PERKG              3.75
    0.961 kg NET @ $3.90/kg
  BUK CHOY 1BUNCH                   5.00
    2 @ $2.50 EACH
ASIAN VEG 2 FOR $4                 -$1.00

Total for 6 items:                $12.80

EFT                               $12.80
GST INCLUDED IN TOTAL              $0.01

            Coles              VIC AU
21/08/19 14:12      31446030   N522B5
***** 7602                       VISA
CREDIT ACCOUNT            Visa Credit
APSN 0001  ATC 0739  A0000000031010
PURCHASE                  AUD$ 12.80
RRN 001150339900          (00)APPROVED
AUTH 035668
NO PIN OR SIGNATURE REQUIRED

          % = Taxable items
```

of the business. Without an operational database, the automatic processes of the business will be severely limited, which was the case prior to the digital era. With the implementation of an operational database, business processes can be made more automatic. For example, in a supermarket situation where a customer purchases goods at the checkout counter, the transactions are recorded in the operational databases, and this leads to several automatic processes, such as accessing the inventory systems, which then leads to the ordering systems, as well as to the sales systems. Figure 1.1 shows a typical supermarket transaction receipt.

The use of an operational database is centred around transactions. In the transaction systems, database correctness, especially in the concurrency setting, is crucial because the database which is centrally located can be accessed simultaneously. An ATM withdrawal is an example of a transaction in the personal banking system, and providing consistency in the personal banking system (as well as in any other system) is critical. With the maturity of Database Management System (DBMS) technology, transactions will always guarantee database consistency, accuracy and correctness. An operational database is also known as *transactional database*, but in this book, we use the term operational database.

In conclusion, an operational database is needed on a daily basis to support the operation of the business. Imagine booking an airline ticket without a database system, which may easily result in a double booking, for example, or a department store without a database, which may result in the need to process all the sales in batches at the end of the day. There are many other examples. In other words, an operational database is pivotal to the business operation. With an operational database, not only can processes be automated, checking or finding things can be done instantly, such as finding prices at the checkout counter or finding a certain product on an online shopping website.

Operational databases are not only used for transactions and querying but are the main source of information for reporting, which is also central to business operations. Business reports can be used to view business operations, such as reports about goods currently on order or reports about sales activities, etc. Because all transactions are recorded in the operational database, reports on this data can easily be produced.

An operational database has quite a structured design, such as an Entity-Relationship (E/R) diagram. Figure 1.2 shows a sample E/R diagram of a product sales system which is used to record sales transactions. It is also straightforward to implement this design into relational tables using Relational DBMS (RDBMS).

An operational database which mainly focuses on aggregation, summarisation, etc. is suitable for decision-making, even though there are aggregate functions and views in RBDMS. To support more efficient decision-making, the pre-computation of aggregation, summarisation, etc. becomes the main aim. Unfortunately, this is not the primary function of an operational database. Hence, we need a data warehouse to address this need.

1.2 Data Warehouses

To address the drawbacks of an operational database and the need for decision-making support data, a data warehouse is needed. A data warehouse is a multi-dimensional view of databases with aggregates and precomputed summaries. In many ways, it is basically doing aggregates in advance; that is, pre-computation is being done at the design level rather than at the query level.

A simple three-dimensional data warehouse can be viewed as a cube, as illustrated in Fig. 1.3. In this cube, the *Sales$* is viewed from three dimensions: Time, Product and Location. The Sales is basically summary data or facts which can be viewed from any of the three dimensions. Therefore, a data warehouse is good for obtaining summarised data or facts. It provides some capabilities to drill down and roll up along the dimensions. The Sales fact is actually a precomputed aggregate to support efficient data analysis. Therefore, a data warehouse is basically a product to change the view of an operational database to a multi-dimensional view. This multi-dimensional view of a database (or a multi-dimensional database) is a data warehouse.

A data warehouse also supports multi-levels of aggregation in the form of a hierarchy of data warehouses of different granularities. This results in a data

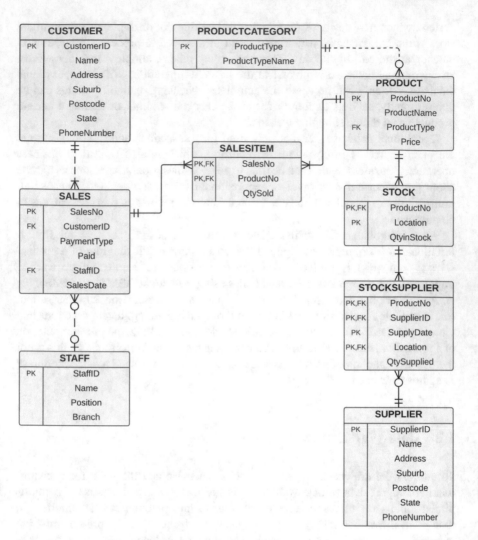

Fig. 1.2 A sample E/R diagram

warehousing architecture. With the availability of data warehouses in different granularities, we can query the data warehouse of different levels, depending on the granularity of the analysis for decision-making. High-level decision-making perhaps only deals with a high-level coarse grain data warehouse, whereas operational decision-making needs to query and analyse data from the highest granularity of a data warehouse. Different levels of data warehousing architecture enable this ability for targeted data analysis, based on the required granularity level. The highest granularity or the most detailed data warehousing is basically a copy of an operational database which is structured in such a way to support data analysis, rather than supporting the daily transaction operations of the business.

Fig. 1.3 A
multi-dimensional view of
data

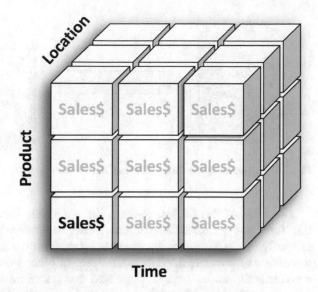

1.3 Building Data Warehouses

There are two important parts in data warehousing: building it and using it. This section discusses how to build data warehouses, whereas the next section discusses how to use data warehouses. A data warehouse is created by transforming an operational database to a data warehouse. The transformation includes a series of steps of data manipulation, such as extracting, cleaning, aggregating, summarising, combining, altering, appending, etc., all of which are called an *Extract-Transform-Load* or an ETL. Figure 1.4 illustrates how an operational database is transformed into a data warehouse.

A data warehouse is basically a multi-dimensional view of the data from the operational database. There are many specialist systems to cater for this multi-dimensional view. Many data warehousing tools have been developed in the last decade, but they come and go.

The most widely used Database Management System (DBMS) to implement operational databases is undoubtedly the Relational DBMS (RDBMS). Hence, it is logical to use the same system to implement data warehouses, which is an RDBMS. The RDBMS has many advantages or features due to its maturity which makes the system robust, stable and backed up by scientific foundation (relational algebra, hence called relational databases). RDBMS is widely used, and most operational databases are implemented in RDBMS. RDBMS is the best choice to implement data warehousing. Implementing data warehousing using a different tool and system is still possible, but RDBMS provides a smoother transition. Therefore, the transformation process transforms RDBMS tables of the operational database into data warehouse tables using SQL.

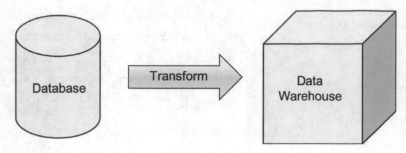

Fig. 1.4 Transforming operational databases to data warehouses

The focus of this book is to learn the systematic transformation approach to transform an operational database into a data warehouse. The implementation will primarily use SQL commands to extract data from the operational database and to create and populate the data warehouse. In other words, a data warehouse is a collection of tables, created, populated and transformed from the tables in the operational database. Therefore, creating a data warehouse is creating tables, and a data warehouse consists of a collection of tables.

1.4   Using Data Warehouses

Once the data warehouse is created, the next step is to use it. Using a data warehouse means to extract data from the data warehouse for further data analysis. The query to extract data from the data warehouse is an *Online Analytical Tool* or OLAP. OLAP is implemented using SQL. Because it uses an SQL command to retrieve data from the data warehouse, the result is a table, and the data is in a relational table format. In short, a data warehouse is a collection of tables, OLAP queries the tables, and the results are also in a table format.

The following is an example of an OLAP to retrieve total sales for all shoes and jeans sales in 2019 and 2020 in Australia. This OLAP uses group by cube which not only gets the total sales as specified in the where clause but also the respective sub-totals, as well as the grand total.

```
select
   T.Year, P.ProductName,
   sum(Total_Sales) as TotalSales
from
   SalesFact S,
   TimeDim T,
   ProductDim P,
   LocationDim L
where S.TimeID = T.TimeID
and S.ProductNo = P.ProductNo
and S.LocationID = L.LocationID
```

```
and T.Year = 2019 and T.Year = 2020
and P.ProductName in ('Shoes', 'Jeans')
and L.Country = 'Australia'
group by cube(T.Year, P.ProductName);
```

The results of the above SQL command are shown in Table 1.1. The blank cells indicate the sub-totals. For example, the line containing 2019 and empty Product Name shows the Total Sales for 2019 (subtotal for 2019), whereas the line containing empty Year followed by Shoes indicates the Total Sales for Shoes (for the year of 2019 and 2020). The last line in the result is the grand total: Total Sales of Shoes and Jeans in both years.

Note that the results only contain the data that satisfies the SQL query, and there is no fancy formatting. The formatting itself is not part of the OLAP. Using SQL commands, OLAP retrieves raw data which can then be later formatted using any Business Intelligence (BI) tools. So, the focus is on the data as the retrieved data is the most important. The BI tool is for further presentation and visualisation.

The data retrieved by the SQL command, as shown in Table 1.1, can be later formatted in a number of ways, depending on the use of the data in the business, as well as the features available in the BI tool. For example, the data can be shown in a matrix-like format, such as in Table 1.2. In this matrix format, the respective sub-totals are more clearly shown.

This can also be shown in various graphs, such as those in Fig. 1.5. The presentation and visualisation is not the focus of OLAP. OLAP only retrieves the required data, that is, the raw data. BI tools which receive this data can present the data in any form: reports, graphs, dashboards, etc. Some BI tools have complex features, whereas others may be simple but adequate for the business. For example, Microsoft Excel is often deemed to be only adequate in presenting some basic graphs, or R also has some visualisation features.

Table 1.1 OLAP raw results

Year	Product name	Total sales
2019	Shoes	1,359,800
2019	Jeans	941,330
2019		2,301,130
2020	Shoes	2,861,450
2020	Jeans	1,827,480
2020		4,688,930
	Shoes	4,221,250
	Jeans	2,768,810
		6,990,060

Table 1.2 Results in a matrix format

	2019	2020	
Shoes	$1,359,800	$2,861,450	$4,221,250
Jeans	$41,330	$1,827,480	$2,768,810
	$2,301,130	$4,688,930	$6,990,060

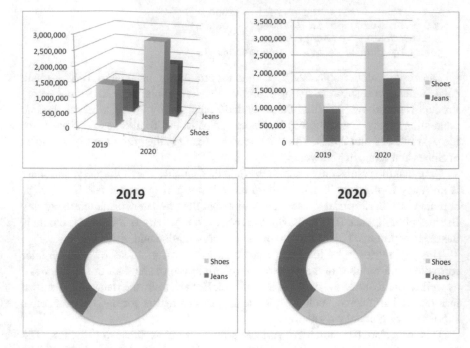

Fig. 1.5 Some graphs produced by a Business Intelligence tool

1.5   The Big Picture

The transformation process from an operational database to a data warehouse is a small part of the big picture where operational data turns into useful and meaningful information in the organisation and each phase in the big picture has a certain role to play in the data journey.

Figure 1.6 shows the entire journey of data in various forms and formats. It starts from an *operational database* (or a *transactional database*), which is the backbone of any information system. The data in the operational database is used to support the daily operations of the business, primarily focusing on transactions. Hence, the primary purpose is to support data integrity, consistency and concurrency. In operational databases, transaction management plays a critical role in guaranteeing data integrity and consistency. Hence, an operational database is often called *OLTP* or *Online Transaction Processing*.

The operational database is then transformed into a *data warehouse* to support systematic data aggregation, which is the foundation of decision support. Various aggregation levels are used in data warehousing, supporting data cubes, which allows managers of an organisation to have understanding of various levels of data aggregates. Transformation from an operational database to a data warehouse is basically changing the structure of the database to support data analysis. Once a

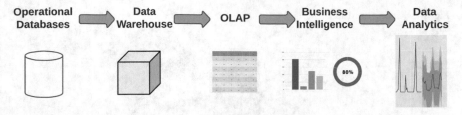

Fig. 1.6   The big picture

data warehouse is built, using the data in the data warehouse is achieved through *OLAP* or *Online Analytical Processing* which is a querying mechanism to obtain the required data from the data warehouse for further analysis. This OLAP query is the foundation of data analytics as it is able to retrieve the required data from the data warehouse which is later retrieved into the data analytics engine. Comparing OLAP and OLTP, OLAP is read-only as it focuses on data retrieval from a data warehouse for further data analysis, whereas OLTP, which is used to support the daily operation of the business, focuses on transactions (e.g. insert, update, delete of transaction data) to maintain data integrity and consistency in the concurrent access environment.

The data retrieved by OLAP is raw data. This raw data needs to be presented in a suitable and attractive format for management to be able to understand various aspects of the organisation. This then becomes *Business Intelligence* which takes the raw data from OLAP and creates various reports, graphs and other data tools for presentation. Using this visual presentation, the data from OLAP can be more appealing and meaningful to the management. This visualisation is often presented as a dashboard in which managers are able to perform some what-if simulations.

The last phase in the big picture is *data analytics*. Data analytics focuses on algorithms to determine the relationship between data to find patterns which otherwise cannot be found and to undertake predictive analysis. The major difference between Business Intelligence and data analytics is that data analytics has predictive capabilities, whereas Business Intelligence mainly focuses on analysing past data to help inform decision-making.

1.6   Fueling the Data Engine

The Economist[1] published an article entitled "The world's most valuable resource is no longer oil, but data" on the 6th of May 2017, highlighting that data is the most valuable resource, and subsequently the term "data is the new oil" was coined. The

---

[1] https://www.economist.com/leaders/2017/05/06/the-worlds-most-valuable-resource-is-no-longer-oil-but-data.

Fig. 1.7 Fueling the data
engine

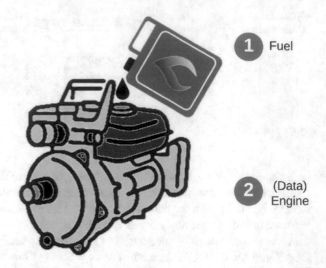

article mentioned the giants that deal with data, such as Google, Amazon, Apple, Facebook and Microsoft. Indeed, data is the largest commodity nowadays.

Oil will only be useful if it is used to fuel an engine, whether it be a vehicle engine, an aircraft engine or a manufacturing engine. In other words, oil fuels the engine. The engine uses fuel and burns it to produce mechanical power. It is designed to covert one form of energy into mechanical energy.

If data is the new oil, then data will only be useful to fuel the data engine, therefore, the term *Fuel for Data Engine*[2] (Fig. 1.7 illustrates for fuel for the data engine). A data engine, like any other engine, produces power that enables the information system to move and operate. The data engine takes the data and processes it to produce useful information and knowledge. This includes reports, decision models, machine learning models, etc. This information and knowledge drives the organisation. The data engine is a data processing software that produces these outcomes. The fuel is the data sources, including operational data sources, data warehouse, data mart, data lake, etc.

The focus of this book is to discuss in depth how data sources, in this case data warehouses, may be used as fuel by data engines to produce reports, dashboards, patterns and other useful information and knowledge for an organisation to assist its decision-making process. The entire process, as discussed in the previous section, is to convert one form of "energy" (e.g. operational databases) to "organisational mechanical energy" (e.g. reports and decision models) through complex processes, covering data transformation, preparation, pre-processing, integration and aggregation, all of which are the central activities of data warehousing, hence the title of this book *Data Warehousing and Analytics: Fueling the Data Engine*.

---

[2] Personal communications with Professor Won Kim who coined the term *Fuel for Data Engine*.

In addition to this book, there are two more books in the Fuel for Data Engine series: (*i*) *Data Lake: Boosting the Data Engine* and (*ii*) *Data Privacy: Protecting the Data Engine*.

1.7 Organisation of the Book

The book is divided into six parts:

   (I) Star Schema
  (II) Snowflake and Bridge Tables
 (III) Advanced Dimension
 (IV) Multi-fact and Multi-input
  (V) Data Warehousing Granularity and Evolution
 (VI) OLAP, Business Intelligence and Data Analytics

Part I—Star Schema describes the foundation of data warehouse design. It consists of two chapters: Chap. 2 introduces the concept of star schema with a particular focus on the transformation process from an operational database to a star schema. Chapter 3 discusses further issues related to the transformation process and star schema creation, including more complex processes in fact measures and dimensions.

Part II—Snowflake and Bridge Tables consists of three chapters. This part expands the concept of a simple star schema by introducing the concept of hierarchy, bridge tables as well as the use of bridge tables in temporal data warehousing. Chapter 4 discusses the concept of hierarchies and the factors for using and not using hierarchies. Chapter 5 introduces the concept of bridge tables and why bridge tables are sometimes necessary in star schemas. Chapter 6 focuses on temporal data warehousing, expanding the use of bridge tables discussed in Chap. 5. Various slowly changing dimensions are covered in the temporal data warehousing chapter.

Part III—Advanced Dimensions consists of four chapters. This part further elaborates various dimension models, namely, Determinant Dimensions (Chap. 7), Junk Dimensions (Chap. 8), Dimension Keys (Chap. 9) and One-Attribute Dimensions (Chap. 10). These dimension models will enrich the semantics of star schema.

Part IV—Multi-fact and Multi-input is divided into three chapters. Multi-fact star schemas are introduced in Chap. 11, where the star schema has multi-fact entities. It discusses when and why a multi-fact is needed in a star schema. A multi-fact can also be created by slicing one fact into multi-facts. This will be discussed in Chap. 12, which covers Vertical and Horizontal Slicing. Chapter 13 introduces the creation of a star schema where an operational database is used as input to the transformation process and consists of multiple operational databases. In conclusion, this part focuses on the multiplicity of the fact entity in the star schema, as well as the operational databases as input to star schema creation.

Part V—Data Warehousing Granularity and Evolution comprises five chapters. Chapter 14 introduces the concept of levels of aggregation in a star schema constellation. These levels of aggregation define the level of granularity of the star schema in the data warehousing architecture. Chapter 15 solely focuses on the lowest-level star schema, including how to design a star schema and why it is needed in data warehousing. Chapter 16 focuses on methods to make the star schema more general or more specific in the data warehousing architecture by adding and removing dimensions. Chapter 17 explains more advanced concepts in data warehousing granularity involving bridge tables. Finally, Chap. 18 introduces the concept of Active Data Warehousing, where the data warehouse is continually updated as the operational databases are being updated. Hence, the data warehouse is active.

Part VI—OLAP, Business Intelligence and Data Analytics is divided into three chapters. Chapter 19 thoroughly explains OLAP—Online Analytical Processing—which is basically a query to the data warehouse. The results obtained by OLAP can then later be plotted in a Business Intelligence reporting tool. Chapter 20 describes two important activities in the data warehousing process, namely, pre-data warehousing and post-data warehousing. Pre-data warehousing includes the exploration of the operational databases with a particular focus on finding and (potentially) removing dirty data. Post-data warehousing covers the exploration of data warehousing, including exploring the extended fact table. The final chapter in this book, Chap. 21, focuses on data analytics, which consists of a suite of methods for data analysis suitable for data warehousing.

1.8 Summary

This chapter introduces the concepts of (i) operational databases (or transactional databases), (ii) data warehouse, (iii) OLAP, (iv) Business Intelligence or BI and (v) data analytics.

The operational databases, which are used to record transactions, are used as inputs to the data warehouse. The transformation process from the operational databases to data warehouses, which may involve data cleaning, filtering, extraction, integration, aggregation, etc., is known as the ETL process. Extracting data from the data warehouse is done by OLAP, which will then present the retrieved data to a BI tool for producing reports and charts. Data analytics uses various data analytics methods to analyse data from the data warehouse. These data analytics methods are specialised methods tailored for data warehouses.

1.9 Exercises

1.1 Using the supermarket checkout example, when you purchase a box of cereal, explain what happens (in a step-by-step manner) to the operational database. This includes an explanation as to what will happen to the tables in the operational database, what piece of information is recorded or updated in the operational database, etc.

1.2 Using the cube example shown in Fig. 1.3, if there are x number of records in the Time Dimension, y records in the Product Dimension and z records in the Location Dimensions, how many Total Sales figures are there in the data warehouse?

1.3 The OLAP retrieves the raw data as shown earlier in Table 1.1. Sort the table based on the Product Name column, and display the results.

1.4 Again, take the results retrieved by OLAP as shown in Table 1.1. Draw a stacked bar chart for this data.

1.5 Refer to the cube example in Fig. 1.3, where the data warehouse has three dimensions, Time, Product and Location, and each small cube represents a Total Sales figure. In this illustration, there are 27 small cubes as there are only three Time instances, three Products and three Locations. Supposing there is a fourth dimension, called Stop Type, which is either supermarkets or convenience stores, what will happen to each of the small cubes?

1.10   Further Readings

Further readings on the operational or transactional databases can be found in many database textbooks. They cover a broad range of topics, from database design, relational theory, SQL, to more advanced transaction management. There are plenty of good textbooks on database systems, such as [1–18] and [19].

There are various good references on data warehousing, covering from star schema design (both logical and physical design), optimisation in both processing and storage, to business adoption of data warehousing: [20–29] and [30].

ETL has been the focus of research for decades. This covers a wide range of areas, such as optimisation, real-time and incremental ETL, semantic ETL and maintenance. Here are some of the references on the topic of ETL: [31–35] and [36].

SQL is the basis for the OLAP implementation. OLAP queries normally use various OLAP functions, such as `cube`, `rollup`, etc., as well as the ranking functions and aggregate functions (e.g. moving aggregates). Most of the database textbooks mentioned above cover SQL in various depth. Specialised resources on SQL are as follows: [37–39] and [40].

In our book, the BI part is mainly used for presenting the OLAP results using graphs, charts and reports. The field of BI actually covers a wide range of topics, such as business requirements, decision-support systems and data integration and analytics. Further readings on BI can be found in the following books: [41–46] and [47].

Data analytics and data mining are matured disciplines which have produced a wealth of techniques and methods. There are excellent textbooks on data mining for further reading in this area: [48–52] and [53].

References

1. C. Coronel, S. Morris, *Database Systems: Design, Implementation, & Management* (Cengage Learning, Boston, 2018)
2. T. Connolly, C. Begg, *Database Systems: A Practical Approach to Design, Implementation, and Management* (Pearson Education, London, 2015)
3. J.A. Hoffer, F.R. McFadden, M.B. Prescott, *Modern Database Management* (Prentice Hall, Englewood Cliffs, 2002)
4. A. Silberschatz, H.F. Korth, S. Sudarshan, *Database System Concepts*, 7th edn. (McGraw-Hill, New York, 2020)
5. R. Ramakrishnan, J. Gehrke, *Database Management Systems*, 3rd edn. (McGraw-Hill, New York, 2003)
6. J.D. Ullman, J. Widom, *A First Course in Database Systems*, 2nd edn. (Prentice Hall, Englewood Cliffs, 2002)
7. H. Garcia-Molina, J.D. Ullman, J. Widom, *Database Systems: The Complete Book* (Pearson Education, London, 2011)
8. P.E. O'Neil, E.J. O'Neil, *Database: Principles, Programming, and Performance*, 2nd edn. (Morgan Kaufmann, Los Altos, 2000)
9. R. Elmasri, S.B. Navathe, *Fundamentals of Database Systems*, 3rd edn. (Addison-Wesley-Longman, Reading, 2000)
10. C.J. Date, *An Introduction to Database Systems*, 7th edn. (Addison-Wesley-Longman, Reading, 2000)
11. R.T. Watson, *Data Management—Databases and Organizations*, 5th edn. (Wiley, London, 2006)
12. P. Beynon-Davies, *Database Systems* (Springer, Berlin, 2004)
13. M. Kifer, A.J. Bernstein, P.M. Lewis, *Database Systems: An Application-Oriented Approach* (Pearson/Addison-Wesley, 2006)
14. W. Lemahieu, S. vanden Broucke, *Principles of Database Management: The Practical Guide to Storing, Managing and Analyzing Big and Small Data* (Cambridge University Press, Cambridge, 2018)
15. S. Bagui, R. Earp, *Database Design Using Entity-Relationship Diagrams*. Foundations of Database Design (CRC Press, Boca Raton, 2003)
16. M.J. Hernandez, *Database Design for Mere Mortals: A Hands-On Guide to Relational Database Design*. For Mere Mortals (Pearson Education, London, 2013)
17. N.S. Umanath, R.W. Scamell, *Data Modeling and Database Design* (Cengage Learning, Boston, 2014)
18. T.J. Teorey, S.S. Lightstone, T. Nadeau, H.V. Jagadish, *Database Modeling and Design: Logical Design*. The Morgan Kaufmann Series in Data Management Systems (Elsevier, Amsterdam, 2011)
19. G. Simsion, G. Witt, *Data Modeling Essentials*. The Morgan Kaufmann Series in Data Management Systems (Elsevier, Amsterdam, 2004)
20. C. Adamson, *Star Schema The Complete Reference* (McGraw-Hill Osborne Media, 2010)
21. R. Laberge, *The Data Warehouse Mentor: Practical Data Warehouse and Business Intelligence Insights* (McGraw-Hill, New York, 2011)
22. M. Golfarelli, S. Rizzi, *Data Warehouse Design: Modern Principles and Methodologies* (McGraw-Hill, New York, 2009)

23. C. Adamson, *Mastering Data Warehouse Aggregates: Solutions for Star Schema Performance* (Wiley, London, 2012)
24. P. Ponniah, *Data Warehousing Fundamentals for IT Professionals* (Wiley, London, 2011)
25. R. Kimball, M. Ross, *The Data Warehouse Toolkit: The Definitive Guide to Dimensional Modeling* (Wiley, London, 2013)
26. R. Kimball, M. Ross, W. Thornthwaite, J. Mundy, B. Becker, *The Data Warehouse Lifecycle Toolkit* (Wiley, London, 2011)
27. W.H. Inmon, *Building the Data Warehouse* ITPro Collection (Wiley, London, 2005)
28. M. Jarke, *Fundamentals of Data Warehouses*, 2nd edn. (Springer, Berlin, 2003)
29. E. Malinowski, E. Zimányi, *Advanced Data Warehouse Design: From Conventional to Spatial and Temporal Applications*. Data-Centric Systems and Applications (Springer, Berlin, Heidelberg, 2008)
30. A. Vaisman, E. Zimányi, *Data Warehouse Systems: Design and Implementation* Data-Centric Systems and Applications (Springer, Berlin-Heidelberg, 2014)
31. X. Liu, N. Iftikhar, H. Huo, P.S. Nielsen, Optimizing ETL by a two-level data staging method. Int. J. Data Warehous. Min. **12**(3), 32–50 (2016)
32. A. Nabli, S. Bouaziz, R. Yangui, F. Gargouri, Two-etl phases for data warehouse creation: design and implementation, in *Advances in Databases and Information Systems—19th East European Conference, ADBIS 2015, Poitiers, France, September 8–11, 2015, Proceedings*, ed. by T. Morzy, P. Valduriez, L. Bellatreche. Lecture Notes in Computer Science, vol. 9282 (Springer, Berlin, 2015), pp. 138–150
33. W. Qu, V. Basavaraj, S. Shankar, S. Dessloch, Real-time snapshot maintenance with incremental ETL pipelines in data warehouses, in *Big Data Analytics and Knowledge Discovery—17th International Conference, DaWaK 2015, Valencia, Spain, September 1-4, 2015, Proceedings*, ed. by S. Madria and T. Hara. Lecture Notes in Computer Science, vol. 9263 (Springer, Berlin, 2015), pp. 217–228
34. A. Simitsis, P. Vassiliadis, T.K. Sellis, Optimizing ETL processes in data warehouses, in *Proceedings of the 21st International Conference on Data Engineering, ICDE 2005, 5–8 April 2005, Tokyo, Japan*, ed. by K. Aberer, M.J. Franklin, S. Nishio (IEEE Computer Society, Silver Spring, 2005), pp. 564–575
35. Z. El Akkaoui, E. Zimányi, J-N. Mazón, J. Trujillo, A BPMN-based design and maintenance framework for ETL processes. Int. J. Data Warehouse. Min. **9**(3), 46–72 (2013)
36. L. Bellatreche, S. Khouri, N. Berkani, Semantic data warehouse design: from ETL to deployment à la carte, in *Database Systems for Advanced Applications, 18th International Conference, DASFAA 2013, Wuhan, China, April 22-25, 2013. Proceedings, Part II*, ed. by W. Meng, L. Feng, S. Bressan, W. Winiwarter, W. Song. Lecture Notes in Computer Science (Springer, Berlin, 2013), pp. 64–83
37. J. Melton, *Understanding the New SQL: A Complete Guide*, vol. I, 2nd edn. (Morgan Kaufmann, Los Altos, 2000)
38. C.J. Date, *SQL and Relational Theory - How to Write Accurate SQL Code*, 2nd edn. Theory in Practice (O'Reilly, 2012)
39. A. Beaulieu, *Learning SQL: Master SQL Fundamentals* (O'Reilly Media, 2009)
40. M.J. Donahoo, G.D. Speegle, *SQL: Practical Guide for Developers*. The Practical Guides. (Elsevier, Amsterdam, 2010)
41. D. Loshin, *Business Intelligence: The Savvy Manager's Guide*. The Morgan Kaufmann Series on Business Intelligence (Elsevier, Amsterdam, 2012)
42. B. Brijs, *Business Analysis for Business Intelligence* (CRC Press, Boca Raton, 2016)
43. R. Sherman, *Business Intelligence Guidebook: From Data Integration to Analytics* (Elsevier, Boca Raton, 2014)
44. L.T. Moss, S. Atre, *Business Intelligence Roadmap: The Complete Project Lifecycle for Decision-support Applications*. Addison-Wesley Information Technology Series (Addison-Wesley, Reading, 2003)
45. C. Vercellis, *Business Intelligence: Data Mining and Optimization for Decision Making*. (Wiley, London, 2011)

46. L. Bulusu, *Open Source Data Warehousing and Business Intelligence* (CRC Press, Boca Raton, 2012)
47. W. Grossmann, S. Rinderle-Ma, *Fundamentals of Business Intelligence*. Data-Centric Systems and Applications (Springer, Berlin-Heidelberg, 2015)
48. I.H. Witten, E. Frank, M.A. Hall, C.J. Pal, *Data Mining: Practical Machine Learning Tools and Techniques*. The Morgan Kaufmann Series in Data Management Systems (Elsevier, Amsterdam, 2016)
49. J. Han, J. Pei, M. Kamber, *Data Mining: Concepts and Techniques*. The Morgan Kaufmann Series in Data Management Systems (Elsevier, Amsterdam, 2011)
50. N. Ye, *Data Mining: Theories, Algorithms, and Examples*. Human Factors and Ergonomics. (CRC Press, Boca Raton, 2013)
51. X. Wu, V. Kumar, *The Top Ten Algorithms in Data Mining*. Chapman & Hall/CRC Data Mining and Knowledge Discovery Series (CRC Press, Boca Raton, 2009)
52. D.J. Hand, H. Mannila, P. Smyth, *Principles of Data Mining*. A Bradford Book (Bradford Book, 2001)
53. K.J. Cios, W. Pedrycz, R.W. Swiniarski, L.A. Kurgan, *Data Mining: A Knowledge Discovery Approach* (Springer, New York, 2007)

Part I
Star Schema

Chapter 2
Simple Star Schemas

This chapter introduces basic data warehouse modelling using *star schemas*. First of all, the notations and processes will be introduced. The best way to learn about data warehouse design and modelling is through case studies. This chapter introduces two simple case studies, focusing on star schema design and implementation through a transformation process from an operational database to a data warehouse. Finally, a simple yet straightforward methodology to check the validity of a star schema, called the *two-column table methodology* will be introduced.

2.1 Notations and Processes

This section describes the basics of data warehouse design notation and processes, including star schema notations, E/R diagram notation and the transformation process from an operational database to a data warehouse.

2.1.1 Star Schema Notation

A data warehouse is basically a multi-dimensional view of the database. A cube, as shown in the previous chapter, is a practical example of a multi-dimensional view of data. Figure 2.1 shows a cube with three dimensions (e.g. the x-axis is *Time*, the y-axis is *Product* and the z-axis is *Location*), where each cell, which is an intersection of these three dimensions, represents a *Sales$*.

This three-dimensional view of data is easily visualised and understood. However, if there are more than three dimensions, it would be more difficult to imagine. For example, it would be quite challenging to imagine how to visualise four dimensions, let alone ten dimensions. However, in real life, a data warehouse

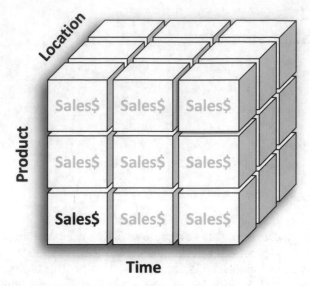

Fig. 2.1 A three-dimensional view of Sales as illustrated in a cube

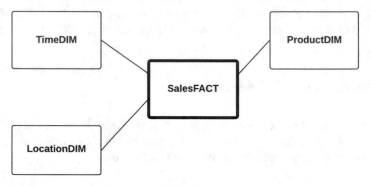

Fig. 2.2 A Sales star schema

has many dimensions (e.g. definitely more than three dimensions in most cases); therefore, there must be a systematic way to represent a multi-dimensional view of data, rather than simply using cubes or a higher degree of cubes, which are more a geometrical representation, rather than a design representation.

This is where *star schema* exists. A star schema is a design representation of a multi-dimensional view. The star schema equivalent to the cube example above is shown in Fig. 2.2. A star schema consists of *Fact*, in the middle, surrounded by *Dimensions* which are connected to the Fact. In the Sales star schema, the three dimensions are Time, Product and Location dimensions. There is no particular order for the dimensions; they are merely a list of dimensions from where we can see the Fact. The Fact in this star schema is Sales. Notation-wise, the Fact uses a bolder line to differentiate between Facts and Dimensions.

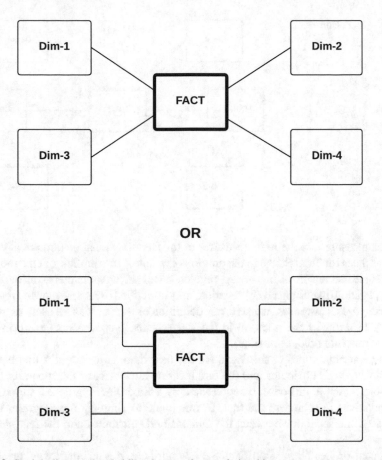

Fig. 2.3 Straight lines vs. bended lines in star schema relationships

The lines that represent a relationship between the fact and dimensions can be straight lines or bended lines (see Fig. 2.3). In this book, we use a straight line notation.

Using the star schema notation, the number of dimensions is unlimited. If there are more dimensions, then we simply add more dimensions linked to the Fact. Figure 2.4 shows that the star schema has six dimensions. It is obvious why a star schema is thus named as it looks like a star. This is especially clear when there are a lot of dimensions connected to the fact.

Similar to entities in the E/R diagram, the fact or dimensions in a star schema also have some details describing them, known as attributes. The fact has DimensionIDs from each of the dimensions. Each DimensionID attribute in the fact is considered to be *Foreign Keys* to the dimension. The DimensionIDs in the fact are printed in bold and italic (e.g. bold indicates a Primary Key, and italics indicates a Foreign Key). Underneath the DimensionID attribute is the *fact measure* attributes.

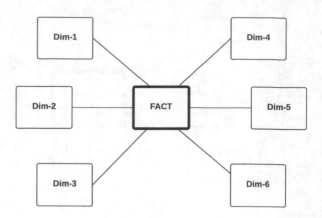

Fig. 2.4 More Dimensions

Fact measures are the main measures of the fact; they must be numerical values, such as Sales (or Total Sales) in the previous example. For simplicity, at the moment, we assume that fact measures are aggregated values (in latter chapters, various levels of aggregation, including no aggregation, may be applied to fact measures. However, for pedagogical purposes, we start our definition of a fact measure as an aggregate value). In terms of the notation in the star schema, fact measures are printed in normal font (not bold, not italics).

There are two types of notations for the fact: there is a separator line between the DimensionID attributes and the fact measure attributes, or the separator line is invisible, which in this case, is represented by a blank line. Figure 2.5 shows these two notations to represent the fact. In this book, to simplify the diagram, we do not use a line separator between the DimensionID attributes and the fact measure attributes.

Dimension has DimensionID as the key attribute (e.g. Primary Key, printed in bold) and potentially additional attributes to describe the dimension. Figure 2.5 shows the fact and dimension.

The Sales star schema shown previously in Fig. 2.2 is an outline star schema as it only shows the fact and dimensions and their relationships; it does not contain the details, such as key attributes, fact measures, etc. The complete Sales star schema, containing the key attributes and fact measures, as well as the cardinality relationship between the dimension and fact, is shown in Fig. 2.6.

In the complete Sales star schema, each of the dimensions has a key attribute printed in bold. Some of the dimensions also have extra attributes, such as Month and Year attributes in the Time Dimension. The Fact has three DimensionIDs, namely, TimeID, LocationID and ProductID. The fact measure in the fact is Total Sales. Also note that the cardinality relationship between the dimensions and the fact is 1-m, indicated by a crow's foot at the fact side. The cardinality relationship is not shown in the outline star schema previously shown in Fig. 2.2.

In this book, sometimes, when a particular dimension is being discussed, the dimension is highlighted in the star schema. This is to assist readers to easily identify the discussed dimension. For example, in the star schema shown in Fig. 2.7, one

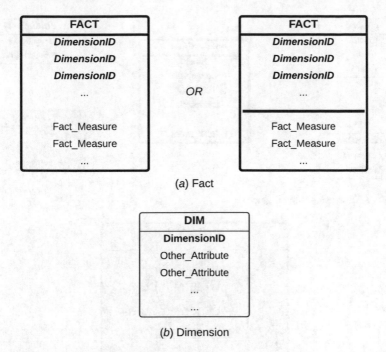

(a) Fact

(b) Dimension

Fig. 2.5 Attributes of Fact and Dimension

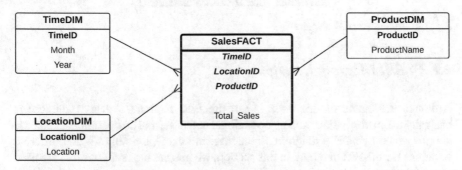

Fig. 2.6 Sales star schema complete with the attributes

more dimension is added to the Sales star schema, namely, the Customer Dimension. Hence, this dimension is highlighted in the star schema, so that the readers are able to focus on this dimension, as this is the dimension which is being discussed. This is the only reason the dimension (or the fact) is highlighted in the star schema. The intention is not to change the semantics of the highlighted dimension or fact.

Other types of dimensions (e.g. determinant dimension) and relationships (e.g. hierarchies and multi-facts) will be introduced later in the subsequent chapters.

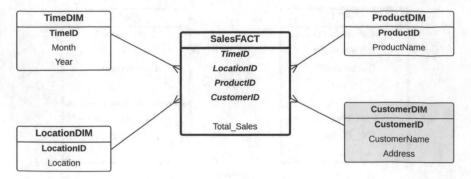

Fig. 2.7 A Dimension is highlighted

ENTITY	
PK	PrimaryKey
	OtherAttribute
	OtherAttribute
	...
FK	ForeignKey

Fig. 2.8 An Entity in E/R diagram

2.1.2 E/R Diagram Notation

Although the theme of this book is not the E/R diagram, a data warehouse is basically a transformation from an operational database, and an operational database is represented by an E/R diagram. Since there are many, possibly slightly different notations for the E/R diagram, in this section, we present the notations that are used as the standard in this book.

Figure 2.8 shows the notation for an Entity. The Entity name is capitalised. The Primary Key is denoted by PK, and the Foreign Key is denoted by FK. The other attributes are shown in a standard font. The entity rectangle has a light-yellow colour as the background.

The relationship notation uses a crow's foot notation. The cardinality of the relationship (e.g. 1-1 and 1-*many*) is shown together with the participation (e.g. mandatory or optional participation). Figure 2.9 shows the cardinality and participation of a relationship. Basically, the cardinality, combined with participation, indicates the min-max of the relationship.

The *many-many* relationship is broken down into 1-*many* and *many*-1 relationships. This is called an *associative relationship* (see Fig. 2.10). The middle entity has

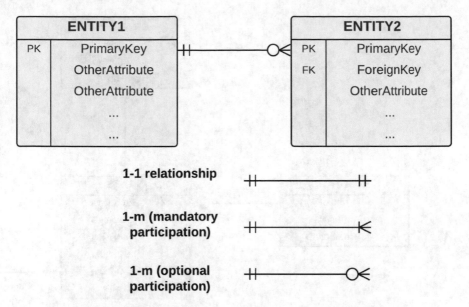

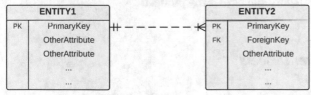

Fig. 2.9 A relationship in E/R diagram

Associative Relationship (m-m)

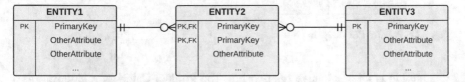

Non-Associative Relationship (1-m)

Fig. 2.10 Associative and non-associative relationship

composite Primary Keys, consisting of the PKs from both sides. A non-associative relationship is where both entities do not share the same PKs. The line representing the relationship is a dotted line (see Fig. 2.10).

Figure 2.11 shows a simple, yet complete E/R diagram for the Sales operational database. The Product-Branch relationship is actually a *many-many* relationship, through the Sales entity, which is why the PK of the Sales entity is a composite key (e.g. ProductNo from Product and BranchID from Branch). Hence, the

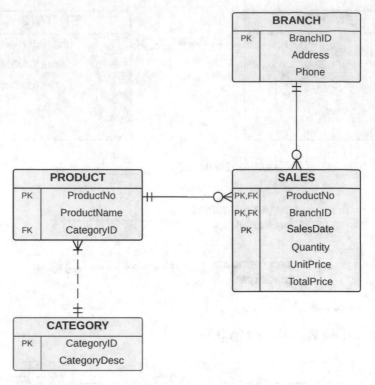

Fig. 2.11 A Sales E/R diagram

relationship is an associative relationship. The Product-Category relationship is a non-associative relationship with a *many*-1 cardinality, where one product belongs to one category, but one category can have many products.

2.1.3   Transformation Process

The process to build a data warehouse is shown in Fig. 2.12. A data warehouse is built by transforming the input operational database. This transformation is a systematic process, which is the focus of this book. The input is one or more operational databases. In the first few chapters of this book, the focus is on using a single operational database as input, and following a systematic transformation process, a data warehouse, represented as a star schema, is then built. In later chapters, case studies showing multiple operational databases are used as the input to the transformation process, and a data warehouse is then created.

   The data warehouse itself can be a simple star schema, as shown in this example, but it can be a complex star schema, with various kinds of dimensions, such as

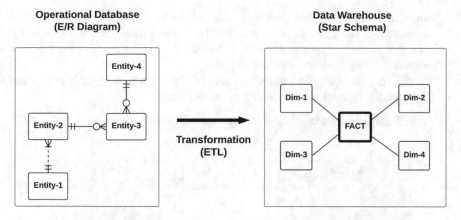

Fig. 2.12 A transformation from an operational database to a data warehouse

hierarchies, determinant, etc., as well as multiple facts, as opposed to single facts and multiple star schemas forming a complex data warehousing architecture.

The main focus of the transformation process is basically an ETL (extract-transform-load), which includes extracting data from the operational databases, transforming the data and then loading it into the new data warehouse. The process consists of a series of SQL commands, which access the tables from the operational databases and perform manipulation, cleaning, transformation and aggregation, which is ingested into the data warehouse.

2.2   First Case Study: A College Star Schema

The easiest way to learn star schema modelling and design is through examples. Hence, the learning method used throughout this book is a case study-based method. The case study will be presented progressively by introducing each concept as it is explained. This section presents the first case study, which is very simple and elementary. The aim is to introduce how to implement a systematic approach to build a star schema from an operational database.

This case study involves an International College. The admissions office handles enrolment, payment and marketing campaigns to international students, often through educational agents located overseas. This admissions office has an operational system that maintains all the details of international students enrolled in the College. Payment details are also handled by this office. The operational system has the following features:

- All student details are kept in the database. This includes the courses in which the students have enrolled.
- As the College is a multi-campus university, some courses are offered at different campuses. The admissions office handles international students on all campuses.

- Some international students coming to the College are handled by an educational
  agent. This is particularly common for the first course in which a student enrols.
  Subsequent courses are not normally handled by an agent because the students
  themselves deal directly with the College.
- International students pay tuition fees several times per year, usually once each
  semester, for each course in which they are enrolled.

An E/R diagram showing the current operational system is presented in Fig. 2.13.

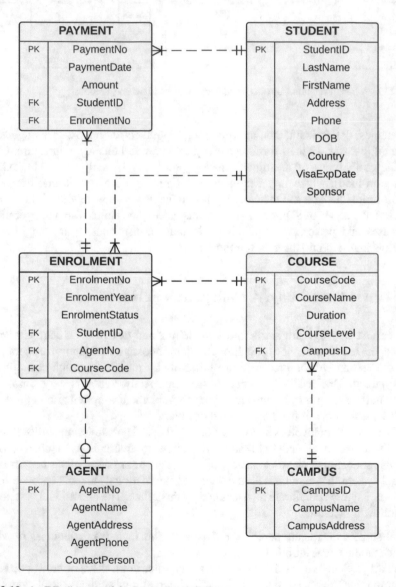

Fig. 2.13 An E/R diagram of the International College

The operational database that maintains the above system has the following tables. Note that the Primary Keys (PK) are underlined and the Foreign Keys (FK) are printed in italic. All the relationships in this E/R diagram are non-associative relationships as the entities do not share any common PK.

- **Student** (StudentID, LastName, FirstName, Address, Phone, DOB, Country, VisaExpDate, Sponsor)
- **Campus** (CampusID, CampusName, CampusAddress)
- **Course** (CourseCode, CourseName, Duration, CourseLevel, *CampusID*)
- **Agent** (AgentNo, AgentName, AgentAddress, AgentPhone, ContactPerson)
- **Enrolment** (EnrolmentNo, EnrolmentYear, EnrolmentStatus, *StudentID, AgentNo, CourseCode*)
- **Payment** (PaymentNo, PaymentDate, Amount, *StudentID, EnrolmentNo*)

The College now requires a data warehouse for analysis purposes. The analysis is needed to answer at least the following questions:

- How many students come from certain countries?
- What is the total income for certain postgraduate courses?
- How many students are handled by certain agents?
- How do enrolment numbers in particular courses fluctuates across the year?

The first question could be used by management to identify countries that may be targeted for future international marketing campaigns. The second question could be used by the financial office for further planning. The third question could be used in conjunction with future international marketing campaigns.

From the above requirements, it is clear that there are two fact measures needed in the star schema. The first one is *Number of Students*, from the "How many students ...?" question. Number of Students is an aggregated numerical value; hence, it satisfies the two requirements of a fact measure. The second fact measure is *Total Income*, from the "What is the total income...?" question.

The dimensions are the perspectives from which the fact measures are viewed. So, "How many students come from certain countries", for example, is to look at the Number of Students from a country point of view; hence *Country* is a dimension. From the other questions, we can deduce other dimensions, such as Course, Agent and Year. As a result, the star schema has four dimensions (e.g. Country, Course, Agent and Year) and two fact measures (e.g. Number of Students and Total Income). The star schema is shown in Fig. 2.14.

For simplicity, some of the dimensions have only one attribute, which is the DimensionID attribute. For example, the Country dimension has only the Country Name attribute and the Year dimension has only the Enrolment Year attribute. In a real case, there will be other attributes in these dimensions. But for this first case study, the scope of the attributes is limited to the information in the tables in the operational database.

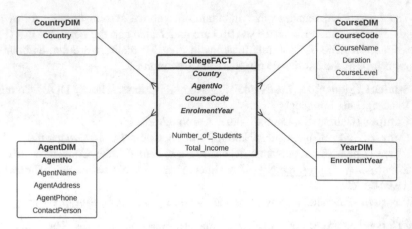

Fig. 2.14 A star schema for the College case study

Once the star schema is drawn, it is time to implement it. The operational database is already implemented in a Database Management System (DBMS). This means that the tables and their records are given as an input to the transformation process. Hence, the data warehouse (or the star schema) will be implemented in a DBMS with tables and records.

The first step in the transformation from an operational database to a data warehouse is to create the dimension tables. Each dimension is basically a new table in the data warehouse. As previously mentioned, there are four dimensions in this star schema; hence, we will need to create four tables. Some of the dimensions are an exact copy of the operational database. Therefore, a direct copy of this table from the operational database to the data warehouse is straightforward and convenient. For example, the Agent dimension is an exact copy of the Agent table in the operational database. The SQL command to copy the Agent table from the operational database is as follows. It uses the `create table as select` command to create the table as well as to copy the records.

```
create table AgentDim as
select * from Agent;
```

The second method to create a dimension table is by selecting certain attributes from the operational database. For example, the Country dimension table only needs the Country attribute from the Student table. The SQL commands to create the three other dimensions (e.g. Country, Course and Semester dimension tables) are as follows:

```
create table CountryDim as
select distinct Country
from Student;
```

```
create table CourseDim as
select CourseCode, CourseName, Duration, CourseLevel
from Course;

create table YearDim as
select distinct EnrolmentYear
from Enrolment;
```

Once all the dimension tables are created, it is now time to create the Fact Table. So, the fact in the star schema is a table in the data warehouse. Firstly, we need to select the attributes from the tables in the operational database. These are Country, AgentNo, CourseCode and EnrolmentYear. The attributes are from the Student and Enrolment tables. The next step is to use the aggregate functions to create the fact measures. In this case, the count and sum functions are used to count the number of students and to calculate the total payment amount. The third step is to join the mentioned tables. Because the amount is from the Payment table, hence, we need to join three tables: Student, Enrolment and Payment. Finally, the fourth step in the SQL is to use the group by clause. The complete SQL command to create the Fact Table is as follows:

```
create table CollegeFact as
select
  S.Country,
  E.AgentNo,
  E.CourseCode,
  E.EnrolmentYear,
  count(S.StudentID) as Number_of_Students,
  sum(P.Amount) as Total_Income
from Student S, Enrolment E, Payment P
where E.EnrolmentNo = P.EnrolmentNo
and E.StudentID = S.StudentID
group by
  S.Country,
  E.AgentNo,
  E.CourseCode,
  E.EnrolmentYear;
```

Our first data warehouse consisting of four dimension tables and one Fact Table has been created and populated. Management is now able to query these tables to answer the questions.

2.3 Another Simple Case Study: A Sales Star Schema

The second case study is similar to the first case study, in terms of the star schema design, but the transformation processes are slightly more complex. The E/R diagram of the second case study is shown in Fig. 2.15. It is a very simplistic Sales operational system with four entities: Product, Sales, Branch and Category. The Product-Sales relationship is a 1-*many* relationship, and the Sales-Branch

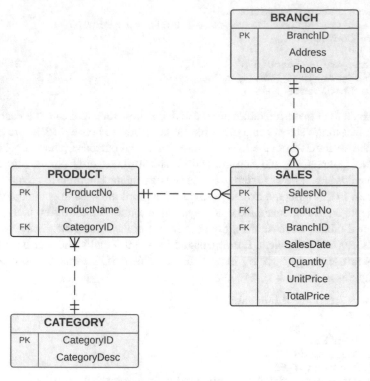

Fig. 2.15 An E/R diagram of the Sales case study

relationship is a *many*-1 relationship. Literally, Product-Branch can be seen as a *many-many* relationship. However, because Sales has its own PK (e.g. SalesNo), the relationships with Branch and Product become non-associative relationships. The Product-Category relationship is a *many*-1 relationship.

Suppose we would like to analyse Total Sales from various points of view, such as Quarter, Branch and Product Category. In other words, we would like to retrieve, for example, the total sales for the first quarter this year, or of a particular branch and product category. Hence, it is obvious that the fact measure is Total Sales and the dimensions are Quarter (or Time), Branch and Product Category. The star schema of the Sales case study is shown in Fig. 2.16, which is slightly different from the star schema presented earlier in Fig. 2.6. In the star schema in Fig. 2.16, we use the Branch Dimension instead of the Location Dimension. For the Product, we use the Product Category Dimension instead of the Product Dimension. For the Time Dimension, we focus on Quarter instead of Month/Year.

The second case study star schema design is quite similar to the first case study. The fact measure is obvious, and the dimensions also make sense. But the SQL processes to create the second data warehouse are slightly more complex than those of the first data warehouse.

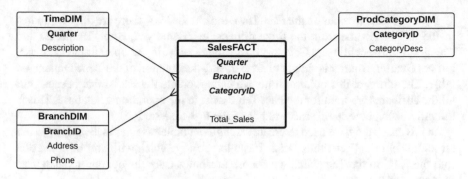

Fig. 2.16 A star schema for the Sales case study

The first step in creating a data warehouse (or implementing a star schema) is to create the dimension tables. To create the Product Category and Branch Dimension tables, we can directly copy from the respective tables in the operational database. The SQL command used is basically a `create table as select * ` command shown as follows:

```
create table ProdCategoryDim as select * from Category;
create table BranchDim as select * from Branch;
```

However, for the Time Dimension table, we need to create the table manually, because we cannot immediately find the two attributes needed by the table, namely, the Quarter and Description attributes. There is a Sales Date attribute in the Sales table in the operational database, but not Quarter. To create a table manually, it is a simple `create table` command is used. Because the table is created manually, a manual insertion must be done through the `insert into` command. Because there are only four quarters, inserting four records into the TimeDim table is not difficult.

```
create table TimeDim
(Quarter number(1),
  Description varchar2(20));

insert into TimeDim values (1, 'Jan-Mar');
insert into TimeDim values (2, 'Apr-Jun');
insert into TimeDim values (3, 'Jul-Sep');
insert into TimeDim values (4, 'Oct-Dec');
```

In summary, there are three ways to create dimension tables:

(a) Use `create table as select *` which is direct copying from the table in the operational database,
(b) Choose selected attributes from the table in the operational database, or
(c) Create the dimension table manually, followed by `insert into` to insert new records into the table.

Now it is time to focus on the Fact. If we look at the Fact, there are three attributes in the fact that come from the three dimensions, namely, Quarter, BranchID and CategoryID. BranchID and CategoryID attributes exist in the operational database, but not Quarter. Quarter is simply created during the creation of the Time Dimension table. Therefore, at this point in time, we cannot create the Fact Table because not all the attributes we want for the Fact Table exist in the operational database. This is the difference between this case study and the first case study.

In this case study, we need to create a *TempFact* table, which is a *Temporary Fact* table. It is not the Fact Table, rather it is a temporary, which will later be converted into the Fact. In the TempFact, we choose attributes from the operational database that we need later for the Fact.

The Fact needs Quarter, but the operational database does not have Quarter—it has the Sales Date attribute. So, the Sales Date attribute is selected. The Fact needs BranchID and CategoryID, which exist in the operational database so we can choose these for the TempFact. Finally, the Fact needs Total Sales, but the operational database has the Total Price attribute; hence, we choose Total Price for the TempFact. The SQL command to create the TempFact Table is as follows. For simplicity, we only use the 2020 Sales records. Note that in the TempFact, there is no aggregation function and no `group by`; it is simply a join query among tables that provide the required attributes (e.g. SalesDate, BranchID, CategoryID and TotalPrice).

```
create table TempFact as
select
  S.SalesDate,
  B.BranchID,
  C.CategoryID,
  S.TotalPrice
from Branch B, Sales S, Product P, Category C
where B.BranchID = S.BranchID
and S.ProductNo = P.ProductNo
and P.CategoryID = C.CategoryID
and to_char(S.SalesDate, 'YYYY') = '2020';
```

Because the Fact requires Quarter but the TempFact has SalesDate, we need to *convert* SalesDate to Quarter. This is done by adding a new attribute called Quarter in TempFact. With this new empty column in TempFact, we need an `update` command to fill in the appropriate quarter number (e.g. 1, 2, 3, or 4). At the end of this process, we have TempFact with five attributes: SalesDate, BranchID, CategoryID, TotalPrice and Quarter.

```
alter table TempFact
add (Quarter number(1));

update TempFact
set Quarter = 1
where to_char(SalesDate, 'MM') >= '01'
and to_char(SalesDate, 'MM') <= '03';
```

```
update TempFact
set Quarter = 2
where to_char(SalesDate, 'MM') >= '04'
and to_char(SalesDate, 'MM') <= '06';

update TempFact
set Quarter = 3
where to_char(SalesDate, 'MM') >= '07'
and to_char(SalesDate, 'MM') <= '09';

update TempFact
set Quarter = 4
where Quarter is null;
```

The TempFact table has all the information we need for the final Fact Table. To create the final Fact Table, we simply pick and choose the attributes from the TempFact Table, namely, Quarter, BranchID and CategoryID. The TempFact Table provides Total Price, so the final Fact Table can simply aggregate the Total Price to get the Total Sales. The SQL command to create the final Fact Table is as follows:

```
create table SalesFact as
select
   Quarter,
   BranchID,
   CategoryID,
   sum(TotalPrice) as Total_Sales
from TempFact
group by Quarter, BranchID, CategoryID;
```

The difference between the first case study and the second case study is very subtle. From a design point of view, there is not much difference. But from the implementation point of view, the second case study introduces the use of TempFact, simply because after the dimension tables are created, there is not enough information to create the final Fact Table immediately; hence, TempFact is needed.

2.4   Two-Column Table Methodology

The main concept of Fact and Dimensions is fact measure and the perspective from which the fact measure is viewed. From a pedagogical point of view, we start by explaining that the fact measure is a statistical numerical value. This restriction will be relaxed in later chapters when discussing advanced topics, such as data warehousing architectures.

When we create a star schema with fact and dimensions, we need to have a systematic method to check if the star schema is correct or not. There is an easy method to check the correctness of a star schema called the *two-column table methodology*. A two-column table is an *imaginary table* of our view to the fact measure from one particular dimension perspective. Using this two-column table, we can easily validate whether the fact measure and its view from a dimension perspective make sense or not.

The following two subsections explain how to use the two-column table methodology to validate the star schemas with the one-fact measure and multiple fact measures.

2.4.1 One-Fact Measure

A two-column table methodology uses tables with two columns. The first column is the *category*, and the second column is the statistical numerical figure, which is the *fact measure*. In the two-column table methodology, we construct several two-column tables. The second column has to be consistent throughout all the two-column tables.

Figure 2.17 shows an example of four two-column tables, each table having two columns. The first table, for example, has columns A and F, the second table columns B and F, etc. Notice several things: firstly, all four tables have column F as their second column. So, the second column is common to all the tables. Secondly, the first column of these tables indicates a category. So, there are different categories for different tables. Finally, as previously mentioned, these tables are imaginary; hence, the records are not real, the tables are not real and they do not exist. We only use these tables to validate our thinking visually to see if the relationship between one category and the fact measure makes sense or not. Hence, we can validate whether a record containing A and F makes sense or not and whether they fulfil the data warehousing requirement or not.

If all these four tables make sense, that is, the fact measure F is correctly viewed from each category (e.g. A, B, C, and D), then we can confidently draw a star schema as shown in Fig. 2.18, where the four categories are the four dimensions and F is the fact measure. Note that when we validate a star schema, we consider each dimension as an independent view to the fact. For example, fact measure F viewed from A alone, fact measure F is viewed from B alone, etc. But of course when the star schema is already done, the star schema itself reflects that fact measure F can

A	F
x	4
y	3

B	F
r	5
s	3

C	F
k	1
m	1

D	F
p	2
q	5

Fig. 2.17 Four two-column tables

Fig. 2.18 A star schema for
one-fact measure two-column
table

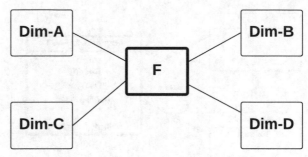

be viewed from any dimension: combined or individually, for example, fact measure
F viewed from A and C.

So basically, using the two-column table methodology, we validate the view of
the fact measure from each individual dimension. The use of the imaginary tables is
to aid us in the visualisation of the fact measure from a certain dimension angle.

Now let us use a case study to illustrate the use of this two-column table
methodology. The case study is an immigration case study, where the Department
of Immigration would like to analyse the number of immigrants in the country. Let
us use Australia as an example. The first step is to create a number of two-column
tables to visualise the number of immigrants from various dimension perspectives.
Note that these tables are imaginary tables, and hence the records are not real.
We only want to see the relationship between the fact, in this case, the number
of immigrants and the dimensions.

The first two-column table could have Year as the first column and Number of
Immigrants as the second column. So, this table gives us a view of the number
of immigrants coming to Australia every year, for example, how many immigrants
came to Australia last year, 2 years ago, etc. The second two-column table could
be Country and Number of Immigrants which gives a view of the number of
immigrants coming to Australia and their country of origin, for example, how many
immigrants came to Australia from the UK, how many from China, etc. Note that
we do not mix and match two pieces of information, namely, Year and Country. We
do not (yet) ask how many immigrants came from China last year. Each two-column
table focuses on one category only.

The third two-column table could be the type of visa the immigrant held, such
as a skilled visa, humanitarian visa, etc. This gives a view of how many immigrants
came to this country on a skilled visa or a humanitarian visa, etc. Again, the view of
the fact measure, which in this case is the number of immigrants, is only from visa
type.

The fourth two-column table could be in which state in Australia the immigrants
settled to give a view of how many immigrants settled in the various states (e.g.
Victoria, South Australia, Western Australia, etc.). This would be useful for state
planning, for example.

If we are satisfied with these four two-column tables, which show the correct
relationship between the fact measure (e.g. Number of Immigrants) and the

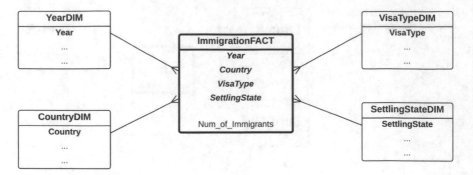

Fig. 2.19 A star schema for the Immigration example

dimensions (e.g. Year, Country, Visa Type, State), then the star schema can be created, which is shown in Fig. 2.19. It has the four dimensions, and the fact measure is Number of Immigrants. Using this star schema, we can ask questions such as how many immigrants came to Australia from the UK last year and settled in the state of Victoria (e.g. three dimensions), or how many immigrants came to Australia on a humanitarian visa last year (e.g. two dimensions)? So, the two-column table methodology is an easy tool to validate the correctness of a star schema.

2.4.2   Multiple Fact Measures

The two-column table, as described above, has a category in the first column (e.g. Column A) and a numerical aggregated value in the second column (e.g. Column F). This second column actually becomes the fact measure in the star schema. If we need to multiple fact measures in a star schema, this means that the second column in the two-column table, namely, Column F, needs to be divided into multiple columns, called Columns $F1$, $F2$, $F3$, etc. Fig. 2.20 shows the two-column tables where the second column is divided into three columns: $F1$, $F2$ and $F3$. The first column is still a single column which is a category column. Note that in this example, the three Column Fs exist in all four tables. We still call this a two-column table methodology, simply because the multiple Column Fs are considered as a single unit, as they must exist in all of the tables.

The star schema that represents the two-column tables (see Fig. 2.20) is shown in Fig. 2.21. There are four dimensions, A, B, C and D, with three fact measures, $F1$, $F2$ and $F3$.

Using a two-column table methodology to validate the star schema with multiple fact measures is very crucial. In this example, all the three Column Fs must exist in all tables. Suppose one of the tables has only Column D as the category and two Column Fs, say $F1$ and $F2$ only. This dimension (say Dimension D) will not be valid for the star schema with three fact measures, because this dimension is only

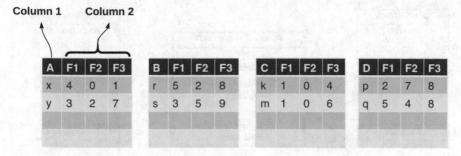

Fig. 2.20 Two-column tables with multiple Column F

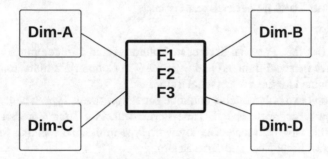

Fig. 2.21 A star schema for multiple fact measure two-column table

applicable to two fact measures and is irrelevant to the third fact measure, because there is no Column $F3$ in the two-column table. Hence, it is important for the data warehouse designer to be able to view every two-column table and validate whether each category is associated with *ALL* of the fact measures. In other words, all of the Column Fs must exist in all of the two-column tables. If not, the dimension is irrelevant to this star schema.

If a two-column table has more fact measures in the second column (e.g. four Column Fs: $F1$, $F2$, $F3$ and $F4$) compared to other tables, this dimension will also be irrelevant to the star schema that has only three fact measures. Hence, once again, the two-column table methodology is shown to be a critical tool to validate each dimension against all the fact measures to ascertain that a star schema is valid.

Let us put this multiple fact measure two-column table into an example. The case study involves a fitness centre. Suppose we want to build a star schema for a fitness centre to analyse their Number of Employees and Total Salary. Therefore, the fact measures (or the second column in the two-column table methodology) are Number of Employees and Total Salary.

There are many kinds of job titles in the fitness centre, for example, Personal Trainer, Gym Instructor, Yoga Teacher, Front Office Assistant, etc. Hence, we could have the first two-column table with Job Title as the first column (or Column A). The second column will comprise two columns: Number of Employees and Total Salary (or Columns $F1$ and $F2$). In this table, we could imagine to have a record

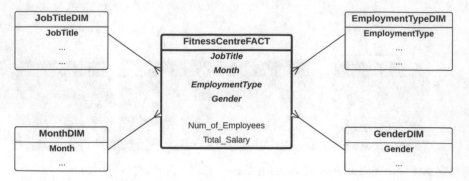

Fig. 2.22 A star schema for the fitness centre example

something like this: Personal Trainer in Column A and 5 in column $F1$ (meaning there are five personal trainers) and then $999 in Column $F2$ indicating the Total Salary combined for the five personal trainers.

The second two-column table could be for Employment Type. There are several employment types, such as Full-Time Permanent, Full-Time Contract and Part-Time. So, this table will have Employment Type in the first column, followed by the Number of Employees and Total Salary.

The third two-column table could be for Gender. So, for each gender, there is Number of Employees and Total Salary. The fourth table could be for Month. For each Month, there is Number of Employees and Total Salary.

Note that both fact measures, Number of Employees and Total Salary, are applicable to all the four categories: Job Title, Employment Type, Gender and Month. Using the two-column table methodology, it is easy to check if all the fact measures are applicable to each particular category. If we could imagine what the four imaginary tables look alike, and if they make sense, then we can confidently draw the following star schema (see Fig. 2.22) with four dimensions (e.g. Job Title, Employment Type, Gender and Month) and two fact measures (e.g. Number of Employees and Total Salary).

2.5 Summary

This chapter introduces star schema, containing fact measures and dimensions. A fact measure is a numerical and aggregated value, whereas a Dimension is the angle where we view the fact measure. To validate whether a star schema is correct or not, we can apply the two-column table methodology, which provides an imaginary view as to how the fact measures are associated with each category (or dimension). This simple yet powerful methodology shows the validity of a star schema.

In terms of the implementation of a star schema, the SQL commands to create the tables required for the data warehouse are used. The data warehouse creation process follows a systematic procedure by extracting data from the tables in the operational database. This chapter presents two case studies: the first case study is a very simple and straightforward process, whereas the second case study introduces a slight complexity in the process through the use of temporary Fact Tables before the final Fact Table is created.

2.6 Exercises

2.1 Using the aforementioned Sales case study, where the operational database E/R diagram is shown in using the Sales case study discussed above, where the operational database E/R diagram is shown in Fig. 2.15 and the star schema is shown in Fig. 2.16, suppose the Time Dimension, which had a granularity of Quarter, now has a Month granularity instead. See the new star schema in Fig. 2.23, and pay particular attention to the Time Dimension, with MonthID as its key identifier. The format for the MonthID attribute is "YYYYMM" which is unique for every Month/Year. Write the SQL commands to create this new star schema (e.g. Dimensions and Fact Tables).

2.2 The Information Technology Services (ITS) department at an International University needs a database to keep track of ITS staff members and the projects on which they are working. The system must satisfy the following requirements:

- For each ITS employee, list the employee number, employee name, job title (position) and the number and name of the ITS group in which he or she works. In addition, for each project to which the employee is assigned, list the project number and name, the percent of the employee's time assignment to the project and the total number of hours the employee has worked on the project so far.
- For each ITS project, list the project number and name (description), the name of the department that requested the project, the name of the contact person in that requesting department, the project type (maintenance, database, etc.), project

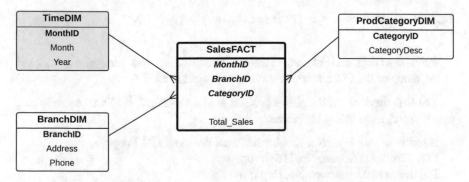

Fig. 2.23 The Sales case study where Time Dimension is Monthly-based

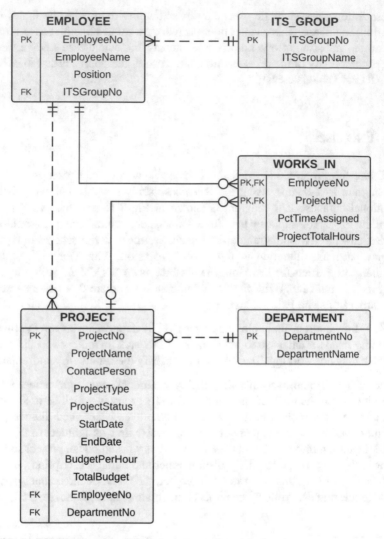

Fig. 2.24 An E/R diagram of the ITS project case study

status, start data, end date, total budgeted person-hours, total budgeted dollars and the name of the ITS employee serving as project leader.

The E/R diagram for the above process is given in Fig. 2.24. The system is now operating with the following tables:

- **Employee**(EmployeeNo, EmployeeName, Position, *ITSGroupNo*)
- **ITS_Group**(ITSGroupNo, ITSGroupName)
- **Department**(DepartmentNo, DepartmentName)

- **Project**(ProjectNo, ProjectName, ContactPerson, ProjectType, ProjectStatus, StartDate, EndDate, BudgetPerHour, TotalBudget, *EmployeeNo, DepartmentNo*)
- **Works_In**(*EmployeeNo, ProjectNo*, PctTimeAssigned, ProjectTotalHours)

The university management would like to analyse the performance of the ITS department, and to assist this process, you are asked to develop a data warehouse for analysis purposes. The analysis is needed to answer at least the following questions:

- What is the total number of projects that are long in duration?
- What is the total budget for a certain project type?
- What is the average budget cost per hour of a certain department?

Based on the above requirements, the fact measures in which management is interested are total budgets, total budget per hour and total projects; and the dimensions are project type, department and project duration. Assume that the management classifies projects into short term (less than 10 days), medium term (between 10 and 30 days) and long term (more than 30 days).

Your tasks are to design a star schema that matches the above requirements and implement the dimensions and Fact Tables using SQL.

2.3 Suppose the Association of Accountants (AoA) would like to analyse its members, who are certified accountants, in a particular city. Assume that the AoA has the full details of its members. Draw at least three two-column tables to visualise the imaginary records of these two-column tables. One table could look at Number of Accountants from an education point of view. Another table could examine the classification of Accountants based on gender. The third could look at the Number of Accountants for each specialist type of the accounting profession.

Based on these two-column tables, draw the star schema.

2.4 A major telecommunication company has implemented a billing system that records all mobile phone calls. A sample bill is shown in Fig. 2.25. Create at least five two-column tables to capture the Number of Calls and Total Duration of mobile phone calls from various aspects, as indicated in the billing above. Based on these two-column tables, draw the star schema.

2.7 Further Readings

This chapter focuses on star schema design and implementation in SQL. Star schema design can be found in various books on data warehousing, such as [1–10] and [11].

Lessons on E/R diagrams and relational modelling can be found in various database and data modelling textbooks, such as [12–32] and [33]. There are a variety of notations used in E/R diagram, but the concepts remain the same, which are entities and relationships, together with a variety of relationship properties

Date	Time	Type	Location	Number	Rate	Duration
15 Jun	05:03pm	National	Strathdale	131008	Peak	01:05
15 Jun	06:15pm	National	Bendigo	0410172913	Peak	04:54
15 Jun	08:48pm	National	Spencer St	0411848821	Off Peak	00:14
15 Jun	08:54pm	National	Spencer St	0411848821	Off Peak	00:10
16 Jun	03:15pm	National	Bundoora	0433762975	Peak	00:04
17 Jun	12:37pm	National	Kew East	0400070985	Peak	00:03
17 Jun	01:57pm	National	Kew East	0410172913	Peak	04:53
17 Jun	02:15pm	National	Kew East	0414861114	Peak	17:53
17 Jun	02:36pm	National	Kew East	0394795753	Peak	12:45
17 Jun	04:09pm	National	Kew East	0426670897	Peak	41:42
17 Jun	05:14pm	National	Kew East	0426670897	Peak	00:54
17 Jun	05:54pm	National	Kew	0424422409	Peak	20:00
20 Jun	01:31pm	National	Macleod	0400070985	Peak	00:03
20 Jun	05:07pm	National	Kew East	0411848821	Peak	00:22
24 Jun	11:42am	National	Hawthorn	0411848821	Peak	01:25
24 Jun	01:07pm	National	Kew East	0411848821	Peak	00:18
24 Jun	01:08pm	National	Kew East	0452609278	Peak	00:03
24 Jun	01:14pm	National	Kew East	0394791241	Peak	04:39
24 Jun	01:58pm	National	Kew East	0394793761	Peak	00:07
24 Jun	02:34pm	National	Kew East	0394791241	Peak	03:32
24 Jun	03:05pm	National	Kew East	0394793761	Peak	00:05
24 Jun	03:05pm	National	Kew East	0394793715	Peak	01:12
24 Jun	03:40pm	National	Kew East	0394791053	Peak	02:46
24 Jun	03:48pm	National	Kew East	0394795753	Peak	05:04
24 Jun	10:06pm	National	Kew East	0435282528	Off Peak	04:01
26 Jun	05:23pm	National	Melb CBD	0411848821	Off Peak	00:02
26 Jun	09:53pm	National	Kew East	0435282528	Off Peak	29:40
27 Jun	06:17pm	National	Heidelberg	0435282528	Peak	06:56
28 Jun	04:48pm	National	Bundoora	0433762975	Peak	01:11
28 Jun	07:01pm	National	Macleod	0433762975	Off Peak	31:24
28 Jun	07:41pm	National	Macleod	0433762975	Off Peak	08:32
29 Jun	05:42pm	National	Heidelberg	0433762975	Peak	06:57
30 Jun	02:48pm	National	Macleod	0390772216	Peak	00:35
30 Jun	04:15pm	National	Heidelberg	0452531299	Peak	01:40
30 Jun	04:26pm	National	Macleod	0425520404	Peak	05:20
30 Jun	06:15pm	National	Macleod	0452531299	Peak	03:10
30 Jun	06:33pm	National	Macleod	0426670897	Peak	06:25
30 Jun	06:47pm	National	Macleod	0468363333	Peak	00:16
30 Jun	07:39pm	National	Heidelberg	0435535358	Off Peak	19:43
01 Jul	03:28pm	National	Kew East	0394795115	Peak	09:16
01 Jul	07:12pm	National	Hawthorn	0411848821	Off Peak	00:21
01 Jul	09:47pm	National	Kew East	0435535358	Off Peak	35:20
02 Jun	05:00pm	National to Telstra Mobiles	Bundoora	0438745117	Peak	16:00
06 Jun	11:09am	National to Telstra Mobiles	Heidelberg	0450605898	Peak	00:15
06 Jun	11:12am	National to Telstra Mobiles	Macleod	0450605898	Peak	02:37
09 Jun	05:39pm	National to Telstra Mobiles	Heidelberg	0429447823	Peak	10:56
09 Jun	08:28pm	National to Telstra Mobiles	Macleod	0422473959	Off Peak	17:00
10 Jun	04:59pm	National to Telstra Mobiles	Kew East	0438745117	Peak	00:29
17 Jun	02:09pm	National to Telstra Mobiles	Kew East	0417305956	Peak	05:22
17 Jun	02:15pm	National to Telstra Mobiles	Kew East	0405562286	Peak	00:03
17 Jun	02:34pm	National to Telstra Mobiles	Kew East	0438745117	Peak	00:09
20 Jun	11:33am	National to Telstra Mobiles	Macleod	0408385054	Peak	16:11
20 Jun	01:32pm	National to Telstra Mobiles	Heidelberg	0455077311	Peak	00:07
20 Jun	01:33pm	National to Telstra Mobiles	Macleod	0422473959	Peak	01:49

Fig. 2.25 A sample billing from a mobile phone user

(e.g. cardinalities, participations, etc.), and entity properties (e.g. weak entities, composite attributes, etc.).

ETL (*Extract-Transform-Load*) is a process of transforming an operational database to a data warehouse. The following papers discussed ETL in various context and details: [34–38] and [39].

SQL commands are used to implement the data warehouse, in the form of dimension and Fact Tables, as well as importing the necessary data from the operational database. Additional resources on SQL include the following books: [40–42] and [43].

This book is the first book that introduces the concept of *two-column table methodology*. However, the concept is similar to the database view. The basic concept of database view is discussed in the following books: [20, 21]. The following papers discuss *materialised view* in more details: [44–49] and [50].

References

1. C. Adamson, *Star Schema The Complete Reference* (McGraw-Hill Osborne Media, 2010)
2. R. Laberge, *The Data Warehouse Mentor: Practical Data Warehouse and Business Intelligence Insights* (McGraw-Hill, New York, 2011)
3. M. Golfarelli, S. Rizzi, *Data Warehouse Design: Modern Principles and Methodologies* (McGraw-Hill, New York, 2009)
4. C. Adamson, *Mastering Data Warehouse Aggregates: Solutions for Star Schema Performance* (Wiley, London, 2012)
5. P. Ponniah, *Data Warehousing Fundamentals for IT Professionals* (Wiley, London, 2011)
6. R. Kimball, M. Ross, *The Data Warehouse Toolkit: The Definitive Guide to Dimensional Modeling* (Wiley, London, 2013)
7. R. Kimball, M. Ross, W. Thornthwaite, J. Mundy, B. Becker, *The Data Warehouse Lifecycle Toolkit* (Wiley, London, 2011)
8. W.H. Inmon, *Building the Data Warehouse*. ITPro Collection (Wiley, London, 2005)
9. M. Jarke, *Fundamentals of Data Warehouses*, 2nd edn. (Springer, Berlin, 2003)
10. E. Malinowski, E. Zimányi, *Advanced Data Warehouse Design: From Conventional to Spatial and Temporal Applications*. Data-Centric Systems and Applications (Springer, Berlin, 2008)
11. A. Vaisman, E. Zimányi, *Data Warehouse Systems: Design and Implementation*. Data-Centric Systems and Applications (Springer, Berlin, 2014)
12. C. Coronel, S. Morris, *Database Systems: Design, Implementation, & Management* (Cengage Learning, Boston, 2018)
13. T. Connolly, C. Begg, *Database Systems: A Practical Approach to Design, Implementation, and Management* (Pearson Education, 2015)
14. J.A. Hoffer, F.R. McFadden, M.B. Prescott, *Modern Database Management* (Prentice Hall, Englewood Cliffs, 2002)
15. A. Silberschatz, H.F. Korth, S. Sudarshan, *Database System Concepts*, 7th edn. (McGraw-Hill, New York, 2020)
16. R. Ramakrishnan, J. Gehrke, *Database Management Systems*, 3rd edn. (McGraw-Hill, New York, 2003)
17. J.D. Ullman, J. Widom, *A First Course in Database Systems*, 2nd edn. (Prentice Hall, Englewood Cliffs, 2002)
18. H. Garcia-Molina, J.D. Ullman, J. Widom, *Database Systems: The Complete Book* (Pearson Education, 2011)

19. P.E. O'Neil, E.J. O'Neil, *Database: Principles, Programming, and Performance*, 2nd edn. (Morgan Kaufmann, Los Altos, 2000)
20. R. Elmasri, S.B. Navathe, *Fundamentals of Database Systems*, 3rd edn. (Addison-Wesley-Longman, 2000)
21. C.J. Date, *An Introduction to Database Systems*, 7th edn. (Addison-Wesley-Longman, 2000)
22. R.T. Watson, *Data Management—Databases and Organizations*, 5th edn. (Wiley, London, 2006)
23. P. Beynon-Davies, *Database Systems* (Springer, Berlin, 2004)
24. E.F. Codd, *The Relational Model for Database Management, Version 2* (Addison-Wesley, Reading, 1990)
25. T.A. Halpin, T. Morgan, *Information Modeling and Relational Databases*, 2nd edn. (Morgan Kaufmann, Los Altos, 2008)
26. J.L. Harrington, *Relational Database Design Clearly Explained*. Clearly Explained Series (Morgan Kaufmann, Los Altos, 2002)
27. M. Kifer, A.J. Bernstein, P.M. Lewis, *Database Systems: An Application-Oriented Approach* (Pearson/Addison-Wesley, 2006)
28. W. Lemahieu, S. vanden Broucke, *Principles of Database Management: The Practical Guide to Storing, Managing and Analyzing Big and Small Data* (Cambridge University Press, Cambridge, 2018)
29. S. Bagui, R. Earp, *Database Design Using Entity-Relationship Diagrams*. Foundations of Database Design (CRC Press, Boca Raton, 2003)
30. M.J. Hernandez, *Database Design for Mere Mortals: A Hands-On Guide to Relational Database Design*. For Mere Mortals (Pearson Education, 2013)
31. N.S. Umanath, R.W. Scamell, *Data Modeling and Database Design* (Cengage Learning, Boston, 2014)
32. T.J. Teorey, S.S. Lightstone, T. Nadeau, H.V. Jagadish, *Database Modeling and Design: Logical Design*. The Morgan Kaufmann Series in Data Management Systems (Elsevier, Amsterdam, 2011)
33. G. Simsion, G. Witt, *Data Modeling Essentials*. The Morgan Kaufmann Series in Data Management Systems (Elsevier, Amsterdam, 2004)
34. X. Liu, N. Iftikhar, H. Huo, P.S. Nielsen, Optimizing ETL by a two-level data staging method. Int. J. Data Warehouse. Min. **12**(3), 32–50 (2016)
35. A. Nabli, S. Bouaziz, R. Yangui, F. Gargouri, Two-ETL phases for data warehouse creation: design and implementation, in *Advances in Databases and Information Systems—19th East European Conference, ADBIS 2015, Poitiers, France, September 8–11, 2015, Proceedings*, ed. by T. Morzy, P. Valduriez, L. Bellatreche. Lecture Notes in Computer Science, vol. 9282 (Springer, Berlin, 2015), pp. 138–150
36. W. Qu, V. Basavaraj, S. Shankar, S. Dessloch, Real-time snapshot maintenance with incremental ETL pipelines in data warehouses, in *Big Data Analytics and Knowledge Discovery—17th International Conference, DaWaK 2015, Valencia, Spain, September 1–4, 2015, Proceedings*, ed. by S. Madria, T. Hara Lecture Notes in Computer Science, vol. 9263 (Springer, Berlin, 2015), pp. 217–228
37. A. Simitsis, P. Vassiliadis, T.K. Sellis, Optimizing ETL processes in data warehouses, in *Proceedings of the 21st International Conference on Data Engineering, ICDE 2005, 5–8 April 2005, Tokyo, Japan*, ed. by K. Aberer, M.J. Franklin, S. Nishio (IEEE Computer Society, Silver Spring, 2005), pp. 564–575
38. Z. El Akkaoui, E. Zimányi, J.N. Mazón, J. Trujillo, A bpmn-based design and maintenance framework for ETL processes. Int. J. Data Warehouse. Min. **9**(3), 46–72 (2013)
39. L. Bellatreche, S. Khouri, N. Berkani, Semantic data warehouse design: From ETL to deployment à la carte, in *Database Systems for Advanced Applications, 18th International Conference, DASFAA 2013, Wuhan, China, April 22–25, 2013. Proceedings, Part II*, ed. by W. Meng, L. Feng, S. Bressan, W. Winiwarter, W. Song. Lecture Notes in Computer Science (Springer, Berlin, 2013), pp. 64–83

40. J. Melton, *Understanding the New SQL: A Complete Guide*, vol. I, 2nd edn. (Morgan Kaufmann, Los Altos, 2000)
41. C.J. Date, *SQL and Relational Theory—How to Write Accurate SQL Code*, 2nd edn. Theory in Practice (O'Reilly, 2012)
42. A. Beaulieu, *Learning SQL: Master SQL Fundamentals* (O'Reilly Media, 2009)
43. M.J. Donahoo, G.D. Speegle, *SQL: Practical Guide for Developers*. The Practical Guides. (Elsevier, Amsterdam, 2010)
44. R. Ahmed, R.G. Bello, A. Witkowski, P. Kumar, Automated generation of materialized views in Oracle. Proc. VLDB Endow. **13**(12), 3046–3058 (2020)
45. J.M.C. Barca, B.V. Sanchez, P.C.G. de Marina, Evaluation of an implementation of cross-row constraints using materialized views. SIGMOD Rec. **48**(3), 23–28 (2019)
46. P.A. Larson, J. Zhou, Efficient maintenance of materialized outer-join views, in *Proceedings of the 23rd International Conference on Data Engineering, ICDE 2007, The Marmara Hotel, Istanbul, Turkey, April 15–20, 2007*, ed. by R. Chirkova, A. Dogac, M.T. Özsu, T.K. Sellis (IEEE Computer Society, Silver Spring, 2007), pp. 56–65
47. G. Luo, Partial materialized views, in *Proceedings of the 23rd International Conference on Data Engineering, ICDE 2007, The Marmara Hotel, Istanbul, Turkey, April 15–20, 2007*, ed. by R. Chirkova, A. Dogac, M.T. Özsu, T.K. Sellis (IEEE Computer Society, Silver Spring, 2007), pp. 756–765
48. J. Zhou, P.A. Larson, J. Goldstein, L. Ding, Dynamic materialized views, in *Proceedings of the 23rd International Conference on Data Engineering, ICDE 2007, The Marmara Hotel, Istanbul, Turkey, April 15–20, 2007*, ed. by R. Chirkova, A. Dogac, M.T. Özsu, T.K. Sellis (IEEE Computer Society, Silver Spring, 2007), pp. 526–535
49. J. Zhou, P.A. Larson, H.G. Elmongui, Lazy maintenance of materialized views, in *Proceedings of the 33rd International Conference on Very Large Data Bases, University of Vienna, Austria, September 23–27, 2007*, ed. by C. Koch, J. Gehrke, M.N. Garofalakis, D. Srivastava, K. Aberer, A. Deshpande, D. Florescu, C.Y. Chan, V. Ganti, C.C. Kanne, W. Klas, E.J. Neuhold (ACM, New York, 2007), pp. 231–242
50. R. Chirkova, C. Li, J. Li, Answering queries using materialized views with minimum size. VLDB J. **15**(3), 191–210 (2006)

Chapter 3
Creating Facts and Dimensions: More Complex Processes

The previous chapter introduced the basic concept of star schema, consisting of Fact and Dimensions. The implementation process to create a star schema uses SQL commands to create the required tables for the data warehouse but following systematic processes consisting of a number of steps in creating fact and dimension tables.

This chapter will discuss slightly more complex processes in creating the fact and dimension tables. The first three sections of this chapter will focus on Fact Tables. A *Fact* table consists of key attributes from each dimension and fact measures. In the previous example, it shows how Fact Tables are created, either through *TempFact* or direct retrieval from the tables in the operational database. The fact measure itself is basically an aggregated value. In the SQL command, the fact measure attribute in the Fact Table is created using an aggregate function, such as count or sum, and the group by operation. In this chapter, we are going to learn about the complexity of aggregate functions, such as different ways of using the count function, as well as the problems of having an average function in the fact measure. A Fact Table is basically created by a join operation, which joins several tables from the operational database. However, in some cases, a Fact Table is created using an outer join operation. This chapter will discuss a case study as an example of the use of an outer join operation.

The previous chapter outlined three ways to create a dimension table: (*i*) direct copying from tables in the operational database, (*ii*) extracting some relevant attributes (or records) from tables in the operational database or (*iii*) manually creating a table for the dimension; hence, the table needs to be inserted with records, also manually. This chapter will discuss more complex processes when creating dimension tables, as the required dimension tables cannot be obtained directly from the operational database and cannot be created manually, especially when a *Temporary Dimension* table needs to be created before the final Dimension table is created.

This chapter will end with a discussion as to why an additional transformation process of tables in the operational database is sometimes necessary. The main

© The Author(s), under exclusive license to Springer Nature Switzerland AG 2021   49
D. Taniar, W. Rahayu, *Data Warehousing and Analytics*, Data-Centric Systems and Applications, https://doi.org/10.1007/978-3-030-81979-8_3

reason for this is that the table in the operational database cannot be readily used to create a Fact (or a TempFact) Table. As a result, the table in the operational database needs to go through a temporary table (*Temporary Operational Database Table*). This temporary operational database table (instead of the original operational database table) will then be used when creating the Fact (or TempFact) Table.

3.1 Use of `count` Function

The `count` function is one of the most common aggregate functions used to create the fact measures. The `count` function basically counts the number of records in a given table. Technically, in SQL, there are three ways of using the `count` function: (*i*) `count(*)`, (*ii*) `count(attribute)` or (*iii*) `count(distinct attribute)`.

The difference between these three `count` functions is often overlooked, especially when they are used in creating a Fact Table. The `count(*)` function counts the number of records in the query result, whereas the `count(attribute)` counts the number of records of that attribute excluding the `null` values. In other words, if the attribute does not contain any `null` values, then both `count(*)` and `count(attribute)` will return the same result. `count(distinct attribute)` on the other hand will count the number of *unique* values in the mentioned attribute.

To fully understand the implication of these different `count` functions, especially when creating the Fact Table, let's use the following case study. This case study is on mobile app repositories, where people publish their apps to be downloaded by other people. A short description of this case study is as follows.

Monalisa University is an international university. It has an online mobile app store, the Monalisa App Store, which allows students from any university in the world to publish their applications and receive feedback. This app store is considered a research environment where applications developed during research and studies can be tested and used by real users. The applications are free and open source. Basically, the operational system has the following features: all user details are kept in the database, which includes the universities in which the students are enrolled. Users can publish their apps or download other apps. They can also give feedback and ratings to other apps. The download and feedback statistics are stored in the database. Apps are organised into different categories, and many authors may have more than one app in different categories. An E/R diagram to show the current operational system is shown in Fig. 3.1.

The E/R diagram shows that an App User belongs to a University. An App User may publish one or more apps (i.e. App_User-Application relationship) and may download one or more apps (i.e. App_User-Download-Application relationship). An app, which belongs to a category (i.e. Application-Category relationship) may have multiple feedbacks (i.e. Application-Feedback relationship). Finally, an App User may give feedback (i.e. App_User-Feedback relationship).

A data warehouse, or a star schema, is needed to analyse the ratings and feedback on different authors and apps, so the management will be able to connect with

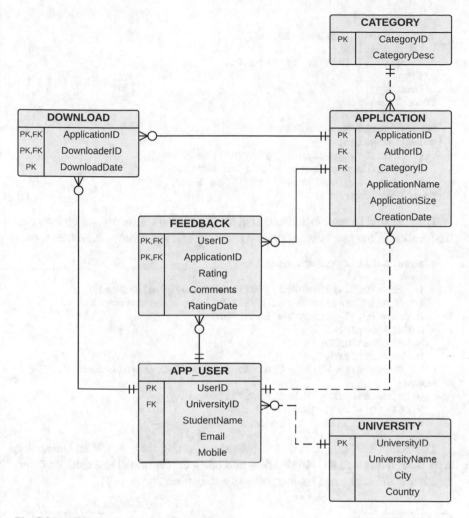

Fig. 3.1 An E/R diagram of apps downloads

talented authors and send them an annual award, for example. The author with the highest average app rating will be named author of the year and will receive an award. Hence, a star schema is created and is shown in Fig. 3.2.

For simplicity, we only focus on Total Downloads. There are four dimensions used in this star schema, namely, Category, University, Location and Time Dimensions. The SQL commands to create the dimensions are as follows:

```
create table CategoryDim as
select * from Category;

create table UniversityDim as
select UniversityID, UniversityName from University;
```

```
create table LocationDim as
select distinct
  Country || City as LocationID,
  City,
  Country
from University;

create table TimeDim as
select distinct
  to_char(DownloadDate, 'YYYYMM') as TimeID,
  to_char(DownloadDate, 'MM') as Month,
  to_char(DownloadDate, 'YYYY') as Year
from Download;
```

The Location Dimension indicates the location of the University, which indicates the location of the user (student) who created the app and downloaded other apps.

```
create table TempFact as
select
  to_char(D.DownloadDate, 'YYYYMM') as DownloadMonth,
  to_char(A.CreationDate, 'YYYYMM') as CreationMonth,
  U.Country || U.City as LocationID,
  A.CategoryID,
  A.ApplicationID,
  U.UniversityID
from University U, App_User R, Download D, Application A
where
  U.UniversityID = R.UniversityID and
  R.UserID = D.DownloaderID and
  D.ApplicationID = A.ApplicationID;
```

The TempFact Table is basically created by joining the four tables, University, App User, Download and Application, and takes the download and creation dates, together with the city and country, category and university.

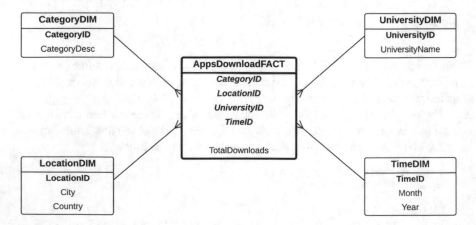

Fig. 3.2 A star schema of the mobile apps store

Now the Fact Table is created using the following SQL command. Note that in the SQL command to create the Fact Table, the Total Downloads fact measure simply uses count(*), because it counts all records, which indicate the number of downloads. As one app may have several downloads, the Total Downloads should count all downloads; each download counts toward the Total Downloads. Therefore, the count(*) function is used.

```
create table AppsDownloadFact as
select
   DownloadMonth as TimeID,
   LocationID,
   CategoryID,
   UniversityID,
   count(*) as TotalDownloads
from TempFact
group by
   DownloadMonth,
   LocationID,
   CategoryID,
   UniversityID;
```

Suppose now we would like to focus on Total Apps, instead of Total Downloads. Hence, the Total Downloads fact measure in the star schema is changed to Total Apps. The Fact Table now uses the following SQL command to create the table.

```
create table AppsFact as
select
   CreationMonth as TimeID,
   LocationID,
   CategoryID,
   UniversityID,
   count(distinct ApplicationID) as TotalApps
from TempFact
group by
   CreationMonth,
   LocationID,
   CategoryID,
   UniversityID;
```

There are a couple of changes in the create table. One is the Creation-Month attribute which is used instead of DownloadMonth. The second one is the count function. The count function now uses distinct ApplicationID, because we are counting the applications. Since the TempFact Table joins with the Download table, there will be duplicates for those apps that have multiple downloads. Therefore, it is necessary to use distinct in the count, so that the same apps will be counted once for Total Applications.

This short and simple case study shows clearly the difference between count(*) and count(distinct attribute). This is particularly important because when a Fact Table is created, it joins several tables from the operational database. Therefore, it is important to understand the results of the join operations so that the correct count function can be utilised.

3.2   Average in the Fact

The `count` and `sum` aggregate functions are the two most common aggregate functions used to create fact measures in star schemas, as it is seen in all the case studies so far. Technically, there are many aggregate and statistical functions that can be used to aggregate values for fact measures. However, there are some pitfalls on how to use aggregate functions correctly in the context of fact measures in star schemas, especially when the fact measures are highly aggregated. In this section, we are going to highlight the inappropriate use of the average function in the fact measures, where we should avoid using an average fact measure.

Consider an example in Table 3.1. Suppose we have the following 16 records as our data source in the operational database. Note that there are nine Semester One and seven Semester Two records. Of the eight Database Unit records, six are in Semester One and two are in Semester Two.

The star schema of the above operational database contains one fact and two dimensions (see Fig. 3.3). The dimensions are Subject and Semester, and the fact measure is average score.

Table 3.1 An operational database

Unit code	Unit title	Semester	Student first name	Score
IT001	Database	1	Mirriam	81
IT001	Database	1	Allan	41
IT001	Database	1	Ben	74
IT001	Database	1	Kate	85
IT001	Database	1	Larry	87
IT001	Database	1	Leonard	75
IT001	Database	2	Juan	64
IT001	Database	2	Andy	32
IT002	Java	1	Ally	65
IT002	Java	1	Menson	47
IT002	Java	2	Mirriam	78
IT002	Java	2	Ben	73
IT002	Java	2	Larry	64
IT003	SAP	1	Ally	63
IT004	Network	2	Juan	53
IT004	Network	2	Menson	52

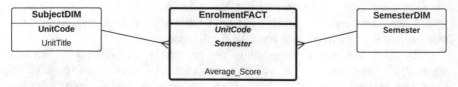

Fig. 3.3 A star schema with average score

Table 3.2 Table: Fact

Unit code	Semester	Average score
IT001	1	73.833
IT001	2	48
IT002	1	56
IT002	2	71.667
IT003	1	63
IT004	2	52.5

Table 3.3 Table: Subject
Dimension

Unit code	Unit title
IT001	Database
IT002	Java
IT003	SAP
IT004	Network

Table 3.4 Table: Semester
Dimension

Semester
1
2

The Fact Table aggregates these score records based on their dimensions, which are Subject and Semester. If we store average score in the Fact Table, this is what the Fact Table will look like (see Table 3.2). The dimension tables are shown in Tables 3.3 and 3.4.

Looking at the Fact Table in Table 3.2, the average score for the Database Unit in Semester One is **73.833** (average of the first six score records); the average score for the Database Unit in Semester Two is **48** (=(64 + 32)/2). These are actually incorrect.

For example, if we want to query the Fact Table to find out the average score of the Database Unit by looking at the above Fact Table, the answer would be (73.833+48)/2= **60.9165**.

The SQL to query the Fact Table is as follows:

```
select avg(Average_Score)
from EnrolmentFact
where UnitCode  = 'IT001';
```

This is incorrect. In the operational database, there are eight records for Database Unit in Semesters One and Two (see the first eight records in the operational database in Table 3.1). If we sum all the scores of these eight records and divide this by eight records, the result will be 539/8=**67.375**, not 60.9165.

Let's make further comparisons. The average score for the Java Unit in Semester One and Two using the above Fact Table is (56+71.667)/2=**63.833**. The actual average score for the Java Unit in Semester One and Two is not 63.833, but **65.4** (see the next five records in the above score list, and sum these scores and then divide by 5, 327/5). So again, the above Fact Table, which stores the average score, will not produce correct results.

Table 3.5 Table: Fact version 2

Unit code	Semester	Total score	Number of students
IT001	1	443	6
IT001	2	96	2
IT002	1	112	2
IT002	2	215	3
IT003	1	63	1
IT004	2	105	2

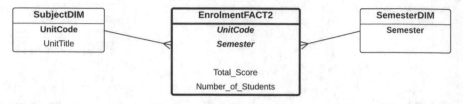

Fig. 3.4 The Correct Star Schema with Total Score and Number of Students

Ok, now let's calculate further. The average score for Semester One using the above fact is $(73.833+56+63)/3=$**64.278**. In the above score list records, there are nine Semester One records, and the average is, in fact, **68.667**.

For Semester Two, using the above Fact Table, the average score for Semester Two is $(48+71.667+52.5)/3=$**57.389**, whereas the actual average score for the seven Semester Two records is **59.4286**.

So, storing the average as a fact measurement is not a good idea. To avoid these problems, in the Fact Table, we should store the *Total Score* and *Number of Students* in each aggregate group. Hence, a correct Fact Table should look like the one depicted in Table 3.5. Note that the dimension tables remain unchanged, because the details of the units and semesters are unchanged (see Tables 3.3 and 3.4).

Using the correct Fact Table above, it is easy to calculate the average score of the Database Unit, which is $(443+96)/(6+2)=$**67.375**. The SQL to query the correct Fact Table is as follows:

```
select
  sum(Total_Score)/sum(Number_of_Students)
  as Average_Score
from EnrolmentFact2
where UnitCode  = 'IT001';
```

The correct star schema is then shown in Fig. 3.4.

The problem of average in the fact is known as the *Average of an Average* problem. This problem is well known in mathematics and statistics. The average of an average will simply produce an incorrect average result (almost all the time). Hence, it is not desirable to have an average measure in the fact, unless the analysis ALWAYS uses all the dimensions. This will be discussed further in Chap. 7.

If average should not be used in the fact, how about Min or Max? Yes, we can. Because Max of Max is always a global Max and Min of Min is always a global

Table 3.6 Table: Fact
version 3

Unit code	Semester	Min score	Max score
IT001	1	41	87
IT001	2	32	64
IT002	1	47	65
IT002	2	64	78
IT003	1	63	63
IT004	2	52	53

Min. For example, using the above sample data, assume we have Max Score and Min Score in the Fact, as shown in the Fact Table in Table 3.6. The Subject and Semester Dimensions are unchanged.

Assuming we want to get the Max Score of IT001, then the max of {87, 64} will produce 87, and 87 is the maximum score of IT001 because 87 is the max in Semester One, which is greater than any max of IT001 (e.g. in Semester Two). In other words, Max of Max is correct. The SQL to retrieve the maximum score of IT001 is as follows:

```
select max(Max_Score)
from EnrolmentFact3
where UnitCode = 'IT001';
```

The same applies to Min of Min. If we want to get the Minimum Score of IT001, the result will be 32, which is the minimum between 41 and 32.

```
select min(Min_Score)
from EnrolmentFact3
where UnitCode = 'IT001';
```

We certainly don't want to mix Min and Max. For example, retrieving the minimum of Max Score would be meaningless as would retrieving the maximum of Min Score.

As a final conclusion:

- Average in the Fact is not desirable, although technically it satisfies the two criteria of the fact (e.g. must be a numerical and aggregate value).
- Min and Max in the Fact can still be used, since Min Score and Max Score are valid fact measures (e.g. they are numerical and aggregated values).
- In general, Count and Sum are more common. Count is the *number of* records, and Sum is the *total of* a certain attribute.

3.3 Outer Join

The inner join operation is normally used when creating a Fact Table or a TempFact Table. The join operation is needed to get records from several tables in the operational database before aggregation is done to calculate the fact measures.

However, in some cases, an *outer join* operation is needed instead of an inner join. In this section, we are going to learn about a case study where an outer join operation is used to create the Fact Table.

This case study is on an Employment Agency which places temporary workers in companies during peak periods. The business process is as follows: The Employment Agency has a file of candidates who are willing to work. They record the candidate's number, name, contact address, contact phone number and maximum hours the candidate is available per week. Each candidate may have several qualifications. The Agency uses special codes to record the candidate's qualification for a job opening. As well as recording the code, the Agency also records the experience of the candidate in each qualification area, expressed as the number of months of experience.

The Agency also has a list of companies that request temporaries. Each company is assigned a company number as an identifier. The company name, type of business and principal contact for employment placements are also recorded. Each time a company requests a temporary employee, the Agency makes an entry in the (job) openings file. This file contains an opening number, the company requesting an employee, the required qualification, start date, anticipated end date and hourly pay. Each opening requires only one specific or main qualification. The Agency may be able to fill the opening from the staff on its books; however, in some circumstances, it cannot fill the request.

When a candidate matches the qualification and is available, he/she is given the job, and an entry is made in the placement record folder. This folder contains an opening number, candidate number, actual start date, total hours worked to date and end date if the placement is completed. The placement record folder is used by the Agency as a source of placement histories for its various temporaries.

An E/R diagram of the current operational system is shown in Fig. 3.5. Note that the relationship between Opening and Placement entities is a 1-1 relationship, where not all Openings have placements, but each Placement comes from an Opening. Over time, a candidate may have several Placements. A Company may also open several Openings, and each Opening has one main qualification requirement.

A star schema is needed, and for simplicity, it focuses on several types of job durations (e.g. short term which is less than 10 days, medium term between 10 and 30 days and long term which is longer than 30 days), qualification requirements and month. The star schema is shown in Fig. 3.6. The fact measures are Total Opening and Total Placement, with three dimensions: Duration, Qualification and Month Dimensions. Using these fact measures and dimensions, we are able to answer questions, such as how many short-term job openings are there, how many jobs are there which require certain qualifications and how many people started working in a certain month? The answers to which are useful in decision-making.

The three dimension tables are created as follows. For simplicity, there are three types of Duration, which are short term, medium term and long term.

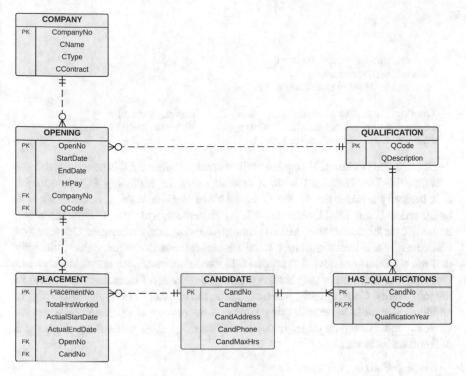

Fig. 3.5 An E/R diagram for the Employment Agency

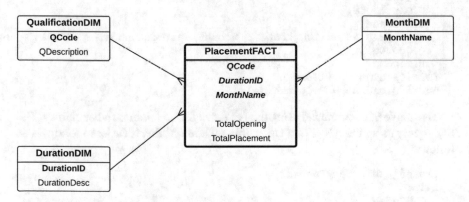

Fig. 3.6 A star schema for the Employment Agency

```
create table QualificationDim as
select * from Qualification;

create table MonthDim as
select
  distinct to_char(ActualStartDate, 'Month')
```

```
  as MonthName
from Placement;

create table DurationDim
(DurationID number,
 DurationDesc varchar2(20));

insert into DurationDim values (1, 'Short-Term');
insert into DurationDim values (2, 'Medium-Term');
insert into DurationDim values (3, 'Long-Term');
```

Because the Duration Dimension table is created manually, it is necessary to have a TempFact. The TempFact Table is created using the following SQL command. It is basically a join between the Opening and Placement tables, because we will later need to count Total Opening and Total Placement, which come from these two tables. However, note that in the E/R diagram, the relationship between Opening and Placement is a 1-1 relationship, but the Placement side is not mandatory, meaning that not all Openings have a Placement. If we do an inner join operation between Opening and Placement, only those Openings that have a Placement will appear in the join results. Consequently, when counting, the number of openings will result in an incorrect result. To remedy this problem, a *left outer join* operation between the Opening and Placement tables is needed so that Openings without any Placement will still be included.

```
create table TempFact as
select
  O.QCode,
  O.StartDate,
  O.EndDate,
  to_char(P.ActualStartDate, 'Month') as MonthName,
  O.OpenNo,
  P.CandNo
from Opening O, Placement P
where O.OpenNo = P.OpenNo (+);
```

The above SQL command using the (+) for *Left Outer Join* is rather clumsy. The SQL query using the ANSI Left Outer Join notation, which is easier to follow, is as follows:

```
create table TempFact as
select
  O.QCode,
  O.StartDate,
  O.EndDate,
  to_char(P.ActualStartDate, 'Month') as MonthName,
  O.OpenNo,
  P.CandNo
from Opening O left outer join Placement P
  on O.OpenNo = P.OpenNo;
```

To continue to the Fact Table, the TempFact Table needs to add a new attribute to store the DurationID. After some proper updates in the TempFact, the Fact Table is now created:

```
alter table TempFact
add (DurationID number);

update TempFact
set DurationID = 1
where EndDate - StartDate < 10;

update TempFact
set DurationID = 2
where EndDate - StartDate >= 10
and EndDate - StartDate <= 30;

update TempFact
Set DurationID = 3
where EndDate - StartDate > 30;

create table AgencyFact as
select
   QCode, DurationID, MonthName,
   count(OpenNo) as TotalOpening,
   count(CandNo) as TotalPlacement
from TempFact
group by QCode, DurationID, MonthName;
```

This simple case study shows the importance of understanding the relationship between entities in the E/R diagram, which may affect the way we join the tables in the star schema. In this example, an outer join operation is needed.

3.4 Creating Temporary Dimension Tables

The previous chapter explained that there are three ways to create a dimension table: (*i*) direct copying from a table in the operational database, (*ii*) selecting certain attributes (or records) from a table in the operational database or (*iii*) creating the dimension manually from scratch and inserting records into the newly created table. The latter method is suitable if the dimension is small, and only a few records need to be inserted.

If all of the dimension tables are created using the first two ways mentioned above, after the dimension tables are created, the Fact Table can immediately be created. However, if any of the dimension tables are created manually, then a temporary Fact Table (e.g. TempFact) needs to be created as an intermediate, before the final Fact Table is created.

However, there are circumstances where the required dimension table cannot be created directly. It requires an intermediate step, where a *Temporary Dimension*

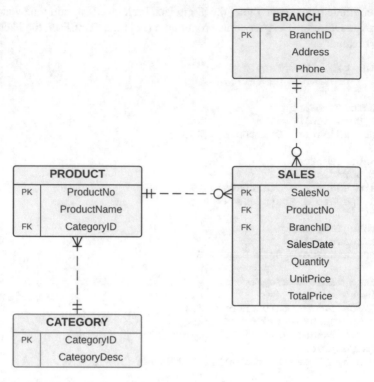

Fig. 3.7 An E/R diagram for the Sales case study

table needs to be created. This section showcases an example where a Temporary Dimension table is needed.

The previous chapter described a Sales case study. Let's use the same case study. The E/R diagram of the Sales operational database is shown in Fig. 3.7. The Category-Product relationship is a 1-*m*. Product has many Sales, and many Sales were made by a Branch.

The star schema for the Sales case study is shown in Fig. 3.8. The main difference between this star schema and the star schema studied in the previous chapter is that in this star schema, the Time Dimension has QuarterID, Quarter and Year attributes. The format for the QuarterID is "YYYYQ", where Q is the Quarter Number between 1 and 4. In other words, QuarterID is a unique combination between year and quarter. So, it is still Quarterly based, but now the Year information is included in the Quarter, whereas in the previous chapter, the Quarter does not have the Year information (or the Year is disregarded).

Let's start with the easy dimensions (e.g. Branch and Product Category Dimensions). The SQL commands to create the Branch and Product Category Dimensions are as follows:

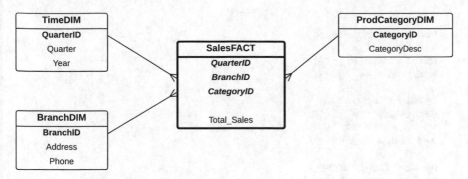

Fig. 3.8 A star schema for the Sales case study

```
create table BranchDim as
select * from Branch;

create table ProdCategoryDim as
select * from Category;
```

The Time Dimension that has QuarterID, Quarter and Year attributes needs to be created manually. But if we need to insert the records one by one, there are two problems. The first problem is that we may not know how many records need to be inserted because we need to check how many years of data there are. The second problem is of course if a large number of records need to be inserted due to the large number of years, then inserting the records manually will not be productive and may introduce human insertion errors. So, in this case, we need to go through additional intermediate steps when creating the required Time Dimension table, and as a result, a Temporary Time Dimension table is needed.

First, we need to create a Temporary Time Dimension table that contains Month and Year; these two attributes will be used in the final Time Dimension table. Then we need to add a new attribute, called QuarterID. After this, a series of updates need to be performed to fill in the QuarterID column with the correct values. Finally, the final Time Dimension table can be created. The SQL commands are as follows:

```
create table TimeDimTemp as
select distinct
  to_char(SalesDate, 'MM') as Month,
  to_char(SalesDate, 'YYYY') as Year
from Sales;

alter table TimeDimTemp add
(QuarterID char(5),
 Quarter   char(1));

update TimeDimTemp
set Quarter = '1'
where Month >= '01'
and Month <= '03';
```

```
update TimeDimTemp
set Quarter = '2'
where Month >= '04'
and Month <= '06';

update TimeDimTemp
set Quarter = '3'
where Month >= '07'
and Month <= '09';

update TimeDimTemp
set Quarter = '4'
where Month >= '10'
and Month <= '12';

update TimeDimTemp
set QuarterID = Year||Quarter;

create table TimeDim as
select distinct QuarterID, Quarter, Year
from TimeDimTemp;
```

The rest of the steps are identical to the steps mentioned in the previous chapters. These are also given here for completeness. The first step is to create a TempFact Table; this is because the Time Dimension is not a direct copy from the operational database. This is then followed by a series of alter table and *update* commands to obtain the correct Quarter and QuarterID attributes in TempFact. Finally, the final Fact Table is created.

```
create table TempFact as
select
  to_char(S.SalesDate, 'YYYY') as Year,
  to_char(S.SalesDate, 'MM') as Month,
  B.BranchID,
  P.CategoryID,
  S.TotalPrice
from Branch B, Sales S, Product P
where B.BranchID = S.BranchID
and S.ProductNo = P.ProductNo;

alter table TempFact
add (Quarter char(1));

update TempFact
set Quarter = '1'
where Month >= '01'
and Month <= '03';

update TempFact
set Quarter = '2'
where Month >= '04'
and Month <= '06';
```

```
update TempFact
set Quarter = '3'
where Month >= '07'
and Month <= '08';

update TempFact
set Quarter = '4'
where Quarter is null;

alter table TempFact
add (QuarterID char(5));

update TempFact
set QuarterID = Year||Quarter;

create table SalesFact as
select
  QuarterID,
  BranchID,
  CategoryID,
  sum(TotalPrice) as Total_Sales
from TempFact
group by QuarterID, BranchID, CategoryID;
```

3.5 Creating Temporary Tables in the Operational Database

In the case studies discussed so far, the Fact Table (or the TempFact Table) uses the tables from the operational database directly, joining them and aggregating the required attributes for the fact measures. This means that the tables from the operational database are ready to be processed and aggregated for the data warehouse. However, this might not be the case in many situations. The tables from the operational database might need to go through further transformation before they are ready to be used to create the Fact Table. In this section, a case study will be used to show how an operational database table is further transformed before being used to create the data warehouse. This case study is on hiring sessional staff in a university.

An International University employs its students to do various jobs, such as tutoring (teaching assistantship), programming, administration, etc. These jobs are called "sessional" jobs (also known as "contract" jobs). For each sessional job, students need to sign a contract. For example, to take a tutoring/teaching assistantship, the student will sign a contract with the university for one semester. The E/R diagram of the operational database is shown in Fig. 3.9. The relationship between Department and Employee is an $m - m$ relationship through the Contract. Each Employee may have several degrees.

A data warehouse (or a star schema) is built to analyse the number of contracts, based on year, department, employee and their most recent degree. Hence, with this data warehouse, we are able to analyse the number of contracts issued per

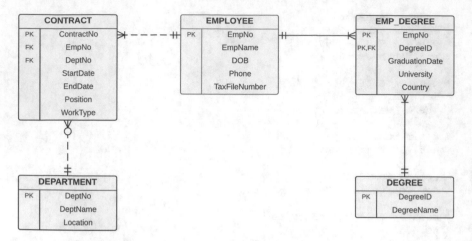

Fig. 3.9 An E/R diagram for the sessional contract jobs

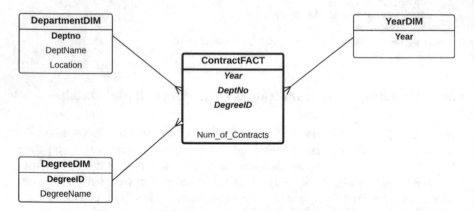

Fig. 3.10 A star schema for the sessional contract jobs

year by a certain department. Management sometimes needs to know how many contracts have been given to those whose most recent degree is a Master's degree (note that current Master's students are required to hold a Bachelor's degree to work as teaching assistants or programmers; current PhD students who have a Master's degree often work as teaching assistants too; hence, it is necessary to find out how many contracts have been signed by those who have a certain degree as their most recent degree). In this case, we do not need to know the entire number of degrees that an employee has—rather, we only need to keep track of their most recent degree.

A star schema for this is shown in Fig. 3.10. It has only one fact measure, which is Number of Contracts, with three dimensions: Department, Year and Degree Dimensions.

The SQL commands to create the dimension tables are as follows:

```
create table DepartmentDim as
select * from Department;

create table DegreeDim as
select * from Degree;

create table YearDim as
select distinct to_char(StartDate, 'YYYY') as Year
from Contract;
```

Note that an Employee (e.g. a record in the Employee Table) may have several degrees. But the data warehouse needs information on their most recent degree. Therefore, the Employee table needs a further transformation to obtain information on only one degree for each employee, this being their most recent degree. Hence, in this case, we create a table called the EmployeeTemp Table:

```
create table EmployeeTemp as
select
  T.EmpNo, T.EmpName, T.DOB,
  T.Phone, T.TaxFileNumber, T.DegreeID
from (
  select
    E.EmpNo, E.EmpName, E.DOB, E.Phone,
    E.TaxFileNumber, D.DegreeID,
    rank() over
      (partition by E.EmpNo
      order by D.GraduationDate desc) as Rank
  from Employee E, Emp_Degree D
  where E.EmpNo = D.EmpNo) T
where T.Rank = 1;
```

The EmployeeTemp Table basically extracts the most recent degree for each employee. This is achieved using the rank() over function with the partition by clause. Once the degrees are ranked according to their Graduation Date, for each Employee, it chooses the highest degree, which is the first in the ranking.

This EmployeeTemp Table will then be used to create the Fact Table instead of the original Employee Table. The SQL commands to create the Fact Table are as follows:

```
create table ContractFact as
select
  E.DegreeID,
  to_char(C.StartDate, 'YYYY') as Year,
  C.DeptNo,
  count(*) as Num_of_Contracts
from EmployeeTemp E, Contract C
where E.EmpNo = C.EmpNo
group by
  E.DegreeID,
```

```
to_char(C.StartDate, 'YYYY'),
C.DeptNo;
```

This case study shows a simple transformation of a table from the operational database. In this example, the Employee Table is further transformed because of the need to extract the employee's most recent degree. This transformation is relatively simple. In other cases, the transformation might need several temporary tables forming a chain of transformation before the final table in the transformation is ready to be used to create the Fact Table.

3.6   Summary

This chapter focuses on more complex processes in creating the Fact and Dimension tables.

(a) The first three sections in this chapter discuss slightly more complex processes to create the Fact Table, including discussions on aggregate functions (e.g. various uses of the `count` functions) and the issues of average in fact measures. It also discusses the use of an outer join operation, instead of the more common inner join operation to construct the Fact (or TempFact) Table. Understanding the relationship among the entities in the E/R diagram that will contribute to the creation of the data warehouse determines the use of which join operation will be used.

(b) Then it discusses steps to create a *Temporary Dimension* table. This is often needed when it is not possible to create a dimension table directly. Hence, we need to apply a series of additional steps when creating a dimension table through the creation of an intermediate table, called the Temporary Dimension table.

(c) Finally, the tables in the operational databases may go through a transformation before they are ready to be processed to create the data warehouse. This is because the tables in the operational database cannot always be used to create the Fact Table. Hence, the table in the operational database needs to go through a transformation phase where an intermediate table, called the *Temporary Operational Database Table*, is created.

3.7   Exercises

3.1 Melbourne International would like to analyse their policy in regard to the English requirements for admission into a course. Melbourne International has the following data (see Tables 3.7, 3.8, and 3.9).

A data warehouse based on the above data has been created. A star schema is shown in Fig. 3.11.

Table 3.7 Table: Student_IELTS

Student ID	Student name	Listening	Reading	Writing	Speaking	Overall
228493	Sooying Tan	6.5	6.5	6.0	7.0	6.5
229094	Xuebing Lu	5.5	5.5	5.5	5.5	5.5
231289	Amandh Kumar	6.0	7.0	6.0	7.0	6.5
234354	Agus Hidayat	5.5	6.0	6.0	6.5	6.0
234355	Budi Rahayu	7.0	7.0	7.0	7.0	7.0
...	...					

Table 3.8 Table: Student_Course

Student ID	Student name	Course	Start year
228493	Sooying Tan	MBIS	2013
229094	Xuebing Lu	MBIS	2013
231289	Amandh Kumar	MIT	2013
234354	Agus Hidayat	MIT	2013
234355	Budi Rahayu	MIT	2013
...	...	...	...

Table 3.9 Table: Student

Student ID	Student name	Address	Suburb	Phone number	Country
228493	Sooying Tan				Singapore
229094	Xuebing Lu				China
231289	Amandh Kumar				India
234354	Agus Hidayat				Indonesia
234355	Budi Rahayu				Indonesia
...	...				...

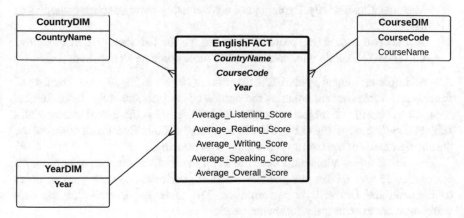

Fig. 3.11 A star schema for the English requirements

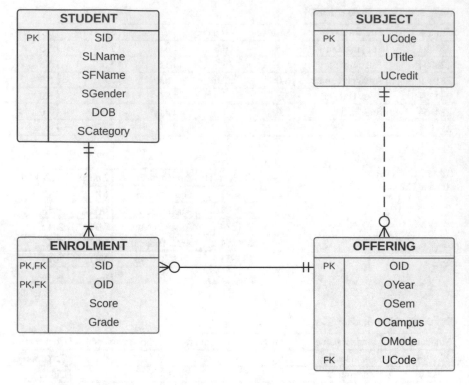

Fig. 3.12 The E/R diagram of an enrolment system

Questions

(a) The star schema in Fig. 3.11 will not produce the correct analysis of the fact measures. Explain why. Explain your answer using more concrete examples or data.

(b) How do you correct this problem by changing the fact measures of the above star schema? Explain your solution using more concrete examples or data.

3.2 A simple enrolment system E/R diagram is shown in Fig. 3.12. It consists of four tables. Table Student contains the details of students, whereas Table Subject contains the details of subjects or units. Table Offering is the actual instance of a subject showing when the subject is offered. Lastly, Table Enrolment contains the enrolment details of the students enrolling in an offering.

The University administrator needs to keep track of the number of enrolments for particular unit or campus and the students' performance each year in order to maintain the University's performance. The main requirements of the data warehouse are to answer the following queries:

(a) How many students are enrolled in the Database Unit offered on the Main Campus?

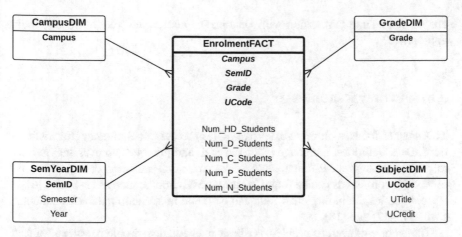

Fig. 3.13 Star schema of the enrolment system

(b) What is the average score of students taking the Database Unit in a particular semester?
(c) How many students received a HD (High Distinction) score in the Java unit offered in a particular semester?

Design a star schema with a number of dimensions and fact measures in the Fact, following the above requirements. Implement the fact and dimension tables using the SQL commands.

3.3 Using the same enrolment system as in the previous question (see the E/R diagram in Fig. 3.12), suppose the star schema is shown in Fig. 3.13. The grading system used in this case study is HD for High Distinction, D for Distinction, C for Credit, P for Pass and N for Fail. This is equivalent to A, B, C, D and F used in many other grading systems. The star schema then has five fact measures: Number of HDs, Ds, Cs, Ps and Ns. The dimensions are Campus, Semester Year, Grade and Subject.

Question Is this star schema correct?

3.4 This exercise uses the Sales case study presented in the previous chapter, as well as in this chapter. In the previous chapter, the star schema of the Sales case study is Quarterly based. The Time Dimension has two attributes: Quarter and Description. Quarter is simply either 1, 2, 3 or 4.

The Sales star schema in this chapter is also Quarterly based. But the attributes in the Time Dimension are different. They are QuarterID ('YYYYQ'), Quarter (1–4) and Year.

What are the differences between these two star schemas? In which situation is the star schema from the previous chapter (e.g. Time Dimension with Quarter and Description attributes) used, and in which situation is the star schema from this chapter (e.g. Time Dimension with QuarterID, Quarter and Year attributes) more appropriate?

3.8 Further Readings

This chapter focuses on more extensive SQL processes, including aggregate functions, Data Definition Language (DDL) of SQL and join query processing. Further references on aggregate functions (e.g. `count`, `sum`) and `create table` and other SQL commands can be found in various SQL books, such as [1–3] and [4].

Join queries, including outer join, can be found in standard database textbooks, such as [5–17] and [18].

The average of average problem is a basic mathematical problem. More advanced statistical functions can be found in various books on statistics, such as [19, 20] and [21].

The basic join query algorithms, such as Nested-Loop, Sort-Merge and Hash Join algorithms, can be found in [22]. There is much less work on outer join algorithms. Some early work on outer join algorithms can be found in [23, 24] and [25].

References

1. J. Melton, *Understanding the New SQL: A Complete Guide*, vol. I, 2nd edn. (Morgan Kaufmann, Los Altos, 2000)
2. C.J. Date, *SQL and Relational Theory—How to Write Accurate SQL Code*, 2nd edn. Theory in Practice (O'Reilly, 2012)
3. A. Beaulieu, *Learning SQL: Master SQL Fundamentals* (O'Reilly Media, 2009)
4. M.J. Donahoo, G.D. Speegle, *SQL: Practical Guide for Developers*. The Practical Guides (Elsevier, Amsterdam, 2010)
5. C. Coronel, S. Morris, *Database Systems: Design, Implementation, & Management*. (Cengage Learning, Boston, 2018)
6. T. Connolly, C. Beggm, *Database Systems: A Practical Approach to Design, Implementation, and Management* (Pearson Education, 2015)
7. J.A. Hoffer, F.R. McFadden, M.B. Prescott, *Modern Database Management* (Prentice Hall, Englewood Cliffs, 2002)
8. A. Silberschatz, H.F. Korth, S. Sudarshan, *Database System Concepts*, 7th edn. (McGraw-Hill, New York, 2020)
9. R. Ramakrishnan, J. Gehrke, *Database Management Systems*, 3rd edn. (McGraw-Hill, New York, 2003)
10. J.D. Ullman, J. Widom, *A First Course in Database Systems*, 2nd edn. (Prentice Hall, Englewood Cliffs, 2002)
11. H. Garcia-Molina, J.D. Ullman, J. Widom, *Database Systems: The Complete Book* (Pearson Education, 2011)
12. P.E. O'Neil, E.J. O'Neil, *Database: Principles, Programming, and Performance*, 2nd edn. (Morgan Kaufmann, Los Altos, 2000)
13. R. Elmasri, S.B. Navathe, *Fundamentals of Database Systems*, 3rd edn. (Addison-Wesley-Longman, 2000)

14. C.J. Date, *An Introduction to Database Systems*, 7th edn. (Addison-Wesley-Longman, 2000)
15. R.T. Watson, *Data Management—Databases and Organizations*, 5th edn. (Wiley, London, 2006)
16. P. Beynon-Davies, *Database Systems* (Springer, London, 2004)
17. M. Kifer, A.J. Bernstein, P.M. Lewis, *Database Systems: An Application-Oriented Approach* (Pearson/Addison-Wesley, 2006)
18. W. Lemahieu, S. vanden Broucke, *Principles of Database Management: The Practical Guide to Storing, Managing and Analyzing Big and Small Data* (Cambridge University Press, Cambridge, 2018)
19. D.A. Forsyth, *Probability and Statistics for Computer Science* (Springer, London, 2018)
20. D.E. Shasha, M. Wilson, *Statistics Is Easy!* Synthesis Lectures on Mathematics and Statistics (Morgan & Claypool Publishers, 2008)
21. D.P. Foster, R.A. Stine, R.P. Waterman, *Basic Business Statistics—A Casebook*. (Springer, Berlin, 1998)
22. D. Taniar, C.H.C. Leung, W. Rahayu, S. Goel, *High Performance Parallel Database Processing and Grid Databases* (Wiley, London, 2008)
23. Y. Xu, P. Kostamaa, Efficient outer join data skew handling in parallel DBMS. Proc. VLDB Endow. **2**(2), 1390–1396 (2009)
24. Y. Xu, P. Kostamaa, A new algorithm for small-large table outer joins in parallel DBMS, in *Proceedings of the 26th International Conference on Data Engineering, ICDE 2010, March 1–6, 2010, Long Beach, California, USA*, ed. by F. Li, M.M. Moro, S. Ghandeharizadeh, J.R. Haritsa, G. Weikum, M.J. Carey, F. Casati, E.Y. Chang, I. Manolescu, S. Mehrotra, U. Dayal, V.J. Tsotras (IEEE Computer Society, Silver Spring, 2010), pp. 1018–1024
25. G. Hill, A. Ross, Reducing outer joins. VLDB J. **18**(3), 599–610 (2009)

Part II
Snowflake and Bridge Tables

Chapter 4
Hierarchies

A simple star schema, such as the one shown in Fig. 4.1, has four dimensions (e.g. Dim1, Dim2, Dim3 and Dim4), and all of the dimensions are connected to the Fact. A *Hierarchy* is formed when a dimension is broken down to two or more dimensions in a hierarchical manner. In Fig. 4.2, Dim4 is broken into a hierarchy, which is then further broken down into Dim4a, Dim4b and Dim4c. Because the original dimension Dim4 is "broken" down into multiple dimensions, the star schema now looks like a *snowflake* because the "branches" from the centre have grown longer and are possibly more spread out which is why this schema is often called a *snowflake* schema.

When studying hierarchical dimensions, it is important to understand the differences between the hierarchical version and the non-hierarchical version. The following comparisons will be explained in this chapter:

1. Hierarchy versus Non-hierarchy Dimensions
2. Hierarchy versus Multiple Independent Dimensions
3. Linked Dimensions
4. Hierarchy Design Considerations

4.1 Hierarchy vs. Non-hierarchy

To understand the concept of hierarchy, we need to compare and contrast two approaches: the hierarchical approach and the non-hierarchical approach. Let's look at the Student Enrolment case study, where the fact of the number of student enrolments is examined from the course, the campus and the semester/year point of view. Figure 4.3 shows a simple star schema with one-fact measure: Number of Students and three dimensions: Course Dimension, Campus Dimension and Semester/Year Dimension.

© The Author(s), under exclusive license to Springer Nature Switzerland AG 2021
D. Taniar, W. Rahayu, *Data Warehousing and Analytics*, Data-Centric Systems
and Applications, https://doi.org/10.1007/978-3-030-81979-8_4

Fig. 4.1 A simple star
schema

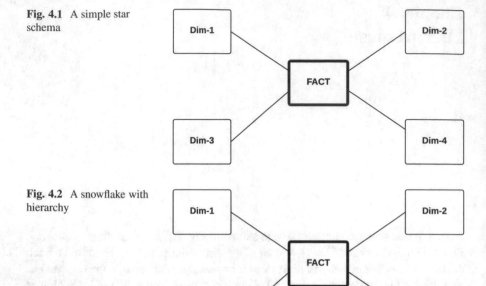

Fig. 4.2 A snowflake with
hierarchy

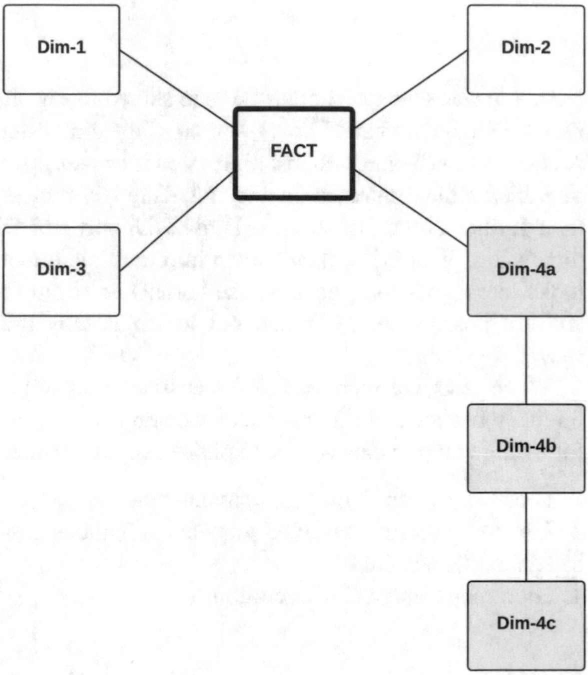

Let's focus on the Campus Dimensions. Table 4.1 shows a snapshot of the Campus Dimension table.

Another option to present this schema is to use a hierarchy on the Campus Dimension. The Campus Dimension, shown in Fig. 4.3, is now broken down into three dimensions (see Fig. 4.4): Campus, City, and Country Dimensions, and these dimensions form a hierarchy, from the Campus to City and then to the Country Dimension. The link to the fact is still through the Campus Dimension, so the fact remains unchanged.

Note that in the Campus Dimension, there is a CityID attribute which is linked to the CityID attribute in the City Dimension through a *many*-1 relationship (as one city may have multiple campuses). The same is applied to the City and Country

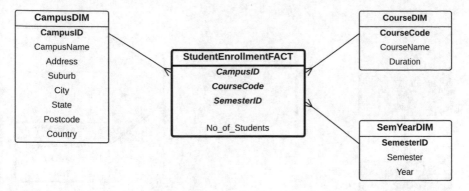

Fig. 4.3 Option 1: A non-hierarchy version

Table 4.1 Campus Dimension table

Campus ID	Campus Name	Address	Suburb	City	State	Postcode	Country
CL	Clayton Campus	Wellington Road	Clayton	Melbourne	Victoria	3800	Australia
CA	Caulfield Campus	Dandenong Road	Caulfield East	Melbourne	Victoria	3145	Australia
PA	Parkville Campus	Royal Parade	Parkville	Melbourne	Victoria	3052	Australia
SY	Sydney Campus	Opera Boulevard	Sydney	Sydney	New South Wales	2001	Australia
MUM	Malaysia Campus	Jalan Lagoon	Bandar Sunway	Kuala Lumpur	Selangor	47500	Malaysia
MSA	South Africa Campus	Peter Street	Johannes-burg	Johannes-burg	Johannes-burg	1725	South Africa

hierarchy, where in the City Dimension, there is a CountryID attribute which is linked to the CountryID attribute in the Country Dimension through a *many*-1 relationship (i.e. one country may have many cities).

Comparing the non-hierarchy version (option 1) and the hierarchy version (option 2), in the hierarchy version, we pull out the city information from the campus (from the non-hierarchy version) and then create a hierarchy between campus and city. We also pull out the country from the campus Dimension (from the non-hierarchy version) and create a hierarchy between city and country. Hence, the hierarchy version is a *normalised* version (i.e. Third Normal Form or NF), whereas the non-hierarchy version is lower than 3NF (e.g. First Normal Form (1NF) or Second Normal Form (2NF)).

The contents of the dimensions in the hierarchy, the Campus, City and Country Dimension tables, are shown in Tables 4.2, 4.3, and 4.4. Note that the contents of the Fact Table and other dimensions are not affected.

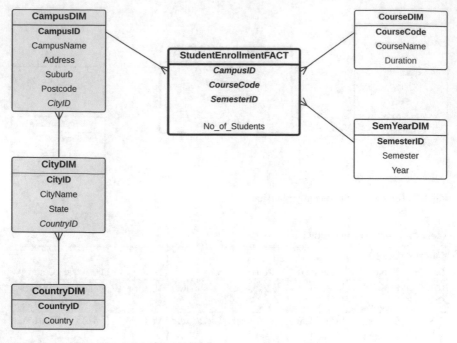

Fig. 4.4 Option 2: A hierarchy version

Table 4.2 Campus Dimension table

Campus ID	Campus name	Address	Suburb	Postcode	CityID
CL	Clayton Campus	Wellington Road	Clayton	3800	MEL
CA	Caulfield Campus	Dandenong Road	Caulfield East	3145	MEL
PA	Parkville Campus	Royal Parade	Parkville	3052	MEL
SY	Sydney Campus	Opera Boulevard	Sydney	2001	SYD
MUM	Malaysia Campus	Jalan Lagoon	Bandar Sunway	47500	KUL
MSA	South Africa Campus	Peter Street	Johannesburg	1725	JNB

Table 4.3 City Dimension table

City ID	City name	State	Country ID
MEL	Melbourne	Victoria	AU
SYD	Sydney	New South Wales	AU
KUL	Kuala Lumpur	Selangor	MA
JNB	Johannesburg	Johannesburg	SA

Table 4.4 Country Dimension table

Country ID	Country name
AU	Australia
MA	Malaysia
SA	South Africa

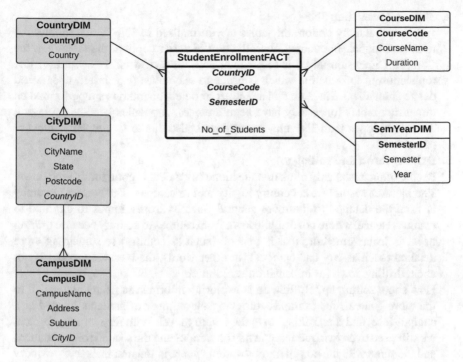

Fig. 4.5 A wrong hierarchy

The hierarchy should start with the highest detail dimension, in this case, the Campus Dimension. This top dimension in the hierarchy is then linked to the fact. The lower detail (e.g. City) becomes the child in the hierarchy, and the lowest detail (e.g. Country) becomes the lowest in the hierarchy. In the context of cardinality, the correct hierarchy should have a *many*-1 relationship, not a 1-*many* relationship. One of the most common mistakes is the wrong ordering of the dimensions in the hierarchy, such as the star schema shown in Fig. 4.5, where the cardinality is 1-*many* instead of *many*-1.

Since there are two options in implementing a star schema, with and without the hierarchy, which option is preferable? Here are some comparisons:

1. **One table vs. many tables**
   Without the hierarchy, there is only one table (e.g. the Campus Dimension table). With the hierarchy, there are three tables (e.g. Campus, City and Country Dimension tables). The consequence is in the join query processing when producing reports. When you want to produce a report involving the fact and campus, with the no hierarchy option, we only need to use two tables: the Campus Dimension table and the Fact Table. Hence, join query processing is simple. With the hierarchy option, we need to join the fact with three tables (Campus, City and Country). Consequently, join query processing will be more complex.

2. **3NF vs. lower than 3NF**

With the hierarchy option, the tables are normalised in 3NF which follows the relational model. In contrast, with the non-hierarchy option, the table (e.g. the Campus Dimension table) is lower than 3NF. In fact, it is 2NF with a visible replication of information which is prone to anomalies (e.g. insert, update and delete anomalies). Since no update, insert or delete operations are performed on dimension tables (once they have been created), anomalies become irrelevant. Therefore, lower than 3NF may be preferred, also due to its simpler join query processing.

3. **Drilling down and rolling up**

People often mistakenly think that the hierarchy model is good for drilling down. The hierarchy is not from country to city and to campus. The correct hierarchy is from the detailed to the more general, such as from campus to city and to country. Hence, we are not drilling down from campus to country because drilling down is from something that is general (such as country) to something more detailed (such as city and campus). In other words, the hierarchy model is not about drilling down for information exploration.

How about rolling up? Rolling up is exploring information from the detailed to the more general, for example, retrieving the number of students located at a campus level and then rolling up to the country level. With the hierarchy option, we still need to do two queries, one for the campus and the second for the country, and the same with the non-hierarchy option, where in the first instance, we query the campus and after that we do another query, but just for the country.

As a summary:

- From the **query point of view,** both versions need two queries.
- From the **query processing point of view**, the non-hierarchy version uses one join operation only because it only needs to join the fact and one dimension.
- From the **conceptual point of view**, the hierarchy model does not actually offer better roll up or drill down features.

If we insist on having a drill down, that is, from country to campus, we still need to do the two queries but we will do the country query first and then the campus query second.

4.2 Hierarchy Versus Multiple Independent Dimensions

In the previous section, we learned how to split a dimension into a hierarchy of dimensions. We also compared the two models: hierarchy vs. non-hierarchy models. In this section, we will compare the hierarchy and multiple independent dimensions. Let's look at the star schema in Fig. 4.6. Let's focus on the two dimensions: Campus and Country. They are independent dimensions, connected to the Fact. This is a simple but valid star schema. There is no relationship whatsoever between

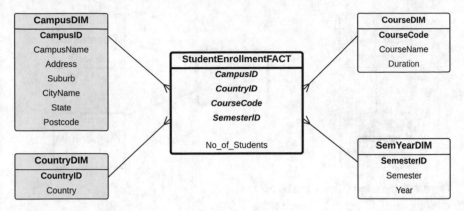

Fig. 4.6 Campus and Country as Separate Dimensions

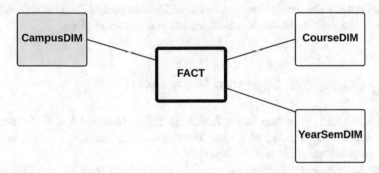

Fig. 4.7 Campus and Country as a Combined Dimension

the Campus and Country Dimensions. The Campus and Country identifiers (e.g. CampusID and CountryID) are also in the Fact.

Actually, there are possibly three versions of this example:

1. **Separate Dimensions**
   In this case study (refer to the star schema in Fig. 4.6), we are focusing on the Campus and Country Dimensions, which is a Separate Dimension model, because we actually split between the Campus and Country Dimensions.
2. **Combined Dimension**
   When the Campus and Country Dimensions are combined into one dimension, as in Fig. 4.7, we call this a Combined Dimension model. For simplicity, the attributes are not shown; only the dimensions and fact are shown in Fig. 4.7
3. **Hierarchy Dimensions**
   In the Hierarchy model, now the Campus and Country Dimensions form a hierarchy, as shown in Fig. 4.8. Again, for simplicity, the attributes are not shown.

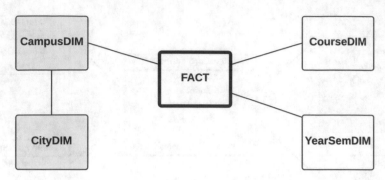

Fig. 4.8 Campus and Country in a Hierarchy

In order to assess which one is the best star schema, we need to compare two things: Separate Dimension versus Combined Dimension and Combined Dimension versus Hierarchy. These will be discussed in the next two sections.

4.2.1   Separate vs. Combined Dimension

We are comparing the two star schemas as shown in Fig. 4.6 (for the Separate Dimension model) and in Fig. 4.7 (for the Combined Dimension model). We need to see some sample records in both models.

For the Separate Dimension model, the sample records are shown in Tables 4.5, 4.6, and 4.7 for the Campus and Country Dimension, and the Fact.

Table 4.5 Separate Dimension Model-Campus Dimension table

Campus ID	Campus name	Address	Suburb	City name	State	Postcode
CL	Clayton Campus	Wellington Road	Clayton	Melbourne	Victoria	3800
CA	Caulfield Campus	Dandenong Road	Caulfield East	Melbourne	Victoria	3145
PA	Parkville Campus	Royal Parade	Parkville	Melbourne	Victoria	3052
SY	Sydney Campus	Opera Boulevard	Sydney	Sydney	New South Wales	2001
MUM	Malaysia Campus	Jalan Lagoon	Bandar Sunway	Kuala Lumpur	Selangor	47500
MSA	South Africa Campus	Peter Street	Johannes-burg	Johannes-burg	Johannes-burg	1725

Table 4.6 Separate
Dimension model-Country
Dimension table

Country ID	Country name
AU	Australia
MA	Malaysia
SA	South Africa

Table 4.7 Separate Dimension model-Fact Table

Campus ID	Country ID	Course code	Semester ID	No of students
CL	AU	A3001	202001	450
CL	AU	A3002	202001	150
CL	AU	A3003	202001	200
CL	AU	...	...	...
CA	AU	B5001	202001	115
CA	AU	B5002	202001	160
CA	AU	...	...	...
PA	AU	C6001	202001	75
PA	AU	C6002	202001	50
PA	AU	...	...	...
SY	AU	A3001	202001	40
SY	AU	B5002	202001	35
SY	AU	...	...	...
MUM	MA	A3001	202001	150
MUM	MA	B5001	202001	80
MUM	MA	B5002	202001	100
MUM	MA	...	...	...
MSA	SA	A3002	202001	25
MSA	SA	A3002	202001	20
MSA	SA	...	...	...

Now look at the sample records for the Combined Dimension model, where the Campus and Country Dimensions are collapsed into one dimension called the Campus Dimension. Tables 4.8 and 4.9 show the Campus Dimension table and the Fact Table, respectively.

From the fact point of view, a comparison of Tables 4.7 and 4.9 shows that there is only a minor difference; that is, in the Combined Dimension model, the Fact has one less attribute because the CountryID attribute is not there anymore. This doesn't change the number of records. However, there is a minor implication. If we want to retrieve, for example, the Number of Students, based on Country (say Australia), using the Separate Dimension model (refer to Table 4.7), we can solely use the Fact Table without the need to join with the dimension. However, if we use the Combined Dimension model (refer to Table 4.7) because the information on the Country is in the Campus Dimension, we then need to join the Fact Table with the Campus Dimension table. So, this is a minor difference.

Table 4.8 Combined Dimension model-Campus Dimension table

Campus ID	Campus name	Address	Suburb	City name	State	Postcode	Country
CL	Clayton Campus	Wellington Road	Clayton	Melbourne	Victoria	3800	Australia
CA	Caulfield Campus	Dandenong Road	Caulfield East	Melbourne	Victoria	3145	Australia
PA	Parkville Campus	Royal Parade	Parkville	Melbourne	Victoria	3052	Australia
SY	Sydney Campus	Opera Boulevard	Sydney	Sydney	New South Wales	2001	Australia
MUM	Malaysia Campus	Jalan Lagoon	Bandar Sunway	Kuala Lumpur	Selangor	47500	Malaysia
MSA	South Africa Campus	Peter Street	Johannes-burg	Johannes-burg	Johannes-burg	1725	South Africa

Table 4.9 Combined Dimension model-Fact Table

Campus ID	Course code	Semester ID	No of students
CL	A3001	202001	450
CL	A3002	202001	150
CL	A3003	202001	200
CL	...	...	...
CA	B5001	202001	115
CA	B5002	202001	160
CA	...	...	...
PA	C6001	202001	75
PA	C6002	202001	50
PA	...	...	...
SY	A3001	202001	40
SY	B5002	202001	35
SY	...	...	...
MUM	A3001	202001	150
MUM	B5001	202001	80
MUM	B5002	202001	100
MUM	...	...	...
MSA	A3002	202001	25
MSA	A3002	202001	20
MSA	...	...	...

If we look at this from the dimension point of view, that is, comparing Tables 4.5 and 4.6, and 4.8, obviously, the Separate Dimension model needs to maintain two tables, whereas the Combined Dimension model only needs to maintain one table. Maintaining one dimension is certainly easier than that of two dimensions.

Additionally, the star schema in the Combined Dimension model (Fig. 4.7) is simpler.

So, if the two pieces of information, e.g. Campus and Country, are often seen as one entity or one piece of information, as in this case, it would be easier if the two pieces of information are present in one dimension. Hence, use the Combined Dimension model as in the star schema shown in Fig. 4.7, where the Campus Dimension incorporates the country and there is no need to separately maintain the Country Dimension.

4.2.2 Combined Dimension vs. Hierarchy

Now let's compare the Combined Dimension model (Fig. 4.7) with the Hierarchy (Fig. 4.8). In the Hierarchy model, Campus Dimension and Country Dimension form a hierarchy. So, the Hierarchy model is slightly different from the Separate Dimension model (Fig. 4.7) because in the Separate Dimension model, the two dimensions, Campus and Country, are independently connected to the Fact.

The sample data of the Combined Dimension model are shown in Tables 4.8 and 4.9. The following tables (see Tables 4.10, 4.11, and 4.12) show the sample data of the Hierarchy model. The Hierarchy model is very similar to the Separate Dimension model. Table 4.10 shows the Campus Dimension; notice that the CountryID is maintained in this table to link to the Country table as the hierarchy is maintained. There is no difference in the Country table, between the Separate Dimension model and the Hierarchy model.

The Fact Table is shown in Table 4.12. It is exactly identical to the Combined Dimension model, because in both the Combined Dimension model and the

Table 4.10 Hierarchy model-Campus Dimension table

Campus ID	Campus name	Address	Suburb	Name name	State	Postcode	Country ID
CL	Clayton Campus	Wellington Road	Clayton	Melbourne	Victoria	3800	AU
CA	Caulfield Campus	Dandenong Road	Caulfield East	Melbourne	Victoria	3145	AU
PA	Parkville Campus	Royal Parade	Parkville	Melbourne	Victoria	3052	AU
SY	Sydney Campus	Opera Boulevard	Sydney	Sydney	New South Wales	2001	AU
MUM	Malaysia Campus	Jalan Lagoon	Bandar Sunway	Kuala Lumpur	Selangor	47500	MA
MSA	South Africa Campus	Peter Street	Johannes-burg	Johannes-burg	Johannes-burg	1725	SA

Table 4.11 Hierarchy model-Country Dimension table

Country ID	Country name
AU	Australia
MA	Malaysia
SA	South Africa

Table 4.12 Hierarchy model-Fact Table

Campus ID	Course code	Semester ID	No of students
CL	A3001	202001	450
CL	A3002	202001	150
CL	A3003	202001	200
CL	...	...	...
CA	B5001	202001	115
CA	B5002	202001	160
CA	...	...	...
PA	C6001	202001	75
PA	C6002	202001	50
PA	...	...	...
SY	A3001	202001	40
SY	B5002	202001	35
SY	...	...	...
MUM	A3001	202001	150
MUM	B5001	202001	80
MUM	B5002	202001	100
MUM	...	...	...
MSA	A3002	202001	25
MSA	A3002	202001	20
MSA	...	...	...

Hierarchy model, there is no CountryID in the Fact. Hence, the main difference is in the Campus Dimension table, where the Foreign Key (FK) which is the CountryID is kept.

In the Hierarchy model, it seems unnecessary to split the Campus Dimension into a Hierarchy. Apart from the unnecessary CountryID attribute in the Campus Dimension table, the reasons why the Hierarchy model is not ideal are quite similar to those of the Separate Dimension model, which are:

• An efficient query processing to query the fact and dimensions,
• The identifier of the Campus Dimension (e.g. CampusID) already covers the child dimension (e.g. Country Dimension), and
• Campus and Country information is often regarded as one entity, at least in this case study.

Therefore, the Combined Dimension model (refer to the star schema in Fig. 4.7) is often seen as the best option. Splitting Campus into two dimensions, Campus and Country, whether they are two independent dimensions or two dimensions in a hierarchy, will not offer much benefit, in query processing or in capturing the model.

4.3 Linked Dimensions

Apart from the aforementioned three models: (*i*) Separate Dimension model, (*ii*) Combined Dimension model and (*iii*) Hierarchy, there is one more possibility, that is, the *Linked Dimension*. This is very often used when people are creating a star schema from an E/R diagram and the "E/R diagram kind of relationship" is used in the star schema, where the entities are connected to other entities. Hence, dimensions are connected to other dimensions. Figure 4.9 shows an example where the Campus and Country Dimensions are connected to each other, and at the same time, they are also linked directly to the Fact. On the surface, this seems reasonable because there is a natural relationship between Campus and Country, and at the same time, they are connected to the Fact to form a star schema. So, this is a kind of combination between the Hierarchy model, and the Separate Dimension model, as previously discussed. The main question which remains is whether it is necessary to link dimensions.

Sample data for the Linked Dimension case is shown in Tables 4.13, 4.14, and 4.15: Campus Dimension, Country Dimension and the Fact Tables. Because the Linked Dimension model is actually a combination of the Hierarchy model and the Separate Dimension model, there is a lot of redundant information. Linking Campus and Country makes sense from an E/R diagram point of view because one country has many campuses. So, technically, it is not wrong. Also, because the two dimensions (e.g. Campus and Country) are connected directly to the Fact, both identifiers (e.g. CampusID and CountryID) exist in the Fact. So, this is also correct, from a star schema standpoint.

However, from the query point of view, the inefficiency problem faced in the Separate Dimension model exists here in the Linked Dimension model, simply

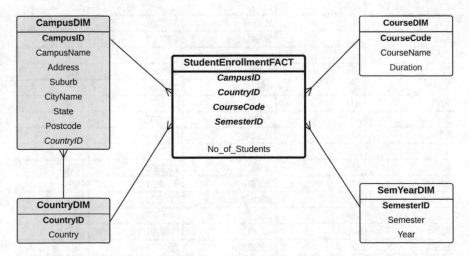

Fig. 4.9 Linked Dimensions

Table 4.13 Linked Dimension model-Campus Dimension table

Campus ID	Campus name	Address	Suburb	City name	State	Postcode	Country ID
CL	Clayton Campus	Wellington Road	Clayton	Melbourne	Victoria	3800	AU
CA	Caulfield Campus	Dandenong Road	Caulfield East	Melbourne	Victoria	3145	AU
PA	Parkville Campus	Royal Parade	Parkville	Melbourne	Victoria	3052	AU
SY	Sydney Campus	Opera Boulevard	Sydney	Sydney	New South Wales	2001	AU
MUM	Malaysia Campus	Jalan Lagoon	Bandar Sunway	Kuala Lumpur	Selangor	47500	MA
MSA	South Africa Campus	Peter Street	Johannes-burg	Johannes-burg	Johannes-burg	1725	SA

Table 4.14 Linked Dimension model-Country Dimension table

Country ID	Country name
AU	Australia
MA	Malaysia
SA	South Africa

Table 4.15 Linked Dimension model-Fact Table

Campus ID	Country ID	Course code	Semester ID	No of students
CL	AU	A3001	202001	450
CL	AU	A3002	202001	150
CL	AU	A3003	202001	200
CL	AU	...	...	...
CA	AU	B5001	202001	115
CA	AU	B5002	202001	160
CA	AU	...	...	...
PA	AU	C6001	202001	75
PA	AU	C6002	202001	50
PA	AU	...	...	...
SY	AU	A3001	202001	40
SY	AU	B5002	202001	35
SY	AU	...	...	...
MUM	MA	A3001	202001	150
MUM	MA	B5001	202001	80
MUM	MA	B5002	202001	100
MUM	MA	...	...	...
MSA	SA	A3002	202001	25
MSA	SA	A3002	202001	20
MSA	SA	...	...	...

because it has two dimensions (e.g. Campus and Country) linked to the Fact. As a result, there are more join operations, as well as unnecessary additional attributes in the Fact.

The problem faced by the Hierarchy model, that is, an inefficient model is maintaining the hierarchy, also exists in this Linked Dimension model. In this case, there is no need to have CountryID in the Campus because the query is from the Fact, and the Fact has already has information about the country. Although technically it is not wrong, the redundant information makes it inefficient.

So, the bottom line is that creating relationships between dimensions, in a form of hierarchy, perhaps will not add any efficient access or schema. All relationships from each dimension must be linked directly to the fact and not among dimensions.

4.4 Hierarchy Design Considerations

Hierarchy is basically a *many*-1 relationship between dimensions, such as the Campus-Country hierarchy. Now let's include the Course Dimension in the hierarchy. A campus may likely have more than one course. Let's assume that each course is only offered on one campus. Hence, the relationship between Course and Campus is also *many*-1 and hence can form a hierarchy.

Therefore, the hierarchy is Course, Campus and then Country. The star schema is shown in Fig. 4.10. Technically, it is not wrong because the hierarchy uses *many*-1 relationships, from Course to Campus and to Country. However, is this hierarchy proper?

There are two aspects from which Course and Campus are not a good hierarchy. Firstly, Campus and Country Dimensions have a similarity, which is the spatial context. However, Course Dimension is a totally different thing. Hence, forming a hierarchy between Course and Campus may not be appropriate, although technically they have a *many*-1 relationship.

Secondly, data access to the data warehouse is always through the Fact, which is that's why the Fact has fact measures, which are aggregated values of the data warehouse. These aggregated values, or fact measures, are the focus of the data retrieval. All dimensions should be linked directly to the fact whenever possible, so that when retrieving the fact measure, we can directly access the required dimension. So, in this example, we can get the number of students per course because CourseID provides a direct link from the Fact to the Course Dimension. If there are a lot of queries relating to the campus, such as retrieving the number of students for each campus (or for any particular campus), because the Campus Dimension is not linked directly to the fact, we need to go through the Course. If this is done too often, data retrieval becomes less efficient.

If we think of collapsing the hierarchy and moving all the attributes of Campus (and Country) into the Course, the same problem still exists; that is, accessing the number of students from the Campus point of view needs to go through the Course. Therefore, in this case, the best solution is to split the Course and Campus

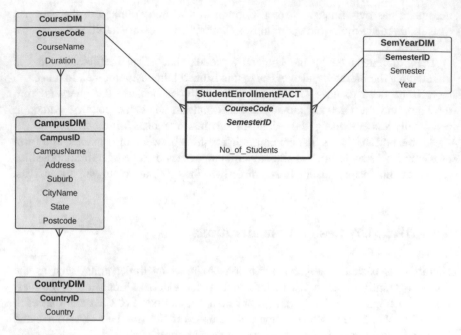

Fig. 4.10 Course Dimension in the Hierarchy

into two independent dimensions, each of these being directly linked to the Fact. Consequently, accessing information from any angle, whether it be Course or Campus, can be done directly from the Fact.

However, if Campus is a minor information, then collapsing Campus (and Country) into the Course Dimension will be more efficient and simpler.

We now look back at the Campus and Country Dimensions issue discussed in the previous section. Why would the Campus and Country Dimensions be better off collapsed into one dimension, called Campus (e.g. the Combined Dimension model)? The main reason is that it is assumed that in this case study, the information about Country is minor, meaning that it is very rare that the retrieval of the number of students is based on the country of the campus. Therefore, merging Country and Campus will be more feasible; and therefore, a Combined Dimension model was adopted.

Therefore, in designing a hierarchy, we need to think about two aspects, in particular, the topic of the information on each dimension involved (whether they are of the same topic) and the data retrieval aspect (whether the information on a particular dimension is minor). This will give a good indication as to whether we need to use a hierarchy or simply collapse the dimensions in the hierarchy into one dimension.

4.5 Summary

This chapter discusses the concept of Dimension Hierarchy in star schemas. A dimension hierarchy is a connection of two or more dimensions in a hierarchical manner, using a *many*-1 relationship. As a result, the dimensions in a hierarchy are normalised, in 3NF (Third Normal Form), using the context of the Relational Database Design.

This chapter also compares and contrasts five different models:

1. Separate Dimension model
2. Combined Dimension model
3. Hierarchy model
4. Linked Dimension model
5. Hierarchy Design Considerations

The Combined Dimension model combines the two dimensions in the Separate Dimension model into one dimension, whereas the Hierarchy model maintains the two dimensions but forms a hierarchy. The Linked Dimension model keeps the hierarchy, which links multiple dimensions in a hierarchical relationship, as well as connecting each of these dimensions directly to the fact, like the Separate Dimension models.

After comparing the star schemas and also looking at some sample data, it can be concluded that the Combined Dimension model is more efficient and gives a simpler model. Hence, keeping the number of dimensions to a minimum would be more beneficial rather than splitting them into more dimensions, either independently or in a hierarchical manner.

If we need to use a Hierarchy model, it has to be designed appropriately. Simply creating a hierarchy by linking dimensions with *many*-1 relationships does not guarantee a good hierarchy model, although technically it is not wrong. However, a hierarchy must show a natural hierarchy, rather than merely a *many*-1 relationship between two dimensions.

4.6 Exercises

4.1 In this chapter, using the Campus-Country Student Enrolment Case Study, we explore four models: (*i*) Separate Dimension model, (*ii*) Combined Dimension model, (*iii*) Hierarchy model and (*iv*) Linked Dimension model.

Explore another case study, and draw the four models of the star schema, highlighting the differences between the four models in the star schema.

4.2 Considering the Campus-Country Student Enrolment Case Study used in this chapter. Figure 4.11 shows a star schema with the Course Dimension now in the Campus hierarchy. In what scenario is this star schema wrong? It is not about the inefficiency of the model or of the query processing; rather, it is simply plain wrong.

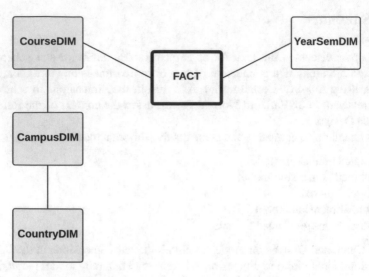

Fig. 4.11 Star schema with Hierarchy

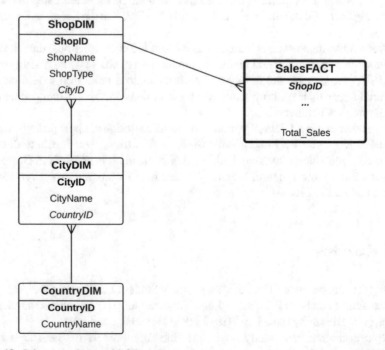

Fig. 4.12 Sales star schema with Hierarchy

4.3 Consider the following star schemas (refer to Figs. 4.12 and 4.13). The first star schema contains a hierarchy in the dimension, whereas the second star schema collapses the hierarchy into one dimension.

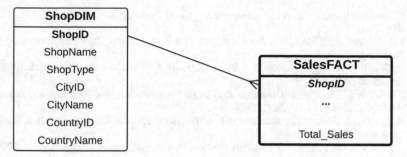

Fig. 4.13 Sales star schema without Hierarchy

Questions

(a) Draw sample table contents of the fact and dimension tables of the two star schemas

(b) Compare and contrast the two star schemas using the sample tables in question (*a*) above. Explain the pros and cons of each star schema.

4.7 Further Readings

This chapter solely focuses on the concept of *Hierarchy*, which is applied to Dimensions. A hierarchy is modelled using a *many*-1 relationship, which exists in a form of Primary Key-Foreign Key (PK-FK) relationships in relational databases. The details of various cardinality relationships can be typically found in many database textbooks, such as [1–15] and [16].

The concept of hierarchy in database design can also be found in many data modelling books, such as [17–21] and [22].

Dimensional hierarchy in star schema is also explained in various data warehousing books, such as [23–32] and [33].

References

1. C. Coronel, S. Morris, *Database Systems: Design, Implementation, & Management* (Cengage Learning, 2018)
2. T. Connolly, C. Begg, *Database Systems: A Practical Approach to Design, Implementation, and Management* (Pearson Education, 2015)
3. J.A. Hoffer, F.R. McFadden, M.B. Prescott, *Modern Database Management* (Prentice Hall, Englewood Cliffs, 2002)
4. A. Silberschatz, H.F. Korth, S. Sudarshan, *Database System Concepts*, 7th edn. (McGraw-Hill, New York, 2020)

5. R. Ramakrishnan, J. Gehrke, *Database Management Systems*, 3rd edn. (McGraw-Hill, New York, 2003)
6. J.D. Ullman, J. Widom, *A First Course in Database Systems*, 2nd edn. (Prentice Hall, Englewood Cliffs, 2002)
7. H. Garcia-Molina, J.D. Ullman, J. Widom, *Database Systems: The Complete Book*. (Pearson Education, 2011)
8. P.E. O'Neil, E.J. O'Neil, *Database: Principles, Programming, and Performance*, 2nd edn. (Morgan Kaufmann, Los Altos, 2000)
9. R. Elmasri, S.B. Navathe, *Fundamentals of Database Systems*, 3rd edn. (Addison-Wesley-Longman, 2000)
10. C.J. Date, *An Introduction to Database Systems*, 7th edn. (Addison-Wesley-Longman, 2000)
11. R.T. Watson, *Data Management—Databases and Organizations*, 5th edn. (Wiley, London, 2006)
12. P. Beynon-Davies, *Database Systems* (Springer, Berlin, 2004)
13. E.F. Codd, *The Relational Model for Database Management, Version 2* (Addison-Wesley, 1990)
14. J.L. Harrington, *Relational Database Design Clearly Explained*. Clearly Explained Series (Morgan Kaufmann, Los Altos, 2002)
15. M. Kifer, A.J. Bernstein, P.M. Lewis, *Database Systems: An Application-Oriented Approach* (Pearson/Addison-Wesley, 2006)
16. W. Lemahieu, S. vanden Broucke, *Principles of Database Management: The Practical Guide to Storing, Managing and Analyzing Big and Small Data* (Cambridge University Press, Cambridge, 2018)
17. T.A. Halpin, T. Morgan, *Information Modeling and Relational Databases*, 2nd edn. (Morgan Kaufmann, Los Altos, 2008)
18. S. Bagui, R. Earp, *Database Design Using Entity-Relationship Diagrams*. Foundations of Database Design (CRC Press, Boca Raton, 2003)
19. M.J. Hernandez, *Database Design for Mere Mortals: A Hands-On Guide to Relational Database Design*. For Mere Mortals (Pearson Education, 2013)
20. N.S. Umanath, R.W. Scamell, *Data Modeling and Database Design* (Cengage Learning, 2014)
21. T.J. Teorey, S.S. Lightstone, T. Nadeau, H.V. Jagadish, *Database Modeling and Design: Logical Design*. The Morgan Kaufmann Series in Data Management Systems (Elsevier, Amsterdam, 2011)
22. G. Simsion, G. Witt, *Data Modeling Essentials*. The Morgan Kaufmann Series in Data Management Systems (Elsevier, Amsterdam, 2004)
23. C. Adamson, *Star Schema The Complete Reference* (McGraw-Hill Osborne Media, 2010)
24. R. Laberge, *The Data Warehouse Mentor: Practical Data Warehouse and Business Intelligence Insights* (McGraw-Hill, New York, 2011)
25. M. Golfarelli, S. Rizzi, *Data Warehouse Design: Modern Principles and Methodologies* (McGraw-Hill, New York, 2009)
26. C. Adamson, *Mastering Data Warehouse Aggregates: Solutions for Star Schema Performance* (Wiley, London, 2012)
27. P. Ponniah, *Data Warehousing Fundamentals for IT Professionals* (Wiley, London, 2011)
28. R. Kimball, M. Ross, *The Data Warehouse Toolkit: The Definitive Guide to Dimensional Modeling* (Wiley, London, 2013)
29. R. Kimball, M. Ross, W. Thornthwaite, J. Mundy, B. Becker, *The Data Warehouse Lifecycle Toolkit* (Wiley, London, 2011)
30. W.H. Inmon, *Building the Data Warehouse*. ITPro Collection (Wiley, London, 2005)
31. M. Jarke, *Fundamentals of Data Warehouses*, 2nd edn. (Springer, Berlin, 2003)
32. E. Malinowski, E. Zimányi, *Advanced Data Warehouse Design: From Conventional to Spatial and Temporal Applications*. Data-Centric Systems and Applications (Springer, Berlin, 2008).
33. A. Vaisman, E. Zimányi, *Data Warehouse Systems: Design and Implementation*. Data-Centric Systems and Applications (Springer, Berlin, 2014)

Chapter 5
Bridge Tables

A *Bridge Table* is a table that links two dimensions, and only one of these two dimensions is linked to the fact. As a result, the star schema becomes a *snowflake* schema. The term snowflake schema was introduced in the previous chapter on *Hierarchy*, where a string of dimensions is linked through a hierarchy relationship, and the parent dimension is linked to the fact. It is called a snowflake because the hierarchy of dimension is like a snowdrop. Remember that the cardinality relationship in the hierarchy from the parent dimension to the child dimension is *many*-1. A Bridge Table is also a snowflake, but the relationship between the two dimensions that are linked through a Bridge Table has a cardinality of 1-*many* and *many*-1.

The following example shows the difference between a normal star schema (Fig. 5.1), a snowflake with a Bridge Table (Fig. 5.2) and a snowflake with a hierarchy (Fig. 5.3). In the snowflake with a Bridge Table, Dim-4b is not linked directly to the fact but is connected to another dimension (e.g. Dim-4a) through a Bridge Table. Dim-4a has a 1-*many* relationship with the Bridge Table, which in turn has a *many*-1 relationship with Dim-4b.

On the other hand, the snowflake with a hierarchy (Fig. 5.3) shows that dimensions 4a, 4b and 4c are linked through a hierarchy relationship with *many*-1 cardinalities between each pair of parent-child dimensions.

A Bridge Table is needed when a dimension cannot be connected directly to the Fact as it has to go through another dimension. For example, Dim-4b cannot be connected directly to the fact without going through Dim-4a. For Dim-4b to be connected to Dim-4a, it has to go through a Bridge Table. This is because the cardinality relationship between Dim-4a and Dim-4b is actually a *many-many* relationship which is why a Bridge Table needs to be an intermediate table between these two dimensions.

© The Author(s), under exclusive license to Springer Nature Switzerland AG 2021   97
D. Taniar, W. Rahayu, *Data Warehousing and Analytics*, Data-Centric Systems
and Applications, https://doi.org/10.1007/978-3-030-81979-8_5

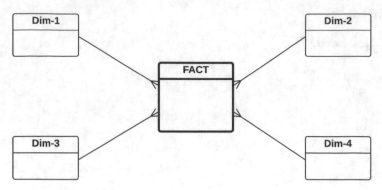

Fig. 5.1 A star schema

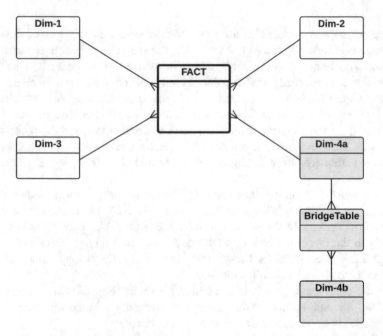

Fig. 5.2 A snowflake with a Bridge Table

There are at least two reasons, why a dimension cannot be connected directly to the Fact:

(a) The Fact Table has a fact measure, and the dimension has a key identity. In order to connect a dimension to the Fact, the dimension's key identity must contribute directly to the calculation of the fact measure. Unfortunately, this cannot happen if the operational database does not have this data.

(b) The operational database does not have this data if the relationship between two entities in the operational database that hold the information about the dimension's key identity and the intended fact measure is a *many-many* relationship.

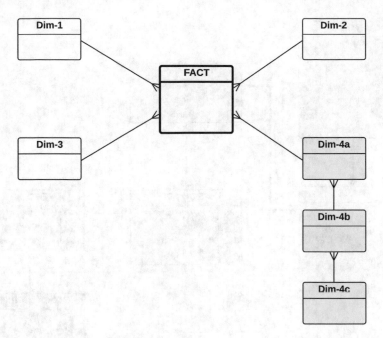

Fig. 5.3 A snowflake with a Hierarchy

In this chapter, we are going to study two case studies that highlight the importance and the use of a Bridge Table. Additional features, including a Weight Factor attribute and a List Aggregate attribute, will also be explained.

5.1   A Product Sales Case Study

To understand the concept of Bridge Tables in data warehousing, let's use a Product Sales example as a case study. The E/R diagram is shown in Fig. 5.4. The Sales entity on the left side of the E/R diagram is a transaction entity, where all sales transactions are recorded. This includes the customers and the staff who performed the sales. The Product entity, in the middle of the E/R diagram, maintains a list of the company's products. The relationship between Sales and Product is a *many-many*, through the SalesItem entity, where the QtySold attribute is recorded. Each product may be stored in several locations, as shown by the Stock entity, where the QtyinStock attribute is recorded. The Supplier entity maintains a list of suppliers. Finally, the relationship between Stock and Supplier is a *many-many*, which basically records each supply, including the supplied quantity.

In this case study, the management of the company would like to analyse the statistics of its product sales history. The analysis is needed to identify popular products, the suppliers who supplied these products, the best time to purchase more

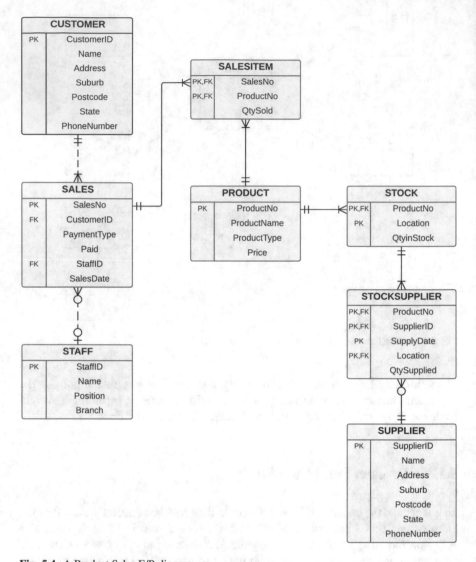

Fig. 5.4 A Product Sales E/R diagram

stock, etc. Hence, a small data warehouse is to be built to keep track of the statistics. The management is particularly interested in analysing the *total sales* (*quantity * price*) by *product*, *customer suburbs*, *sales time periods* (month and year) and *supplier*.

We start the data warehouse by designing two-column tables (see Tables 5.1, 5.2, and 5.3). Here are some possible scenarios (note that the total sales figures are imaginary).

The imaginary data in Tables 5.1, 5.2, and 5.3 seem to be logical, and hence a star schema for the Product Sales case study is shown in Fig. 5.5.

Table 5.1 Product point of view

ProductNo	TotalSales
A1	$130,000
B2	$15,900
C3	$2,500,000
...	...

Table 5.2 Time point of view

TimeID	TotalSales
201801	$25,000
201802	$4,700
201803	$3,500
...	...

Table 5.3 Suburb point of view

Suburb	TotalSales
Caulfield	$6,500
Chadstone	$12,000
Clayton	$1,800
...	...

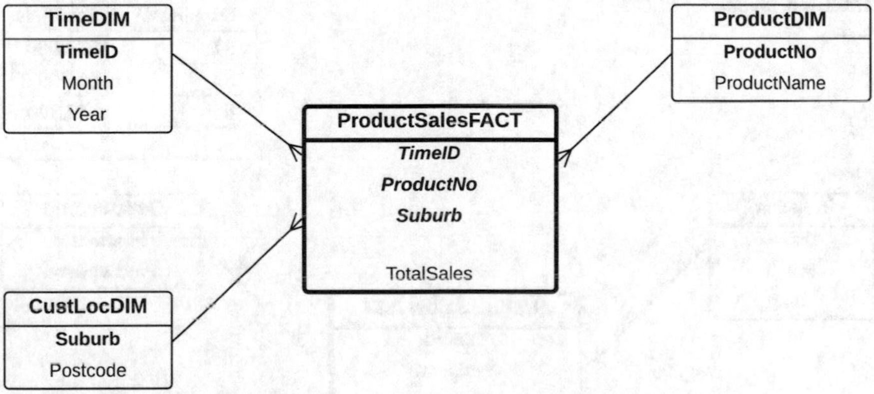

Fig. 5.5 An initial star schema

The contents of the Product Sales Fact table are shown in Table 5.4.

However, the management would also like information about the supplier which is still missing from the report in Table 5.4. Therefore, one could have another two-column table for suppliers, which may look like the one in Table 5.5, and a revised star schema is shown in Fig. 5.6 in which a new supplier dimension is added to the star schema. Therefore, the contents of the new Product Sales Fact are modified to include information about the supplier (see Table 5.6).

The revised ProductSalesFact table (with the addition of the Supplier Dimension) seems to be accurate. Actually, it is impossible to produce such a report because in the E/R diagram (in the operational database), it shows that one supplier may supply many products (or one product is supplied by many suppliers). However, the sales

Table 5.4 ProductSalesFact table

TimeID	Suburb	ProductNo	TotalSales
201801	Caulfield	A1	$450
201801	Caulfield	B2	$100
201801	Caulfield	C3	$320
201801	Caulfield	...	...
201801	...	...	...
201801	Chadstone	A1	$75
201801	Chadstone	B2	$600
201801	Chadstone	C3	$55
201801	Chadstone	...	...
201801	...	...	...
201801	Clayton	A1	$130
201801	...	...	...
201802	Caulfield	A1	$500
201802	Caulfield	B2	$430
201802	Caulfield	C3	$120
...	...	...	...

Table 5.5 Supplier point of view

SupplierID	TotalSales
S1	$77,000
S2	$5,700
S3	$12,500
...	...

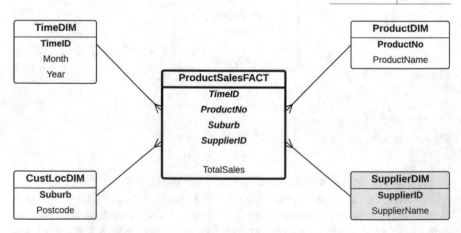

Fig. 5.6 A revised star schema

do not record which supplier supplied a product. In other words, there is no direct relationship between suppliers and total sales. Therefore, the two-column table for the supplier as illustrated in Table 5.5, although seeming to be correct, is actually wrong.

The problem will even be bigger if we add the date the supplies were delivered (e.g. the date attribute from the StockSupplier table; see the E/R diagram in Fig. 5.4).

Table 5.6 ProductSalesFact
table with Supplier

TimeID	Suburb	ProductNo	SupplierID	TotalSales
201801	Caulfield	A1	S1	...
201801	Caulfield	A1	S2	...
201801	Caulfield	A1	S3	...
201801	Caulfield	A1	...	...
201801	Caulfield	B2	S1	...
201801	Caulfield	B2	S2	...
201801	Caulfield	B2	S3	...
201801	Caulfield	B2	...	...
201801	Caulfield	C3	S1	...
201801	Caulfield	C3	S2	...
201801	Caulfield	C3	S3	...
201801	Caulfield	C3	...	...
201801	...	...	...	...
201801	Chadstone	A1	S1	...
201801	Chadstone	A1	S2	...
201801	Chadstone	A1	S3	...
201801	Chadstone	A1	...	...
201801	...	...	...	...
201802	Caulfield	A1	S1	...
201802	Caulfield	A1	S2	...
201802	Caulfield	A1	S3	...
201802	Caulfield	A1	...	...
...	...	...	...	

Imagine that from the ProductSales Fact, we would like to produce a report that takes all attributes from the Fact Table and adds them to one additional column next to the SupplierID, called SupplyDate. This will create a conflict between the TimeID column and the SupplyDate column. In addition, it is also impossible to include a SupplyDate column because a one product-supplier pair may have multiple supply dates.

The reasons why this problem occurs are the following: (i) there is no direct relationship between supplier and product sales, and (ii) supplier information is not available in the sales of a particular product.

One solution is to move the Supplier Dimension from being connected to the Fact and to create a relationship with the Product Dimension, resulting a **Bridge Table** between Product Dimension and Supplier Dimension (see Fig. 5.7). Note that three additional pieces of information are added: location, date and quantity of supplies.

The aim of using a Bridge Table is to "drill down" for information on products. For example, after examining the fact report, management might be interested in drilling down further on a particular product. Perhaps the product performs extremely well in sales or for any other reasons. Hence, we can drill down for information on that product and find out, in this example, details of the supply history and the suppliers as well.

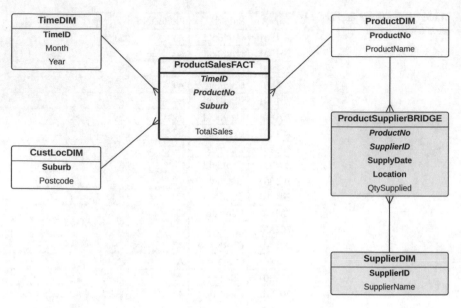

Fig. 5.7 A snowflake schema with a Bridge Table

The SQL commands to create fact and dimension tables are quite straightforward, which are as follows:

```
-- Time Dimension
create table TimeDim as
select
    distinct to_char(SalesDate, 'YYYYMM') as TimeID,
    to_char(SalesDate, 'YYYY') as Year,
    to_char(SalesDate, 'MM') as Month
from Sales;

-- Customer Location Dimension
create table CustLocDim as
select distinct Suburb, Postcode
from Customer;

-- Product Dimension
create table ProductDim as
select distinct ProductNo, ProductName
from Product;

-- Bridge Table
create table ProductSupplierBridge as
select *
from StockSupplier;

-- Supplier Dimension
create table SupplierDim as
```

```
Select SupplierID, Name as SupplierName
from Supplier;

-- Fact Table
create table ProductSalesFact as
Select
    to_char(S.SalesDate, 'YYYYMM') as TimeID,
    P.ProductNo,
    C.Suburb,
    sum(SI.QtySold*P.Price) as TotalSales
from Sales S, Product P, Customer C, SalesItem SI
where S.SalesNo = SI.SalesNo
and SI.ProductNo= P.ProductNo
and C.CustomerID = S.CustomerID
group by
    to_char(S.SalesDate, 'YYYYMM'),
    P.ProductNo,
    C.Suburb;
```

If, for example, the operational database does not maintain a master list of all suppliers (assume that each supplier may supply any product in an ad hoc manner), we will not have the Supplier Dimension in the schema (see Fig. 5.8). Instead, we will only have the Bridge Table, but for each Product-Supplier pair, we will have Location and Date implemented as a weak entity, because for each product-supplier, there are multiple supplies. Therefore, the Bridge Table acts as a *temporal dimension*. This means, for each product, there is a list of the history of supplies.

The Product Dimension table and the Bridge Table may look something like the ones in Tables 5.7 and 5.8.

In this case, the Product Supplier Bridge Table acts as a *temporal dimension* since the history of supplies is recorded.

If the history is not needed (in a non-temporal data warehouse), then the Product Supplier Bridge Table will only have ProductNo and SupplierName attributes, without the history of supplies (see Fig. 5.9).

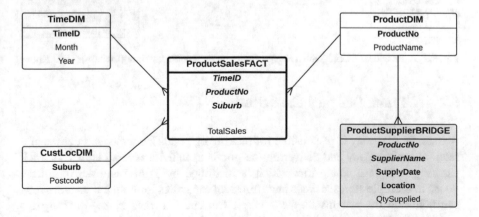

Fig. 5.8 A snowflake schema with a Bridge Table, but without a Supplier Dimension

Table 5.7 ProductDim table

ProductNo	ProductName
A1	Fandy Handbag
B2	Mercer Women Shoes
C3	Plain T-Shirt
...	...

Table 5.8 ProductSupplierBridge table

ProductNo	SupplierName	DateSupplied	Location	QtySupplied
A1	Cheap Goods Pty	21-May-2018	Caulfield	100
A1	Cheap Goods Pty	15-Aug-2018	Caulfield	150
A1	Cheap Goods Pty	19-Dec-2018	Clayton	50
A1	Cheap Goods Pty	...	...	...
A1	Just Bags	21-May-2018	Chadstone	200
A1	Just Bags	30-Jun-2018	Clayton	80
A1	JustBags	...	...	...
A1	Baggy	...	...	...
A1	...	...	...	...
B2	Cheap Goods Pty	...	...	...
...	...	...	...	...

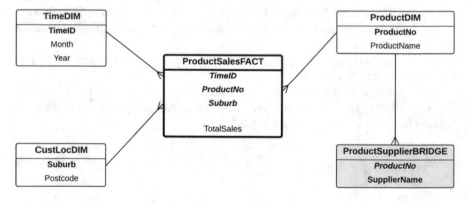

Fig. 5.9 A snowflake schema with a Bridge Table, but without maintaining the history of supplies

5.2 A Truck Delivery Case Study

A trucking company is responsible for picking up goods from the warehouses of a retail chain company and delivering the goods to individual retail stores. A truck carries goods on a single trip, which is identified by TripID and delivers these goods to multiple stores. Trucks have different capacities for both the volumes they can hold and the weights they can carry. Currently, a truck makes several trips each week. An operational database is being used to keep track of the deliveries,

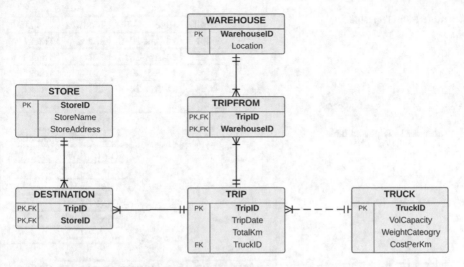

Fig. 5.10 An E/R diagram of the truck delivery system

including the scheduling of trucks to provide timely deliveries to stores. The E/R diagram of the truck delivery system is shown in Fig. 5.10.

Based on the E/R diagram, (*i*) a trip may pick up goods from many warehouses (i.e. a *many-many* relationship between Warehouse and Trip); (*ii*) a trip uses one truck only, and obviously a truck may have many trips in the history (i.e. a *many-1* relationship between Trip and Truck); and (*iii*) a trip delivers goods (e.g. TVs, fridges, etc.) potentially to several stores (a *many-many* relationship between Trip and Store, which is represented by the Destination table).

The tables from the operational database are as follows:

- **Warehouse**(WarehouseID, Location)
- **Trip**(TripID, Date, TotalKm, *TruckID*)
- **TripFrom**(*TripID*, *WarehouseID*)
- **Truck**(TruckID, VolCapacity, WeightCategory, CostPerKm)
- **Store**(StoreID, StoreName, Address)
- **Destination**(*TripID*, *StoreID*)

Some sample data in the operational database are shown in Tables 5.9, 5.10, 5.11, 5.12, 5.13, and 5.14.

Table 5.9 Warehouse table

WarehouseID	Location
W1	Warehouse1
W2	Warehouse1
W3	Warehouse1
...	...

Table 5.10 Trip table

TripID	Date	TotalKm	TruckID
Trip1	14-Apr-2018	370	Truck1
Trip2	14-Apr-2018	570	Truck2
Trip3	14-Apr-2018	250	Truck3
Trip4	15-Apr-2018	450	Truck1
...	...	...	...

Table 5.11 TripFrom table

TripID	WarehouseID
Trip1	W1
Trip1	W2
Trip1	W3
Trip2	W1
Trip2	W2
...	...

Table 5.12 Truck table

TruckID	VolCapacity	WeightCategory	CostPerKm
Truck1	250	Medium	$1.20
Truck2	300	Medium	$1.50
Truck3	100	Small	$0.80
Truck4	550	Large	$2.30
Truck5	650	Large	$2.50
...	...	...	...

Table 5.13 Store table

StoreID	StoreName	Address
M1	MyStore City	Melbourne
M2	MyStore Chaddy	Chadstone
M3	MyStore HiPoint	High Point
M4	MyStore Westfield	Doncaster
M5	MyStore North	Northland
M6	MyStore South	Southland
M7	MyStore East	Eastland
M8	MyStore Knox	Knox City
...	...	...

The management of this trucking company would like to analyse the delivery cost based on *trucks*, *time period* and *store*.

5.2.1   Solution Model 1: Using a Bridge Table

The fact measure to be included in the Fact Table is "Total Delivery Cost", which is calculated by *distance* (total kilometres in the Trip table) and *cost per kilometre*

Table 5.14 Destination table

TripID	StoreID
Trip1	M1
Trip1	M2
Trip1	M4
Trip1	M3
Trip1	M8
Trip2	M4
Trip2	M1
Trip2	M2
...	...

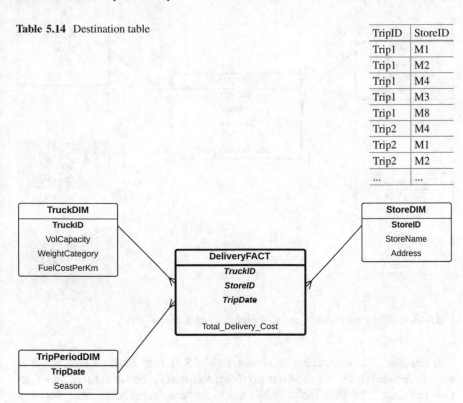

Fig. 5.11 A possible truck delivery star schema

(from the Truck table). The dimensions are Truck, Trip Period and Store. A possible star schema is shown in Fig. 5.11.

Based on the sample data as shown previously in Tables 5.9, 5.10, 5.11, 5.12, 5.13, and 5.14:

- From the Truck point of view, Truck1 makes two trips (e.g. Trip1 and Trip4), travelling a total of 820 km (370 km + 450 km). The cost for Truck1 is $1.20/km. Hence, calculating the cost for Truck1 is straightforward. The cost of the other trucks can also be calculated this way.
- From the Period point of view (say from a date point of view), three trips were made on 14 April 2018 (e.g. Trip1, Trip2 and Trip3). Trip1 (370 km) is delivered by Truck1 which costs $1.20/km. Trip2 and Trip 3 on the same day can be calculated in the same way. Hence, on 14 April 2018, the total cost can be calculated.
- From the Store point of view, the cost is calculated based on Trip, but a trip delivers goods to many stores. Therefore, the delivery cost for each store cannot be calculated. The delivery cost is for the trip, not for the store.

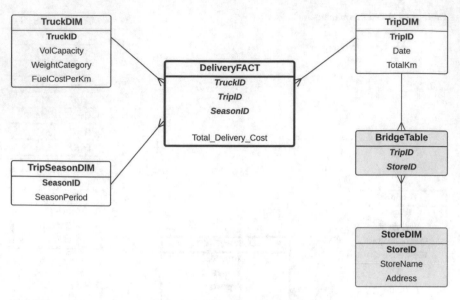

Fig. 5.12 A correct truck delivery snowflake schema with a Bridge Table

Therefore, the star schema as shown in Fig. 5.11 may look correct, but it is actually **incorrect**, because there is **no** direct relationship between Store and Total Delivery Cost in the Fact Table. This is due to the *many-many* relationship between the Trip entity and Store entity in the E/R diagram of the operational database. To solve this problem, a Bridge Table can be used, as shown in Fig. 5.12.

5.2.2 Solution Model 2: Add a Weight Factor Attribute

The Solution Model 1 shown in the previous section demonstrates that there is no direct relationship between Total Delivery Cost and Store. Total delivery cost is calculated at a Trip level and not at a Store level. For example, The length of Trip 1 was 370 km on 14 April using Truck1, and the cost per kilometre for Truck1 is $1.20. Hence, Trip1 delivered goods to five stores at a cost of $444. Using the current data that we have, it is impossible to calculate the delivery cost for each of these five stores for Trip1. What we can say is that the cost for Trip1 is x and Trip1 delivered to y number of stores. However, we can estimate the total delivery cost per store, if we want to. This can be estimated through the "Weight Factor" (see the Weight Factor attribute in the Trip Dimension in Fig. 5.13).

A *weight factor* is a proportion of the trip that goes to each store for that particular trip. For example, if Trip1 went to five stores, then the Weight Factor is 0.2 (or 20%). This implies that each store "contributes" 20% of the total delivery cost for that trip. This is certainly inaccurate, but this is the only estimate that we can make, based on the data that we have.

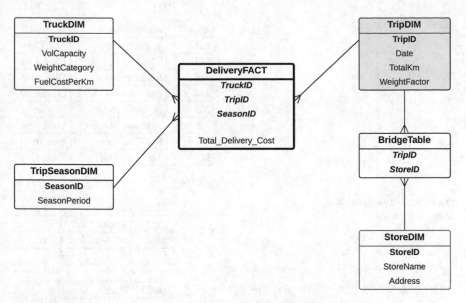

Fig. 5.13 A truck delivery snowflake schema with a Weight attribute

Table 5.15 Trip Dimension table

TripID	Date	TotalKm	WeightFactor
Trip1	14-Apr-2018	370	0.20
Trip2	14-Apr-2018	570	0.33
...	...	...	...

Table 5.16 Bridge Table

TripID	StoreID
Trip1	M1
Trip1	M2
Trip1	M4
Trip1	M3
Trip1	M8
Trip2	M4
Trip2	M1
Trip2	M2
...	...

Not all snowflake schemas with Bridge Tables must have a weight factor. A weight factor is only needed if we want to estimate the contribution that a dimension made to the fact. Using the star schema in Fig. 5.11, if a weight factor is not used, then we will not be able to estimate the total delivery cost per store. Is missing this information very critical? It depends. After all, the information on total delivery cost per store is an estimated figure anyway.

The data in the Trip, Bridge Table and Store Dimensions are shown in Tables 5.15, 5.16, and 5.17.

Table 5.17   Store table

StoreID	StoreName	Address
M1	MyStore City	Melbourne
M2	MyStore Chaddy	Chadstone
M3	MyStore HiPoint	High Point
M4	MyStore Westfield	Doncaster
M5	MyStore North	Northland
M6	MyStore South	Southland
M7	MyStore East	Eastland
M8	MyStore Knox	Knox City
...	...	...

Table 5.18   Trip Dimension
table

TripID	Date	TotalKm
Trip1	14-Apr-2018	370
Trip2	14-Apr-2018	570
...	...	...

Table 5.19   Bridge Table

TripID	StoreID	WeightFactor
Trip1	M1	0.20
Trip1	M2	0.20
Trip1	M4	0.20
Trip1	M3	0.20
Trip1	M8	0.20
Trip2	M4	0.33
Trip2	M1	0.33
Trip2	M2	0.33
...	...	...

The 0.2 weight factor for Trip1 does not mean that Trip1 has 20% of the total delivery cost; rather, Trip1 delivers to five stores, and each store contributes 20% (or 0.20) to the total delivery cost. If this is what it means, why not have the Weight Factor attribute in the Bridge Table, as shown in Tables 5.18 and 5.19, to indicate, for example, that store M1 (from Trip 1) contributes 20% (0.20) to the total delivery cost of Trip1? The answer is yes, we could do that, but then the Bridge Table will contain redundant information, because all the stores in Trip1 should have a weight factor of 20%. So, instead of storing the 0.20 in each of the stores for Trip1, we store the 0.20 in the Trip1 record itself in the Trip Dimension table (refer to Table 5.15).

Creating the Trip Dimension table with the WeightFactor attribute is done using the following SQL:

```
create table TripDim2 as
select T.TripID, T.TripDate, T.TotalKm,
    1.0/count(*) as WeightFactor
from Trip T, Destination D
where T.TripID = D.TripID
group by T.tripid, T.tripdate, T.totalkm;
```

5.2.3   Solution Model 3: A List Aggregate Version

In the *ListAgg* (short for *List Aggregate*) option, we have one additional attribute in the parent dimension (e.g. the Trip Dimension) which holds the information on the group of each record in the parent dimension table (e.g. stores within each trip). Figure 5.14 includes an attribute called "StoreGroupList" in the Trip Dimension. In this attribute, all stores for each trip are concatenated to become a store group list.

The contents of the Trip Dimension and the Bridge Table are shown in Tables 5.20 and 5.21.

To create the Trip Dimension table with a StoreGroupList attribute, we can use the `listagg` function in SQL. The `listagg` function has the following format:

```
listagg (Attr1, '_') within group
    (order by Attr1) as ColumnName
```

where Attr1 is the StoreID and the "_" indicates that the StoreIDs are concatenated with the "_" symbol (e.g. M1_M2_M3_M4_M8). If we want to list the stores listed in descending order (e.g. M8_M4_M3_M2_M1), then we use `within`

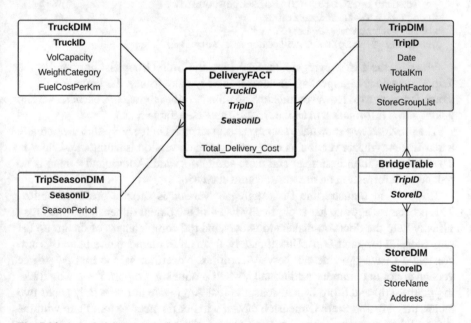

Fig. 5.14   A truck delivery snowflake schema with a StoreGroupList attribute

Table 5.20   Trip Dimension table

TripID	Date	TotalKm	WeightFactor	StoreGroupList
Trip1	14-Apr-2018	370	0.20	M1_M2_M3_M4_M8
Trip2	14-Apr-2018	570	0.33	M1_M2_M4
...	...	...	...	...

Table 5.21 Bridge Table

TripID	StoreID
Trip1	M1
Trip1	M2
Trip1	M4
Trip1	M3
Trip1	M8
Trip2	M4
Trip2	M1
Trip2	M2
...	...

`group (order by Attr1 Desc)`. The SQL to create the Trip Dimension table becomes:

```
create table TripDim3 as
select T.TripID, T.TripDate, T.TotalKm,
    1.0/count(D.StoreID) as WeightFactor,
    listagg (D.StoreID, '_') within group
    (order by D.StoreID) as StoreGroupList
from Trip T, Destination D
where T.TripID = D.TripID
group by T.TripID, T.TripDate, T.TotalKm;
```

Visually, the List Aggregate attribute (e.g. the StoreGroupList attribute in the Trip Dimension) is appealing because it is very easy to see the complete list of stores for each trip. However, this information is rather redundant because we can get the same information from the Bridge Table (see Table 5.21).

The previous two snowflake schemas for the Truck Delivery System are actually a non-List Aggregate version. A non-List Aggregate version is simpler and cleaner. The implementation is simpler and more straightforward. Additionally, there is no redundant information on the stores within each trip.

However, in industry, the List Aggregate version is often a preferred option. This is because a group list is physically listed in the parent dimension, which may visually help the decision-makers to understand the completeness of the group list (e.g. StoreID for each Trip). Unfortunately, from an implementation point of view, creating a List Aggregate can be very complex. Nevertheless, the List Aggregate version is not uncommon. Additionally, when producing a report, if we would like to join the Trip and Store Dimensions, the List Aggregate attribute only needs two tables, the Trip and Store Dimension tables, without the need to join them with the Bridge Table.

```
select *
from TripDim3 T, StoreDim3 S
where T.StoreGroupList like '%'||S.StoreID||'%';
```

In this SQL, the joining is based on the StoreGroupList attribute in the Trip Dimension table and the StoreID in the Store Dimension table. However, the join condition is not a simple `T.StoreGroupList = S.StoreID` because the join condition is not a simple match. Instead, we use a `like` operator to check if the StoreID value exists in the StoreGroupList.

In contrast, without the StoreGroupList attribute in the Trip Dimension, we need to join three tables, the Trip Dimension, the Bridge Table and the Store Dimension tables, using the following SQL:

```
select *
from TripDim3 T, BridgeTable3 B, StoreDim3 S
where T.TripID = B.TripID
and B.StoreID = S.StoreID;
```

5.3   Summary

In principle, a Bridge Table is used:

(a) When it is impossible to have a dimension connected directly to the Fact Table, because simply there is no relationship between this dimension and the Fact Table (e.g. in the Product Sales case study, it is impossible to have a direct link from SupplierDim to ProductSalesFact)
(b) When an entity (which will become a dimension) has a *many-many* relationship with another entity (dimension) in the E/R schema of the operational database (e.g. Supplier and Stock has a *many-many* relationship).
(c) When the temporality aspect (data history) is maintained in the operational database and the Bridge Table can be used to accommodate a dimension that has temporal attributes (e.g. product supply history is maintained in the second snowflake schema example).

When a Bridge Table is used in the schema, there are two additional options:

(a) A Weight Factor is used to estimate the contribution of a dimension in the calculation of the fact measure. Because this is only an estimate, a weight factor is an option.
(b) Every snowflake schema (whether it has Weight Factor or not) can be implemented in two ways: a List Aggregate version and a non-List Aggregate version.

5.4   Exercises

5.1 Figure 5.15 is an E/R diagram of an operational database. The system stores information about books, including the authors, publishers, book categories as well as the reviews that each book has received. The "stars" attribute in the Review entity records the star rating for each review (e.g. 5 stars for excellent to 1 star for poor, etc.). One book may receive many reviews. For simplicity, it is assumed, as also shown in the E/R diagram, that a book will only have one category.

The E/R diagram also includes entities related to the sale of books and the stores which sell the books. Each store has many sales transactions (i.e. the Sales

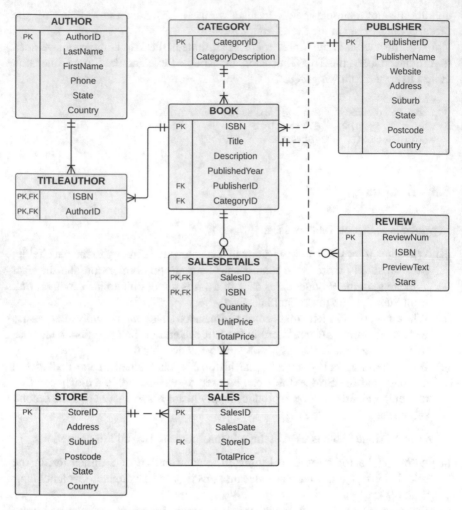

Fig. 5.15   A book sales system

entity), and each sales transaction may include several books (i.e. the SalesDetails entity). The TotalPrice attribute in the SalesDetails entity is quantity multiplied by UnitPrice, whereas the TotalPrice attribute in the Sales entity is the total price for each sales transaction.

You are required to design a small data warehouse for analysis purposes. The analysis is needed to answer at least the following questions:

- What are the total sales for each store in a month?
- How many books are sold in each category?
- Which book category has the highest sales?
- Which author has sold the highest number of books?

Write the SQL commands to create the dimensions and Fact Tables.

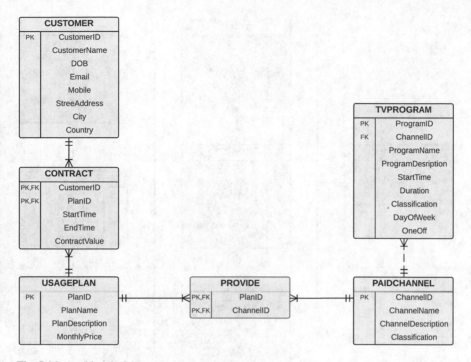

Fig. 5.16 A cable television system

5.2 The Program Manager of a Cable Television company is interested in analysing the statistics on the revenue it receives from paid channel subscriptions. The analysis is needed to identify which paid TV channels are more attractive to users than the others. The results of the analysis will be useful to determine which channels to purchase in the following years. You are required to design a small Data Warehouse to keep track of the statistics. The director is particularly interested in analysing the *total number of contracts* and *total revenue* (total contract value) by channels, location, year and month (as in contract start time). The E/R diagram of the operational database is shown in Fig. 5.16.

Based on the Program Manager's requirements, develop a snowflake schema. If you are using a Bridge Table, make sure you include a Weight Factor attribute and a List Aggregate attribute in your design. Write the queries to create (and populate) the dimension tables and the Fact Table.

5.3 The Rural University of Victoria (RUV) has a number of campuses in several cities and towns in the state of Victoria. Each campus has several departments. Staff from one campus may need to travel to a different campus for various reasons, for example to teach, to attend administrative meetings or to conduct collaborative research with staff on another campus.

To facilitate the travel between campuses, some departments have vehicles which staff may borrow. In addition to this, the university also has a fleet of vehicles which staff may borrow if the department's cars are booked.

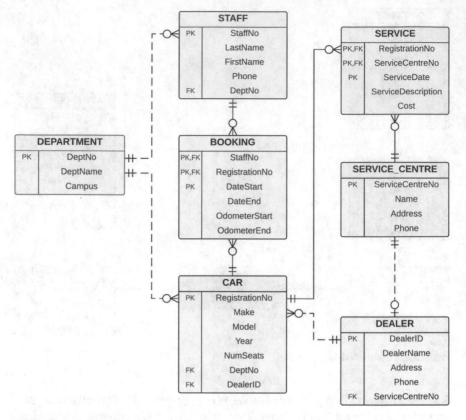

Fig. 5.17 A car booking system

These cars need a regular service, and from time to time, they also require non-regular maintenance, such as new tires, replacement batteries, etc.

The E/R diagram of the operational database is shown in Fig. 5.17.

The university now requires a data warehouse to analyse the cost of car services and service costs and to answer at least the following questions:

- How much is the total service cost in each month/year?
- How many times is a car serviced each month?
- How much is the total service cost for each department?
- Approximately, how much of the total service cost is incurred by each user?

Based on the above requirements, develop a snowflake schema with a Bridge Table. The snowflake schema must also include a Weight Factor attribute and a List Aggregate attribute. Write the queries to create (and populate) the dimension tables and the Fact Table.

5.4 This question is related to the Car Service System, shown in Fig. 5.18. Every time a service is conducted, a record is entered into the database. The information

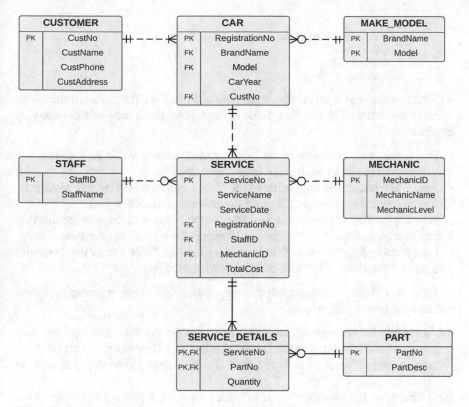

Fig. 5.18 A Car Service System

recorded includes service number, service name, car registration number, the name of the staff member who booked the service and liaised with the customer, the mechanic who performed the repair and the total cost of the repair.

Questions

(a) A simple star schema for the Car Service System has been designed and is shown in Fig. 5.19. It is a simple star schema with four dimensions: Time, Brand, Mechanic and Part Dimensions. The fact measure is the Number of Services. The Fact Table is created by a join operation of the Service, Service Details and Car tables in the operational database.

```
create table CarServiceFact as
select
  to_char(S.ServiceDate, 'YYYYMM') as TimeID,
  C.BrandName,
  S.MechanicID,
  D.PartNo,
  count(S.ServiceNo) as Number_of_Services
from Service S, Service_Details D, Car C
where S.ServiceNo = D.ServiceNo
and S.RegistrationNo = C.RegistrationNo
```

```
group by
  to_char(S.ServiceDate, 'YYYYMM'),
  C.BrandName,
  S.MechanicID,
  D.PartNo;
```

Your task is to create all the dimension tables. Is the Fact Table created correctly? Analyse the contents of the Fact Table by analysing the results of the following queries:

- Pick a Part, and then retrieve the Number of Services for this Part. Check this query result manually against the operational database.
- Pick a Mechanic, and then retrieve the Number of Services by this Mechanic. Again, check the query result against the operational database.
- Now use the Part and Mechanic, and retrieve the Number of Services for this Part and Mechanic. Check your query result against the operational database.
- Lastly, retrieve the Number of Services from the Fact Table. Check this manually against the number of services in the operational database

If the Fact Table is incorrect, explain why the star schema is incorrect, and then redesign the correct star schema.

(b) Build another star schema, but this time to analyse the Total Service Cost, also based on each month, car brand, mechanic and part. Note that every service may use several different parts. Write the SQL commands to create the dimension and Fact Tables.

(c) Looking at the above two star schemas, one for Number of Services (see Fig. 5.19) and the other for Total Service Cost (your star schema from question (b)), what is the main difference, and what/where was the mistake?

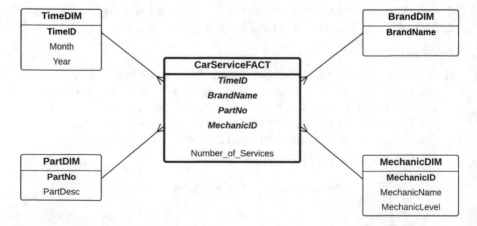

Fig. 5.19 A star schema for the Car Service System

5.5 Further Readings

This chapter focuses solely on the concept of *Bridge Tables* in star schemas. A Bridge Table is actually a *many-many* relationship between two dimension tables. The *many-many* relationships have been used extensively in relational database design and modelling. They can be found in typical database and data modelling textbooks, such as [1–21] and [22]

Aggregate functions, such as `listagg` and other aggregate functions, can be found in many SQL books, such as [23–25] and [26].

String subset join and other non-equi join queries are explained in various SQL books and database textbooks, such as [7–11, 13–15, 23–25] and [26].

Basic join algorithms, such as Nested-Loop, Sort-Merge and Hash Join algorithms, are explained in [27]. Non-equi join algorithms are less often discussed. Some earlier work on non-equi join and band join algorithms can be found in [28, 29] and [30].

References

1. T.A. Halpin, T. Morgan, *Information Modeling and Relational Databases*, 2nd edn. (Morgan Kaufmann, Los Altos, 2008)
2. S. Bagui, R. Earp, *Database Design Using Entity-Relationship Diagrams*. Foundations of Database Design (CRC Press, Boca Raton, 2003)
3. M.J. Hernandez, *Database Design for Mere Mortals: A Hands-on Guide to Relational Database Design*. For Mere Mortals (Pearson Education, 2013)
4. N.S. Umanath, R.W. Scamell, *Data Modeling and Database Design* (Cengage Learning, 2014)
5. T.J. Teorey, S.S. Lightstone, T. Nadeau, H.V. Jagadish, *Database Modeling and Design: Logical Design*. The Morgan Kaufmann Series in Data Management Systems (Elsevier, Amsterdam, 2011)
6. G. Simsion, G. Witt, *Data Modeling Essentials*. The Morgan Kaufmann Series in Data Management Systems (Elsevier, Amsterdam, 2004)
7. C. Coronel, S. Morris, *Database Systems: Design, Implementation, & Management* (Cengage Learning, 2018)
8. T. Connolly, C. Begg, *Database Systems: A Practical Approach to Design, Implementation, and Management* (Pearson Education, 2015)
9. J.A. Hoffer, F.R. McFadden, M.B. Prescott, *Modern Database Management* (Prentice Hall, 2002)
10. A. Silberschatz, H.F. Korth, S. Sudarshan, *Database System Concepts*, 7th edn. (McGraw-Hill, New York, 2020)
11. R. Ramakrishnan, J. Gehrke, *Database Management Systems*, 3rd edn. (McGraw-Hill, New York, 2003)
12. J.D. Ullman, J. Widom, *A First Course in Database Systems*, 2nd edn. (Prentice Hall, Englewood Cliffs, 2002)
13. H. Garcia-Molina, J.D. Ullman, J. Widom, *Database Systems: The Complete Book* (Pearson Education, 2011)
14. P.E. O'Neil, E.J. O'Neil, *Database: Principles, Programming, and Performance*, 2nd edn. (Morgan Kaufmann, Los Altos, 2000)

15. R. Elmasri, S.B. Navathe, *Fundamentals of Database Systems*, 3rd edn. (Addison-Wesley-Longman, 2000)
16. C.J. Date, *An Introduction to Database Systems*, 7th edn. (Addison-Wesley-Longman, 2000)
17. R.T. Watson, *Data Management—Databases and Organizations*, 5th edn. (Wiley, London, 2006)
18. P. Beynon-Davies, *Database Systems* (Springer, Berlin, 2004)
19. E.F. Codd, *The Relational Model for Database Management, Version 2* (Addison-Wesley, Reading, 1990).
20. J.L. Harrington, *Relational Database Design Clearly Explained*. Clearly Explained Series (Morgan Kaufmann, Los Altos, 2002)
21. M. Kifer, A.J. Bernstein, P.M. Lewis, *Database Systems: An Application-oriented Approach* (Pearson/Addison-Wesley, 2006)
22. W. Lemahieu, S. vanden Broucke, *Principles of Database Management: The Practical Guide to Storing, Managing and Analyzing Big and Small Data* (Cambridge University Press, Cambridge, 2018)
23. J. Melton, *Understanding the New SQL: A Complete Guide*, vol. I 2nd edn. (Morgan Kaufmann, Los Altos, 2000)
24. C.J. Date, *SQL and Relational Theory—How to Write Accurate SQL Code*, 2nd edn. Theory in Practice (O'Reilly, 2012)
25. A. Beaulieu, *Learning SQL: Master SQL Fundamentals* (O'Reilly Media, 2009)
26. M.J. Donahoo, G.D. Speegle, *SQL: Practical Guide for Developers*. The Practical Guides (Elsevier, Amsterdam, 2010)
27. D. Taniar, C.H.C. Leung, W. Rahayu, S. Goel, *High Performance Parallel Database Processing and Grid Databases* (Wiley, London, 2008)
28. J. Van den Bercken, B. Seeger, P. Widmayer, The bulk index join: a generic approach to processing non-equijoins, in *Proceedings of the 15th International Conference on Data Engineering, Sydney, Australia, March 23–26, 1999*, ed. by M. Kitsuregawa, M.P. Papazoglou, C. Pu (IEEE Computer Society, 1999), p. 257
29. P. Bouros, K. Lampropoulos, D. Tsitsigkos, N. Mamoulis, M. Terrovitis, Band joins for interval data, in *Proceedings of the 23rd International Conference on Extending Database Technology, EDBT 2020, Copenhagen, Denmark, March 30 - April 02, 2020*, ed. by A. Bonifati, Y. Zhou, M.A.V Salles, A. Böhm, D. Olteanu, G.H.L. Fletcher, A. Khan, B. Yang (2020), pp. 443–446. OpenProceedings.org
30. H. Lu, K.L Tan, On sort-merge algorithm for band joins. IEEE Trans. Knowl. Data Eng. **7**(3), 508–510 (1995)

Chapter 6
Temporal Data Warehousing

A *temporal data warehousing* is a warehousing technique where the temporal (or historical) aspect of records is incorporated into the warehouse. For example, if a book is a dimension (in a bookshop setting), the book price changes over time. If we would like to keep track of the book prices in the data warehouse, then a temporal data warehouse technique must be used. Without this temporal or historical aspect of the book price, the analysis made from the data warehouse is only based on the current book price.

This chapter focuses on temporal data warehousing covering star schemas, which maintain a temporal aspect of the data. Temporal data warehousing is also known as *slowly changing dimensions* (*SCD*). This chapter will also describe various techniques that can be used to deal with SCD.

6.1 A Bookshop Case Study

To learn the concept of temporal data warehousing, let's use the Bookshop example as a case study. The management of a bookshop with a number of branches would like to build a data warehouse to analyse their book sales. They have already stored all book sales transactions in an operational database. The management would particularly like to analyse their book sales performance from various perspectives, such as from a monthly basis, from a book basis and from a branch basis. The first task to do this is to define a star schema.

A simple star schema for this Bookshop case study is shown in Fig. 6.1. It has three dimensions: Time, Branch and Book Dimensions, and the fact has one fact measure, called Number of Books Sold. With this star schema, the management is able to retrieve the number of books sold based on the month, based on the branch and based on the book.

© The Author(s), under exclusive license to Springer Nature Switzerland AG 2021   123
D. Taniar, W. Rahayu, *Data Warehousing and Analytics*, Data-Centric Systems
and Applications, https://doi.org/10.1007/978-3-030-81979-8_6

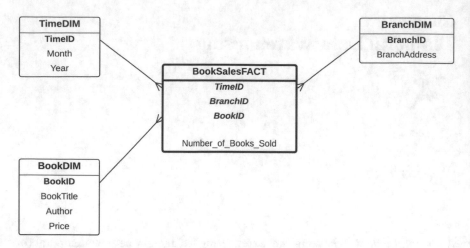

Fig. 6.1 The Bookshop star schema

The Fact Table would probably look like the following (see Table 6.1). Note that the sales figures are for illustrative purposes only, and the following entries in the tables are not necessarily comprehensive.

The Fact Table can be joined with the dimensions to produce a more comprehensive report. In the example below, the Fact Table is joined with the Book Dimension table, and consequently the attributes in the Book Dimension table will appear in the report. Assume the Book Dimension table has the following data (see Table 6.2).

The new report is shown in Table 6.3. Note the three additional attributes: Book Title, Author and Price from the Book Dimension table. Pay attention to the Price column in Report 1. In this case, the Price column contains the *Current Price* of each book, and it is likely to be the *Original Price*. However, from time to time, a book is sold at a discounted price. The above sales report does not reflect this; therefore, the analysis from this report can be misleading.

To incorporate the temporal values of the Price, we need to add a *Bridge Table* to the Book Dimension to store the history of book prices (which is the Book Price Dimension). Note that the Book Price Dimension (the Bridge Table) is implemented as a *Weak Entity*, with a composite primary key of BookID, Start Date and End Date (refer to Fig. 6.2).

The Book Dimension and Book Price Dimension tables are as follows (refer to Tables 6.4 and 6.5). In the Book Price Dimension table, a *temporal* attribute Price is managed, where for each price, the period (Start and End Dates) and the remarks are recorded. Note that we use Dec9999 to indicate that the end date is still open until now.

Table 6.1 BookSalesFact
Table

TimeID	BranchID	BookID	Number of books sold
Mar2008	City	C1	5
Mar2008	City	H6	15
Mar2008	City	DV	23
Mar2008	City	...	...
Mar2008	Chadstone	C1	15
Mar2008	Chadstone	H6	3
Mar2008	Chadstone	DV	2
Mar2008	Chadstone	...	...
Mar2008	Camberwell	C1	1
Mar2008	Camberwell	H6	1
Mar2008	Camberwell	DV	2
Mar2008	Camberwell	...	...
Mar2008	...	...	...
...	...	...	...
...	...	...	...
Dec2007	City	C1	15
Dec2007	City	H6	6
Dec2007	City	DV	6
Dec2007	City	...	...
Dec2007	Chadstone	C1	10
Dec2007	Chadstone	H6	8
Dec2007	Chadstone	DV	1
Dec2007	Chadstone	...	...
Dec2007	Camberwell	C1	18
Dec2007	Camberwell	H6	3
Dec2007	Camberwell	DV	2
Dec2007	Camberwell	...	...
Dec2007	...	...	...
...	...	...	...

Table 6.2 Book Dimension
table

BookID	Book title	Author	Price
C1	CSIRO Diet	CSIRO Team	$45.95
H6	Harry Potter 6	Rowling	$30.95
DV	Da Vinci Code	Dan Brown	$27.95
...	...	...	...

Table 6.3 Report 1 (Book Sales Fact with Book Dimension)

TimeID	BranchID	BookID	Book title	Author	Price	Number of books sold
Mar2008	City	C1	CSIRO Diet	CSIRO Team	$45.95	5
Mar2008	City	H6	Harry Potter 6	Rowling	$30.95	15
Mar2008	City	DV	Da Vinci Code	Dan Brown	$27.95	23
Mar2008	City	...	...	...	...	...
Mar2008	Chadstone	C1	CSIRO Diet	CSIRO Team	$45.95	15
Mar2008	Chadstone	H6	Harry Potter 6	Rowling	$30.95	3
Mar2008	Chadstone	DV	Da Vinci Code	Dan Brown	$27.95	2
Mar2008	Chadstone	...	...	...	...	...
Mar2008	Camberwell	C1	CSIRO Diet	CSIRO Team	$45.95	1
Mar2008	Camberwell	H6	Harry Potter 6	Rowling	$30.95	1
Mar2008	Camberwell	DV	Da Vinci Code	Dan Brown	$27.95	2
Mar2008	Camberwell	...	...	...	...	...
Mar2008	...	...	...	...	...	...
...	...	...	...	...	...	...
...	..	...	...	...	...	...
Dec2007	City	C1	CSIRO Diet	CSIRO Team	$45.95	15
Dec2007	City	H6	Harry Potter 6	Rowling	$30.95	6
Dec2007	City	DV	Da Vinci Code	Dan Brown	$27.95	6
Dec2007	City	..	...	...	...	...
Dec2007	Chadstone	C1	CSIRO Diet	CSIRO Team	$45.95	10
Dec2007	Chadstone	H6	Harry Potter 6	Rowling	$30.95	8
Dec2007	Chadstone	DV	Da Vinci Code	Dan Brown	$27.95	1
Dec2007	Chadstone	..	...	...	...	...
Dec2007	Camberwell	C1	CSIRO Diet	CSIRO Team	$45.95	18
Dec2007	Camberwell	H6	Harry Potter 6	Rowling	$30.95	3
Dec2007	Camberwell	DV	Da Vinci Code	Dan Brown	$27.95	2
Dec2007	Camberwell	..	...	...	...	...
Dec2007	...	...	...	...	...	...
...	..	...	...	...	...	...

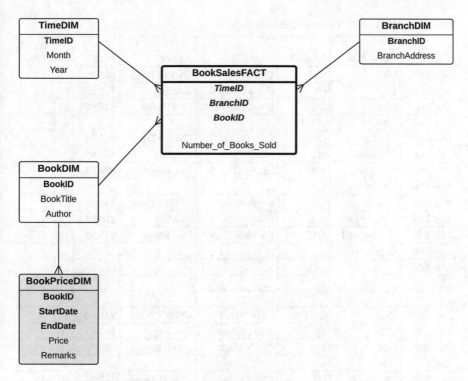

Fig. 6.2 A temporal data warehousing star schema

Table 6.4 Book Dimension table

BookID	Book title	Author
C1	CSIRO Diet	CSIRO Team
H6	Harry Potter 6	Rowling
DV	Da Vinci Code	Dan Brown
...	...	...

Table 6.5 Book Price Dimension table

BookID	Start date	End date	Price	Remarks
C1	Jan2007	July2007	$45.95	Full Price
C1	Aug2007	Oct2007	$36.75	20% Discount
C1	Nov2007	Jan2008	$23.00	Half Price
C1	Feb2008	**Dec9999**	$45.95	Full Price
H6	Jan2007	Mar2007	$21.95	Launching
H6	Apr2007	Jan2008	$30.95	Full Price
H6	Feb2008	**Dec9999**	$10.00	End of Product Sale
DV	Jan2007	**Dec9999**	$27.95	Full Price
...	...	...	...	...

Table 6.6 Report 2 with the correct Book Price

TimeID	BranchID	BookID	Book title	Author	Price	Number of books sold
Mar2008	City	C1	CSIRO Diet	CSIRO Team	$45.95	5
Mar2008	City	H6	Harry Potter 6	Rowling	**$10.00**	15
Mar2008	City	DV	Da Vinci Code	Dan Brown	$27.95	23
Mar2008	City	...	...	...	...	...
Mar2008	Chadstone	C1	CSIRO Diet	CSIRO Team	$45.95	15
Mar2008	Chadstone	H6	Harry Potter 6	Rowling	**$10.00**	3
Mar2008	Chadstone	DV	Da Vinci Code	Dan Brown	$27.95	2
Mar2008	Chadstone	...	...	...	...	...
Mar2008	Camberwell	C1	CSIRO Diet	CSIRO Team	$45.95	1
Mar2008	Camberwell	H6	Harry Potter 6	Rowling	**$10.00**	1
Mar2008	Camberwell	DV	Da Vinci Code	Dan Brown	$27.95	2
Mar2008	Camberwell	...	...	...	...	...
Mar2008	...	...	...	...	...	...
...	...	...	...	...	...	...
...	...	...	...	...	...	...
Dec2007	City	C1	CSIRO Diet	CSIRO Team	**$23.00**	15
Dec2007	City	H6	Harry Potter 6	Rowling	$30.95	6
Dec2007	City	DV	Da Vinci Code	Dan Brown	$27.95	6
Dec2007	City	..	...	...	...	...
Dec2007	Chadstone	C1	CSIRO Diet	CSIRO Team	**$23.00**	10
Dec2007	Chadstone	H6	Harry Potter 6	Rowling	$30.95	8
Dec2007	Chadstone	DV	Da Vinci Code	Dan Brown	$27.95	1
Dec2007	Chadstone	..	...	...	...	...
Dec2007	Camberwell	C1	CSIRO Diet	CSIRO Team	**$23.00**	18
Dec2007	Camberwell	H6	Harry Potter 6	Rowling	$30.95	3
Dec2007	Camberwell	DV	Da Vinci Code	Dan Brown	$27.95	2
Dec2007	Camberwell	..	...	...	...	...
Dec2007	...	...	...	...	...	...
...	...	...	...	...	...	...

Since each book has a temporal feature (i.e. temporal attribute), the previous Report 1 can be revised to incorporate the correct sale price of the book. Hence, the new report (Report 2) is shown in Table 6.6. Note that now we can understand why more copies of CSIRO Diet were sold in December 2007 as it was half price. Also note that the sale of Harry Potter is coming to an end (e.g. not much demand). The Da Vinci Code has always been sold for the full price.

6.2   Implementation of Temporal Data Warehousing

This section will discuss the SQL implementation details of the temporal data warehousing. Consider an E/R diagram of the operational database as shown in Fig. 6.3.

The star schema was previously shown in Fig. 6.2. The SQL commands to create the dimension tables are as follows:

```
create table BranchDim as
select * from Branch;

create table BookDim as
select * from Book;

create table TimeDim as
select distinct
  to_char(TransactionDate, 'MonYYYY') as TimeID,
  to_char(TransactionDate, 'Mon') as Month,
  to_char(TransactionDate, 'YYYY') as Year
from BookTransaction;

create table BookPriceDim as
select * from BookPriceHistory;
```

The SQL to create the Fact Table is as follows:

```
create table BookSalesFact1 as
select
  to_char(T.TransactionDate, 'MonYYYY') as TimeID,
  BK.BookID,
  BR.BranchID,
  sum(T.Quantity) as Number_of_Books_Sold
from BookTransaction T, Book BK, Branch BR
where T.BranchID = BR.BranchID
and T.BookID = BK.BookID
group by
  to_char(T.TransactionDate, 'MonYYYY'),
  BK.BookID,
  BR.BranchID;
```

Calculating Number of Books Sold is quite straightforward, that is, the sum of the attribute quantity from the Book Transaction table in the operational database.

Suppose we would like to have one additional fact measure, called *Total Sales*. Therefore, the new star schema is shown in Fig. 6.4.

The SQL command to create the second Fact Table (we call it *BookSalesFact2*) is as follows:

```
create table BookSalesFact2 as
select * from BookSalesFact1;

alter table BookSalesFact2
add (Total_Sales number);
```

```
declare
  cursor PriceCursor is
    select *
    from BookPriceDim;
begin
  for Item in PriceCursor loop
    -- update value for Total_Sales in BookSalesFact2
    update BookSalesFact2
    set Total_Sales = Number_Of_Books_Sold * Item.Price
    where BookID = Item.BookID
    and to_date(TimeID, 'MonYYYY') >=
      to_date(Item.StartDate, 'MonYYYY')
    and to_date(TimeID, 'MonYYYY') <=
      to_date(Item.EndDate, 'MonYYYY');
  end loop;
end;
/
```

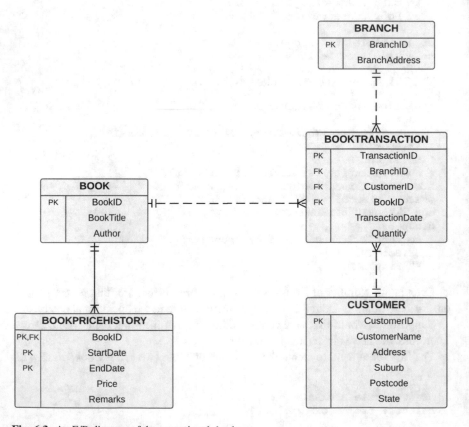

Fig. 6.3 An E/R diagram of the operational database

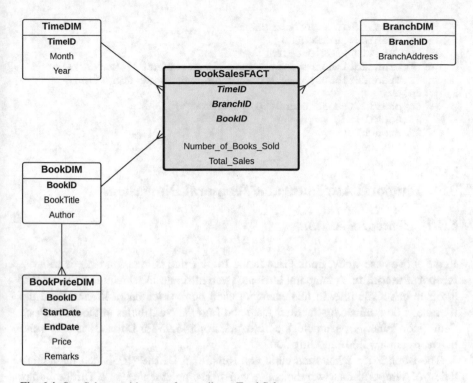

Fig. 6.4 Star Schema with a new fact attribute: Total Sales

For each record in the BookPriceDim table, update the BookSalesFact2 table and calculate the total sales by multiplying Number of Books Sold in the BookSalesFact2 and the "correct" price of the BookPriceDim (where the TimeID in the fact should be between the start date and end date of the book price).

The aforementioned method assumes that BookSalesFact1 has been created. BookSalesFact2 is created by copying from BookSalesFact1, and then update it by adding new column called Total Sales. However, if we do not assume that BookSalesFact1 has already been created, we can create BookSalesFact2 from scratch, using the following create table statement:

```
create table BookSalesFact2 as
select
    to_char(T.TransactionDate, 'MonYYYY') as TimeID,
    BK.BookID,
    BR.BranchID,
    sum(T.Quantity) as Number_Of_Books_Sold,
    sum(T.Quantity * BP.Price) as Total_Sales
from
    BookTransaction T,
    Book BK,
    Branch BR,
    BookPriceHistory BP
```

```
where T.BranchID = BR.BranchID
and T.BookID = BK.BookID
and BK.BookID = BP.BookID
and T.TransactionDate >= to_date(BP.StartDate, 'MonYYYY')
and T.TransactionDate <= to_date(BP.EndDate, 'MonYYYY')
group by
  to_char(T.TransactionDate, 'MonYYYY'),
  BK.BookID,
  BR.BranchID;
```

6.3   Temporal Attributes and Temporal Dimensions

6.3.1   Temporal Attributes

In the above case study, Book Price in the Book Price Dimension table is called a **temporal attribute**. A temporal attribute is an attribute in which the value of that attribute has a lifetime. In this example, each book price has a lifetime, and the lifetime is determined by the Start Date and End Date attributes in the Book Price Dimension table. For example, the book price of $45.95 for Book C1 is only valid between January 2007 and July 2007.

The Book Price Dimension table is a Relational DBMS (RDBMS) implementation of temporal data warehousing, which is implemented as a Bridge Table. The reason why the Book Price Dimension is separated from the original Book Dimension is because one book may have many prices in different time periods. Because the relational model does not permit a nested table, the information about book prices has to be separated into another table. The Book Price Dimension table was shown in Table 6.5.

As stated in the previous section, when we produce a report that joins the Fact Table with the dimension tables (e.g. joining the BookSalesFact Table with the Book Dimension and Book Price Dimension tables), we can produce Report 2 (refer to Table 6.6) which contains all of the attributes from the BookSalesFact Table with the addition of Book Title and Author attributes from the Book Dimension table and the Price attribute from the Book Price Dimension table. This report is produced by the following SQL statement:

```
select
  F.TimeID,
  F.BranchID,
  F.BookID,
  B.BookTitle,
  B.Author,
  P.Price,
  F.Number_of_Books_Sold
from BookSalesFact F, BookDim B, BookPriceDim P
where F.BookID = B.BookID
and B.BookID = P.BookID
```

```
and to_date(F.TimeID, 'MonYYYY') >=
  to_date(P.StartDate, 'MonYYYY')
and to_date(F.TimeID, 'MonYYYY') <=
  to_date(P.EndDate, 'MonYYYY');
```

This SQL command joins the three mentioned tables (i.e. Book Sales Fact, Book Dimension and Book Price Dimension) and checks if TimeID (which is Month and Year) in the BookSalesFact Table falls in the period of the book price as stated by the Start Date and End Date in the Book Price Dimension table. As a result, the correct price will be displayed in the report which matches with the month (i.e. TimeID) of the fact record.

However, a problem will arise if the granularity of time that is used by the book price and the fact is different. For example, if the price of BookID C1 is $23.00 but not from November 2007 to January 2008, but from November 2007 to 15 January 2008 (consequently the price increased to $45.95 as stated from 16 January 2008 instead of February 2008; refer to Table 6.7), then the where clause condition in the above SQL becomes:

```
and to_date(F.TimeID, 'MonYYYY') >=
  to_date(P.StartDate, 'MonYYYY')
and to_date(F.TimeID, 'MonYYYY') <=
  to_date(P.EndDate, 'MonYYYY');
```

This where clause condition will however produce an incorrect report because TimeID 200701 for January 2007 will match with two records in the Book Price Dimension table, one with a price of $23.00 and the other with a price of $45.95. Consequently, in the report, there will be two records for book C1 in January 2008, in each branch, one with the price of $23.00 and the other with the price of $45.95, simply because there are two book prices for the month of January 2008 (see Report 3 in Table 6.8).

This problem can only be solved if the granularity of TimeID in the fact is the same as the Start Date and End Date in the Book Price Dimension table. If TimeID is on the month level, then the period of the book price must be exclusively based on month. On the other hand, if the period of the book price is at a date granularity

Table 6.7 BookPriceDim table

BookID	Start date	End date	Price	Remarks
C1	Jan2007	Jul2007	$45.95	Full Price
C1	Aug2007	Oct2007	$36.75	20% Discount
C1	Nov2007	**15Jan2008**	$23.00	Half Price
C1	**16Jan2008**	Dec9999	$45.95	Full Price
H6	Jan2007	Mar2007	$21.95	Launching
H6	Apr2007	Jan2008	$30.95	Full Price
H6	Feb2008	Dec9999	$10.00	End of Product Sale
DV	Jan2007	Dec9999	$27.95	Full Price
...	...	...	...	...

Table 6.8 Report 3: an incorrect report

TimeID	BranchID	BookID	Book title	Author	Price	Number of books sold
Jan2008	**City**	**C1**	**CSIRO Diet**	**CSIRO Team**	**$23.00**	**25**
Jan2008	**City**	**C1**	**CSIRO Diet**	**CSIRO Team**	**$45.95**	**25**
Jan2008	City	H6	Harry Potter 6	Rowling	$30.95	10
Jan2008	City	DV	Da Vinci Code	Dan Brown	$27.95	7
Jan2008	City	...	...	...	...	...
Jan2008	**Chadstone**	**C1**	**CSIRO Diet**	**CSIRO Team**	**$23.00**	**30**
Jan2008	**Chadstone**	**C1**	**CSIRO Diet**	**CSIRO Team**	**$45.05**	**30**
Jan2008	Chadstone	H6	Harry Potter 6	Rowling	$30.95	15
Jan2008	Chadstone	DV	Da Vinci Code	Dan Brown	$27.95	5
Jan2008	Chadstone	...	...	...	...	...
Jan2008	**Camberwell**	**C1**	**CSIRO Diet**	**CSIRO Team**	**$23.00**	**20**
Jan2008	**Camberwell**	**C1**	**CSIRO Diet**	**CSIRO Team**	**$45.05**	**20**
Jan2008	Camberwell	H6	Harry Potter 6	Rowling	$30.95	5
Jan2008	Camberwell	DV	Da Vinci Code	Dan Brown	$27.95	5
Jan2008	Camberwell	...	...	...	...	...
Jan2008	...	...	...	...	...	...
...	...	...	...	...	...	...

level, then the TimeID in the fact must also be at a date granularity level. If the level of granularity is different, the aforementioned problem will occur. Another way to "solve" this problem is by not allowing the report to join with the Book Price Dimension table. In other words, the Book Price Dimension table is purely treated as additional information in the data warehouse in case the management wants to drill down for information on certain books.

The above problem of mismatched TimeID granularity between the Book Price Dimension table and the Fact can be solved if we display "two" prices on the same record in the report (see Table 6.9). For example, in the first record in Table 6.9, it has two book prices because for that month (e.g. January 2008), there was a change in the book price from $23.00 to $45.95.

Report 4 can be created by the following SQL command:

```
select
   F.TimeID,
   F.BranchID,
   F.BookID,
   B.BookTitle,
   B.Author,
   listagg(P.Price, ';') within group (order by P.Price)
     as Price,
   F.Number_of_Books_Sold
from BookSalesFact F, BookDim B, BookPriceDim P
where F.BookID = B.BookID
and B.BookID = P.BookID
and to_date(F.TimeID, 'MonYYYY') >=
  to_date(P.StartDate, 'MonYYYY')
```

```
and to_date(F.TimeID, 'MonYYYY') <=
  to_date(P.EndDate, 'MonYYYY')
group by
  F.TimeID,
  F.BranchID,
  F.BookID,
  B.BookTitle,
  B.Author,
  F.Number_of_Books_Sold;
```

The above problem in Report 3 occurs because we joined the Fact Table with the Book Price Dimension table; there is no problem with the records in the Fact Table itself. One might ask what happened with the Total Sales attribute in the Fact Table. The calculation for the Total Sales is still correct, as shown in the following SQL statement:

```
create table BookSalesFact2 as
select
  to_char(T.TransactionDate, 'MonYYYY') as TimeID,
  BK.BookID,
  BR.BranchID,
  sum(T.Quantity) as Number_of_Books_Sold,
  sum(T.Quantity * BP.Price) as Total_Sales
from
  BookTransaction T,
  Book BK,
  Branch BR,
  BookPriceHistory BP
where T.BranchID = BR.BranchID
and T.BookID = BK.BookID
and BK.BookID = BP.BookID
and T.TransactionDate >= BP.StartDate
and T.TransactionDate <= BP.EndDate
group by
  to_char(T.TransactionDate, 'MonYYYY'),
  BK.BookID,
  BR.BranchID;
```

This SQL joins the four tables in the operational database, namely, Book Transaction, Book, Branch and Book Price History. While joining these tables, it obtains the correct book price by comparing Transaction Date with Start Date and End Date of the book price. In this case, we assume that Start Date and End Date, as well as Transaction Date, have a date granularity. Once these tables are joined, the grouping is then based on month. The individual book total sales are correct, and the grouping is then based on month. Hence, the problem of an incorrect report is not due to the incorrect fact but because of the join between the fact and the dimension that stores the temporal attribute.

Table 6.9 Report 4: multiple book prices on one month

TimeID	BranchID	BookID	Book title	Author	Price	Number of books Sold
Jan2008	City	C1	CSIRO Diet	CSIRO Team	**$23.00;$45.95**	25
Jan2008	City	H6	Harry Potter 6	Rowling	$30.95	10
Jan2008	City	DV	Da Vinci Code	Dan Brown	$27.95	7
Jan2008	City	...	...	...	...	...
Jan2008	Chadstone	C1	CSIRO Diet	CSIRO Team	**$23.00;$45.95**	30
Jan2008	Chadstone	H6	Harry Potter 6	Rowling	$30.95	15
Jan2008	Chadstone	DV	Da Vinci Code	Dan Brown	$27.95	5
Jan2008	Chadstone	...	...	...	...	...
Jan2008	Camberwell	C1	CSIRO Diet	CSIRO Team	**$23.00;$45.95**	20
Jan2008	Camberwell	H6	Harry Potter 6	Rowling	$30.95	5
Jan2008	Camberwell	DV	Da Vinci Code	Dan Brown	$27.95	5
Jan2008	Camberwell	...	...	...	...	...
Jan2008	...	...	...	...	...	...
...	...	...	...	...	...	...

6.3.2 Temporal Dimensions

If a temporal attribute is an attribute where the value has a lifetime, a *temporal dimension* is a dimension where the record of the dimension has a lifetime. For example, if BookID C1 appeared in 2007, disappeared in 2008 and then reappeared again in 2009, then the book dimension needs a temporal dimension. In another situation, if a branch opens and closes several times, then the branch dimension needs a temporal dimension. However, these two examples are not realistic because it is very rare that books appear and disappear and branches open and close. So, in order to highlight the temporal dimensions, we are going to use a more realistic case study of a Calendar shop.

A *Calendar shop* has a number of branches. It sells different types of calendars and diaries. Some branches are seasonal; they open a stall in major shopping malls, and these stalls are only open for certain months (for instance, some stalls open from October to February, while others may open from November to January). This company also has a small number of permanent shops which are open all year round.

This Calendar shop company wants to analyse their sales, similar to the Bookshop case study. The star schema for the Calendar shop has three dimensions: Time, Branch and Merchandise Dimensions (instead of Book Dimension in the Bookshop case study). For the Merchandise Dimension, if we want to keep track of the changes in prices, like we did with the Book Price Dimension, we could have the merchandise price as a temporal attribute in the Merchandise Price Dimension.

For the Branch Dimension, because some branches have a certain lifetime, Branch Dimension needs a temporal dimension called the Branch History Dimension. In the Branch History Dimension, the entire branch record has a temporal

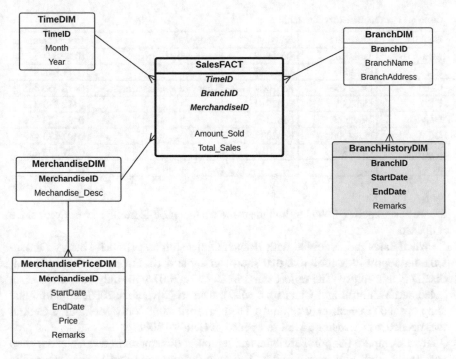

Fig. 6.5 Star schema for the Calendar shop case study

Table 6.10 Branch Dimension table	BranchID	Branch name	Branch address
	MEL	Melbourne Central	88 Lonsdale St
	CH	Chadstone Mall	109 Dandenong Rd
	DOW	Doncaster Westfield	75 Doncaster Rd
	...	...	...

element, indicated by the Start and End Dates. The star schema for the Calendar shop is shown in Fig. 6.5.

The sample records of Branch Dimension and Branch History Dimension are shown in Tables 6.10 and 6.11.

In the Branch History Dimension table, Start Date and End Date refer to the BranchID or the entire record, rather than to a specific attribute. This means, for example, the Chadstone Branch (BranchID CH) opened several times in the last few years, as indicated by the Start Date and End Date, whereas the Melbourne Central shop (BranchID MEL) is open all year round. The Remarks attribute is only used for remarks for each occurrence of the branch entry in the Branch History Dimension table. The Contact Number attribute shows the contact number of a particular branch at a particular lifetime. In this example, it shows that the Chadstone branch (CH) had a different contact phone number (a mobile number) every year, whereas the

Table 6.11 BranchHistoryDim table

BookID	Start date	End date	Remarks	Contact number
MEL	Jan0000	Dec9999	Main shop	(03) 9859 8070
CH	Oct2007	Mar2008		0411 848 821
CH	Oct2008	Feb2009	Under re-construction	0413 356 665
CH	Oct2009	Feb2010		0412 313 313
DOW	Nov2007	Feb2008		0427 123 456
DOW	Nov2008	Feb2009		0427 123 456
DOW	Oct2009	Feb2010		0427 123 456
...	...	...	...	...

Doncaster branch (DOW) has had the same contact mobile number every year since it opened.

When Sales fact is joined with Branch Dimension and Branch History Dimension, the report may look like that shown in Table 6.12. For simplicity, MerchandiseID is left empty. The report starts in October 2007 with only two shops open (Melbourne Central and Chadstone Mall), then in November 2007, an additional shop opened (Doncaster Westfield). Then in April 2008, only Melbourne Central was opened, as Chadstone Mall re-opened in October 2008.

If we compare temporal attribute and temporal dimension in this Calendar shop case study, both look very similar. The only difference is that in the case of the temporal attribute, there is a temporal attribute in the history dimension table (e.g. book price in the Book Price Dimension table), whereas in the temporal dimension case, there is no such attribute. The attributes in the temporal dimension are only the key identifier of the parent dimension and possibly any remarks attributes. Or in other words, we can say that a temporal dimension is where all attributes in the dimension are all, collectively, temporal attributes.

6.3.3 Another Temporal Dimension

Temporal dimensions or dimensions with temporal attributes are normally implemented as a Bridge Table, as discussed in the previous sections. In this section, we are going to see an example where a temporal dimension is not implemented as a Bridge Table, rather as a normal dimension.

Suppose there is a simple star schema that maintains a list of courses/degrees (e.g. Bachelor of Computer Science, Master of Information Technology, etc.). A course or a degree structure changes from time to time, simply due to the nature of the discipline, especially those which are moving at a reasonably high pace, like information technology. Each course has a course code. When the course changes significantly, the university or the faculty usually introduces a new course (or degree) and phases out the old course (or degree) rather than "updating" the

Table 6.12 Report: SalesFact joined with Branch Dimension and Branch History Dimension

Time ID	Branch ID	Branch name	Branch address	Start date	End date	Remarks	Contact number	Merchandise	Amount sold	Total sales
Oct2007	MEL	Melbourne Central	88 Lonsdale Street	Jan0000	Dec9999	Main shop₂	...	...	...	...
Oct2007	CH	Chadstone Mall	109 Dandenong Road	Oct2007	Mar2008		...	...	...	...
Nov2007	MEL	Melbourne Central	88 Lonsdale Street	Jan0000	Dec9999	Main shop₂	...	...	...	...
Nov2007	CH	Chadstone Mall	109 Dandenong Road	Oct2007	Mar2008		...	...	...	...
Nov2007	DOW	Doncaster Westfield	75 Doncaster Road	Nov2007	Feb2008		...	...	...	...
Dec2007	MEL	Melbourne Central	88 Lonsdale Street	Jan0000	Dec9999	Main shop	...	...	...	...
Dec2007	CH	Chadstone Mall	109 Dandenong Road	Oct2007	Mar2008		...	...	...	...
Dec2007	DOW	Doncaster Westfield	75 Doncaster Road	Nov2007	Feb2008		...	...	...	...
Jan2008	MEL	Melbourne Central	88 Lonsdale Street	Jan0000	Dec9999	Main shop	...	...	...	...
Jan2008	CH	Chadstone Mall	109 Dandenong Road	Oct2007	Mar2008		...	...	...	...
Jan2008	DOW	Doncaster Westfield	75 Doncaster Road	Nov2007	Feb2008		...	...	...	...

(continued)

Table 6.12 (continued)

Time ID	Branch ID	Branch name	Branch address	Start date	End date	Remarks	Contact number	Merchandise	Amount sold	Total sales
Feb2008	MEL	Melbourne Central	88 Lonsdale Street	Jan0000	Dec9999	Main shop	...	...	...	...
Feb2008	CH	Chadstone Mall	109 Dandenong Road	Oct2007	Mar2008		...	...	...	...
Feb2008	DOW	Doncaster Westfield	75 Doncaster Road	Nov2007	Feb2008		...	...	...	...
Mar2008	MEL	Melbourne Central	88 Lonsdale Street	Jan0000	Dec9999	Main shop	...	...	...	...
Mar2008	CH	Chadstone Mall	109 Dandenong Road	Oct2007	Mar2008		...	...	...	...
Apr2008	MEL	Melbourne Central	88 Lonsdale Street	Jan0000	Dec9999	Main shop	...	...	...	...
May2008	MEL	Melbourne Central	88 Lonsdale Street	Jan0000	Dec9999	Main shop	...	...	...	...
...	...	...	...	...	...	...	...	...	...	...
Oct2008	MEL	Melbourne Central	88 Lonsdale Street	Jan0000	Dec9999	Main shop	...	...	...	...
Oct2008	CH	Chadstone Mall	109 Dandenong Road	Oct2008	Feb2009	Under re-construction	...	...	...	...
...	...	...	...	...	...	...	...	...	...	...

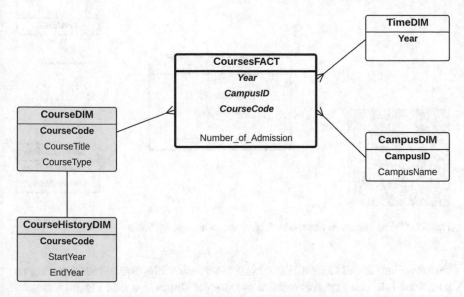

Fig. 6.6 Course History as a temporal dimension: using a Bridge Table

existing degree. When this happens, students entering this course/degree will be admitted to the new course (rather than the old course), while existing students will still be taught using the old degree structure until there are no more students in the old course. The new course will have a new course code, even though the course name might remain the same. Hence, it is common that a course or a degree has a number of course codes, but only one of them can admit new students.

Figure 6.6 shows a simple star schema with a temporal dimension, called Course History Dimension, with a Course Code attribute as the identifier. The year when the course is introduced is stored in the Start Year attribute. When the course no longer admits new students (in other words, the course is being phased out or closed), the End Year is recorded. Course without End Year means that the course is an active and current course. The star schema in Fig. 6.6 is able to analyse the number of "new" students (admission) to enrol in courses, for each year and each campus. Because a course has only "one" history, the relationship between Course and Course History Dimension is a 1-1 relationship.

This case study is rather different from the Calendar Shop case study in the previous section where a branch (or a shop) is open for several months a year and is closed for the rest of the year. However, the shop opens again next year. This means that the shop opens and closes, hence, a shop has many instances in the history of the shop.

On the other hand, when a course is closed, it never re-opens. Hence, a course has only one instance in the history. If the End Year is not specified, the course is still active. If there is an End Year (which is normally a past year), the course is already closed and does not admit any new students. Because the relationship cardinality

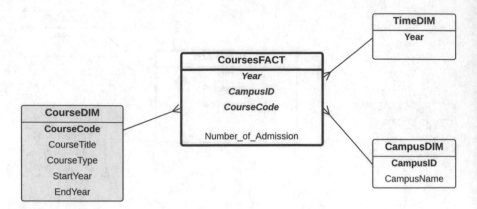

Fig. 6.7 Course History as a temporal dimension: not as a Bridge Table

between Course and Course History is a 1-1 relationship, both dimensions can be combined into one dimension called the Course Dimension (see Fig. 6.7). Hence, the temporal dimension, which is in this case the Course Dimension, is directly linked to the Fact, so this temporal dimension is actually a "Normal" dimension, even though there is a lifetime for each record in this Course Dimension table.

6.4   Slowly Changing Dimensions

Temporal data warehousing is often known as *slowly changing dimensions* or *SCD*. Slowly changing dimensions are dimensions where the records of these dimensions change slowly over a period of time. Using the Bookshop case study earlier in this chapter, the Book Dimension has price information and it is common that the price of a book changes "slowly" over time, which is why the Book Dimension is an example of a slowly changing dimension. It is called "slowly" because changes to price do not occur, for example, every hour or even every week; rather, book prices change over a longer period of time, such as months or even years.

The topic of slowly changing dimensions is different from attributes or records that are "rapidly" changing, such as the location attribute of a moving taxi, which changes very frequently (e.g. changes to the location coordinate are recorded every minute in the operational database), or the price of shares in the stock market, which may change every few seconds, etc. The topic of these rapid changes of records or values is often studied in the area of Real-Time Data Warehousing or Stream Data Warehousing, which is not the scope of this chapter.

There are three basic types of SCD called Type 1, Type 2 and Type 3. Each of these types treats an implementation of SCD differently. However, recently, data warehousing practitioners have added new types, called Type 0, Type 4 and Type 6, which enrich the implementation options for SCD.

6.4.1 SCD Type 0 and Type 1

SCD Type 0 and Type 1 are quite similar; they do not actually record the history of changes in the dimension.

In **Type 0**, the dimension stores the "Original or Initial" value of the records when the data warehouse is built. If the value of the dimension attributes changes, the changes are not recorded. The records remain the same as when the records were first inserted into the data warehouse.

Using the Bookshop case study, these values are the "Full Price". For many books, the original or initial price was the full price, but for some books, the price listed initially may not be the full price. So, the Book Dimension table using SCD Type 0 which lists the full price is shown in Table 6.13. If the book prices change after this, the new price will not be recorded in the data warehouse.

For other systems, it is more desirable to store the original or initial value rather than the so-called full price as in this example.

Because SCD Type 0 does not record the history of the book price, the star schema does not have a temporal dimension. The star schema for the Bookshop case study is shown in Fig. 6.1.

Type 1 also does not record the history of changes; rather it only records the latest value of the record. Using the Book Dimension example, the book price in the Book Dimension will be the latest price. This means that when there is a change of price, the old price will be overwritten by the new price in the Book Dimension table. Table 6.14 shows the contents of the Book Dimension table using SCD Type 1. In this case, the $10.00 price of BookID H6 is the latest price. Note that the other two books in this example have the same price as the full price in SCD Type 0, but this does not mean that the prices were never changed. This Book Dimension table only tells us that these are the current price of the books.

Similar to SCD Type 0, the SCD Type 1 star schema does not maintain a temporal dimension. The star schema is the one shown in Fig. 6.1.

Because SCD Type 0 and Type 1 do not maintain the history of book prices, if we need to produce a report that joins the Book Dimension table with the Book Sales

Table 6.13 Book Dimension table (SCD Type 0)

BookID	Book title	Author	Price
C1	CSIRO Diet	CSIRO Team	$45.95
H6	Harry Potter 6	Rowling	$30.95
DV	Da Vinci Code	Dan Brown	$27.95
...	...	...	...

Table 6.14 Book Dimension Table (SCD Type 1)

BookID	Book title	Author	Price
C1	CSIRO Diet	CSIRO Team	$45.95
H6	Harry Potter 6	Rowling	**$10.00**
DV	Da Vinci Code	Dan Brown	$27.95
...	...	...	...

Fact Table, we need to be careful not to draw an association between the book price from Book Dimension and the TimeID from the Book Sales Fact, because the book price in the report may not necessarily be the book price at that particular TimeID. In such a report, the column title for the book price can be changed to "original book price" (for SCD Type 0) or "latest book price" (for SCD Type 1) to avoid any misunderstanding in interpreting the report.

6.4.2 SCD Type 2

SCD Type 2 keeps track of the history but does not separate the history from the main dimension; instead, the new records are continually added to the dimension. Using the Book Dimension as an example, when the price of a book is changed, it creates "another book" with the same details but with the new BookID and of course the new price. In addition to the Start Date and End Date, it also has a Current Flag to indicate whether a record is the current or a past record. Any additional information, such as remarks, may also be included. The Book Dimension table for SCD Type 2 is shown in Table 6.15.

Note that the same book has a different BookID for a different book price and period. Usually, the BookID attribute is implemented as a Surrogate Key, but in this example, we just added a sequence number to the original BookID to differentiate the same book in a different time period. Because of these multiple BookIDs for the same book, the report that joins the Book Sales Fact and the Book Dimension tables will appear as follows (see report 3 SCD Type 2 in Table 6.16).

Note that the first record in Report 3, the BookID, is C1_4, because this was the BookID of the CSIRO Diet book in March 2008. Because SCD Type 2 only changes the original Book Dimension table, the star schema looks similar to that of Fig. 6.1, but the Book Dimension table has additional attributes (see Fig. 6.8).

Table 6.15 Book Dimension table (SCD Type 2)

BookID	Book title	Author	Start Date	End Date	Price	Remarks	Current Flag
C1_1	CSIRO Diet	CSIRO Team	Jan2007	Jul2007	$45.95	Full Price	N
C1_2	CSIRO Diet	CSIRO Team	Aug2007	Oct2007	$36.75	20% Discount	N
C1_3	CSIRO Diet	CSIRO Team	Nov2007	Jan2008	$23.00	Half Price	N
C1_4	CSIRO Diet	CSIRO Team	Feb2008	Dec9999	$45.95	Full Price	Y
H6_1	Harry Potter 6	Rowling	Jan2007	Mar2007	$21.95	Launching	N
H6_2	Harry Potter 6	Rowling	Apr2007	Jan2008	$30.95	Full Price	N
H6_3	Harry Potter 6	Rowling	Feb2008	Dec9999	$10.00	End of Product Sale	Y
DV_1	Da Vinci Code	Dan Brown	Jan2007	Dec9999	$27.95	Full Price	Y
...	...	...	...	...	...	...	...

Table 6.16 Report 3 (SCD Type 2)

TimeID	BranchID	BookID	Book title	Author	Price	Number of books sold
Mar2008	City	C1_4	CSIRO Diet	CSIRO Team	$45.95	5
Mar2008	City	H6_3	Harry Potter 6	Rowling	$10.00	15
Mar2008	City	DV_1	Da Vinci Code	Dan Brown	$27.95	23
Mar2008	City	...	...	...	...	...
Mar2008	Chadstone	C1_4	CSIRO Diet	CSIRO Team	$45.95	15
Mar2008	Chadstone	H6_3	Harry Potter 6	Rowling	$10.00	3
Mar2008	Chadstone	DV_1	Da Vinci Code	Dan Brown	$27.95	2
Mar2008	Chadstone	...	...	...	...	...
Mar2008	Camberwell	C1_4	CSIRO Diet	CSIRO Team	$45.95	1
Mar2008	Camberwell	H6_3	Harry Potter 6	Rowling	$10.00	1
Mar2008	Camberwell	DV_1	Da Vinci Code	Dan Brown	$27.95	2
Mar2008	Camberwell	...	...	...	...	...
Mar2008	...	...	...	...	...	...
...	...	...	...	...	...	...
...	...	...	...	...	...	...
Dec2007	City	C1_3	CSIRO Diet	CSIRO Team	$23.00	15
Dec2007	City	H6_2	Harry Potter 6	Rowling	$30.95	6
Dec2007	City	DV_1	Da Vinci Code	Dan Brown	$27.95	6
Dec2007	City	..	...	...	...	...
Dec2007	Chadstone	C1_3	CSIRO Diet	CSIRO Team	$23.00	10
Dec2007	Chadstone	H6_2	Harry Potter 6	Rowling	$30.95	8
Dec2007	Chadstone	DV_1	Da Vinci Code	Dan Brown	$27.95	1
Dec2007	Chadstone	..	...	...	...	...
Dec2007	Camberwell	C1_3	CSIRO Diet	CSIRO Team	$23.00	18
Dec2007	Camberwell	H6_2	Harry Potter 6	Rowling	$30.95	3
Dec2007	Camberwell	DV_1	Da Vinci Code	Dan Brown	$27.95	2
Dec2007	Camberwell	..	...	...	...	...
Dec2007	...	...	...	...	...	...
...	...	...	...	...	...	...

6.4.3   SCD Type 3

SCD Type 3 is a simplification of Type 2. Unlike Type 2 which maintains multiple records of the same book, Type 3 does not have multiple records for the same book. One book has one entry in the Book Dimension table. For the price, it only records the current price and the previous price. In other words, it does not maintain the entire history of price changes; rather, it only keeps the last two prices. The Book Dimension table for SCD Type 3 is shown in Table 6.17. Note that for a book which does not have any previous price, a Null value will be recorded in the Previous Price column.

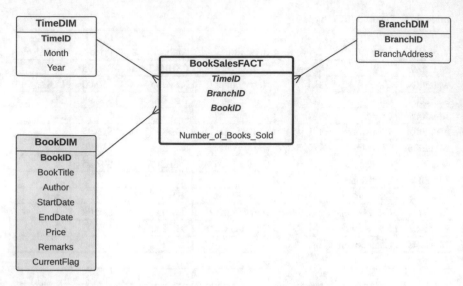

Fig. 6.8 The Bookshop star schema with SCD Type 2 for Book Dimension

Table 6.17 Book Dimension table (SCD Type 3)

BookID	Book title	Author	Current price	Previous price
C1	CSIRO Diet	CSIRO Team	**$45.95**	**$23.00**
H6	Harry Potter 6	Rowling	**$10.00**	**$30.95**
DV	Da Vinci Code	Dan Brown	**$27.95**	**Null**
...	...	...	...	...

The main rationale for adopting SCD Type 3 is that most of the analysis will be based on the current price and, at most, one past price, this being the previous price. Perhaps, this is so it can be compared with the pricing trend. It is assumed that analysing the complete history is not necessary.

Additionally, although we can store information on when the price was changed, the original SCD Type 3 does not record this. It only records the current and the previous prices. Without keeping the date when the price was changed, it is not possible to correlate the book price with the TimeID information in the Book Sales Fact Table. Consequently, the report that can be produced will need to pay particular attention to the potential mismatch between the information in the book price column (from the Book Dimension table) and the TimeID column (from the Book Sales Fact Table).

The star schema for SCD Type 3 is shown in Fig. 6.9.

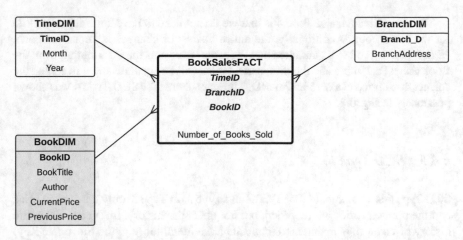

Fig. 6.9 The Bookshop star schema with SCD Type 3 for Book Dimension

Table 6.18 Book Dimension table

BookID	Book title	Author
C1	CSIRO Diet	CSIRO Team
H6	Harry Potter 6	Rowling
DV	Da Vinci Code	Dan Brown
...	...	...

Table 6.19 Book Price Dimension table

BookID	Start date	End date	Price	Remarks
C1	Jan2007	Jul2007	$45.95	Full Price
C1	Aug2007	Oct2007	$36.75	20% Discount
C1	Nov2007	Jan2008	$23.00	Half Price
C1	Feb2008	Dec9999	$45.95	Full Price
H6	Jan2007	Mar2007	$21.95	Launching
H6	Apr2007	Jan2008	$30.95	Full Price
H6	Feb2008	Dec9999	$10.00	End of Product Sale
DV	Jan2007	Dec9999	$27.95	Full Price
...	...	...	...	...

6.4.4   SCD Type 4

SCD Type 4 is a new way to treat SCD that was not in the original SCD theory. In SCD Type 4, we create a new dimension to maintain the history of attribute value changes. Basically, the temporal dimension that is described in the beginning of this chapter is the SCD Type 4. In Type 4, the original Book Dimension table is kept without the price attribute. The price attribute, Start Date and End Date are separated into another table, which is the Book Price Dimension table. The Book Dimension and the Book Price Dimension tables are shown in Tables 6.18 and 6.19.

The main advantage of Type 4 is that we do not need to have a different BookID for the same book. Additionally, the entire history of changes is kept. As shown earlier, this method will guarantee that the report that joins the information from the Book Sales Fact Table and the dimension tables will be accurate and will reflect the correct book price at a certain TimeID. The star schema for SCD Type 4 was shown previously in Fig. 6.2.

6.4.5   SCD Type 6

SCD Type 6 is a combination of Type 2 and Type 3. In Type 3, only the current price and the previous price are recorded, but not the entire history. In Type 2, the entire history of changes is maintained, but a separate identifier (e.g. the Surrogate Key) is needed. In Type 6, a separate identifier for the same book is not needed (as in Type 3), but the entire history is kept (similar to Type 2). SCD Type 6 for the Book Dimension table is shown in Table 6.20.

In Type 6, there is no need to maintain a separate history table. The history itself is kept in the original dimension table. The star schema for SCD Type 6 is shown in Fig. 6.10. Note that the Book Dimension table has a composite key comprising BookID, Start Date and End Date.

Table 6.20 Book Dimension (SCD Type 6)

Book ID	Book Title	Author	Start Date	End Date	Current Price	Previous Price	Remarks	Current Flag
C1	CSIRO Diet	CSIRO Team	Jan2007	Jul2007	$45.95	Null	Full Price	N
C1	CSIRO Diet	CSIRO Team	Aug2007	Oct2007	$36.75	$45.95	20% Discount	N
C1	CSIRO Diet	CSIRO Team	Nov2007	Jan2008	$23.00	$36.75	Half Price	N
C1	CSIRO Diet	CSIRO Team	Feb2008	Dec9999	$45.95	$23.00	Full Price	Y
H6	Harry Potter 6	Rowling	Jan2007	Mar2007	$21.95	Null	Launching	N
H6	Harry Potter 6	Rowling	Apr2007	Jan2008	$30.95	$21.95	Full Price	N
H6	Harry Potter 6	Rowling	Feb2008	Dec9999	$10.00	$30.95	End of Product	Y
DV	Da Vinci Code	Dan Brown	Jan2007	Dec9999	$27.95	Null	Full Price	Y
...	...	...	...	...	...	...	...	...

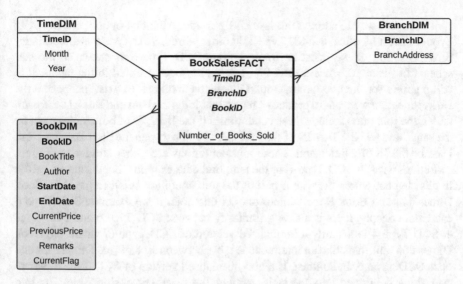

Fig. 6.10 The Bookshop star schema with SCD Type 6 for Book Dimension

If we examine this star schema (and the Book Dimension table), the Book Dimension table has a composite key, which comprises BookID, Start Date and End Date, whereas the Fact Table uses only BookID as the reference to the Book Dimension table. Taking BookID C1 as an example, there are four records of C1 in the Book Dimension table, and there are many records of C1 in the Fact Table. The cardinality relationship between the Book Dimension table and Fact Table is no longer 1-m, but m-m. Normally, there are three possible options to address this issue:

1. Add a new surrogate key to the Book Dimension table, and consequently the Fact Table will have this surrogate key from the Book Dimension as a reference. This surrogate key can also simply be a concatenation between BookID, Start Date and End Date. Or, the surrogate key is BookID with a sequence number, like in SCD Type 2.
2. If we do not want a surrogate key in the Book Dimension table, then the Fact Table must also include the Start Date and End Date, in addition to the BookID. But this will be messy because the Fact already had TimeID, referencing the Time Dimension table.
3. The last option is to have an associative table (or a bridge table) between Book Dimension and Fact, because the cardinality between Book Dimension and Fact is m-m. This middle table that links Book Dimension and Fact will have a composite key comprising the identifier from Book Dimension and Fact. This is also a messy option, because we are trying to use a relational theory to solve this m-m issue.

The easiest way to address this issue is by adopting the first option, making SCD Type 6 almost identical to SCD Type 2. In other words, SCD Type 6 is unnecessary or useless. But, if we insist on implementing SCD Type 6 as shown in the star schema in Fig. 6.10, since no PK-FK is explicitly implemented in the tables, it is still possible for the two tables, Book Dimension and Fact, to exist independently. However, when we want to produce a report that joins the Fact and Book Dimension tables, the join must include these conditions: (*i*) the BookID of both tables must be the same, and (*ii*) the TimeID of the Fact must be in between the Start Date and End Date of the Book Dimension. These join conditions are rather similar to the join conditions used in SCD Type 4, or the temporal data warehousing discussed earlier in this chapter, where the date is part of the join condition between the Fact, Book Dimension and Book Price Dimension. On this note, if we examine SCD Type 6 even more deeply, there is a strong similarity between SCD Type 6 and SCD Type 4. SCD Type 4 is actually a normalised version of SCD Type 6, where the Book Dimension table acts like an intermediate table between Book Price Dimension and Fact. SCD Type 6, in contrast, is a non-normalised version of SCD Type 4, where everything is lumped into one table, namely, the Book Dimension table. The join conditions to join the Book Dimension and the Fact are identical.

6.4.6 Implementation of SCD in SQL

The concepts of SCD are very easy to digest. However, implementing SCD in SQL has several challenging technical aspects. In this section, we are going to examine the SQL to create the six SCD types.

The source tables in the operational database are the Book table and the Book Price History table, as shown in the E/R diagram (refer to Fig. 6.1). The structures of these two tables are:

- **Book** (BookID, BookTitle, Author)
- **BookPriceHistory** (BookID, StartDate, EndDate, Price, Remarks)

SCD Type 0 is a non-temporal dimension. The Book Dimension table using SCD Type 0 is shown in Table 6.13. Note that the "original price" may not be the "initial price" because some books have a launching price which can be heavily discounted. Therefore, the Book Dimension table using SCD Type 0 checks for the Remarks "Full Price".

```
create table SCD0 as
select distinct
  B.BookID, B.BookTitle,
  B.Author, H.Price as OriginalPrice
from Book B, BookPriceHistory H
where B.BookID = H.BookID
and H.Remarks = 'Full Price';
```

SCD Type 1 uses the latest price (refer to Table 6.14). We could choose a book where the End Date is "Dec9999". Alternatively, we can sort by the Start Date and choose the "latest" Start Date. In the following SQL, the inner query ranks the book records based on the Start Date in descending order, and the outer query chooses those books which have a rank of 1.

```
create table SCD1 as
select
  T.BookID, T.BookTitle,
  T.Author, T.Price as CurrentPrice
from (
  select
    B.BookID, B.BookTitle, B.Author,
    to_date(H.StartDate, 'MonYYYY'), H.Price,
    rank() over( partition by B.BookID
       order by to_date(H.StartDate, 'MonYYYY') desc)
       as Rank
  from Book B, BookPriceHistory H
  where B.BookID = H.BookID) T
where T.Rank = 1;
```

SCD Type 2 is a join between the Book table and Book Price History table, so that we can get the book details as well as the Start Date and End Date. However, a book with the same BookID should be added with a sequence number, as previously shown in Table 6.15.

The SQL for SCD Type 2 uses the rank function as well as the partition clause, so the same book will be ranked according to their Start Date. The Current Flag column will identify whether the book record has the current price or not.

```
create table SCD2 as
select   B.BookID  || '_' ||
   rank() over(partition by B.BookID
      order by to_date(H.StartDate, 'MonYYYY') asc)
      as BookID,
  B.BookTitle, B.Author, H.StartDate,
  H.EndDate, H.Price, H.Remarks,
  case H.EndDate when 'Dec9999' then 'Y' else 'N'
  end as CurrentFlag
from Book B, BookPriceHistory H
where B.BookID = H.BookID;
```

SCD Type 3 is quite complex as it not only has the Current Price (similar to SCD Type 2) but also the Previous Price. If SCD Type 2 uses rank 1, SCD Type 3 needs rank 2 (for the Previous Price) as well. Hence, SCD Type 3 joins the table with rank 1 and the table with rank 2. Since some books do not have the Previous Price, we need to use an outer join instead of an inner join.

```
create table SCD3 as
select
  T1.BookID, T1.BookTitle, T1.Author,
  T1.CurrentPrice, T2.CurrentPrice as PreviousPrice
from (
```

```
select
  T.BookID, T.BookTitle,
  T.Author, T.Price as CurrentPrice
from (
   select
     B.BookID, B.BookTitle,
     B.Author, to_date(H.StartDate, 'MonYYYY'),
     H.Price,
     rank() over( partition by B.BookID
       order by to_date(H.StartDate, 'MonYYYY') desc)
       as Rank
   from Book B, BookPriceHistory H
   where B.BookID = H.BookID) T
 where T.Rank = 1) T1,
 (select
   T.BookID, T.BookTitle,
   T.Author, T.Price as CurrentPrice
 from (
   select
     B.BookID, B.BookTitle, B.Author,
     to_date(H.StartDate, 'MonYYYY'), H.Price,
     rank() over( partition by B.BookID
       order by to_date(H.StartDate, 'MonYYYY') desc)
       as Rank
   from Book B, BookPriceHistory H
   where B.BookID = H.BookID) T
  where T.Rank = 2) T2
where T1.BookID = T2.BookID(+);
```

SCD Type 4 is actually the temporal data warehousing discussed in the first part of this chapter. The SQL to create the Book Dimension is an entire extraction from the Book Price History table from the operational database.

```
create table SCD4 as
select * from BookPriceHistory;
```

SCD Type 6 is a combination of SCD Type 2 and SCD Type 3. However, the implementation of SCD Type 6 in SQL can be very challenging. We cannot simply join SCD Type 2 and SCD Type 3 because SCD Type 2 has the complete history of prices, whereas SCD Type 3 has only the current and previous prices. Additionally, joining between SCD Type 2 and SCD Type 3 cannot use an equi-join. Instead, we need to check if the BookID of SCD Type 3 is part of the BookID of SCD Type 2, using the like operator.

Creating SCD Type 6 from SCD Type 2 and SCD Type 3 needs a couple of steps. The first step is to join SCD Type 2 and SCD Type 3 using the like operator. The join results use the BookID attribute from SCD Type 3, and the rest of the attributes are from SCD Type 2. Additionally, an Order Number attribute will have a sequence number for each group of BookID. The results of this join query are shown in Table 6.21.

Table 6.21 Join between SCD Type 2 and SCD Type 3

Book ID	Book Title	Author	Start Date	End Date	Price	Remarks	Current Flag	Order Number
C1	CSIRO Diet	CSIRO Team	Jan2007	Jul2007	$45.95	Full Price	N	1
C1	CSIRO Diet	CSIRO Team	Aug2007	Oct2007	$36.75	20% Discount	N	2
C1	CSIRO Diet	CSIRO Team	Nov2007	Jan2008	$23.00	Half Price	N	3
C1	CSIRO Diet	CSIRO Team	Feb2008	Dec9999	$45.95	Full Price	Y	4
DV	Da Vinci Code	Dan Brown	Jan2007	Dec9999	$27.95	Full Price	Y	1
H6	Harry Potter 6	Rowling	Jan2007	Mar2007	$21.95	Launching	N	1
H6	Harry Potter 6	Rowling	Apr2007	Jan2008	$30.95	Full Price	N	2
H6	Harry Potter 6	Rowling	Feb2008	Dec9999	$10.00	End of Product	Y	3
...	...	...	...	...	...	...	...	...

```
select
  SCD3.BookID,
  SCD2.BookTitle, SCD2.Author,
  SCD2.StartDate, SCD2.EndDate,
  SCD2.Price, SCD2.Remarks, SCD2.CurrentFlag,
  row_number() over
    (partition by SCD3.BookID
     order by SCD3.BookID, SCD2.StartDate) as
     OrderNumber
  from SCD2, SCD3
where SCD2.BookID like SCD3.BookID||'_%'
order by SCD3.BookID, SCD2.StartDate;
```

The next step is to do a self-join of Table 6.21. The Price from the first table will become the Current Price, and the Price from the second table will become the Previous Price. There are two join conditions for this self-join query. The first condition is that the BookID must be the same. In order for the Price from the second table to become the Previous Price, we need to add the Order Number in the second table by one, and then the second join condition is to match the Order Number from these two tables. The SQL for the self-join is as follows, whereas the result is shown in Table 6.22.

```
select
  T1.BookID, T1.BookTitle, T1.Author,
  T1.StartDate, T1.EndDate,
  T1.Price as CurrentPrice,
  T2.Price as PreviousPrice,
  T1.Remarks, T1.CurrentFlag
from (
  select SCD3.BookID,
    SCD2.BookTitle, SCD2.Author, SCD2.StartDate,
```

```
      SCD2.EndDate, SCD2.Price, SCD2.Remarks,
      SCD2.CurrentFlag,
      row_number() over
        (partition by SCD3.BookID
         order by SCD3.BookID, SCD2.StartDate) as
         OrderNumber
      from SCD2, SCD3
    where SCD2.BookID like SCD3.BookID||'_%'
    order by SCD3.BookID, SCD2.StartDate) T1,
    (
    select SCD3.BookID,
      SCD2.BookTitle, SCD2.Author, SCD2.StartDate,
      SCD2.EndDate, SCD2.Price, SCD2.Remarks,
      SCD2.CurrentFlag,
      row_number() over
        (partition by SCD3.BookID
         order by SCD3.BookID, SCD2.StartDate) + 1 as
         OrderNumber
      from SCD2, SCD3
    where SCD2.BookID like SCD3.BookID||'_%'
    order by SCD3.BookID, SCD2.StartDate) T2
  where T1.BookID = T2.BookID
  and T1.OrderNumber = T2.OrderNumber
  order by T1.BookID, T1.StartDate;
```

However, the self-join results excluded the first record of each book, because
there is no Previous Price for the first record of each book. In order to fix this
mistake, the self-join query needs to use an outer join, instead of an inner join.
The correct SQL query for SCD Type 6 that uses an outer join is very similar to the
above SQL query that uses an inner join, with only one difference, that is, the join

Table 6.22 Self-join results

Book ID	Book Title	Author	Start Date	End Date	Current Price	Previous Price	Remarks	Current Flag
C1	CSIRO Diet	CSIRO Team	Aug2007	Oct2007	$36.75	$45.95	20% Discount	N
C1	CSIRO Diet	CSIRO Team	Nov2007	Jan2008	$23.00	$36.75	Half Price	N
C1	CSIRO Diet	CSIRO Team	Feb2008	Dec9999	$45.95	$23.00	Full Price	Y
H6	Harry Potter 6	Rowling	Apr2007	Jan2008	$30.95	$21.95	Full Price	N
H6	Harry Potter 6	Rowling	Feb2008	Dec9999	$10.00	$30.95	End of Product	Y
...	...	...	...	...	...	...	...	...

Table 6.23 The correct self-join results (SCD Type 6)

Book ID	Book Title	Author	Start Date	End Date	Current Price	Previous Price	Remarks	Current Flag
C1	CSIRO Diet	CSIRO Team	Jan2007	Jul2007	$45.95	Null	Full Price	N
C1	CSIRO Diet	CSIRO Team	Aug2007	Oct2007	$36.75	$45.95	20% Discount	N
C1	CSIRO Diet	CSIRO Team	Nov2007	Jan2008	$23.00	$36.75	Half Price	N
C1	CSIRO Diet	CSIRO Team	Feb2008	Dec9999	$45.95	$23.00	Full Price	Y
DV	Da Vinci Code	Dan Brown	Jan2007	Dec9999	$27.95	Null	Full Price	Y
H6	Harry Potter 6	Rowling	Jan2007	Mar2007	$21.95	Null	Launching	N
H6	Harry Potter 6	Rowling	Apr2007	Jan2008	$30.95	$21.95	Full Price	N
H6	Harry Potter 6	Rowling	Feb2008	Dec9999	$10.00	$30.95	End of Product	Y
...	...	...	...	...	...	...	...	...

condition is now an outer join (with a (+) sign at the end of each join condition), which is as follows:

```
where T1.BookID = T2.BookID (+)
and T1.OrderNumber = T2.OrderNumber (+)
```

The final results are shown in Table 6.23.

6.4.7 Creating the Fact Tables

After tackling the implementation challenges of SCD, the next step is to examine the SQL commands to create the Fact Tables. For SCD Types 0, 1, 3 and 4 (see the sample records in Tables 6.13, 6.14, 6.17, 6.18 and 6.19), the BookID attributes are exactly the same as the BookID in the "non-temporal" version of data warehousing (refer to Table 6.2 at the beginning of this chapter). Consequently, the Fact Table will not be affected when we use SCD Types 0, 1, 3 or 4. The original Book Dimension, as shown in the non-temporal version, has three books with BookIDs: C1 (CSIRO Diet), H6 (Harry Potter) and DV (Da Vinci Code). The Book Dimensions in SCD Types 0, 1, 3 and 4 also have the same BookIDs as the original Book Dimension, and the number of records in these tables is the same.

However, when producing a report, by joining the Fact Table and the Book Dimension table, when all the attributes from the Book Dimension are present (including the Price), we cannot draw a correlation between the TimeID in the Fact and the Book Price in the Book Dimension table, because the Book Price displayed on the report might not be the price on the TimeID. This problem was illustrated in Table 6.3. This is why instead of using SCD Types 0, 1 or 3, we use SCD Type 4, which was discussed earlier in this chapter. With SCD Type 4, we not only maintain the complete history of Book Price; we are also able to produce correct reports when joining with the Fact Table.

For SCD Type 2 (see the sample records in Table 6.15), the number of records is not the same as the original Book Dimension table. This is because every time the Book Price is changed, a "new" book is inserted into the dimension table. This has an impact on the contents of the Fact Table. For example, in the Fact Table, it should not be BookID C1 for CSIRO Diet Book, but C1_1 or C1_2, etc., to refer to different Book instance in the history. Therefore, when SCD Type 2 is used, the contents of the Fact Table must contain the correct BookID. The SQL command to create the Fact Table for SCD Type 2 is as follows:

```
create table BookSalesFactWithSCD2 as
select
  to_char(T.TransactionDate, 'MonYYYY') as TimeID,
  BK.BookID,
  BR.BranchID,
  sum(T.Quantity) as Number_of_Books_Sold
from BookTransaction T, SCD2 BK, Branch BR
where T.BranchID = BR.BranchID
and BK.BookID like T.BookID||'_%'
and to_date(BK.StartDate, 'MonYYYY') <= T.TransactionDate
and T.TransactionDate <= to_date(BK.EndDate, 'MonYYYY')
group by
  to_char(T.TransactionDate, 'MonYYYY'),
  BK.BookID,
  BR.BranchID;
```

There are three things to note: (i) It joins with the Book Dimension SCD Type 2. The reason for this is that the Book Table in the operational database does not have the new BookIDs. The new BookIDs only exist in SCD Type 2, because the new BookIDs are created when the Book Price is changed. (ii) The join condition uses a LIKE and a wildcard to match the new BookID in SCD Type 2 and the BookID in the Boo Transaction table from the operational database. (iii) The join condition must include checking the dates so that the correct record in the Book Dimension SCD Type 2 is used when joining with the records in the Book Transaction table from the operational database. The Fact Table is shown in Table 6.24.

SCD Type 6 is similar to SCD Type 2, that is, it contains the same number of records. The only difference is that in SCD Type 6, BookID does not change, and the original BookID is used. Therefore, the SQL to create the Fact Table is slightly simpler. In the join condition, it simply compares the BookID from SCD Type 6 and the Book Transaction table (note that in SCD Type 2 Fact Table, it uses the LIKE and wildcard). The SQL command to create the Fact Table for SCD Type 6 is as follows:

```
create table BookSalesFactWithSCD6 as
select
  to_char(T.TransactionDate, 'MonYYYY') as TimeID,
  BK.BookID,
  BR.BranchID,
  sum(T.Quantity) as Number_of_Books_Sold
from BookTransaction T, SCD6 BK, Branch BR
where T.BranchID = BR.BranchID
```

Table 6.24 Fact Table (SCD Type 2)

TimeID	BranchID	BookID	Number of books sold
Mar2008	City	C1_4	5
Mar2008	City	H6_3	15
Mar2008	City	DV_1	23
Mar2008	City	...	...
Mar2008	Chadstone	C1_4	15
Mar2008	Chadstone	H6_3	3
Mar2008	Chadstone	DV_1	2
Mar2008	Chadstone	...	...
Mar2008	Camberwell	C1_4	1
Mar2008	Camberwell	H6_3	1
Mar2008	Camberwell	DV_1	2
Mar2008	Camberwell	...	...
Mar2008	...	...	...
...	...	...	...
...	...	...	...
Dec2007	City	C1_3	15
Dec2007	City	H6_2	6
Dec2007	City	DV_1	6
Dec2007	City	..	...
Dec2007	Chadstone	C1_3	10
Dec2007	Chadstone	H6_2	8
Dec2007	Chadstone	DV_1	1
Dec2007	Chadstone	..	...
Dec2007	Camberwell	C1_3	18
Dec2007	Camberwell	H6_2	3
Dec2007	Camberwell	DV_1	2
Dec2007	Camberwell	..	...
Dec2007	...	...	...
...	...	...	...

```
and BK.BookID = T.BookID
and to_date(BK.StartDate, 'MonYYYY') <= T.TransactionDate
and T.TransactionDate <= to_date(BK.EndDate, 'MonYYYY')
group by
  to_char(T.TransactionDate, 'MonYYYY'),
  BK.BookID,
  BR.BranchID;
```

The Fact Table is shown in Table 6.25. It has almost the same records as Table 6.24, except that here the original BookID is kept.

It is worthwhile to note that when we join SCD Type 2 and the Fact Table, or when we join SCD Type 6 and the Fact Table, the correct Book Price will be shown (as Book Price matches with the TimeID), and therefore, there will not be any incorrect correlation between these two attributes in the report. The reason for

Table 6.25 Fact Table (SCD Type 6)

TimeID	BranchID	BookID	Number of books sold
Mar2008	City	C1	5
Mar2008	City	H6	15
Mar2008	City	DV	23
Mar2008	City	...	...
Mar2008	Chadstone	C1	15
Mar2008	Chadstone	H6	3
Mar2008	Chadstone	DV	2
Mar2008	Chadstone	...	...
Mar2008	Camberwell	C1	1
Mar2008	Camberwell	H6	1
Mar2008	Camberwell	DV	2
Mar2008	Camberwell	...	...
Mar2008	...	...	...
...	...	...	...
...	...	...	...
Dec2007	City	C1	15
Dec2007	City	H6	6
Dec2007	City	DV	6
Dec2007	City	..	...
Dec2007	Chadstone	C1	10
Dec2007	Chadstone	H6	8
Dec2007	Chadstone	DV	1
Dec2007	Chadstone	..	...
Dec2007	Camberwell	C1	18
Dec2007	Camberwell	H6	3
Dec2007	Camberwell	DV	2
Dec2007	Camberwell	..	...
Dec2007	...	...	...
...	...	...	...

this is that both SCD Type 2 and Type 6 maintain the complete history of Book Prices; they keep the StartDate and EndDate of each Book Price.

SCD Type 4 (our temporal data warehousing version) also keeps the complete history of Book Prices, and therefore, the report when joining the Fact Table and the Book Dimension Table is also correct, as shown previously in Table 6.6. SCD Type 2 and SCD Type 6 will also produce the same report. The only difference is the way the Book Price history is maintained. SCD Type 4 uses two tables (i.e. Book Dimension and Book Price Dimension), whereas SCD Type 2 and SCD Type 6 use one table only, where the Book Price history is maintained in the Book Dimension.

6.5 Summary

In this chapter, we focus on incorporating historical data in the data warehouse. This is called *temporal data warehousing*. A temporal data warehousing uses the concept of the Bridge Table (or a Weak Entity), where the history is maintained in a bridge table. Maintaining the history of certain attributes is important in order to make associative analysis more accurate when analysing the reports produced by the fact and dimensions. However, a certain degree of caution is needed when joining the Fact Table and the temporal dimension, especially when the level of granularity of time between the fact and the temporal dimension is not the same.

Temporal data warehousing is also known as *slowly changing dimensions (SCD)*. In this chapter, various treatments and types for SCD are presented. Different types will serve different purposes of data warehousing.

6.6 Exercises

6.1 It is very common for professors in a university to embark on a consulting project with a company. This consulting job is often regarded as a research activity, as the professor's research expertise is being utilised by a company in a particular project. The research office in the university keeps track of all consulting jobs that professors undertake. A simple E/R diagram to do this is shown in Fig. 6.11. Each consulting project has Project Number and Title, as well as the Company who provided the money, the amount and the Year when the consulting was conducted. For research reporting purposes, the research office at the university assigned a research field for each consulting job. Examples of some research fields are AI, Big Data, Visualisation, Web Development, Mobile Apps, etc.

Also note that the professors' position history is also maintained. This is particularly useful to see the trend of the amount of money that the professor secures from the company as he/she moves up the ladder of the professor ranks (e.g. Assistant Professor, Associate Professor and Professor).

The task is to design a star schema to keep track of the Total Consulting Amount for each year, each research field as well as each professor. Write the SQL commands to create the dimensions and Fact Tables.

6.2 Employee Dimension is a *slowly changing dimension (SCD)*. For simplicity, the dimension contains only these attributes: EmployeeID, Name and Salary. SCD has the six types: Types 0, 1, 2, 3, 4 and 6. Draw sample tables for Employee Dimension using each of the SCD types. Assume that the Salary attribute is the *Temporal attribute*, which may change from time to time, due to promotion, increase of salary, etc. Add more attributes to the table, whenever required by the SCD. For each type, you need to have at least two employee records, and these are the two employees:

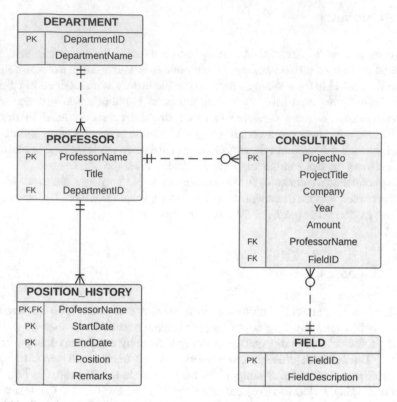

Fig. 6.11 Consulting E/R diagram

- The first employee is *Adam* with an EmployeeID: A1. He started his job in 1 March 2016 with a salary of $3900. After his probation ended in 31 May 2016, his salary was increased to $4300. At the beginning of 2017, he received a pay increase, and his salary became $5000. From 1 July 2017 (until now), his salary has been $5500.
- The second employee is *Ben* with an EmployeeID: B2. He started his job in 1 February 2017 with an initial salary of $4000. After his probation ended in 31 May 2017, his salary became $4750 which is his current salary.

6.7   Further Readings

Slowly changing dimension (SCD) has been discussed and explained in various data warehousing books, such as [1–7] and [8]. There have been discussions to extend the SCD from Type 6 to more complex SCDs, by combining several basic SCDs.

[9] discusses an object-relational implementation of temporal data warehouses, using multi-valued attributes, objects and classes. Further details about object-

relational implementation of an RDBMS, such as Oracle, can be found in the following book: [10].

Very early work on temporal databases can be found in [11–14] and [15].

References

1. C. Adamson, *Star Schema The Complete Reference* (McGraw-Hill Osborne Media, 2010)
2. R. Laberge, *The Data Warehouse Mentor: Practical Data Warehouse and Business Intelligence Insights* (McGraw-Hill, New York, 2011)
3. M. Golfarelli, S. Rizzi, *Data Warehouse Design: Modern Principles and Methodologies* (McGraw-Hill, New York, 2009)
4. R. Kimball, M. Ross, *The Data Warehouse Toolkit: The Definitive Guide to Dimensional Modeling* (Wiley, London, 2013)
5. R. Kimball, M. Ross, W. Thornthwaite, J. Mundy, B. Becker, *The Data Warehouse Lifecycle Toolkit* (Wiley, London, 2011)
6. W.H. Inmon, *Building the Data Warehouse*. ITPro Collection (Wiley, London, 2005)
7. M. Jarke, *Fundamentals of Data Warehouses*, 2nd edn. (Springer, Berlin, 2003)
8. A. Vaisman, E. Zimányi, *Data Warehouse Systems: Design and Implementation*. Data-Centric Systems and Applications (Springer, Berlin, 2014)
9. E. Malinowski, E. Zimányi, *Advanced Data Warehouse Design: From Conventional to Spatial and Temporal Applications*. Data-Centric Systems and Applications (Springer, Berlin, 2008)
10. J.W. Rahayu, D. Taniar, E. Pardede, *Object-oriented Oracle*. Solutions For IT Professionals (IRM Press, 2006)
11. R.T. Snodgrass, I. Ahn, Temporal databases. Computer **19**(9), 35–42 (1986)
12. R.T. Snodgrass, Temporal databases, in *Theories and Methods of Spatio-Temporal Reasoning in Geographic Space, International Conference GIS—From Space to Territory: Theories and Methods of Spatio-Temporal Reasoning, Pisa, Italy, September 21–23, 1992, Proceedings*, ed. by A.U. Frank, I. Campari, U. Formentini. Lecture Notes in Computer Science, vol. 639 (Springer, Berlin, 1992), pp. 22–64
13. C.S. Jensen, R.T. Snodgrass, Temporal data management. IEEE Trans. Knowl. Data Eng. **11**(1), 36–44 (1999)
14. R.T. Snodgrass, I. Ahn, A taxonomy of time in databases, in *Proceedings of the 1985 ACM SIGMOD International Conference on Management of Data, Austin, Texas, USA, May 28–31, 1985*, ed. by S.B. Navathe (ACM Press, New York, 1985), pp. 236–246
15. R.T. Snodgrass, *Developing Time-Oriented Database Applications in SQL* (Morgan Kaufmann, Los Altos, 1999)

Part III
Advanced Dimension

Chapter 7
Determinant Dimensions

In star schema, a dimension is the angle from where you view the fact. Normally, a dimension consists of a key identifier attribute and additional informative attributes, whereas the fact contains the fact measure attributes and the key identifier from each dimension. This key identifier from each dimension forms an implicit link between each dimension and the fact. When we say that we query a data warehouse, this means that we query the fact measures in the Fact Table, because fact measures store the aggregated values which are the central focus of the data warehouse.

When we query the Fact Table (or in other words, when we retrieve the interested fact measures attributes from the Fact Table), a join between the Fact Table and dimension tables is often needed to retrieve details from each dimension. In circumstances where the additional attributes from the dimensions do not need to be retrieved, the join with the dimensions can be avoided. Nevertheless, the key identifier from the dimension which is already stored in the fact must be used to select from which dimension angle we want to view the fact measures. Therefore, although joining the dimension tables may not be needed in some instances, we still say that the dimensions are used to retrieve the fact measures. For simplicity, when we refer to a dimension being used in data retrieval, it is applicable to both joining the required dimension or simply using the key identifier of the dimension in the Fact Table.

However, when retrieving data from a data warehouse (that is retrieving the fact measures from the Fact Table), we are not forced to use all the dimensions (or all the dimensions' key identifiers in the Fact Table) as the selector. We may use as little as one dimension (or the key identifier of one dimension) only. We do not have to use all the dimensions (or the key identifier of all dimensions), simply because the star schema includes dimensions. In other words, each dimension (or key identifier of a dimension in the Fact Table) may or may not be used in data retrieval.

In certain circumstances, a star schema may have a special dimension (or a key identifier) which must be used for all data retrieval, or else the retrieved data will be meaningless. This dimension (or its key identifier) has a special role in data retrieval

© The Author(s), under exclusive license to Springer Nature Switzerland AG 2021
D. Taniar, W. Rahayu, *Data Warehousing and Analytics*, Data-Centric Systems
and Applications, https://doi.org/10.1007/978-3-030-81979-8_7

because it has to be used. This dimension is called a *Determinant Dimension*, and subsequently, the key identifier of this attribute is called a *Determinant Attribute*.

In this chapter, we are going to learn about scenarios when a determinant dimension is used and when it is not necessarily to use a determinant dimension. We are going to study various case studies to explore the complexities of determinant dimensions: (*i*) the *Petrol Station* case study, which illustrates the use of a determinant dimension; (*ii*) the *Olympic Games* case study, which compares and contrasts determinant and non-determinant dimensions; (*iii*) the *PTE Academic Test* case study, which explores the technical aspects and the complexities of both determinant and non-determinant dimensions; (*iv*) the *University's Student Population* case study to illustrate the use of year in the determinant dimension; and (*v*) the *Private Taxi Company* case study to explore the impact of complex dimensions on pivot tables.

7.1   Introducing a Determinant Dimension: Petrol Station Case Study

The Petrol Station case study is about daily petrol prices in Victoria, Australia. Unlike petrol prices in other parts of the world, petrol prices in Victoria (Australia) fluctuate every day. Usually, petrol prices change once a day, and the changes generally occur at midnight. Also, petrol stations are not monopolised by one company; in fact, there are many different petrol companies which own many petrol stations. Hence, the prices may differ from one company to another. Additionally, petrol prices in different geographical locations might differ as well. This is well understood because country towns may have higher petrol prices due to many factors, such as higher transportation costs from the depot to the petrol station. It is also known that petrol stations in more well-off suburbs may have higher petrol prices. Generally, customers want to know where to find the cheapest petrol price.

A company maintains a database of all petrol prices from all petrol stations in Victoria. The following is a sample operational database on petrol prices (for different kinds of petrol) at petrol stations over a period of time (see Fig. 7.1). It is not our purpose to inspect the data in details, but at least we can see the kind of information the table provided. This includes petrol station detail (e.g. name, company and address), day/date and petrol type and price. As previously noted, the various petrol companies may change the price of different types of petrol on different days.

In this section, we are going to examine two versions of the star schema for this Petrol Station case study.

Petrol station	Company	Address	Fuel Type	Price	Date	Day of Week	Suburb
Caltex Star Mart Box Hill	Caltex	793-797 Whitehorse Road, Box Hill	Unleaded	120.9	1/08/16	Monday	Box Hill
Caltex Star Mart Box Hill	Caltex	793-797 Whitehorse Road, Box Hill	Diesel	118.9	1/08/16	Monday	Box Hill
Caltex Star Mart Box Hill	Caltex	793-797 Whitehorse Road, Box Hill	LPG	58.9	1/08/16	Monday	Box Hill
Caltex Star Mart Box Hill	Caltex	793-797 Whitehorse Road, Box Hill	U95	132.9	1/08/16	Monday	Box Hill
Caltex Star Mart Box Hill	Caltex	793-797 Whitehorse Road, Box Hill	Premium Unl	139.9	1/08/16	Monday	Box Hill
7-Eleven Clayton	7-Eleven	187-191 Clayton Road, Clayton	unleaded	122.7	1/08/16	Monday	Clayton
7-Eleven Clayton	7-Eleven	187-191 Clayton Road, Clayton	LTG	56.7	1/08/16	Monday	Clayton
7-Eleven Clayton	7-Eleven	187-191 Clayton Road, Clayton	diesel	115.9	1/08/16	Monday	Clayton
United Fitzroy	United	390 Nicholson Street, Fitzroy	Unleaded	119.9	1/08/16	Monday	Fitzroy
United Fitzroy	United	390 Nicholson Street, Fitzroy	Diesel	109.9	1/08/16	Monday	Fitzroy
United Fitzroy	United	390 Nicholson Street, Fitzroy	LPG	59.7	1/08/16	Monday	Fitzroy
7-Eleven Hawthorn	7-Eleven	747-755 Toorak Road, Hawthorn East	Unleaded	121.9	1/08/16	Monday	Hawthorn East
7-Eleven Hawthorn	7-Eleven	747-755 Toorak Road, Hawthorn East	Diesel	115.9	1/08/16	Monday	Hawthorn East
7-Eleven Hawthorn	7-Eleven	747-755 Toorak Road, Hawthorn East	LPG	57.9	1/08/16	Monday	Hawthorn East
7-Eleven Hawthorn	7-Eleven	747-755 Toorak Road, Hawthorn East	U95	132.9	1/08/16	Monday	Hawthorn East
7-Eleven Hawthorn	7-Eleven	747-755 Toorak Road, Hawthorn East	Premium Unl	137.9	1/08/16	Monday	Hawthorn East
United Point Cook	United	1 Wallace Avenue, Point Cook	Unleaded	123.7	1/08/16	Monday	Point Cook
United Point Cook	United	1 Wallace Avenue, Point Cook	Diesel	118.9	1/08/16	Monday	Point Cook
United Point Cook	United	1 Wallace Avenue, Point Cook	LPG	55.7	1/08/16	Monday	Point Cook
United Murrumbeena	United	90-92a Kangaroo Road, Carnegie	E10	114.9	1/08/16	Monday	Carnegie
United Murrumbeena	United	90-92a Kangaroo Road, Carnegie	Unleaded	119.9	1/08/16	Monday	Carnegie
United Murrumbeena	United	90-92a Kangaroo Road, Carnegie	Diesel	118.7	2/08/16	Monday	Carnegie
United Murrumbeena	United	90-92a Kangaroo Road, Carnegie	LPG	58.9	2/08/16	Monday	Carnegie
United Murrumbeena	United	90-92a Kangaroo Road, Carnegie	Premium Unl	124.9	2/08/16	Monday	Carnegie
Caltex Star Mart Box Hill	Caltex	793-797 Whitehorse Road, Box Hill	unleaded	120.9	2/08/16	Tuesday	Box Hill
Caltex Star Mart Box Hill	Caltex	793-797 Whitehorse Road, Box Hill	Diesel	118.9	2/08/16	Tuesday	Box Hill
Caltex Star Mart Box Hill	Caltex	793-797 Whitehorse Road, Box Hill	LPG	58.9	2/08/16	Tuesday	Box Hill
Caltex Star Mart Box Hill	Caltex	793-797 Whitehorse Road, Box Hill	U95	132.9	2/08/16	Tuesday	Box Hill
Caltex Star Mart Box Hill	Caltex	793-797 Whitehorse Road, Box Hill	Premium Unl	139.9	2/08/16	Tuesday	Box Hill
7-Eleven Box Hill	7-Eleven	786 Whitehorse Road& Elgar Road, Box Hill	unleaded	120.9	2/08/16	Tuesday	Box Hill
7-Eleven Box Hill	7-Eleven	786 Whitehorse Road& Elgar Road, Box Hill	Diesel	115.9	7/08/16	Tuesday	Box Hill
7-Eleven Box Hill	7-Eleven	786 Whitehorse Road& Elgar Road, Box Hill	LPG	58.9	2/08/16	Tuesday	Box Hill
7-Eleven Box Hill	7-Eleven	786 Whitehorse Road& Elgar Road, Box Hill	U95	131.9	2/08/16	Tuesday	Box Hill
7-Eleven Box Hill	7-Eleven	786 Whitehorse Road& Elgar Road, Box Hill	Premium Unl	136.9	2/08/16	Tuesday	Box Hill
United Murrumbeena	United	90-92a Kangaroo Road, Carnegie	E10	114.9	2/08/16	Tuesday	Carnegie
United Murrumbeena	United	90-92a Kangaroo Road, Carnegie	Unleaded	118.9	2/08/16	Tuesday	Carnegie
United Murrumbeena	United	90-92a Kangaroo Road, Carnegie	Diesel	118.7	2/08/16	Tuesday	Carnegie
United Murrumbeena	United	90-92a Kangaroo Road, Carnegie	LPG	58.9	2/08/16	Tuesday	Carnegie
United Murrumbeena	United	90-92a Kangaroo Road, Carnegie	Premium Unl	124.9	2/08/16	Tuesday	Carnegie
BP AA Richmond - Church Street	BP	581 Church Street, Richmond	unleaded	121.9	2/08/16	Tuesday	Richmond
BP AA Richmond - Church Street	BP	581 Church Street, Richmond	LPG	58.9	2/08/16	Tuesday	Richmond
BP AA Richmond - Church Street	BP	581 Church Street, Richmond	Diesel	115.9	2/08/16	Tuesday	Richmond
Coles Oakleigh	Coles	1388 Dandenong Road, Oakleigh	Unleaded	122.9	2/08/16	Tuesday	Oakleigh
Coles Oakleigh	Coles	1388 Dandenong Road, Oakleigh	Diesel	115.9	2/08/16	Tuesday	Oakleigh
Coles Oakleigh	Coles	1388 Dandenong Road, Oakleigh	LPG	56.9	2/08/16	Tuesday	Oakleigh
Coles Oakleigh	Coles	1388 Dandenong Road, Oakleigh	U95	136.9	3/08/16	Tuesday	Oakleigh
Coles Oakleigh	Coles	1388 Dandenong Road, Oakleigh	Premium Unl	142.9	3/08/16	Tuesday	Oakleigh
7-Eleven Point Cook	7-Eleven	Cnr. Boardwalk Blvd. & Tom Roberts Parade, P(	Unleaded	119.7	3/08/16	Wednesday	Point Cook
7-Eleven Point Cook	7-Eleven	Cnr. Boardwalk Blvd. & Tom Roberts Parade, P(	Diesel	115.9	3/08/16	Wednesday	Point Cook
7-Eleven Point Cook	7-Eleven	Cnr. Boardwalk Blvd. & Tom Roberts Parade, P(	LPG	55.7	3/08/16	Wednesday	Point Cook
7 Eleven Point Cook	7-Eleven	Cnr. Boardwalk Blvd. & Tom Roberts Parade, P(	Premium Unl	135.7	3/08/16	Wednesday	Point Cook
7-Eleven Point Cook	7-Eleven	Cnr. Boardwalk Blvd. & Tom Roberts Parade, P(	U95	130.7	3/08/16	Wednesday	Point Cook
7-Eleven Point Cook	7-Eleven	Cnr. Boardwalk Blvd. & Tom Roberts Parade, P(	E10	117.7	3/08/16	Wednesday	Point Cook
Coles Express Ashburton	Coles	High & Johnston Streets, Ashburton	Unleaded	117.9	3/08/16	Wednesday	Ashburton
Coles Express Ashburton	Coles	High & Johnston Streets, Ashburton	Diesel	115.9	3/08/16	Wednesday	Ashburton
Coles Express Ashburton	Coles	High & Johnston Streets, Ashburton	Premium Unl	138.9	3/08/16	Wednesday	Ashburton
AA Glen Iris	AA	44-56 High street, Glen Iris	Unleaded	117.9	3/08/16	Wednesday	Glen Iris
AA Glen Iris	AA	44-56 High street, Glen Iris	Diesel	115.9	3/08/16	Wednesday	Glen Iris
AA Glen Iris	AA	44-56 High street, Glen Iris	LPG	58.9	3/08/16	Wednesday	Glen Iris
AA Glen Iris	AA	44-56 High street, Glen Iris	U95	125.9	3/08/16	Wednesday	Glen Iris
AA Glen Iris	AA	44-56 High street, Glen Iris	Premium Unl	133.9	3/08/16	Wednesday	Glen Iris
7-Eleven Black Rock	7-Eleven	583-589 Balcombe Road, Black Rock	Unleaded	117.9	3/08/16	Wednesday	Black Rock
7-Eleven Black Rock	7-Eleven	583-589 Balcombe Road, Black Rock	Diesel	116.9	3/08/16	Wednesday	Black Rock
7-Eleven Black Rock	7-Eleven	583-589 Balcombe Road, Black Rock	LPG	56.9	3/08/16	Wednesday	Black Rock
7-Eleven Black Rock	7-Eleven	583-589 Balcombe Road, Black Rock	Premium Unl	133.9	3/08/16	Wednesday	Black Rock
7-Eleven Black Rock	7-Eleven	583-589 Balcombe Road, Black Rock	U95	128.9	3/08/16	Wednesday	Black Rock
7-Eleven Black Rock	7-Eleven	583-589 Balcombe Road, Black Rock	E10	115.9	3/08/16	Wednesday	Black Rock
Coles Express Essendon North	Coles	249 Keilor Road, Essendon North	Unleaded	118.7	3/08/16	Wednesday	Essendon North
Coles Express Essendon North	Coles	249 Keilor Road, Essendon North	Diesel	116.9	4/08/16	Wednesday	Essendon North
Coles Express Essendon North	Coles	249 Keilor Road, Essendon North	Premium Unl	138.7	4/08/16	Wednesday	Essendon North
Coles Express Essendon North	Coles	249 Keilor Road, Essendon North	LPG	59.9	4/08/16	Wednesday	Essendon North

Fig. 7.1 Sample data from petrol stations

7.1.1   Petrol Station Star Schema Version 1

The requirements for the data warehouse are to answer questions related to (*a*) average petrol price, (*b*) minimum petrol price and (*c*) maximum petrol price. These are the three prices of interest to most customers.

From this sample data, we choose the following three dimensions: (*i*) day of week, (*ii*) suburb and (*iii*) petrol company. From a *Day of Week* point of view, a sensible question to answer is which day of the week has the lowest petrol price? We use a two-column table methodology to visualise the conceptual model of the star schema. The two-column table for the day of week is shown in Table 7.1.

There are a couple of things to note from Table 7.1.

1. The first column (say, column *A*, which is Day of Week) is the *category*. The second columns (say, columns *B*1, *B*2, *B*3 and *B*4) are the *fact measures*.
2. One of the requirements for the fact measure is average petrol price. Since the Fact Table should not include average as a fact measure, two attributes, total petrol price and number of petrol stations, are used instead.

The visualisation of the above two-column table for Day of Week is rather incomplete because there is no data in the fact measure columns (i.e. columns Total Petrol Price, Number of Petrol Station, Minimum Petrol Price and Maximum Petrol Price). Take a look at the Minimum Petrol Price column. What kind of value should be in that column? Suppose Monday has a value in this column. What does this value mean? It does not specify which petrol type, that is, whether it is Unleaded, or Premium 95, or Premium 98, for example. Therefore, the Minimum Petrol Price column, although it seems correct, does not make sense at all. How can this problem be solved?

One solution is to have one Minimum Petrol Price for each petrol type (e.g. Unleaded, Premium 95, Premium 98, E10, Diesel and LPG). Therefore, we have six attributes to represent the Minimum Petrol Price, namely Min Price Unleaded, Min Price Premium 95, Min Price Premium 98, Min Price E10, Min Price Diesel and Min Price LPG (see Table 7.2).

We need to do the same for the Total Price and Maximum Petrol Price, each of which should be divided into six columns. The new two-column table for Day of Week is shown in Table 7.3.

Since not all petrol stations sell Premium 98, for instance, hence, the number of petrol stations for each petrol type will differ. Consequently, calculating the average

Table 7.1 Petrol prices based on Day of Week

Day of week	Total petrol price	Number of petrol stations	Min Petrol price	Max Petrol price
Monday				
Tuesday				
...				

Table 7.2 Minimum petrol prices

Day of week	Total petrol price	Number of petrol stations	Min petrol price							Max petrol price
			Unleaded	P95	P98	E10	Diesel	LPG		
Monday			119.9c	132.9c	139.9c	114.9c	115.9c	55.7c		
Tuesday			118.9c	131.9c	136.9c	114.9c	115.9c	56.9c		
...										

Table 7.3 Petrol prices of each petrol type

Day of week	Total petrol price						Number of petrol stations	Min petrol price						Max petrol price					
	U	95	98	E	D	LPG		U	95	98	E	D	LPG	U	95	98	E	D	LPG
Monday																			
Tuesday																			

price for Unleaded Petrol or Premium 98 on a Monday is not a matter of simply dividing Total Price Unleaded by the Number of Petrol Stations, or dividing Total Price Premium 98 by the Number of Petrol Stations, because the base line, which is the Number of Petrol Station for different fuel types, is different. Therefore, we need to split the Number of Petrol Station column into six columns, one for each petrol type. As a result, we will end up with 24 columns to store the fact measures, instead of only four columns, as initially thought.

Using the same principle, we can have a two-column table for suburb and another one for company (see Tables 7.4 and 7.5).

Based on these three two-column tables (e.g. Day of Week, Suburb and Company), the star schema for the Petrol Station case study is shown in Fig. 7.2. There are three dimensions—Day of Week, Suburb and Company—and there are 24 fact measures and six different petrol types for total price, number of stations and minimum and maximum prices.

7.1.2   Petrol Station Star Schema Version 2

The star schema has 24 fact measures, as shown in Fig. 7.2, because there are six different kinds of petrol. Is there another way to reduce the number of fact measures? The answer is yes!

We can have a new dimension called *Petrol Type Dimension*, which stores the different kinds of petrol. With this new dimension, the fact measure is reduced to four, which is the number of original fact measures we had in the beginning. The new star schema is shown in Fig. 7.3.

With this new star schema, we can retrieve records that show the Min Petrol Price of Premium 98 on Monday or to get the Average Petrol Price (which is Total Petrol Price/Number of Petrol Stations) for Unleaded Petrol in the 7-Eleven petrol station located in the suburb of Clayton.

Remember that when retrieving records from a data warehouse (e.g. from the Fact Table), there is no obligation that the query must use all dimensions (or the key identifier from all dimensions in the Fact Table). For the Monday Premium 98 example, only two dimensions were used: Day of Week Dimension and Petrol Type Dimension (or Day of Week and PetrolType key identifier attributes in the Fact Table). In the second example, only three dimensions were used: Petrol Type Dimension, Company Dimension and Suburb Dimension (or PetrolType, Company and Suburb key identifier attributes in the Fact Table). So to retrieve data from a data warehouse (or the Fact Table), we can use as few as one key identifier of a dimension, but we can also use all dimensions. We are not restricted to use a key identifier of any particular dimension.

However, this poses a new problem. For example, the star schema as shown in Fig. 7.3 allows us to retrieve Min Petrol Price of Monday, which uses only the Day of Week Dimension (or its key identifier: Day of Week attribute). The SQL would look like the following, and the results are shown in Table 7.6. In this example, for

Table 7.4 Petrol prices for each suburb

Suburb	Total petrol price						Number of petrol stations						Min petrol price						Max petrol price					
	U	95	98	E	D	L	U	95	98	E	D	L	U	95	98	E	D	L	U	95	98	E	D	L
Box Hill																								
Clayton																								

Table 7.5 Petrol prices for each company

Company	Total petrol price						Number of petrol stations						Min petrol price						Max petrol price					
	U	95	98	E	D	L	U	95	98	E	D	L	U	95	98	E	D	L	U	95	98	E	D	L
Caltex																								
7-Eleven																								

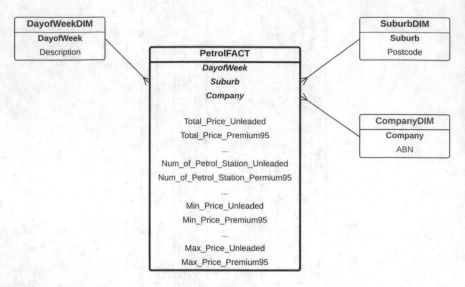

Fig. 7.2 Petrol station star schema (version 1)

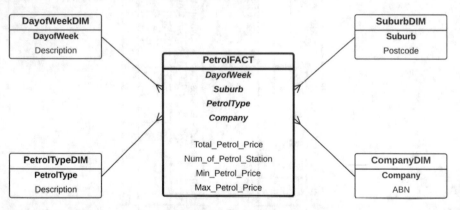

Fig. 7.3 Petrol station star schema with Petrol Type Dimension (version 2)

simplicity, a join between the Fact Table and the two dimension tables (e.g. Day of Week Dimension and Petrol Type Dimension) is not used because the query only uses the key identifier attribute of the dimensions: Day of Week and PetrolType attributes from the Fact Table.

```
select DayofWeek, Min_Petrol_Price
from PetrolFact
where DayofWeek = 'Monday';
```

The data retrieved by the SQL as shown in Table 7.6 does not make any sense because it is not clear which petrol type we are referring to. There are different minimum petrol prices for different petrol types. Based on this example, it is clear that the Petrol Type Dimension holds the key to any data retrieval. We must use

Table 7.6 Minimum Petrol
Price on Monday

Day of week	Min petrol price
Monday	117.9 cents
Monday	56.7 cents
Monday	109.9cents
Monday	125.9 cents
Monday	133.9 cents
Monday	114.9 cents

Table 7.7 Minimum
Unleaded petrol price on
Monday

Day of week	Min petrol price
Monday	117.9 cents

Table 7.8 Minimum Petrol
Price on Monday

Day of week	PetrolType	Min petrol price
Monday	Unleaded	117.9 cents
Monday	LPG	56.7 cents
Monday	Diesel	109.9cents
Monday	E10	125.9 cents
Monday	Premium 95	133.9 cents
Monday	Premium 98	114.9 cents

the Petrol Type Dimension (or its key identifier: Petrol Type attribute) for all data retrieval from the Fact Table. The following are examples of data retrieval that use Petrol Type: the Min Petrol Price of Premium 98 on Monday, the Average Petrol Price of Unleaded in the 7-Eleven petrol station located in the suburb of Clayton, etc.

The SQL to retrieve the minimum Unleaded petrol price on Monday is as follows, and the result is shown in Table 7.7. Note that the petrol type is used in the where clause in the SQL command to filter out the records.

```
select DayofWeek, Min_Petrol_Price
from PetrolFact
where DayofWeek = 'Monday'
and PetrolType = 'Unleaded';
```

Or if Petrol Type is used in the select clause rather than in the where clause, then all petrol types will be retrieved, but at least we will see from the list which petrol type we are interested in (see Table 7.8).

```
select DayofWeek, PetrolType, Min_Petrol_Price
from PetrolFact
where DayofWeek = 'Monday';
```

Petrol Type Dimension, in this example, is called a *Determinant Dimension*, a dimension which must be used in the data retrieval because the fact measures are determined by this dimension, and its key identifier (e.g. Petrol Type attribute) is called a *Determinant Attribute*. Again, when we are referring to a dimension,

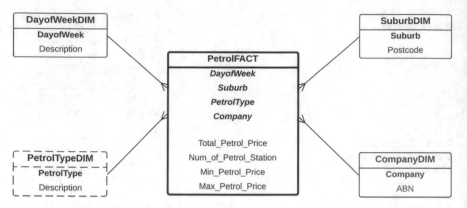

Fig. 7.4 A Determinant Dimension version

such as the Petrol Type Dimension, we often refer to its key identifier. Hence, the terms Determinant Dimension and Determinant Attribute are used interchangeably. But most of the time, we will use the term Determinant Dimension to refer to Determinant Attribute.

In the star schema shown in Fig. 7.4, the Determinant Dimension is indicated with a dotted box to differentiate it from the other normal dimensions. This star schema shows that if we want to retrieve any of the fact measures from the Fact Table, we must use the Petrol Type attribute in the SQL.

In the previous chapter, we learned not to use average as the fact measure due to the average of average problem. This is why, in this example, we keep the fact two attributes, Total Petrol Price and Number of Petrol Station, to calculate the Average Petrol Price. The average of average problem occurs because we are not using all the dimensions in the data retrieval; hence, we need to calculate yet another average of the average fact measure. However, if all dimensions are used in the data retrieval, the average fact measure is perfectly fine because there is no second average function to be used anymore. This means that if all dimensions are determinant dimensions, then average fact measure can be used (see Fig. 7.5).

So we may end up with two star schemas, one for Min and Max Petrol Prices and the other for the Avg Petrol Price. The former uses the Petrol Type Dimension as a Determinant Dimension, whereas the latter uses all dimensions as Determinant Dimensions. Otherwise, we will use the star schema in Fig. 7.4, which includes all the four fact measures, in which the average petrol prices can be calculated when retrieving the records from the Fact Table.

The concept of Determinant Dimension is quite straightforward as explained above, but in practice, it may impose some technical challenges. A normal dimension is created using the create table statement in SQL. The dimension table can either be extracted from any existing tables from the operational database or must be created manually. If the latter, records must be inserted manually into the dimension table after the table is created.

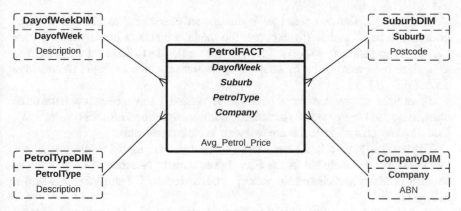

Fig. 7.5 Determinant Dimensions for the Average Petrol Price

The Determinant Dimension is also created this way, that is, using a normal `create table` statement. The question is now how to ensure that this dimension (or its key identifier in the Fact Table) is used during data retrieval. Note from the above examples, the Petrol Type key attribute is either used in the `where` clause to restrict a certain petrol type or in the `select` clause to at least display the information on all petrol types so that the minimum petrol price for each petrol type can easily be identified. However, in SQL, how can we ensure that a certain attribute is used? In this context, how can we ensure that the Determinant Dimension or Determinant Attribute is used in data retrieval? This is the main challenge.

There are two ways to address this issue. The first one is to covert the star schema that features a Determinant Dimension (like the one shown in Fig. 7.4) to a star schema where all the required information is captured in the fact measures (like the one shown in Fig. 7.2). The latter is basically a star schema with a *Pivoted Fact Table*. More details about this conversion will be given in the next sections.

The second method is to enforce the use of the Determinant Dimension or Determinant Attribute at the *User Interface Level*, where the users are forced to choose value(s) from the required Determinant Attribute. Using the Petrol Price case study, a web form contains an entry of Petrol Type which must be chosen by the user, without which the form will be invalid. The star schema design enforces the constraints at the user-interface level.

7.2   Determinant vs. Non-determinant Dimensions: The Olympic Games Case Study

In this section, we are going to examine a case study which has a Type Dimension to see if it is a Determinant Dimension or not. We use an Olympic Games case study.

The Olympic Games committee maintains an operational database that stores information on all matches, games and the medal winners of the Olympic Games over the years (https://www.olympic.org/). We would like to build a data warehouse to analyse the medal count by country, sport and at which Olympic Games they occurred.

There are two possible star schemas. The version 1 star schema contains three dimensions, and the version 2 star schema contains four dimensions. For simplicity, a minimal number of attributes are included in each dimension.

The version 1 star schema has Country, Sport and Olympic Dimensions. Three fact measures are included in the Fact Table, namely Number of Gold Medals, Number of Silver Medals and Number of Bronze Medals. The star schema is shown in Fig. 7.6.

The version 2 star schema has four dimensions, with the additional of the Medal Type Dimension (which is either Gold, Silver or Bronze), but there is only one fact measure in the Fact Table: Number of Medals. The star schema is shown in Fig. 7.7.

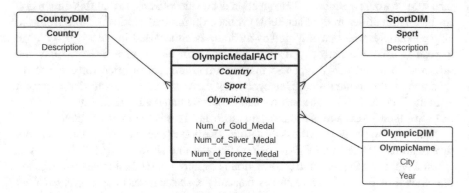

Fig. 7.6 The Olympic Games star schema (version 1)

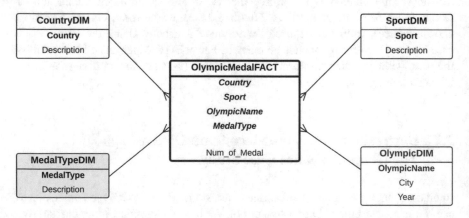

Fig. 7.7 The Olympic Games star schema with MedalTypeDim (version 2)

We are going to answer two questions relating to these two star schemas: What is the difference between these two versions of star schemas? Is the Medal Type Dimension a Determinant Dimension? The answer is next.

7.2.1   Star Schema Version 1 (Without Medal Type Dimension)

Firstly, we need to visualise two-column tables for each dimension category. The first two-column table is from the Country point of view (refer to Table 7.9).

These are the Number of Gold, Silver and Bronze medals that these countries won for all Olympic Games (several Olympic Games) recorded in the operational database. Note that this two-column table methodology helps the data warehouse designer to visualise the view of the fact measures from each dimension.

The second two-column table is from the Sport point of view. Assuming that the operational database records 20 past Olympic Games and at each Olympic Games, there is only one gold for the Men's 100m Butterfly (see Table 7.10). The third two-column table is from the Olympic Name point of view (see Table 7.11).

These two-column tables are known as *Pivot Tables* as they are actually two-dimensional pivot tables. A two-dimensional pivot table is a 2D matrix with two categories, as denoted by the columns and rows. The intersecting cells between the columns and rows are respective value of the column and row categories. Taking Table 7.9 as an example, the row category is Country, and the column category is Medals.

Table 7.9 Olympic Medals—Country point of view

Country	Num of Gold	Num of Silver	Num of Bronze
USA	733	602	488
China	199	143	133
Australia	167	170	189

Table 7.10 Olympic medals—Sport point of view

Sport	Num of Gold	Num of Silver	Num of Bronze
Swimming 100 m Butterfly Men	20	20	20
Swimming 400 m Freestyle Women	20	20	20
Swimming 4 × 100 m Medley Relay Men	20	20	20

Table 7.11 Olympic medals—Olympic Name point of view

Olympic name	Num of Gold	Num of Silver	Num of Bronze
London 2012	302	304	356
Beijing 2008	302	303	353
Athens 2004	301	301	327

All of the above two-column tables seem to be reasonably correct. The first columns are the categories, and the other columns are the fact measures, which are numeric and aggregate values. Because these three two-column tables make sense, star schema version 1 is correct.

7.2.2 Star Schema Version 2 (With Medal Type Dimension)

The two-column tables for the first three dimensions, namely Country, Sport and Olympic Names, are in Tables 7.12, 7.13, and 7.14.

The question is whether these two-column tables make sense. If we look at the Country, it makes sense to see how many medals Australia has received at all Olympic Games. Finding how many medals (regardless of the medal types) for each Country, for each Sport and for each Olympic Games seems to be reasonable as well.

The fourth two-column table for star schema version 2 is the Medal Type, as shown in Table 7.15.

This two-column table on Medal Type seems to be reasonable too. Hence, star schema version 2 (with Medal Type Dimension) is correct.

So in conclusion, both star schemas (with or without Medal Type Dimension) are correct. Now let's go back to the original question: is Medal Type Dimension a Determinant Dimension? Let's read the next section.

Table 7.12 Olympic medals—Country point of view

Country	Num of medals
USA	1823
China	475
Australia	526

Table 7.13 Olympic medals—Sport point of view

Sport	Num of medals
Swimming 100m Butterfly Men	60
Swimming 400m Freestyle Women	60
Swimming 4x100m Medley Relay Men	60

Table 7.14 Olympic medals—Olympic Name point of view

Sport	Num of medals
London 2012	962
Beijing 2008	958
Athens 2004	929

Table 7.15 Olympic medals—Medal Type point of view

Medal type	Num of medals
Gold	4115
Silver	4095
Bronze	4474

7.2.3   Determinant or Non-Determinant Dimensions

A *Determinant Dimension* is a dimension on which the fact measure relies on, and consequently, all data retrieval from the data warehouse must include this dimension (or the key identifier from this dimension). If the data retrieval from the data warehouse does not include this Determinant Dimension (or Determinant Attribute), the retrieved data will not make sense at all.

In the previous case study on petrol stations, the Petrol Type Dimension is a Determinant Dimension (refer to the star schema in Fig. 7.4). All fact measures— Total Petrol Price, Number of Petrol Stations, Min Petrol Price and Max Petrol Price—depend on Petrol Type, which is indicated by Petrol Type Dimension. This means that analysing the minimum petrol price from a Day of Week point of view must include petrol type. Otherwise, it will not make sense to retrieve data to show that on Monday, the lowest (minimum) petrol price is, for example, 117.9 cents. As there is no indication as to which petrol type it is, this information on the lowest petrol price is meaningless. Therefore, it is better to retrieve a record to show that, for example, on Monday, the minimum *Unleaded* petrol price is 117.9 cents (e.g. Unleaded is a petrol type obtained from the Petrol Type Dimension).

Now let's go back to the Olympic Games case study (refer to version 2 star schema with the Medal Type Dimension as shown in Fig. 7.7. Is the Medal Type Dimension a Determinant Dimension?

The answer to this question can be answered by another question. To retrieve the data from the version 2 star schema, must we have the information from the Medal Type Dimension? The answer is clearly no because we can simply retrieve a record from the Fact Table to show that at the London 2012 Olympic Games, Australia received ten medals in swimming. This covers three dimensions, namely Country (Australia), Olympic Name (London 2012) and Sport (Swimming). In this example, the Medal Type Dimension is not involved, and the information retrieved still makes sense. So to answer the question as to whether the Medal Type Dimension is a Determinant Dimension or not, the answer is clearly no!

The next question is: "what is the difference between the Olympic Games case study and the Petrol Station case study?" Both are very similar, but in the Olympic Games case study, the Medal Type Dimension is not a Determinant Dimension, whereas in the Petrol Station case study, the Petrol Type Dimension is a Determinant Dimension. The answer may not be straightforward. But in this case, the aggregate functions used in the fact measure in both case studies are different, and this may indicate whether other dimensions are needed in the data retrieval.

In the Olympic Games case study, the fact measure function is count, which is the medal count. The breakdown of the medals is Gold, Silver and Bronze; but the main aggregate function of the fact measure is the number of medals, which is count. We can still analyse the fact measure which is the total medals from the other dimensions, without the Medal Type Dimension. Consequently, the Medal Type Dimension is not a Determinant Dimension.

On the other hand, the Petrol Station case study relies on min and max as the aggregation functions, and the information of these functions relies on a particular information from a dimension, which is the Petrol Type Dimension. Consequently, the Petrol Type Dimension is a Determinant Dimension.

In short, having a "Type Dimension", such as Petrol Type or Medal Type, does not directly indicate whether it is a Determinant Dimension or not. We need to check if all the data retrieved from the Fact Table actually need information from this dimension or its key identifier. If yes, then it is a Determinant Dimension; if not, then it is a normal dimension.

7.2.4 Version 1 (Without Medal Type Dimension) vs. Version 2 (With Medal Type Dimension)

As both versions in the Olympic Games case study are correct (see Figs. 7.6 and 7.7), let's compare these two versions. To do this, let's have a look at the records in the respective Fact Tables.

The Fact Table for the version 1 star schema (without the Medal Type Dimension) consists of six attributes: three from the dimensions and the other three for the fact measures. The contents of the Fact Table are shown in Table 7.16.

The Fact Table for the version 2 star schema (with the Medal Type Dimension) consists of five columns: four from the dimension, but only one fact measure (see Table 7.17).

There are basically three main differences:

Table 7.16 Fact (star schema version 1)

Country	Sport	Olympic Name	Num of Gold	Num of Silver	Num of Bronze
USA	Swimming	London 2012	16	9	6
China	Swimming	London 2012	5	1	4
Australia	Swimming	London 2012	1	6	3

Table 7.17 Fact (star schema version 2)

Country	Sport	Olympic name	Medal type	Num of medals
USA	Swimming	London 2012	Gold	16
USA	Swimming	London 2012	Silver	9
USA	Swimming	London 2012	Bronze	6
China	Swimming	London 2012	Gold	5
China	Swimming	London 2012	Silver	1
China	Swimming	London 2012	Bronze	4
Australia	Swimming	London 2012	Gold	1
Australia	Swimming	London 2012	Silver	6
Australia	Swimming	London 2012	Bronze	3

1. Storage point of view
   It is clear that version 1 is the winner. It has only three records, whereas in version 2, the same information is represented in nine records.
2. Modelling point of view
   Version 2 with an additional dimension makes the model more concise and compact because there are fewer fact measures in the fact. When the number of fact measures is reasonably large, as in the Petrol Station case study, the star schema with a determinant dimension looks very slim and compact; hence, it is easy to understand. But consequently, the storage requirement increases as well. In contrast, with many different petrol types, if the star schema does not use a determinant dimension, the number of attributes in the fact will dramatically increase, and the schema looks more complex and crowded, but the storage cost is lower.
3. Query Processing point of view
   With an additional dimension, as in version 2, the query processing to join the Fact Table with the dimension tables requires an additional join operation because of the increase in the number of dimension tables, whereas in version 1, the dimension values are actually incorporated into the fact measures, and hence, the join processing between the dimension tables and the Fact Table reduces by one dimension.

7.2.5   Technical Challenges

The concepts of Determinant Dimensions, Pivot Tables and Non-determinant Dimensions are quite easy to digest. Implementation-wise, they may not be so. Using the Olympic Games case study from the previous section, Table 7.18 shows the medal tally of some selected countries which received medals in swimming at the Rio Olympic 2016.

The star schema shown in Fig. 7.7 has three dimensions: Country, Sport and Olympic with three fact measures: Number of Gold, Silver and Bronze medals. Technically, there are two challenges:

1. Some entries in the fact measures might be zero. For example, Italy did not receive any Silver medals in swimming at the Rio Olympic. Regardless of what the operational database (and its tables) looks like, a Fact Table is generally

Table 7.18   Rio Olympic 2016 swimming medal tally

Country	Sport	Olympic name	Num of Gold	Num of Silver	Num of Bronze
USA	Swimming	Rio 2016	16	8	9
Australia	Swimming	Rio 2016	3	4	2
China	Swimming	Rio 2016	1	2	3
Italy	Swimming	Rio 2016	1	0	2

Table 7.19 Rio Olympic 2016 swimming medal tally

Country	Sport	Olympic name	Medal type	Num of medals
USA	Swimming	Rio 2016	Gold	16
USA	Swimming	Rio 2016	Silver	8
USA	Swimming	Rio 2016	Bronze	9
Australia	Swimming	Rio 2016	Gold	3
Australia	Swimming	Rio 2016	Silver	4
Australia	Swimming	Rio 2016	Bronze	2
China	Swimming	Rio 2016	Gold	1
China	Swimming	Rio 2016	Silver	2
China	Swimming	Rio 2016	Bronze	3
Italy	Swimming	Rio 2016	Gold	1
Italy	Swimming	Rio 2016	Bronze	2

created by join and aggregate operations. The join operation is usually an inner join, which only finds matched records, and subsequently, the aggregation functions will not have any zeroes. So how are zeroes recorded in the Fact Table.

2. The Fact Table as shown in Table 7.18 is a *Pivoted Fact Table*. Usually, a Fact Table looks like Table 7.19, which has the same data as the Pivoted Fact Table in Table 7.18. However, note that Table 7.19 does not include Italy's zero silver medal in swimming because the Fact Table does not contain zero values as the inner join was used when creating the Fact Table:

```
select
  Country,
  Sport,
  OlympicName,
  MedalType,
  count(*) as Num_of_Medals
from <Table 1> T1, <Table 2> T2, ..., <Table n> Tn
where T1.attribute = T2.attribute
and T2.attribute = T3.attribute
and ...
group by Country, Sport, OlympicName, MedalType;
```

These two challenges will be addressed in the next section using the PTE Academic Test case study.

7.3   Determinant Dimensions vs. Pivoted Fact Table: The PTE Academic Test Case Study

The PTE Academic Test (administered by Pearson Inc.) is the world's leading computer-based test of English for study abroad and immigration, which is recognised by universities worldwide and is approved by the Australian Government for

Table 7.20 PTE band scale

Proficiency level	Grade	Test component	Test score
Functional	4.5	Overall score	30–35
Vocational	5	Listening	36–49
		Reading	
		Writing	
		Speaking	
Competent	6	Listening	50–64
		Reading	
		Writing	
		Speaking	
Proficient	7	Listening	65–78
		Reading	
		Writing	
		Speaking	
Superior	8–9	Listening	79–90
		Reading	
		Writing	
		Speaking	

visa applications. The PTE Test consists of four components: listening, reading, speaking and writing. A student taking a PTE test will have one score for each of these four components as well as one overall score. The score is a numerical score, ranging from 10 to 90. The Australian Immigration Department uses PTE scores to identify an applicant's English language proficiency level. The band scale is shown in Table 7.20.

The PTE Academic Test has many test venues (all major cities in Australia and overseas). Each country may have different price settings. For example, taking a PTE test in China is cheaper than taking the same test in Melbourne. The PTE Academic Test is a computer-based test. During the test, one or more supervisors oversee the test. The PTE Academic Test results are usually released 5 days after the test date. A typical score card contains the student's details and the five test scores: Listening, Reading, Writing, Speaking and Overall.

Generally, universities and the Australian Government look at each component score rather than the overall score. For example, in order for an applicant to be able to apply for a Permanent Residency in Australia, they must have at least a score of 65 in all of the four components (e.g. proficient in all components, listening, reading, writing and speaking), regardless of their overall score.

It is common for students to take a PTE Academic Test several times to improve their score. A student is identified by their RegistrationID. The RegistrationID for the same student stays the same and does not change when the student sits for a new test.

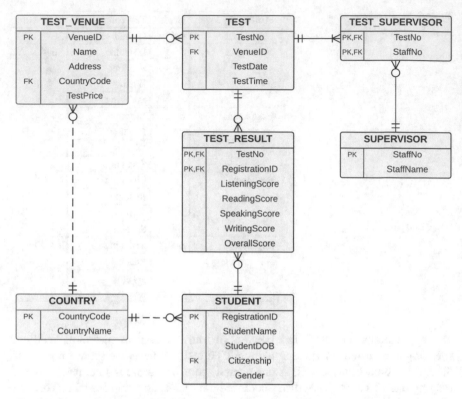

Fig. 7.8 PTE Academic Test E/R diagram

Table 7.21 Table student

RegistrationID	Student name	DOB	Gender	Citizenship
318678895	Ying Chang	27/AUG/89	F	CN
312124682	Ryou Kato	23/MAR/94	M	JP
117609895	Susi Bambang	19/FEB/91	F	ID
891186588	Su-yeon Park	07/JUL/94	F	KR
387160595	Hong Lei	16/DEC/90	M	CN
312905800	Tim Yeoh	08/OCT/91	M	SG
311765711	Berta Ferrari	27/DEC/87	F	IT
392837402	Wenbo Lin	27/AUG/92	M	CN

The E/R diagram of the operational system is shown in Fig. 7.8.

Sample records from the Student, Test, Test Venue and Test Result tables are shown in Tables 7.21, 7.22, 7.23, and 7.24. For simplicity, only eight students from various countries are shown. Note that some students repeated the test. For example, the student with a RegistrationID 318678895 has increased her score from 74 to 78 in two tests (the two rows are highlighted in Table 7.24).

Table 7.22 Table test

Test no	Test date	Test time	Venue ID
1	01/AUG/17	10:00 a.m.	1
2	01/AUG/17	11:00 a.m.	1
3	10/AUG/17	10:00 a.m.	2
4	10/AUG/17	11:00 a.m.	2
5	17/AUG/17	10:00 a.m.	3
6	17/AUG/17	11:00 a.m.	3
7	17/AUG/17	10:00 a.m.	1
8	25/AUG/17	10:00 a.m.	1

Table 7.23 Table test venue

Venue ID	Name	Address	Price	Country Code
1	RMIT English Worldwide	Level 10 235-251 Bourke Street Melbourne Victoria 3000	300 AUD 300 AUD	AU AU
2	Pearson Professional Centers San Francisco	Suite 200 201 Filbert Street San Francisco, California	200 USD	US
3	PLT at Cliftons Sydney	Level 12 60 Margaret Street Sydney New South Wales 2000	300 AUD	AU

Table 7.24 Table test result

Test No	RegistrationID	Listening score	Reading score	Speaking score	Writing score	Overall score
1	318678895	69	77	78	70	**74**
1	312124682	55	57	60	75	60
2	117609895	68	68	67	69	68
2	891186588	62	63	57	68	61
3	387160595	67	72	78	66	72
3	312905800	66	73	71	67	70
4	311765711	57	55	59	60	57
7	117609895	75	68	78	78	75
7	891186588	69	70	76	75	72
8	318678895	80	79	78	78	**78**
8	312124682	55	45	47	50	47

We would like to analyse two versions of star schemas for the data warehouse of the PTE Academic Test.

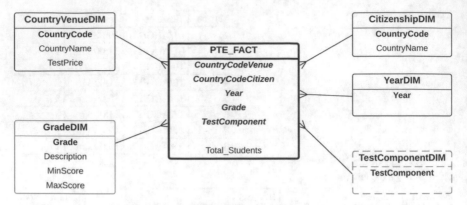

Fig. 7.9 PTE star schema—a Determinant Dimension version

7.3.1 A Determinant Dimension Version

For simplicity, five dimensions are chosen: Country Venue, Citizenship, Year, Grade and Test Component dimensions. The Test Component dimensions will only have five records: Listening, Reading, Writing, Speaking and Overall. The Test Component dimension is a *Determinant Dimension*. There is only one fact measure in the fact: Total Students. Therefore, we could ask questions such as: how many Chinese students received a proficient grade in the Listening part in 2017? This question involves Citizenship, Year, Grade, and, of course, Test Component dimensions. The star schema is shown in Fig. 7.9.

Assuming the operational database, as shown in the E/R diagram in Fig. 7.8, exists, the next step is to implement the star schema in SQL. The first step is to create a table for each of the five dimensions.

The Country Venue Dimension can be created by joining two tables from the operational database: Test Venue and Country, to get the Country Code, Country Name and Test Price attributes. For the Citizenship Dimension, it is also a join between Student and Country tables from the operational database to get the Citizenship (which is the Country Code) and Country Name. For the Year Dimension, it is basically an extraction from the Test Date attribute in the Test table. The SQL to create these three dimensions is as follows:

```
create table CountryVenueDim as
select distinct C.CountryCode, C.CountryName, T.TestPrice
from Test_Venue T, Country C
where T.CountryCode = C.CountryCode;

create table CitizenshipDim as
select distinct
  C.CountryCode as Citizenship,
  C.CountryName
from Student S, Country C
where S.Citizenship = C.CountryCode;
```

```
create table YearDim as
select distinct to_char(TestDate, 'YYYY') as Year
from Test;
```

The Grade and Test Component Dimensions cannot be extracted from the operational database; rather, they have to be created manually. The create table and insert into statements for these two dimensions are as follows. For simplicity, the Test Component Dimension includes only one attribute.

```
create table GradeDim
(Grade        varchar2(3),
 Description  varchar2(20),
 MinScore     number,
 MaxScore     number);

insert into GradeDim values ('4.5', 'Functional', 30, 35);
insert into GradeDim values ('5',   'Vocational', 36, 49);
insert into GradeDim values ('6',   'Competent',  50, 64);
insert into GradeDim values ('7',   'Proficient', 65, 78);
insert into GradeDim values ('8-9', 'Superior',   79, 90);

create table TestComponentDim
(TestComponent varchar2(20));

insert into TestComponentDim values ('Listening');
insert into TestComponentDim values ('Reading');
insert into TestComponentDim values ('Writing');
insert into TestComponentDim values ('Speaking');
insert into TestComponentDim values ('Overall');
```

As mentioned in the earlier chapters, when a dimension is created manually (instead of extracted from the operational database), we need to create a temporary Fact Table called TempFact. In this TempFact, we will not only get Country Code from Test Venue, Citizenship from Student and Test Date from Test tables, but also the five test scores (e.g. Listening, Reading, Writing, Speaking and Overall scores), as well as the Student's RegistrationID.

```
create table TempFact as
select
   TV.CountryCode,
   S.Citizenship,
   to_char(T.TestDate, 'YYYY') as Year,
   TR.ListeningScore,
   TR.ReadingScore,
   TR.WritingScore,
   TR.SpeakingScore,
   TR.OverallScore,
   TR.RegistrationID
from Test_Venue TV, Test T, Student S, Test_Result TR
where TV.VenueID = T.VenueID
and T.TestNo = TR.TestNo
and TR.RegistrationID = S.RegistrationID;
```

Creating the final Fact Table will be tricky because the final Fact Table has only one fact measure, namely Number of Students. Number of students will need to be broken down into five different test components. This is why in the TempFact, we store the five scores as well as the RegistrationID.

The next step is to alter the TempFact Table by adding five new attributes to store the "Grade" of each Test Component.

```
alter table TempFact
add (GradeOverall    varchar2(3),
     GradeListening  varchar2(3),
     GradeReading    varchar2(3),
     GradeWriting    varchar2(3),
     GradeSpeaking   varchar2(3)
);
```

Next, we "convert" each score (e.g. Listening score, etc.) to a grade using the band scale as shown in Table 7.20. Because there are five different scores, we need five grades, one score category for a grade category (e.g. Listening score to Listening grade). Hence, the five update statements are as follows:

```
update TempFact
set GradeOverall =
  (case
    when OverallScore >= 30 and OverallScore <= 35 then '4.5'
    when OverallScore >= 36 and OverallScore <= 49 then '5'
    when OverallScore >= 50 and OverallScore <= 64 then '6'
    when OverallScore >= 65 and OverallScore <= 78 then '7'
    when OverallScore >= 79 and OverallScore <= 90 then '8-9'
  end);

update TempFact set GradeListening = ...;
update TempFact set GradeReading = ...;
update TempFact set GradeWriting = ...;
update TempFact set GradeSpeaking = ...;
```

A case when condition is used in the set clause of the update statement, in order to check for an appropriate conversion from a score to a grade. For simplicity, the repeated code in the other four updates is shown in three dots, instead of the full case when conditions.

After the updates are completed, the TempFact Table has the correct grades for each student. The problem is now to break down the grades for a student into multiple test components. This can be done by creating yet another temporary fact for each of the test components. For example, for the Overall Score, we need a temporary fact, called "OverallFact", and for the Listening Score, we will have "ListeningFact". In other words, we will have five temporary Fact Tables, one for each test component. The structure of the OverallFact, for example, is Country Code, Citizenship, Year, GradeOverall (which we obtained from the above update statement) and the string "Overall" which represents the Test Component and the count of RegistrationID for Total Students. This process is repeated for the other four temporary facts, called ListeningFact, ReadingFact, WritingFact and SpeakingFact.

```
create table OverallFact as
select
  CountryCode,
  Citizenship,
  Year,
  GradeOverall As Grade,
  'Overall' as TestComponent,
  count(RegistrationID) as Total_Students_Overall
from TempFact
group by
  CountryCode,
  Citizenship,
  Year,
  GradeOverall,
  'Overall';

create table ListeningFact as select ...;
create table ReadingFact as select ...;
create table WritingFact as select ...;
create table SpeakingFact as select ...;
```

Using the sample data shown earlier in Table 7.24, the contents of the OverallFact and ListeningFact Tables are shown in Tables 7.25 and 7.26. There were eleven students taking the test as shown in Table 7.24, which are now grouped into nine records in Table OverallFact (e.g. two students from Indonesia (ID) received an Overall grade of 7, and two students from China (CN) also received an Overall grade of also 7; however, it is easy to cross-check with the Test Result table to identify these students). Note that according to the band scale as shown in Table 7.20, grade 7 is equivalent to a score of 65–78, and there are seven students who achieved this grade as shown in the OverallFact Table (Table 7.25).

For the ListeningFact, there are also nine records (two students from Indonesia (ID) received a grade of 7 in the Listening part, and two students from Japan (JP) received a grade of 6 in the Listening part).

There are also three other Fact Tables: ReadingFact, WritingFact and SpeakingFact, which are not shown here, but the structures of the five Fact Tables are

Table 7.25 OverallFact table

Country code	Citizenship	Year	Grade	Test component	Total students overall
AU	JP	2017	6	Overall	1
AU	KR	2017	6	Overall	1
AU	ID	2017	7	Overall	2
US	IT	2017	6	Overall	1
AU	JP	2017	5	Overall	1
US	CN	2017	7	Overall	1
US	SG	2017	7	Overall	1
AU	KR	2017	7	Overall	1
AU	CN	2017	7	Overall	2

Table 7.26 ListeningFact Table

Country code	Citizenship	Year	Grade	Test component	Total students listening
AU	CN	2017	7	Listening	1
AU	ID	2017	7	Listening	2
US	IT	2017	6	Listening	1
AU	KR	2017	6	Listening	1
US	CN	2017	7	Listening	1
US	SG	2017	7	Listening	1
AU	JP	2017	6	Listening	2
AU	CN	2017	8–9	Listening	1
AU	KR	2017	7	Listening	1

identical. These five Fact Tables can now be "combined" to form one final Fact Table. This final Fact Table will be the *union* of the five Fact Tables; hence, the union operator is used in the following SQL:

```
create table FinalFact as
select
   CountryCode,
   Citizenship,
   Year,
   Grade,
   TestComponent,
   Total_Students_Overall as Total_Students
from OverallFact
union
   select * from ListeningFact
union
   select * from ReadingFact
union
   select * from WritingFact
union
   select * from SpeakingFact;
```

Note that the column names (attribute names) of the FinalFact will be based on the first fact in the Union, which is the OverallFact. Table 7.27 shows a snapshot of the table. It contains 45 records, which is a comnbination of the five Fact Tables. The first seven records are two students from China (CN) who sat the test in Australia (AU); the sum of the Total Students column is ten, representing five test component grades for 2 students. On the other hand, the last five records pertain to one student from Singapore (SG) who sat the test in the USA and achieved a score of 7 in all of the test components.

As a conclusion, when the input record in the operational database contains the scores of all the test components as one record, it will be challenging to break down the records into multiple records in the final Fact Table because there is a Test Component dimension. In the above example, not only the TempFact needs to convert the score into a grade, it is also necessary to create five temporary facts, one for each test component. Finally, these are union-ed to form the final Fact Table.

Table 7.27 Final Fact Table

Country code	Citizenship	Year	Grade	Test component	Total students
AU	CN	2017	7	Listening	1
AU	CN	2017	7	Overall	2
AU	CN	2017	7	Reading	1
AU	CN	2017	7	Speaking	2
AU	CN	2017	7	Writing	2
AU	CN	2017	8–9	Listening	1
AU	CN	2017	8–9	Reading	1
...	...	...	...	...	...
...	...	...	...	...	...
US	SG	2017	7	Listening	1
US	SG	2017	7	Overall	1
US	SG	2017	7	Reading	1
US	SG	2017	7	Speaking	1
US	SG	2017	7	Writing	1

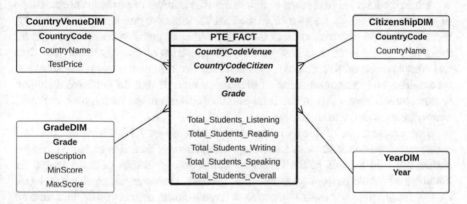

Fig. 7.10 PTE star schema—a Non-determinant Dimension version

7.3.2 A Non-determinant Dimension Version or the Pivoted Fact Table Version

The star schema for the non-determinant dimension is shown in Fig. 7.10. The Test Component dimension is removed, but there is a fact measure for each of the test components (e.g. Total Students Listening, Total Students Reading, etc.). This means that the Fact Table is a *Pivoted Fact Table* in which each record in the Fact Table can be seen as a 2D matrix between the dimension key identifiers and the five fact measures. This Non-determinant Dimension version can be seen as shifting the Test Component into the fact measure. Consequently, the Determinant Dimension is removed from the star schema.

The four dimension tables that have been created in the previous section can be re-used here, so there is no need to separately re-create each dimension table.

In general, when a Fact Table is created, it may not contain all possible combinations from all records of all dimensions. The reason is that the Fact Table is created by a join (an inner join) operation. Consequently, combinations of records from the dimension tables that do not have any value for the fact measure will not be included in the Fact Table. Simply put, there are no zero values in the fact measure.

Using the Test Result table as an example (refer to Table 7.20), look at the second last student's (318678895) scores. Of the eleven test results, this is the only student who received a "Superior" grade (score between 79 and 90, inclusive) in the Listening and Reading components. No one received this Superior grade in the Speaking or Writing components, and obviously no one received a Superior grade as an overall grade. Using the normal Fact Table, there will not be an entry in the Fact Table for Superior-Speaking or for Superior-Writing. In other words, there will be no zero values in Total Students where the Grade column is "Superior" and the Test Component column is either "Speaking" or "Writing". This is why the number of records in the Fact Table is equal to or less than the number of possible combinations of all dimension tables. As a proof, in the Fact Table shown previously in Table 7.27, there are only 45 records, whereas the total number of possible combinations is far more (e.g. two countries of test venues, six countries of citizenship, 1 year, five test components and five grades). This is due to two things: the test results data are incomplete, and the combination of dimension records that do not have a student will not have a zero value in the Total Students column in the fact; rather, they will simply be excluded automatically.

If we create a Fact Table as in the star schema in Fig. 7.10, which has five fact measures, Total Students with Listening, with Reading, etc., it is expected that we should have 60 records in the final Fact Table (two countries of test venues, six countries of citizenship, 1 year and five grades). However, since our test result data is incomplete (or even if there is a large volume of test result data, not all possible combinations from the dimension are covered), we will have problems with the final fact because in the Non-determinant Dimension version as shown in Fig. 7.10, we need to keep track of the zeroes. If not, the student with RegistrationID 318678895 who received a grade of Superior in the Listening and Reading tests but only Proficient in the other three tests (Writing, Speaking and Overall) will not be included in the Fact Table because none of the students received a grade of Superior in the last three test components (e.g. Writing, Speaking and Overall). What we want is to have zero values in the Total Students Writing, Total Students Speaking and Total Students Overall in the Fact Table. To achieve this, first, we need to get all possible combinations from all dimensions, which is a Cartesian Product between all dimensions.

```
create table AllDimensions as
select
  CO.CountryCode,
  CI.Citizenship,
  Y.Year,
  G.Grade
```

```
from
    CountryVenueDim CO,
    CitizenshipDim CI,
    YearDim Y,
    GradeDim G;
```

The next step is to re-create the five temporary Fact Tables, one for each test component. These temporary Fact Tables are created by an *Outer Join* operation between the AllDimensions table and each of the temporary Fact Tables created in the previous section. An outer join operation is used so that we preserve all records from the AllDimensions table. The Total Students will be zero when the record in the AllDimensions table does not match any of the records from the temporary fact (Note: an `nvl` function is used). The following shows the first two temporary Fact Tables only (e.g. OverallFactNew and ListeningFactNew). Notice also how the OverallFact and ListeningFact, which are created previously in the Determinant version, are now used to create the respective Fact in the Non-determinant version. The three other tables (e.g. ReadingFactNew, WritingFactNew and SpeakingFactNew) will be created in the same way.

```
create table OverallFactNew as
select
    A.CountryCode,
    A.Citizenship,
    A.Year,
    A.Grade,
    nvl(O.Total_Students_Overall, 0)
        as Total_Students_Overall
from AllDimensions A, OverallFact O
where A.CountryCode = O.CountryCode(+)
and A.Citizenship = O.Citizenship(+)
and A.Year = O.Year(+)
and A.Grade = O.Grade(+);

create table ListeningFactNew as
select
    A.CountryCode,
    A.Citizenship,
    A.Year,
    A.Grade,
    nvl(O.Total_Students_Listening, 0)
        as Total_Students_Listening
from AllDimensions A, ListeningFact O
where A.CountryCode = O.CountryCode(+)
and A.Citizenship = O.Citizenship(+)
and A.Year = O.Year(+)
and A.Grade = O.Grade(+);

create table ReadingFactNew as select ...;
create table WritingFactNew as select ...;
create table SpeakingFactNew as select ...;
```

Once the five temporary Fact Tables are created, the final Fact Table can be created. Note that the structure of the five temporary tables is identical, that is, Country Code, Citizenship, Year, Grade and Total Students—there is no Test Component. The Total Students column is the total students for the particular temporary fact; this means for the ReadingFactNew table, the Total Students column is the Total Students for the Reading Test Component. This is why we have five temporary Fact Tables, one for each test component.

To create the final Fact Table, we use the `join` operation. Remember that for the determinant version of the Fact Table, as discussed in the previous section, a `union` operator was used. But for the non-determinant version, here, we use a `join` operator to join the five temporary facts based on Country, Citizenship, Year and Grade. We will still keep each Total Students column from each of the five temporary facts.

```
create table FinalFact2 as
select
   O.CountryCode,
   O.Citizenship,
   O.Year,
   O.Grade,
   O.Total_Students_Overall,
   L.Total_Students_Listening,
   R.Total_Students_Reading,
   W.Total_Students_Writing,
   S.Total_Students_Speaking
from
   OverallFactNew O,
   ListeningFactNew L,
   ReadingFactNew R,
   WritingFactNew W,
   SpeakingFactNew S
where O.CountryCode = L.CountryCode
and L.CountryCode = R.CountryCode
and R.CountryCode = W.CountryCode
and W.CountryCode = S.CountryCode
and O.Citizenship = L.Citizenship
and L.Citizenship = R.Citizenship
and R.Citizenship = W.Citizenship
and W.Citizenship = S.Citizenship
and O.Year = L.Year
and L.Year = R.Year
and R.Year = W.Year
and W.Year = S.Year
and O.Grade = L.Grade
and L.Grade = R.Grade
and R.Grade = W.Grade
and W.Grade = S.Grade;
```

Because our sample data is small, there are many records with zero values in the Total Students columns. In order to exclude those records, we can simply delete records where Total Students in all of the five test components are equal to zero.

```
delete from FinalFact2
where Total_Students_Overall = 0
and Total_Students_Listening = 0
and Total_Students_Reading = 0
and Total_Students_Writing = 0
and Total_Students_Speaking = 0;
```

Table 7.28 shows 11 records. Look at the second last record with grade 8-9. This refers to the student with RegistrationID 318678895 who received two Superior grades (one for Listening and the other for Reading), and none of the students in the operational database received a Superior grade in the other three text components (Writing, Speaking and Overall), which is why the zero values appear in the Fact Table.

In summary, to create the Non-determinant version (or the Pivoted Fact version), the Fact Tables from the Determinant version are used. As stated earlier, one of the ways to implement a Determinant Dimension version is by converting the Determinant version into a Pivoted Fact version. This section shows how to do this using the SQL commands.

7.4   Non-type as a Determinant Dimension: University Enrolment Case Study

In the previous sections, we learned that a Type Dimension may be a candidate for a Determinant Dimension, Petrol Type Dimension is a Determinant Dimension in the Petrol Price case study, whereas Medal Type Dimension in the Olympic Games case study is not a Determinant Dimension. Nevertheless, a Type Dimension is a good candidate for a Determinant Dimension, but it needs further examination to check if it is really a Determinant Dimension.

In this section, we are going to learn that other dimensions can potentially be a Determinant Dimension. In this case, we use the Year as a Determinant Dimension. Let's have a look at the University Enrolment case study.

Figure 7.11 shows a simple star schema with four dimensions. The Faculty Dimension lists all faculties of a university, and the Class Type Dimension contains Bachelor, Master and PhD as the class levels. The other two dimensions, Gender and Year, are obvious. The fact measure is the Number of Students. With this star schema, we can query the number of students per Faculty, per Year, per Gender and per Class Type, or any combinations of these. Note that in this star schema, there is a Type Dimension which is Class Type (e.g. Bachelor, Master or PhD).

The Class Type Dimension in this example is not a Determinant Dimension because we are not able to get the Number of Students without referring to any of the class types. For example, the number of students in the Science Faculty last year is certainly useful information, even without the Class Type information. Therefore, the Class Type Dimension is not a Determinant Dimension.

Table 7.28 Final Fact Table

Country code	Citizenship	Year	Grade	Total students overall	Total Stud listening	Total Stud reading	Total Stud writing	Total Stud speaking
AU	KR	2017	6	1	1	1	0	1
US	IT	2017	6	1	1	1	1	1
AU	CN	2017	7	2	1	1	2	2
AU	JP	2017	6	1	2	1	1	1
AU	ID	2017	7	2	2	2	2	2
AU	KR	2017	7	1	1	1	2	1
US	CN	2017	7	1	1	1	1	1
AU	JP	2017	5	1	0	1	0	1
US	SG	2017	7	1	1	1	1	1
AU	CN	2017	8-9	0	1	1	0	0
AU	JP	2017	7	0	0	0	1	0

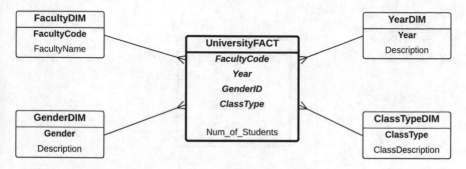

Fig. 7.11 University Enrolment star schema—with four dimensions

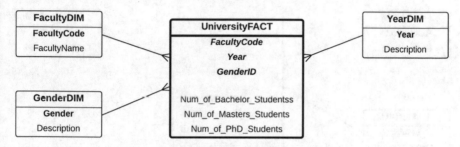

Fig. 7.12 University Enrolment star schema—with three dimensions

Using the knowledge learned from the previous sections, we could shift the Class Type information from the dimension to the fact. As a result, the fact measures become Number of Bachelor Students, Number of Master's Students and Number of PhD students. The star schema is now reduced by one dimension, as shown in Fig. 7.12.

Using this star schema, you can still get the Total Number of Students from the Science Faculty last year simply by combining the Number of Bachelor Students with the Number of Master's Students and Number of PhD Students. So the star schema is still valid, even though we have shifted the Class Type information from the dimension to the fact. Hence, for simplicity, we keep the initial star schema with four dimensions, including the Class Type Dimension, because it is not a Determinant Dimension anyway.

Now let's examine the Year Dimension. In the university context, the analysis of the number of students is usually conducted on a yearly basis. It is not sensible if the year information is omitted. For example, there are 100,000 students in the Science Faculty, which is the total of all years. This is not very meaningful. Or there are 60,000 female students in the Science Faculty, which shows the total number of female students so far (from the reference period of the star schema), but this would be rather unusual. Usually, the information is dependent on the year. For example, the Number of Students in the Science Faculty "last year" was 3000, or the Number of Female Students in the Science Faculty "last year" was 2000, which

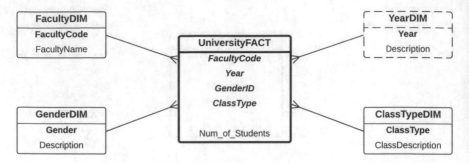

Fig. 7.13 University Enrolment star schema—with Year as a Determinant Dimension

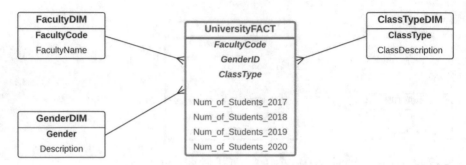

Fig. 7.14 University Enrolment star schema—a Non-determinant Dimension

would be more meaningful. Hence, in this scenario, the information is dependent on the Year Dimension. In other words, the Year Dimension is a Determinant Dimension. The star schema with the Year Dimension as a Determinant Dimension is shown in Fig. 7.13. This example shows that a Type Dimension is not always a Determinant Dimension. Other dimensions, like the Year Dimension, can be a Determinant Dimension, as shown in this case study. Business rules must be used to determine if a dimension is a Determinant Dimension or not.

Converting a Determinant Dimension to a Pivoted Fact Table is done by shifting the information about the Determinant Dimension to become the fact measures. In this case, the Non-determinant Dimension version as shown in Fig. 7.14 has more fact measures, each of which indicates the number of students for that particular year.

In terms of implementation, converting the star schema with a Determinant Dimension (Fig. 7.13) to a star schema with a Pivoted Fact Table (Fig. 7.14) can be done similarly to the PTE Academic Test case study detailed in the previous section.

Two questions might arise from the star schema in Fig. 7.14.

1. What happens if the number of records in the Determinant Dimension is big? In this example, for each year, we need to have one fact measure. In the example in Fig. 7.14, there were only four years, hence four fact measures. This is

manageable. But if the number of years is, for example, 20, then the number of fact measures will be 20. What if, in another scenario that the Determinant Dimension is something else, and has for example one thousand records. It would not be practical to have 1000 fact measures. Hence, in this case, the Determinant Dimension can only be enforced through the User-Interface level, where users are forced to enter/choose a criterion from the Determinant Dimension. In filling out a web-based form, this enforcement is usually indicated by a star (*) next to the text entry to highlight that this text entry must be filled in before the form can proceed.

2. What if there are a lot of attributes in the Determinant Dimension? If we shift the Determinant Dimension to the fact measures, only the key identifier is kept in the fact measure. All other attributes from the Determinant Dimension will be lost from the star schema. This issue will be discussed in the next section.

7.5 Multiple Relationship Between a Dimension and the Fact: Private Taxi Case Study

This section will explore a case study where the dimension (or a Determinant Dimension) to be shifted to the fact measure (e.g. Pivoted Fact Table) has many attributes. In the previous sections, when we shift a dimension (or a Determinant Dimension) to the fact measure, only the key identifier of the dimension (or the Determinant Dimension) is used as the fact measure. None of the other attributes of the dimension (or the Determinant Dimension) exist in the star schema. Let's discuss this issue using the following Private Taxi Company case study as an example.

Figure 7.15 shows a star schema for a Private Taxi Company. For simplicity, it only includes three dimensions: Car, Driver and MonthYear Dimensions, each with some attributes about the dimension. The fact has three fact measures: Total Kilometers, Total Fuel Used and Total Income. With this simple star schema, we can retrieve the Total Kilometers for a particular car (e.g. Car 5) for a particular month (e.g. August 2019), for example. Or we can get the Total Income generated by a particular driver for a particular month.

```
select MonthYear, sum(Total_Kilometers)
from PrivateTaxiFact1
where CarNo = 5
and MonthYear = '1908'
group by MonthYear;

select DriverNo, MonthYear, sum(Total_Income)
from PrivateTaxiFact1
where DriverNo = '01'
and MonthYear = '1908'
group by DriverNo, MonthYear;
```

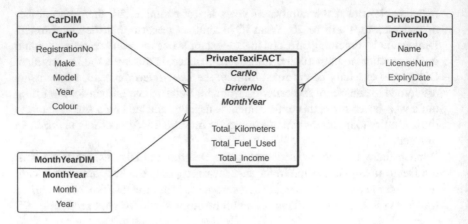

Fig. 7.15 A Private Taxi Company star schema

In this case study, we assume that the business rules state that there is no Determinant Dimension. Even without a Determinant Dimension, we could potentially shift a dimension to become fact measures. In this example, assume that the company has only a limited number of cars, say, five cars, as it is a small private taxi company. So in this case, in the Fact Table, instead of having three fact measures, we will have 15 fact measures, as there are five cars. Hence, the Total Kilometers would become Total Kilometers Car 1, Total Kilometers Car 2,..., Total Kilometers Car 5. The same process is applied to Total Fuel Used and Total Income. As a result, the three original fact measures become 15 fact measures (three fact measures times five cars). In doing this, we reduce the number of dimensions to two, Driver and MonthYear Dimensions.

The same data retrieval is applied to this new star schema. For example, to retrieve Total Kilometers for a particular car (say, Car 5) for a particular month, we will only use the MonthYear Dimension and display the Total Kilometers Car 5 column in the Fact Table. Or to retrieve the Total Income generated by a particular driver for a particular month, this can be done by filtering the Driver Dimension table and the MonthYear Dimension table and then summing the five Total Income (e.g. Total Income Car 1 + Total Income Car 2 + ··· + Total Income Car 5). So there will be no issues with data retrieval.

```
select MonthYear, sum(Total_Kilometers_Car5)
from PrivateTaxiFact2
where MonthYear = '1908'
group by MonthYear;

select
   DriverNo,
   MonthYear,
   sum(Total_Income_Car1 +
   Total_Income_Car2 +
   Total_Income_Car3 +
```

```
  Total_Income_Car4 +
  Total_Income_Car5)
from PrivateTaxiFact2
where DriverNo = '01'
and MonthYear = '1908'
group by DriverNo, MonthYear;
```

However, there is one problem in shifting the Car Dimension to the fact measures. The Car Dimension, as shown in Fig. 7.15, has many attributes, such as Registration No, Make, Model, Year and Colour. By shifting the Car Dimension to the fact measure, we can only keep the key identifier which is Car No. Hence, we won't be able to include other details of the car in the data retrieval as the Car attributes are no longer kept in the star schema.

In this case study, because the Car Dimension is not a Determinant Dimension, we can simply keep the original star schema as shown in Fig. 7.15. Hence, if we want to get the Car's Make and Model, for instance, we can simply join the Fact Table with the Car Dimension table.

```
select
  F.MonthYear,
  D.Make,
  D.Model,
  sum(F.Total_Kilometers)
from PrivateTaxiFact1 F, CarDim D
where F.CarNo = D.CarNo
and F.CarNo = 5
and F.MonthYear = '1908'
group by F.MonthYear, D.Make, D.Model;
```

However, if we insist that the fact measures are broken down into five cars, then we need to have five key identifiers for the Car Dimensions in the fact (e.g. CarNo1, CarNo2, etc.), along with 15 fact measures (see Fig. 7.16). For clarity and simplicity, we only show one relationship between Car Dimension and the fact. But as a matter of fact, there are five relationships, one for each Car No.

When we retrieve the details of the five cars, we need to join the five instances of the Car Dimension tables. For example, the following retrieves the Total Kilometers of the five cars (e.g. Make and Model) in August 2019. If we only need a certain car (e.g. Car 5), then we only need to join once (instead of five times as in the example below) with the Car Dimension table.

```
select
  F.MonthYear,
  D1.Make, D1.Year,
  D2.Make, D2.Year,
  D3.Make, D3.Year,
  D4.Make, D4.Year,
  D5.Make, D5.Year,
  sum(Total_Kilometers)
from PrivateTaxiFact2 F,
  CarDim D1, CarDim D2,
  CarDim D3, CarDim D4,
  CarDim D5
```

```
where F.CarNo1 = D1.CarNo
and F.CarNo2 = D2.CarNo
and F.CarNo3 = D3.CarNo
and F.CarNo4 = D4.CarNo
and F.CarNo5 = D5CarNo
and MonthYear = '1908'
group by
  F.MonthYear,
  D1.Make, D1.Year,
  D2.Make, D2.Year,
  D3.Make, D3.Year,
  D4.Make, D4.Year,
  D5.Make, D5.Year;
```

Although conceptually the dimension has already shifted to the fact measure, we still need to keep the original dimension in the star schema. This is the only way to preserve all the other attributes of that dimension. This relationship replication is applicable to shifting a Determinant Dimension to the fact measure.

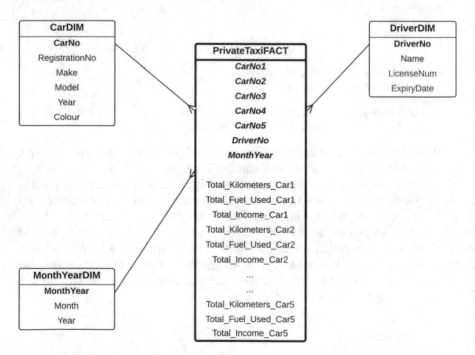

Fig. 7.16 A Private Taxi Company star schema—with a Non-connected Car Dimension

7.6 Summary

This section explores a new concept of a dimension, called a *Determinant Dimension* (or a *Determinant Attribute*). A Determinant Dimension (or a Determinant Attribute) must be used in retrieving the fact to make the retrieved data more meaningful. In other words, it is critical to use the Determinant Attribute in the query, either as a filtering mechanism (in the `where` clause) or in the display (in the `select` clause) in the SQL command.

There are two alternatives to enforce the implementation of a star schema that features a Determinant Dimension. One is to enforce this through the user interface where the user has to fill in or choose the Determinant Attribute during the search. The second is to shift the Determinant Dimension into a Pivoted Fact Table where the key identifier of the Determinant Dimension is incorporated into the fact measure.

7.7 Exercises

7.1 In the Determinant version of the PTE Academic Test case study (refer to Sect. 7.3.1), the Fact Table which contains 45 records is shown in Table 7.27. For example, if the combination of all dimension keys does not have any students, in this example, it will not show as a zero; rather, it will simply be excluded from the Fact Table.

Question: If we also want to show zeroes in the Total Students column (for those combinations of all Dimensions' key that do not have any students), what is the SQL command to produce this? How many records will there be in this new Fact Table?

7.2 Using the University Enrolment case study (refer to Sect. 7.4), in what circumstance will the Year Dimension NOT be considered as a Determinant Dimension? Explain your reason and draw the new star schema as well.

7.3 A medical clinic employs four general practitioners (doctors): Dr. Adele, Dr. Ben, Dr. Kate and Dr. Chris. Some of the doctors do not practise every day. For example, Dr. Adele practises on Mondays and Wednesdays only, whereas Dr. Ben is there only on Thursdays. When a patient comes to the clinic and has a consultation with a doctor, the patient pays a certain consultation fee, depending on the type of consultation the patient had. For example, a general consultation fee (code 113) is $37.50. Because of the nature of medical practice, there are more than 100 different codes for different types of consultations.

The clinic maintains an operational database that records every payment for each consultation by every doctor. A data warehouse is needed for reporting purposes. There are two versions of star schema for this clinic (refer to Figs. 7.17 and 7.18).

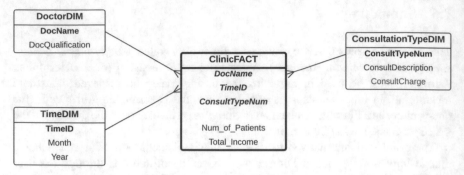

Fig. 7.17 Clinic star schema version 1

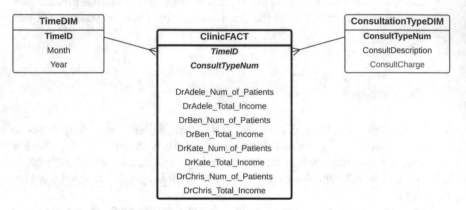

Fig. 7.18 Clinic star schema version 2

Questions (*i*) Is there a determinant dimension in any of the two star schemas? If yes, state which dimension in which star schema and explain your reason. If no, also explain your reason; (*ii*) compare and contrast both star schemas.

7.8   Further Readings

This book is the first book that introduces the concept of *Determinant Dimensions*. Determinant Dimensions are an important concept in data warehousing, which has an impact on how we query the star schema. General star schema design methods, including dimension and fact modelling, can be found in various data warehousing books, such as [1–10] and [11].

Pivot tables are related to whether or not the fact measures are pushed to the dimensions. In the context of star schema design, pivot tables are related to fact and dimension design, especially determinant dimensions. Pivot tables have been used extensively in spreadsheets. Some of the references on pivot tables are as follows: [12] and [13].

References

1. C. Adamson, *Star Schema the Complete Reference* (McGraw-Hill Osborne Media, 2010)
2. R. Laberge, *The Data Warehouse Mentor: Practical Data Warehouse and Business Intelligence Insights* (Mcgraw-Hill, New York, 2011)
3. M. Golfarelli, S. Rizzi, *Data Warehouse Design: Modern Principles and Methodologies* (Mcgraw-Hill, New York, 2009)
4. C. Adamson, *Mastering Data Warehouse Aggregates: Solutions for Star Schema Performance* (Wiley, London, 2012)
5. P. Ponniah, *Data Warehousing Fundamentals for IT Professionals* (Wiley, London, 2011)
6. R. Kimball, M. Ross, *The Data Warehouse Toolkit: The Definitive Guide to Dimensional Modeling* (Wiley, London, 2013)
7. R. Kimball, M. Ross, W. Thornthwaite, J. Mundy, B. Becker, *The Data Warehouse Lifecycle Toolkit* (Wiley, London, 2011)
8. W.H. Inmon, *Building the Data Warehouse*. ITPro Collection (Wiley, London, 2005)
9. M. Jarke, *Fundamentals of Data Warehouses*, 2nd edn. (Springer, Berlin, 2003)
10. E. Malinowski, E. Zimányi, *Advanced Data Warehouse Design: From Conventional to Spatial and Temporal Applications*. Data-Centric Systems and Applications (Springer, Berlin, 2008)
11. A. Vaisman, E. Zimányi, *Data Warehouse Systems: Design and Implementation*. Data-Centric Systems and Applications (Springer, Berlin, 2014)
12. B. Jelen, M. Alexander, *Microsoft Excel 2019 Pivot Table Data Crunching*. Business Skills (Pearson Education, 2018)
13. M. Golfarelli, S. Graziani, S. Rizzi, Shrink: An OLAP operation for balancing precision and size of pivot tables. Data Knowl. Eng. **93**, 19–41 (2014)

Chapter 8
Junk Dimensions

A *Junk Dimension* is a dimension that combines several low-cardinality dimensions. A low-cardinality dimension is a dimension which has a small number of records. A Junk Dimension is a *Cartesian product* of these low-cardinality dimensions. In other words, instead of maintaining the low cardinality separately, we use a Junk Dimension instead.

Tables 8.1 and 8.2 are examples of low-cardinality tables. In this case, they only have three and two records, respectively. A Junk Dimension table that combines (through a Cartesian product) these two tables is shown in Table 8.3.

As clearly shown, the concept of the Junk Dimension is fairly easy to understand. But in this chapter, we are going to learn how a Junk Dimension is used in a star schema and to compare and to contrast two approaches: the Non-junk Dimension approach and the Junk Dimension approach.

8.1 A Real-Estate Case Study

To learn the concept of Junk Dimensions, let's use a real-estate example as a case study. An established real-estate agent started a business many years ago and implemented a very simple database system. The simple database system consists of one large table with the following attributes as shown in Table 8.4.

The manager of the real-estate agency requires a data warehouse for analysis purposes. In particular, the manager would like to analyse a number of properties using several variables, such as properties with a pool or a spa, or properties that face a certain aspect (e.g. north facing). Thus, a small data warehouse needs to be built.

There are two options for the desired star schema: Option 1 is to use a **normal** (or **Non-junk**) **dimension** and option 2 is to use a **Junk Dimension**.

© The Author(s), under exclusive license to Springer Nature Switzerland AG 2021   207
D. Taniar, W. Rahayu, *Data Warehousing and Analytics*, Data-Centric Systems
and Applications, https://doi.org/10.1007/978-3-030-81979-8_8

Table 8.1 Dimension-1 table

attr1	attr2	attr3
a	..	..
b	..	..
c	..	..

Table 8.2 Dimension-2 table

attr4	attr5
x	..
y	..

Table 8.3 Junk Dimension table

attr1	attr2	attr3	attr4	attr5
a	..	..	x	..
a	..	..	y	..
b	..	..	x	..
b	..	..	y	..
c	..	..	x	..
c	..	..	y	..

8.2   Option 1: The Non-junk Dimension Version

This version is the normal star schema with Non-junk Dimensions. For the sake of the discussion of this topic, we call this option 1: a Non-junk Dimension option. Many dimensions can be derived from the aforementioned operational database. However, for simplicity, only four dimensions are chosen—Ensuite Dimension, Pool Dimension, Spa Dimension and AspectFacing Dimension—and only two fact measures are chosen: Number of Property and Total Price. The star schema is shown in Fig. 8.1.

From the Property table (the operational database), attributes such as ensuite, pool, spa and aspect facing have low cardinality. For example, ensuite, pool and spa are only yes or no, whereas aspect facing has four options. In a Non-junk Dimension schema, low-cardinality attributes are stored as individual dimension tables, such as Ensuite Dimension and Pool Dimension.

The SQL statement to create the Ensuite Dimension table is as follows:

```
create table EnsuiteDim as
select distinct Ensuite
from Property
order by Ensuite;

alter table EnsuiteDim
add (EnsuiteID number(1));

update EnsuiteDim
set EnsuiteID = 1
where Ensuite = 'no';
```

```
update EnsuiteDim
set EnsuiteID = 2
where Ensuite = 'yes';
```

Because Ensuite Dimension needs an identifier called EnsuiteID, which does not exist in the operational database (i.e. the Property table), table Ensuite Dimension needs to be altered, and a new attribute needs to be added: EnsuiteID. In order to populate the EnsuiteID column with the correct value, SQL update statements need to be invoked with one update statement for each EnsuiteID value.

The same dimension table creation method is applied to the other three dimensions: Pool Dimension, AspectFacing Dimension and Spa Dimension. Their SQL statements are as follows:

Table 8.4 Property table

Attribute name	Description
Key	Unique key
Date_offered	Date property offered to the public
Summary	Short description of the property
Adtext	Longer description of the property
URL	The URL of the advertisement
Address	Property address
Suburb	Property suburb name
Postcode	Property postcode
Longitude	Longitude of address
Latitude	Latitude of address
Category	Residential or commercial
Zoning	Commercial zoning type
Property_type	Residential property type: house, apartment or lot
Houseprice	Price of property
Num_bedrooms	Number of bedrooms
Lot_size	Size of the lot
Heating	Ducted, gas, open fireplace or wood
Garage	Type of garage
Ensuite	Yes or no
Balcony	Yes or no
Pool	Yes or no
Tennis_court	Yes or no
Spa	Yes or no
Aspect_facing	North, south, east or west
School_distance	Distance to the nearest school (in km)
Shop_distance	Distance to the nearest shops (in km)
Train_distance	Distance to the nearest train station (in km)
Bus_distance	Distance to the nearest bus stop (in km)
Hospital_distance	Distance to the nearest hospital (in km)
Major_road_distance	Distance to the nearest major road (in km)

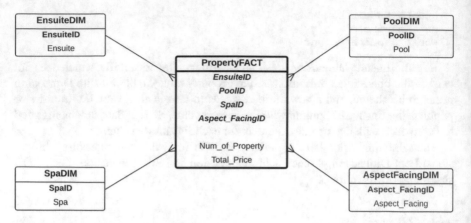

Fig. 8.1 A Non-junk Dimension version

```
-- PoolDim
create table PoolDim as
select distinct Pool
from Property
order by Pool;

alter table PoolDim
add (PoolID number(1));

update PoolDim
set PoolID = 1
where Pool = 'no';

update PoolDim
set PoolID = 2
where Pool = 'yes';

-- AspectFacingDim
create table AspectFacingDim as
select distinct Aspect_Facing
from Property
order by Aspect_Facing;

alter table AspectFacingDim
add (Aspect_FacingID number(1));

update AspectFacingDim
set Aspect_FacingID = 1
where Aspect_Facing = 'East';

update AspectFacingDim
set Aspect_FacingID = 2
where Aspect_Facing = 'North';
```

```
update AspectFacingDim
set Aspect_FacingID = 3
where Aspect_Facing = 'South';

update AspectFacingDim
set Aspect_FacingID = 4
where Aspect_Facing = 'West';

-- SpaDim
create table SpaDim as
select distinct Spa
from Property
order by Spa;

alter table SpaDim
add (SpaID number(1));

update SpaDim
set SpaID = 1
where Spa = 'no';

update SpaDim
set SpaID = 2
where Spa = 'yes';
```

The contents of the four dimension tables are shown in Tables 8.5, 8.6, 8.7, and 8.8.

Because the dimension identifiers (e.g. EnsuiteID, PoolID, Aspect_FacingID and SpaID) do not exist in the operational database (i.e. the Property table), we need to

Table 8.5 EnsuiteDim table

EnsuiteID	Ensuite
1	no
2	yes

Table 8.6 PoolDim table

PoolID	Pool
1	no
2	yes

Table 8.7 AspectFacingDim table

Aspect_FacingID	Aspect_Facing
1	East
2	North
3	South
4	West

Table 8.8 SpaDim table

SpaID	Spa
1	no
2	yes

create a temporary Fact Table. The following is the SQL statement to create the PropertyTempFact1 table:

```
create table PropertyTempFact1 as
select Ensuite, Pool, Aspect_Facing, Spa, Houseprice
from Property;
```

After that, the PropertyTempFact1 table is added with the appropriate dimension identifiers and populated with the correct values. The SQL statements are as follows:

```
-- EnsuiteID
alter table PropertyTempFact1
add (EnsuiteID number(1));

update PropertyTempFact1
set EnsuiteID = 1
where Ensuite = 'no';

update PropertyTempFact1
set EnsuiteID = 2
where Ensuite = 'yes';

-- PoolID
alter table PropertyTempFact1
add (PoolID number(1));

update PropertyTempFact1
set PoolID = 1
where Pool = 'no';

update PropertyTempFact1
set PoolID = 2
where Pool = 'yes';

-- AspectFacingID
alter table PropertyTempFact1
add (Aspect_FacingID number(1));

update PropertyTempFact1
set Aspect_FacingID = 1
where Aspect_Facing = 'East';

update PropertyTempFact1
set Aspect_FacingID = 2
where Aspect_Facing = 'North';

update PropertyTempFact1
set Aspect_FacingID = 3
where Aspect_Facing = 'South';
```

```
update PropertyTempFact1
set Aspect_FacingID = 4
where Aspect_Facing = 'West';

-- SpaID
alter table PropertyTempFact1
add (SpaID number(1));

update PropertyTempFact1
set SpaID = 1
where Spa = 'no';

update PropertyTempFact1
set SpaID = 2
where Spa = 'yes';
```

Finally, a Fact Table (called the PropertyFact1 table) is created using the following SQL statement:

```
create table PropertyFact1 as
select
  EnsuiteID,
  PoolID,
  Aspect_FacingID,
  SpaID,
  count(*) as Num_of_Property,
  sum(Houseprice) as Total_Price
from PropertyTempFact1
group by
  EnsuiteID,
  PoolID,
  Aspect_FacingID,
  SpaID;
```

The contents of the PropertyFact1 table is shown in Table 8.9.

8.3   Option 2: The Junk Dimension Version

A *Junk Dimension* is a type of dimension that consolidates all the low-cardinality attributes or many small dimension tables into a single dimension table. Low-cardinality attributes are attributes with a small range of values, such as male/female, yes/no, 1/2/3 and North/South/East/West. Therefore, the content of a Junk Dimension is a Cartesian product of the values of all its attributes.

An example of a Junk Dimension is a Cartesian product of the four low-cardinality dimensions in Fig. 8.1, namely Ensuite Dimension, Pool Dimension, Aspect_Facing Dimension and Spa Dimension. Now, they are all consolidated into a *Junk Dimension*. The contents of the Junk Dimension table are shown in Table 8.10.

Table 8.9 PropertyFact1 table

EnsuiteID	PoolID	Aspect FacingID	SpaID	Num of property	Total price
1	1	1	1	1824	747,790,330
1	1	1	2	1911	779,736,773
1	1	2	1	832	332,080,307
1	1	2	2	863	348,569,427
1	1	3	1	3801	1,522,896,359
1	1	3	2	3697	1,502,161,380
1	1	4	1	1857	744,817,268
1	1	4	2	1903	766,424,589
1	2	1	1	384	207,206,147
1	2	1	2	375	206,016,159
1	2	2	1	176	92,994,827
1	2	2	2	190	101,747,332
1	2	3	1	640	353,440,961
1	2	3	2	764	415,248,088
1	2	4	1	358	196,980,885
1	2	4	2	331	187,747,229
2	1	1	1	1900	767,864,421
2	1	1	2	1953	780,174,960
2	1	2	1	940	385,170,902
2	1	2	2	950	370,731,351
2	1	3	1	3730	1,474,579,328
2	1	3	2	3741	1,523,864,755
2	1	4	1	1805	741,317,849
2	1	4	2	1854	745,630,764
2	2	1	1	331	185,814,783
2	2	1	2	387	216,223,145
2	2	2	1	175	95,384,840
2	2	2	2	145	82,550,189
2	2	3	1	671	367,434,044
2	2	3	2	716	395,235,770
2	2	4	1	334	182,518,099
2	2	4	2	334	189,319,617

It is clear that a junk table combines (through a Cartesian product) all the four small dimensions into one Junk Dimension. In other words, a Junk Dimension can hold more than one low-cardinality attribute that has no correlation with one another.

Compared to the Non-junk Dimension schema shown in Fig. 8.1, all of the low-cardinality attributes are stored into a dimension named Junk Dimension, as shown in Fig. 8.2. Note that in this example, there are no other dimensions, apart from the Junk Dimension which combines the four small dimensions. We could have other normal dimensions in the schema, if necessary (e.g. a dimension for the property type, another dimension for the suburb, etc.). But for simplicity, the star schema in

Table 8.10 JunkDim table

JunkID	Ensuite	Pool	Aspect_Facing	Spa
1	no	no	East	no
2	no	no	East	yes
3	no	no	North	no
4	no	no	North	yes
5	no	no	South	no
6	no	no	South	yes
7	no	no	West	no
8	no	no	West	yes
9	no	yes	East	no
10	no	yes	East	yes
11	no	yes	North	no
12	no	yes	North	yes
13	no	yes	South	no
14	no	yes	South	yes
15	no	yes	West	no
16	no	yes	West	yes
17	yes	no	East	no
18	yes	no	East	yes
19	yes	no	North	no
20	yes	no	North	yes
21	yes	no	South	no
22	yes	no	South	yes
23	yes	no	West	no
24	yes	no	West	yes
25	yes	yes	East	no
26	yes	yes	East	yes
27	yes	yes	North	no
28	yes	yes	North	yes
29	yes	yes	South	no
30	yes	yes	South	yes
31	yes	yes	West	no
32	yes	yes	West	yes

Fig. 8.2 has only one dimension, which is the Junk Dimension. Note that because there is only one dimension in the star schema, PropertyFact has only one dimension attribute (i.e. JunkID).

The SQL statements to create the Junk Dimension are as follows:

```
create table JunkDim as
select distinct Ensuite, Pool, Aspect_Facing, Spa
from Property
order by Ensuite, Pool, Aspect_Facing, Spa;
```

```
alter table JunkDim
add (JunkID number(2));

drop sequence Seq_ID;

create sequence Seq_ID
start with 1
increment by 1
maxvalue 99999999
minvalue 1
nocycle;

update JunkDim
set JunkID = Seq_ID.nextval;
```

The create table statement creates a Junk Dimension table by extracting the four attributes from the operational database (i.e. the Property table). Because the Junk Dimension table needs an identifier (e.g. JunkID), a new attribute needs to be added to the Junk Dimension table. We use a sequence number to generate a unique number for each junk record.

After the Junk Dimension table is created, a PropertyTempFact2 can now be created. Like the Junk Dimension table, the PropertyTempFact2 table also extracts the four attributes from the operational database (i.e. the Property table). The main difference is that when creating the PropertyTempFact2, all the records from the Property table are retrieved, whereas for the Junk Dimension, only the distinct records from the Property table are retrieved.

```
create table PropertyTempFact2 as
select Ensuite, Pool, Aspect_Facing, Spa, Houseprice
from Property;
```

The next step is to add a column, called JunkID, to the PropertyTempFact2 table:

```
alter table PropertyTempFact2
add (JunkID number(2));
```

Then TempFact2's JunkID attribute must be filled in with the correct values, which correspond to the values of Ensuite, Pool, Aspect_Facing and Spa. There are

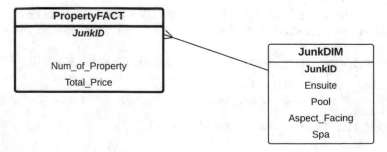

Fig. 8.2 A Junk Dimension version

two ways to do this. The first way is to do an update one by one. Since there are 32 junk records, we need to do 32 updates:

```
update PropertyTempFact2
set JunkID = 1
where Ensuite = 'no'
and Pool = 'no'
and Aspect_Facing = 'East'
and Spa = 'no';

update PropertyTempFact2
set JunkID = 2
where Ensuite = 'no'
and Pool = 'no'
and Aspect_Facing = 'East'
and Spa = 'yes';

...

update PropertyTempFact2
set JunkID = 32
where Ensuite = 'yes'
and Pool = 'yes'
and Aspect_Facing = 'West'
and Spa = 'yes';
```

A better option is to put the update statement in a loop. What it does is as follows: For each record in the Junk Dimension, it updates the PropertyTempFact2 table with the current JunkID, but the four attributes of the PropertyTempFact2 table must be equal to the corresponding attributes in the Junk Dimension table.

```
declare
  cursor JunkCursor is
    select * from JunkDim;
begin
  for JunkCursorRec in JunkCursor loop
    update PropertyTempfact2
    set JunkID = JunkCursorRec.JunkID
    where Ensuite = JunkCursorRec.Ensuite
    and Pool = JunkCursorRec.Pool
    and Aspect_Facing = JunkCursorRec.Aspect_Facing
    and Spa = JunkCursorRec.Spa;
  end loop;
end;
/
```

An alternative solution without a cursor is to have one update command using a nested query approach. For each record in the PropertyTempFact2, update the JunkID with the JunkID from the Junk Dimension table where the four attributes (e.g. Ensuite, Pool, Aspect_Facing and Spa) from both tables must be equal.

```
update PropertyTempFact2 TF
set TF.JunkID =
```

```
(select j.JunkID
 from JunkDim J
 where J.Ensuite= TF.Ensuite
 and J.Pool = TF.Pool
 and J.Aspect_Facing = TF.Aspect_Facing
 and J.Spa = TF.Spa );
```

Finally, the final PropertyFact2 table can now be created:

```
create table PropertyFact2 as
select
  JunkID,
  count(*) as Num_of_Property,
  sum(Houseprice) as Total_Price
from PropertyTempFact2
group by JunkID;
```

The contents of the PropertyFact2 table are shown in Table 8.11.

8.4   Non-junk Dimension Versus Junk Dimension

A comparison of the two versions of star schema, the Non-junk Dimension (Fig. 8.1) and Junk Dimension versions (Fig. 8.2), shows that the Junk Dimension version has a simpler design as it collapses the four dimensions into one dimension. Consequently, when we need to produce a report from the star schema, the Junk Dimension version only needs one join operation, between the Fact Table and the Junk Dimension table. On the other hand, in the Non-junk Dimension version, if all of the dimensions are needed in the report, the query needs four join operations; that is, the Fact Table joins with each of the four dimensions. Having more join operations will degrade query performance.

In order to further digest the differences between the Non-junk Dimension and Junk Dimension versions, we will look at this from the query point view, one from a simple join query point of view and the other from a more nested query point of view.

8.4.1   Simple Join Queries

An example of a simple join query for discussion is "to display the average property price for properties that have pool and spa". For the Non-junk version, the query needs two dimensions only: Pool Dimension and Spa Dimension. Hence, the query

Table 8.11 PropertyFact2 table

JunkID	Num_of_Property	Total_Price
1	1824	747,790,330
2	1911	779,736,773
3	832	332,080,307
4	863	348,569,427
5	3801	1,522,896,359
6	3697	1,502,161,380
7	1857	744,817,268
8	1903	766,424,589
9	384	207,206,147
10	375	206,016,159
11	176	92,994,827
12	190	101,747,332
13	640	353,440,961
14	764	415,248,088
15	358	196,980,885
16	331	187,747,229
17	1900	76,786,4421
18	1953	780,174,960
19	940	385,170,902
20	950	370,731,351
21	3730	1,474,579,328
22	3741	1,523,864,755
23	1805	741,317,849
24	1854	745,630,764
25	331	185,814,783
26	387	216,223,145
27	175	95,384,840
28	145	82,550,189
29	671	367,434,044
30	716	395,235,770
31	334	182,518,099
32	334	189,319,617

only joins PropertyFact1 with Pool Dimension and then with Spa Dimension, which is two join operations. The SQL statement is as follows.

```
--a Non-Junk Dimension version
select
  P.Pool,
  S.Spa,
  (PF.Total_Price/PF.Num_of_Property) as Average_Price
from PropertyFact1 PF, PoolDim P, SpaDim S
where P.Pool = 'yes'
and S.Spa = 'yes'
and PF.PoolID = P.PoolID
```

```
and PF.SpaID = S.SpaID
order by Average_Price;
```

For the Junk Dimension version, it only needs one join between PropertyFact2 and Junk Dimension. The SQL statement is as follows:

```
--a Junk Dimension version
select
  J.Pool,
  J.Spa,
  (PF.Total_Price/PF.Num_of_Property) as Average_Price
from PropertyFact2 PF, JunkDim J
where J.Pool = 'yes'
and J.Spa = 'yes'
and PF.JunkID = J.JunkID
order by Average_Price;
```

The difference between the two versions is probably small, but if the query needs more dimensions, the complexity of the Non-junk Dimension version proportionally increases with the number of dimensions involved in the query. In contrast, for the Junk Dimension version, it only needs one dimension, the Junk Dimension.

8.4.2 Nested Queries

In this section, we will be looking at more complex queries, namely nested queries. A query example is to answer the following question: "What are the features of the most expensive (maximum average price) property?" What we are interested in is to find out whether the maximum average price property has a spa or not, has a pool or not and has an ensuite or not and which aspect does it face (e.g. North, South, East or West).

To answer this question, the query needs several steps. Firstly, we need to know the maximum average price of a property, and secondly, for each property category, we need to check if the average price is the maximum average price. For this, we need to have a nested query.

For the Non-junk Dimension version, the query to retrieve the features of the property category that has the maximum average price is as follows:

```
--a Non-Junk Dimension version
select E1.Ensuite, P1.Pool, A1.Aspect_facing, S1.Spa
from
  PropertyFact1 PF1,
  EnsuiteDim E1,
  PoolDim P1,
  AspectFacingDim A1,
  SpaDim S1
where PF1.Total_Price/PF1.Num_of_Property =
  (select max(P.Total_Price/P.Num_of_Property)
   from PropertyFact1 P)
```

```
and PF1.EnsuiteID = E1.EnsuiteID
and PF1.PoolID = P1.PoolID
and PF1.Aspect_FacingID = A1.Aspect_FacingID
and PF1.SpaID = S1.SpaID ;
```

The inner query retrieves the maximum average property price. Note that in this example, there are only 32 records in PropertyFact1 and this inner query will find which one has the maximum average price. The outer query goes through each record in the PropertyFact1 table and checks whether the average price is equal to the maximum average property price obtained by the inner query. Because the details of each feature are stored in each dimension, a join operation is needed between PropertyFact1 table and all the four dimension tables.

For the Junk Dimension version, the SQL statement to retrieve the property category that has the maximum average price is as follows:

```
--a Junk Dimension version
select *
from  PropertyFact2
where Total_Price/Num_of_Property =
   (select max(P.Total_Price/P.Num_of_Property)
    from   PropertyFact2 P);
```

Note that the above SQL uses `select *` to display the results. Since table PropertyFact2 has only three attributes—JunkID, Num_of_Property and Total_Price—the result of the above query will show the JunkID to indicate which price is the maximum. It does not show the features (e.g. Ensuite, Pool, Aspect_Facing and Spa) of the most expensive property. If we need to get these features, the query needs to join with the Junk Dimension table, as these features are stored in that table. A better SQL is as follows:

```
--a Junk Dimension version (a better version)
select J.Ensuite, J.Pool, J.Aspect_Facing, J.Spa
from PropertyFact2 PF, JunkDim J
where Total_Price/Num_of_Property =
   (select max(P.Total_Price/P.Num_of_Property)
    from PropertyFact2 P)
and PF.JunkID= J.JunkID;
```

It is clear that the complexity of the query increases proportionally with the number of dimensions involved. In this case, in the Non-junk Dimension version, four dimensions are used, and hence, the join operation will need to use all the four dimension tables and the Fact Table. On the other hand, in the Junk Dimension version, only one dimension is used.

8.5   Is Combined Dimension a Junk Dimension?

A Junk Dimension is a Cartesian product of a number of dimensions with low cardinalities. In the previous chapter on Hierarchy, there is a case study on Student

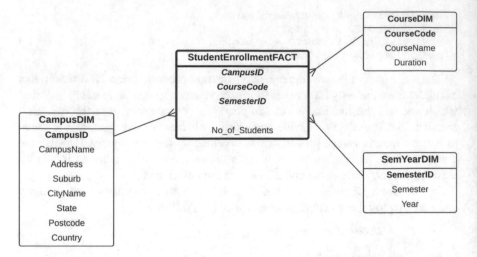

Fig. 8.3 Student Enrolment star schema

Table 8.12 Semester Year
Dimension table

SemesterID	Semester	Year
201701	1	2017
201702	2	2017
201801	1	2018
201802	2	2018
201901	1	2019
201902	2	2019

Enrolment. The star schema is shown in Fig. 8.3. It has three dimensions: Campus, Course and Semester Year Dimensions.

Let's focus on the Semester Year Dimension. Table 8.12 shows what the Semester Year Dimension table might look like. In this example, there are three years and each year has two semesters. The SemesterID is simply a combination between year and month to make it a unique identifier.

The Student Enrolment Fact table might look like that shown in Table 8.13.

In Chap. 4, we discussed the differences between one dimension and separate dimensions in both design and implementation. The Semester Year Dimension could have been implemented in a separate dimension: one dimension for Semester and the other dimension for the Year. Figure 8.4 shows a revised star schema whereby the Semester Year Dimension is split into a Semester Dimension and a Year Dimension.

Both star schemas (Figs. 8.3 and 8.4) are valid star schemas. Now look at what the new Semester Dimension and Year Dimension tables might look like and how they impact the Fact Table. Tables 8.14 and 8.15 show the Semester and Year Dimension tables, and Table 8.16 shows the new Fact Table. The Semester Dimension table has two records, semesters 1 and 2, and the Year Dimension table has three records, representing the 3-year data in this case study. Note what the Fact Table looks like.

Table 8.13 Student
Enrolment Fact table

CampusID	CourseCode	SemesterID	No of students
CL	A3001	201701	450
CL	A3002	201701	150
CL	...	...	...
CL	A3001	201702	275
CL	A3002	201702	105
CL	...	...	...
CL	A3001	201802	525
CL	A3002	201801	270
CL	...	...	...
CL	A3001	201802	350
CL	A3002	201802	160
CL	...	...	...
...	...	...	...

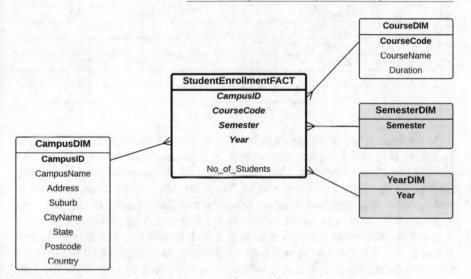

Fig. 8.4 Student Enrolment star schema where Semester and Year are separate dimensions

Table 8.14 Semester
Dimension table

Semester
1
2

The Semester Year column is now split into two columns: Semester and Year—it does not change the number of records in the Fact Table. A comparison of both Fact Tables shows that the difference is negligible. Because Semester and Year are often considered one unit of information, in this context, it would be easier to have one dimension for Semester and Year, as shown above.

Table 8.15 Year Dimension
table

Year
2017
2018
2019

Table 8.16 Student Enrolment Fact table

CampusID	CourseCode	Semester	Year	No of students
CL	A3001	1	2017	450
CL	A3002	1	2017	150
CL	...	...	...	...
CL	A3001	2	2017	275
CL	A3002	2	2017	105
CL	...	...	...	...
CL	A3001	1	2018	525
CL	A3002	1	2018	270
CL	...	...	...	...
CL	A3001	2	2018	350
CL	A3002	2	2018	160
CL	...	...	...	...
...	...	...	...	...

If we look at the Semester Dimension and Year Dimension tables, as shown in Tables 8.14 and 8.15, and compare them with the Semester Year Dimension table (a combined version), as shown in Table 8.12, the Semester Year Dimension table looks like a *Junk Dimension* table. There are two semesters and three years, and therefore, the combination of the Semester and Year Dimension tables has six records, which is the number of records in the Semester Year Dimension table. In fact, the Semester Year Dimension table is a Junk Dimension. The only difference is that in the Real Estate case study above, the JunkID in the Junk Dimension is a sequence number (e.g. 1, 2, 3, . . .), whereas the SemesterID in the Semester Year Dimension table, which can be considered the JunkID, uses a combination between Year and Semester to make it a unique identifier. We could have used a sequence number to represent the SemesterID attribute, if we had wanted to.

Tables 8.17 and 8.18 show two different versions of the Semester Year Dimension, SemesterID using the Year and Semester combination and SemesterID using a sequence number. If we use a sequence number in the SemesterID, the SemesterID in the Fact Table must also be the sequence number referencing the StudentID in the Semester Year Dimension table.

Now look at another example, the Campus Dimension table, also in the star schema in Fig. 8.3. The Campus and Country information in this star schema is implemented in one dimension, the Campus Dimension. This could have been split into two dimensions: Campus Dimension and Country Dimension. Chapter 4 describes two implementations: One uses a hierarchy between Campus and Country

Table 8.17 Semester Year
Dimension table

SemesterID	Semester	Year
201701	1	2017
201702	2	2017
201801	1	2018
201802	2	2018
201901	1	2019
201902	2	2019

Table 8.18 Semester Year
Dimension table

SemesterID	Semester	Year
1	1	2017
2	2	2017
3	1	2018
4	2	2018
5	1	2019
6	2	2019

Dimensions, and the other simply uses two separate individual dimensions. These two possible star schemas are shown in Fig. 8.5.

The same question arises from this Campus-Country Dimension example. The Campus Dimension in the previous star schema in Fig. 8.3 is a combined dimension from the Campus and Country Dimensions in the star schema in Fig. 8.5 (whether it be two distinct dimensions or in a hierarchy). The question is whether the Campus Dimension, which is the combined dimension, is a Junk Dimension. This is quite a similar case to the Semester Year Dimension, but the Campus Dimension (although it is a combined dimension) is slightly different.

Let's look at some sample data in order to understand this matter. Table 8.19 shows the sample records of the combined Campus Dimension; there are six records or six campuses. The Fact Table is shown in Table 8.20.

Now look at the sample data of the Campus and Country Dimensions, as two separate tables (see Tables 8.21 and 8.22).

Notice the difference between the one dimension model as shown in Table 8.19 and the separate dimension models in Tables 8.21 and 8.22 for the separate dimension model. The one dimension model is clearly not a Junk Dimension because it is not a Cartesian product of the two dimensions in the separate dimension model. In other words, the Campus Dimension in the star schema in Fig. 8.3 is not a Junk Dimension. The number of records in the Campus Dimension as a combined dimension is not a multiplication of the number of records in the two separate dimensions, Campus and Country. Therefore, not all the combined dimensions are Junk Dimensions. Junk Dimensions are Cartesian products of all the smaller dimensions that construct the Junk Dimension.

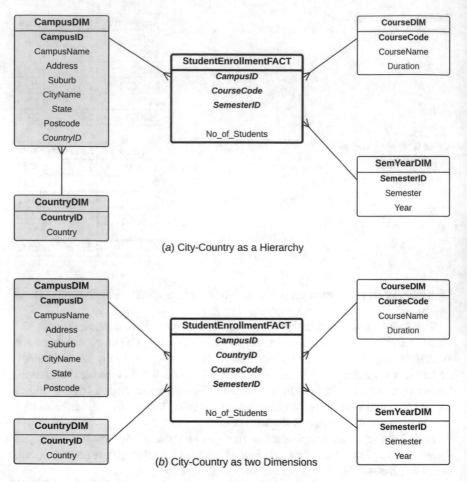

Fig. 8.5 Campus and Country Dimensions as separate dimensions

8.6   Summary

A Junk Dimension is useful when there are a number of low-cardinality dimensions which are then combined to form a Junk Dimension. Compared to the Non-junk Dimension version, the Junk Dimension version offers two advantages: (i) The star schema is simpler, and (ii) the query is less complex.

The Junk Dimension (e.g. the Junk Dimension table) needs a surrogate key, called JunkID, and this is implemented as a sequence of integers. The size of the Junk Dimension table is slightly larger than the original dimension tables because the Junk Dimension table is a Cartesian product of all the original dimension tables. However, since the original dimension tables are small (e.g. low cardinality), the Cartesian product of a number of small tables will not be excessively large. The only thing that we need to be careful about is that the number of records in the Junk Dimension table is not more than the number of records in the Fact Table.

Table 8.19 Campus Dimension table—the combined version

Campus ID	Campus name	Address	Suburb	City	State	Postcode	Country
CL	Clayton Campus	Wellington Road	Clayton	Melbourne	Victoria	3800	Australia
CA	Caulfield Campus	Dandenong Road	Caulfield East	Melbourne	Victoria	3145	Australia
PA	Parkville Campus	Royal Parade	Parkville	Melbourne	Victoria	3052	Australia
SY	Sydney Campus	Opera Boulevard	Sydney	Sydney	New South Wales	2001	Australia
MUM	Malaysia Campus	Jalan Lagoon	Bandar Sunway	Kuala Lumpur	Selangor	47500	Malaysia
MSA	South Africa Campus	Peter Street	Johannes-burg	Johannes-burg	Johannes-burg	1725	South Africa

Table 8.20 Student Enrolment Fact table with CampusID

CampusID	CourseCode	SemesterID	No of students
CL	A3001	201701	450
CL	A3002	201701	150
CL	A3003	201701	200
CL	...	...	...
CA	B5001	201701	115
CA	B5002	201701	160
CA	...	...	...
PA	C6001	201701	75
PA	C6002	201701	50
PA	...	...	...
SY	A3001	201701	40
SY	B5002	201701	35
SY	...	...	...
MUM	A3001	201701	150
MUM	B5001	201701	80
MUM	B5002	201701	100
MUM	...	...	...
MSA	A3002	201701	25
MSA	A3002	201701	20
MSA	...	...	...

Table 8.21 Separate dimension model—Campus Dimension table

Campus ID	Campus name	Address	Suburb	City	State	Postcode
CL	Clayton Campus	Wellington Road	Clayton	Melbourne	Victoria	3800
CA	Caulfield Campus	Dandenong Road	Caulfield East	Melbourne	Victoria	3145
PA	Parkville Campus	Royal Parade	Parkville	Melbourne	Victoria	3052
SY	Sydney Campus	Opera Boulevard	Sydney	Sydney	New South Wales	2001
MUM	Malaysia Campus	Jalan Lagoon	Bandar Sunway	Kuala Lumpur	Selangor	47500
MSA	South Africa Campus	Peter Street	Johannes-burg	Johannes-burg	Johannes-burg	1725

Table 8.22 Separate dimension model—Country Dimension table

CountryID	Country name
AU	Australia
MA	Malaysia
SA	South Africa

8.7 Exercises

8.1 Explore the Real Estate property table in the operational database.

(a) Using the Real Estate property table, write the SQL command to find out how many attributes there are in the property table (Hint: use `all_tab_columns`).

(b) Using the Real Estate property table, write the SQL command to list all attributes that have only small number of values (e.g. attributes that have a maximum of four different values) (Hint: use `all_tab_columns`).

8.2 The JunkID attribute in Junk Dimension, as explained in this chapter, is created using a sequence number: use `create sequence` and then use the `nextval` property of the sequence variable, as shown below:

```
create sequence Seq_ID
start with 1
increment by 1
maxvalue 99999999
minvalue 1
nocycle;

update JunkDim
set JunkID = Seq_ID.nextval;
```

JunkID attribute is basically a sequence number. Another way to create a sequence number is by using the row_number function in SQL.

Question Write the SQL command to use the row_number function to fill in the JunkID attribute, instead of using create sequence.

8.8 Further Readings

Junk Dimensions have been discussed in various data warehousing books, such as [1–10] and [11].

A Junk Dimension is basically a Cartesian product of several low-cardinality dimensions. A Cartesian product is one of the eight relational algebra operators. Further information about the Cartesian product operator and other relational algebra operators can be found in a number of relational database books, such as [12–14] and [15]. The relational model is also explained thoroughly in the following books: [16, 17] and [18]. The Junk Dimension, which is a result of a Cartesian product operation, is therefore not in the third normal form. A complete reference on normal forms can be found in [19].

References

1. C. Adamson, *Star Schema The Complete Reference* (McGraw-Hill Osborne Media, 2010)
2. R. Laberge, *The Data Warehouse Mentor: Practical Data Warehouse and Business Intelligence Insights* (Mcgraw-Hill, New York, 2011)
3. M. Golfarelli, S. Rizzi, *Data Warehouse Design: Modern Principles and Methodologies* (Mcgraw-Hill, New York, 2009)
4. C. Adamson, *Mastering Data Warehouse Aggregates: Solutions for Star Schema Performance* (Wiley, London, 2012)
5. P. Ponniah, *Data Warehousing Fundamentals for IT Professionals* (Wiley, London, 2011)
6. R. Kimball, M. Ross, *The Data Warehouse Toolkit: The Definitive Guide to Dimensional Modeling* (Wiley, London, 2013)
7. R. Kimball, M. Ross, W. Thornthwaite, J. Mundy, B. Becker, *The Data Warehouse Lifecycle Toolkit* (Wiley, London, 2011)
8. W.H. Inmon, *Building the Data Warehouse*. ITPro Collection (Wiley, London, 2005)
9. M. Jarke, *Fundamentals of Data Warehouses*, 2nd edn. (Springer, Berlin, 2003)
10. E. Malinowski, E. Zimányi, *Advanced Data Warehouse Design: From Conventional to Spatial and Temporal Applications*. Data-Centric Systems and Applications (Springer, Berlin, 2008)
11. A. Vaisman, E. Zimányi, *Data Warehouse Systems: Design and Implementation*. Data-Centric Systems and Applications (Springer, Berlin, 2014)
12. A. Silberschatz, H.F. Korth, S. Sudarshan, *Database System Concepts*, 7th edn. (McGraw-Hill, New York, 2020)
13. R. Ramakrishnan, J. Gehrke, *Database Management Systems*, 3rd edn. (McGraw-Hill, New York, 2003)
14. R. Elmasri, S.B. Navathe, *Fundamentals of Database Systems*, 3rd edn. (Addison-Wesley-Longman, 2000)

15. C.J. Date, *An Introduction to Database Systems*, 7th edn. (Addison-Wesley-Longman, 2000)
16. E.F. Codd, *The Relational Model for Database Management, Version 2* (Addison-Wesley, Reading, 1990)
17. H. Darwen, *An Introduction to Relational Database Theory* (Ventus Publishing, 2009)
18. C.J. Date, *The New Relational Database Dictionary: Terms, Concepts, and Examples* (O'Reilly Media, 2015)
19. C.J. Date, *Database Design and Relational Theory: Normal Forms and All That Jazz*. Oreilly and Associate Series (O'Reilly Media, 2012)

Chapter 9
Dimension Keys

In this chapter, two topics related to *Dimension Keys* are going to be discussed: (*i*) *Surrogate Keys* and (*ii*) *Dimension-less Keys*.

In the previous chapters, the identifier for a dimension is normally the original identifier, which is the primary key attribute from the operational database. In practice, however, it is sometimes desirable to have a new identifier for a dimension. This is a *Surrogate Key*. The first section of this chapter will primarily discuss this, briefly covering the concept of the surrogate key and how it is created and used in practice.

Throughout the book, we often encounter dimensions that have only one attribute. As explained earlier, normally in the real-life scenarios, these dimensions have other attributes. So one-attribute dimensions do not exist. However, in this book, because the operational databases used as the case studies are simple databases, often the dimensions created from the operational databases are simple dimensions with one attribute only. We keep these one-attribute dimensions with an understanding that there should be other attributes in these dimensions, where the information of other attributes might be taken from more complex operational databases or other data sources. However, in some cases, one-attribute dimensions solely have one attribute. In this case, these one-attribute dimensions are not needed in the star schema, but the dimension keys are still kept in the Fact Table. These attributes are called *Dimension-less Keys*. The second section of this chapter will discuss this concept and present an example.

9.1 Surrogate Keys

A *Surrogate Key* is a unique identification for each record in a dimension table. It is a primary key for the dimension table, and it is normally implemented as a sequence number.

If we could use the primary key from the operational database, why are surrogate keys needed in some cases? In the real world, input operational databases are not only one system; it can be multiple operational databases. The same table may exist in various forms in these multiple operational databases.

For example, the table Employee may exist in a Payroll database as well as in an Academic database and a Library database, all of which are different database systems. Another example is the table Product, which may exist in an Inventory database as well as in a Manufacturing database, a Sales database, etc.

A data warehouse is normally built from multiple operational databases and when a dimension is created, ideally the primary key, which is the ID of the table, is unique across all operational database systems. For example, if we are going to have EmployeeID as a PK in the Employee Dimension in our data warehouse, we expect that EmployeeID is unique across all the input operational databases, which might likely be the case. Also, if we are going to have a Product Dimension with ProductID as a PK, we expect that ProductID is unique across all database systems.

However, in the real world, a unique identification which is valid across all systems does not always happen. Different databases may have different identification systems. For example, ProductID "AB335" in one system may be coded as "AB3351" in another system. Therefore, we cannot assume that ProductID is unique across all database systems. This is the reason why practitioners often prefer to use surrogate keys in dimension tables.

Using ProductID as an example, instead of using the original ProductID from the operational databases (whether it be "AB335" or "AB3351"), in data warehousing, we use a *Surrogate Key*, which is a new key normally implemented as a sequence number. Hence, the new ProductID will be 1, 2, 3, etc.

So far, in our case studies, because the operational database is simple, surrogate keys are not enforced; we simply adopt the primary keys from the operational databases. If, for example, the operational database has ProductID (primary key) and ProductName, the data warehouse will simply use these attributes in a dimension. There is no need to create a sequence number-based surrogate key for ProductID because it is already a primary key in the operational database.

Another example is a dimension called Suburb Dimension, which consists of Suburb and Postcode. These attributes are simply extracted from the operational database. Because postcode is unique across all operational databases, it can easily become the identifier (or key) of that dimension, without the need to create another surrogate key for Suburb Dimension.

In all of the previous case studies throughout this book, the primary key of the table in the operational database is used as the key of the dimension, if this attribute is included in the star schema. Hence, a surrogate key is not specifically used.

Having said this, however, surrogate keys have actually been used in some of the previous case studies earlier in the book, even though there was no mention about surrogate keys at all. For example, in the Sales case study, there is a Quarter Dimension that has only four records (e.g. four quarters): 1, 2, 3 and 4. This is a surrogate key. The reason that these kinds of surrogate keys were used in the previous case studies is due to the absence of an identifier in the operational database. For example, in the operational database, there is Sales Date, but not Quarter. In the data warehouse, we need to convert the Sales Date into the respective Quarter.

Another example is Project Duration which has three categories: 1 for short term, 2 for medium term and 3 for long term. This is also a surrogate key. The reason for this is that in the operational database, there might be the Start Date and End Date of the project, but not duration. Or if there is duration (e.g. in days), this needs to be converted to a categorical duration in the data warehouse. Hence, we will have 1, 2 or 3 as Project Duration in the data warehouse.

If we look at a university enrolment where there is SemesterID in the Semester Dimension, potentially, we will only have "S1" and "S2" in the SemesterID field of the Semester Dimension table. Although they are not a numerical sequence number, arguably, we can use 1 and 2 instead. So this becomes a surrogate key too.

In a Sales case study, there is a dimension called Sales Method Dimension which records the methods of sales, such as "Online" and "In-Shop". These values are extracted from the operational database. In this case, for simplicity, the data warehouse simply uses these values in the dimension; a surrogate key is not implemented. However, we could have something like 1 for "Online" and 2 for "In-Shop" if we insist on having a surrogate key.

All surrogate keys that have been used in the previous case studies are used in a dimension where the number of records in that dimension table is very small, such as three records in the Project Duration Dimension (e.g. 1 for short term, 2 for medium term and 3 for long term). These records can simply be created manually using three `insert into` statements in SQL. In the TempFact, when it is necessary to populate the TempFact Table with the appropriate project duration value, we can do it with three `update` statements in SQL. Therefore, there is never an issue.

In this chapter, the number of records in the dimension table, such as Product Dimension, is large. If we want to use ProductID as a surrogate key, it is impossible to insert or update one product record at a time, manually. We must have a mechanism to do this. This kind of surrogate key has been used in the previous chapter on Junk Dimension, where the JunkID attribute in the Junk Dimension is a surrogate key. Although the number of records in Junk Dimension is relatively small (e.g. 32 records), it is impractical to do an insert/update manually one by one. In that case, a `cursor` with a `for loop` is used.

9.1.1 An Example

The concept of a *Surrogate Key* is very simple as explained above. It is basically a primary key attribute of a dimension which uses a sequence number. The first record of a dimension table has surrogate key 1, the second record 2, the third record 3, etc. But implementing a surrogate key-based sequence number, in terms of coding, can be challenging.

Table 9.1 Student table (from an operational database)

StudentID	Student name	Suburb	Postcode	Sex
21001	Adam	Caulfield	3162	M
21003	Ben	Caulfield	3162	M
21008	Christine	Chadstone	3148	F
21019	Daisy	Caulfield	3162	F
21033	Edward	Clayton	3168	M
21122	Fred	Caulfield	3162	M
21123	Greg	Chadstone	3148	M

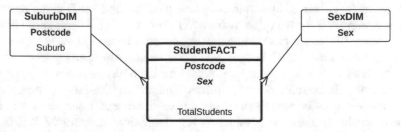

Fig. 9.1 A Non-Surrogate Key version

Table 9.2 SuburbDim table

Suburb	Postcode
Caulfield	3162
Chadstone	3148
Clayton	3168

Suppose an operational database contains one table, called Student (refer to Table 9.1). For simplicity and clarity, only a few records are used.

Suppose the star schema contains only two dimensions: Suburb Dimension and Sex Dimension (refer to Fig. 9.1). The Fact Table has one fact measure, called TotalStudents.

Suburb Dimension table can be created using the following SQL command:

```
create table SuburbDim as
select distinct Suburb, Postcode
from Student;
```

The contents of table Suburb Dimension are shown in Table 9.2.

Table 9.3 Sex Dimension
table

Sex
M
F

Table 9.4 Fact Table

Suburb	Sex	Total students
Caulfield	M	3
Caulfield	F	1
Chadstone	M	1
Chadstone	F	1
Clayton	M	1

Table 9.5 Suburb
Dimension table with
SuburbID as a surrogate key

SuburbID	Suburb	Postcode
1	Caulfield	3162
2	Chadstone	3148
3	Clayton	3168

The Sex Dimension table can be created the same way:

```
create table SexDim as
select distinct Sex
from Student;
```

The contents of table Sex Dimension have only two records (refer to Table 9.3).
The Fact Table can be created using the following SQL command:

```
create table Fact as
select Suburb, Sex, count(*) as TotalStudents
from Student
group by Suburb, Sex;
```

Fact Table will have the following records (see Table 9.4).

In this case, we assume that the attribute Suburb in Suburb Dimension table
becomes the primary key. For Sex Dimension, because it has only one attribute,
namely Sex, this attribute automatically becomes the ID of this table.

Now, suppose we want to implement a surrogate key in the Suburb Dimension,
which we call SuburbID, for example. Hence, the table Suburb Dimension is shown
in Table 9.5. Note that apart from the two original attributes, namely Suburb and
Postcode, the table Suburb Dimension now has one additional attribute, SuburbID,
as the surrogate key.

The SQL to create Suburb Dimension is as follows:

```
create table SuburbDim as
select distinct Suburb, Postcode
from Student;

alter table SuburbDim
add (SuburbID number(2));
```

```
drop sequence Suburb_seq_ID;

create sequence Suburb_seq_ID
start with 1
increment by 1
maxvalue 99999999
minvalue 1
nocycle;

update SuburbDim
set SuburbID = Suburb_seq_ID.nextval;
```

The next question is: how to create the fact (or TempFact) table? First, we need to create a TempFact Table that includes Suburb Name using the following SQL statement:

```
create table TempFact as
select Suburb, Sex
from Student;
```

The next step is to add a column, called SuburbID, in TempFact Table:

```
alter table TempFact
add (SuburbID Number(2));
```

Then Tempfact's SuburbID attribute must be filled in with the correct values, which correspond to the values of Suburb and Postcode. There are two ways to do this. The first way is to do an update one by one. In this case, we only have three suburb records, so we need three updates:

```
update TempFact
set SuburbID = 1
where Suburb = 'Caulfield';

update TempFact
set SuburbID = 2
where Suburb = 'Chadstone';

update TempFact
set SuburbID = 3
where Suburb = 'Clayton';
```

However, if the number of suburbs in the Suburb Dimension is unknown or relatively large, it is undesirable to do one update for each suburb. A better solution is to have one update command as follows:

```
update TempFact TF
set TF.SuburbID =
  (select S.SuburbID
   from SuburbDim S
   where S.Suburb = TF.Suburb);
```

Table 9.6 Fact Table with
SuburbID as a surrogate key

SuburbID	Sex	Total students
1	M	3
1	F	1
2	M	1
2	F	1
3	M	1

Table 9.7 Sex Dimension
table with SexID as a
surrogate key

SexID	Sex
1	M
2	F

The SQL command to create the Fact Table is as follows. The contents of the
new Fact Table are shown in Table 9.6.

```
create table Fact as
select SuburbID, Sex, count(*) as TotalStudents
from TempFact
group by SuburbID, Sex;
```

Obviously, when you want to produce a report, you need to join between Fact
Table and table Suburb Dimension to get the real Suburb/Postcode, rather than just
1, 2 and 3 from SuburbID.

If we do the same for Sex Dimension, Sex Dimension will have two attributes:
SexID as the surrogate key and Sex as an additional attribute. The Fact Table will
then have SexID instead of Sex. The new star schema with surrogate keys for both
dimensions is shown in Fig. 9.2.

The Sex Dimension table and the final Fact Table are shown in Tables 9.7 and 9.8,
respectively.

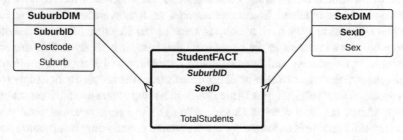

Fig. 9.2 A Surrogate Key version

Table 9.8 Final Fact Table
with Surrogate Keys

SuburbID	SexID	Total students
1	1	3
1	2	1
2	1	1
2	2	1
3	1	1

9.2   Dimension-Less Keys

Throughout the book, we've seen some dimensions which only have one attribute. Since this attribute also appears in the Fact Table, the question often asked is why do we keep these one-attribute dimensions in the star schema since this attribute also appears in the Fact Table. Can we simply delete the dimension but keep the dimension key attribute in the Fact Table?

There are two answers to this. The first answer is to explain why these one-attribute dimensions are used in the book. Note that dimensions (and also Fact Tables) are created by extracting the relevant information from the operational database. The operational databases used in this book are quite simple. Hence, when the necessary information is extracted to populate the dimension tables, we often end up with dimensions having one attribute only, which then becomes the identifier attribute for that dimension. In real-life scenarios, the operational databases that are used as inputs to the data warehouse are very complex and contain all the information about each dimension. As a result, the one-attribute dimensions often seen in this book will usually be dimensions with more than one attribute. Therefore, the existence of these dimensions can be justified.

As we have learned in Chap. 13, the input to the data warehouse can be from multiple operational databases. Therefore, it is very likely that the dimensions will not be one-attribute dimensions. Furthermore, some additional information for a dimension can be obtained from other sources, such as websites. For example, a Campus Dimension may have a campus name as the identifier. Even if the other information is not present in the operational database, it is likely that information such as campus address and campus contact number can be obtained from non-operational databases, such as official documents from the university or simply from the website. As a result, Campus Dimension will not only have one attribute, namely campus name, but also other attributes, although the operational database tables where the information is extracted for the dimension contain only information about campus name.

The second answer is to explain that some one-attribute dimensions are genuinely dimensions with one attribute only. For example, Sex Dimension may only have one attribute, called Sex, which has a value of either M or F. Arguably, there are at least three possibilities for Sex Dimension, as shown in Tables 9.9, 9.10, and 9.11. However, if it is universally accepted that M means "Male" and F means "Female", in the context of Sex, then the one-attribute Sex Dimension table as

Table 9.9 Sex Dimension
table with one attribute

Sex
M
F

Table 9.10 Sex Dimension
table with description

Sex	Sex description
M	Male
F	Female

Table 9.11 Sex Dimension
table with a surrogate key

SexID	Sex	Sex description
1	M	Male
2	F	Female

shown in Table 9.9 can be used, and consequently, in the Fact Table, there will be an attribute called Sex which has a reference to the Sex Dimension table.

Because the Sex Dimension table has only one attribute and this attribute also appears in the Fact Table, the existence of the Sex Dimension table becomes trivial. When querying the Fact Table, if we need the Sex attribute, we do not need to get it from the Sex Dimension table; instead, we can get it from the Fact Table, hence avoiding the need to join with the Sex Dimension table. Therefore, in the star schema, there is no need to have Sex Dimension at all.

The Sex attribute in the Fact Table, however, is still needed, and this kind of attribute that has no reference to the dimension is called a *Dimension-less Key*. In the star schema, dimension-less keys are still printed in bold because they (with other dimension keys) are still part of the composite identifier of the Fact Table, but not in italic, because they do not reference any dimensions.

Using the Student Fact Table as in the previous section, the star schema that uses the Sex attribute as a dimension-less key is shown in Fig. 9.3. The fact in the star schema is divided into three areas: (i) dimension keys (bold and italic), (ii) dimension-less keys (bold only) and (iii) fact measures (non-bold, non-italic). The three areas can be explicitly divided by a line or a blank line. To make the notation simpler, in this book, we adopt a blank line.

9.3 Summary

A *surrogate key* is a key used locally in each dimension. The main reason for doing this is because in many instances, the primary keys of the operational databases may not be unique across different systems; consequently, surrogate keys are needed in the data warehouse, as the data warehouse integrates multiple operational database systems and delivers integrated data for more efficient analysis. However, if the primary key in the operational database is already unique across the systems in the operational databases, surrogate keys become optional in the data warehouse.

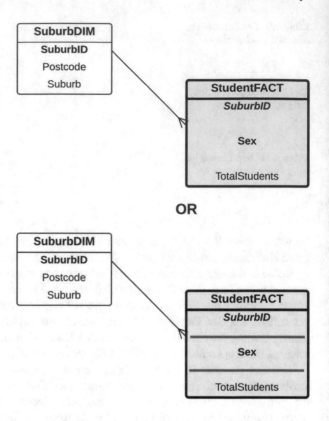

Fig. 9.3 A Dimension-less
Key version

A *dimension-less key* is an attribute in the Fact Table that does not refer to any dimensions. This happens because the dimension that has one attribute is deleted from the star schema. This kind of dimension contains singular information which generally are widely accepted categorical codes.

9.4   Exercises

9.1 *Surrogate Keys*

(a) Continuing the example in this chapter, where SuburbID is created to be a surrogate key in the Suburb Dimension, apply the same method to have SexID as a surrogate key in the Sex Dimension. Then re-create the Fact Table.

(b) Write the SQL command to display the number of students who live in the suburb of Caulfield. Apply this command to the first schema that does not use a surrogate key in Suburb Dimension. Then re-write the SQL command to be applied to the second schema where SuburbID is the key in Suburb Dimension. Compare and contrast the two methods to retrieve the same information.

(c) A new way to create a surrogate key is to use the `row_number()` function instead of the `create sequence` method. Write the SQL command to create a new Suburb Dimension table with SuburbID as the surrogate key using the `row_number()` function instead of the `create sequence` method.

9.2 *Dimension-Less Keys*

Write an SQL statement to display all records from the Fact Table. Do this in two versions: version 1 is the star schema with two dimensions—Suburb Dimension and Sex Dimension—and version 2 is the star schema with a dimension-less key of Sex. Compare and contrast these two versions to retrieve the same information.

9.5   Further Readings

This chapter focuses on two issues of star schema design, namely surrogate keys and dimension-less keys. The concept of surrogate keys has been discussed in many data warehousing books, such as [1–10] and [11].

Surrogate keys are often implemented as sequence numbers. Further details about various sequence number implementations in SQL can be found in various SQL books, such as [12–14] and [15].

Dimension-less keys is a design issue in star schema design. This issue is also found in various data modelling problems. Further readings on data modelling can be found in [16–21] and [22].

References

1. C. Adamson, *Star Schema The Complete Reference* (McGraw-Hill Osborne Media, 2010)
2. R. Laberge, *The Data Warehouse Mentor: Practical Data Warehouse and Business Intelligence Insights* (McGraw-Hill, New York, 2011)
3. M. Golfarelli, S. Rizzi, *Data Warehouse Design: Modern Principles and Methodologies* (McGraw-Hill, New York, 2009)
4. C. Adamson, *Mastering Data Warehouse Aggregates: Solutions for Star Schema Performance* (Wiley, London, 2012)
5. P. Ponniah, *Data Warehousing Fundamentals for IT Professionals* (Wiley, London, 2011)
6. R. Kimball, M. Ross, *The Data Warehouse Toolkit: The Definitive Guide to Dimensional Modeling* (Wiley, London, 2013)
7. R. Kimball, M. Ross, W. Thornthwaite, J. Mundy, B. Becker, *The Data Warehouse Lifecycle Toolkit* (Wiley, London, 2011)
8. W.H. Inmon, *Building the Data Warehouse*. ITPro Collection (Wiley, London, 2005)
9. M. Jarke, *Fundamentals of Data Warehouses*, 2nd edn. (Springer, Berlin, 2003)
10. E. Malinowski, E. Zimányi, *Advanced Data Warehouse Design: From Conventional to Spatial and Temporal Applications*. Data-Centric Systems and Applications (Springer, Berlin, 2008)
11. A. Vaisman, E. Zimányi, *Data Warehouse Systems: Design and Implementation*. Data-Centric Systems and Applications (Springer, Berlin, 2014)
12. J. Melton, *Understanding the New SQL: A Complete Guide*, vol. I, 2nd edn. (Morgan Kaufmann, Los Altos, 2000)

13. C.J. Date, *SQL and Relational Theory - How to Write Accurate SQL Code*, 2nd edn. Theory in Practice (O'Reilly, 2012)
14. A. Beaulieu, *Learning SQL: Master SQL Fundamentals* (O'Reilly Media, 2009)
15. M.J. Donahoo, G.D. Speegle, *SQL: Practical Guide for Developers*. The Practical Guides. (Elsevier, Amsterdam, 2010)
16. G. Simsion, G. Witt, *Data Modeling Essentials*. The Morgan Kaufmann Series in Data Management Systems (Elsevier, Amsterdam, 2004)
17. T.J. Teorey, S.S. Lightstone, T. Nadeau, H.V. Jagadish, *Database Modeling and Design: Logical Design*. The Morgan Kaufmann Series in Data Management Systems (Elsevier, Amsterdam, 2011)
18. N.S. Umanath, R.W. Scamell, *Data Modeling and Database Design* (Cengage Learning, 2014)
19. M.J. Hernandez, *Database Design for Mere Mortals: A Hands-On Guide to Relational Database Design*. For Mere Mortals (Pearson Education, 2013)
20. S. Bagui, R. Earp, *Database Design Using Entity-Relationship Diagrams*. Foundations of Database Design (CRC Press, Boca Raton, 2003)
21. J.L. Harrington, *Relational Database Design Clearly Explained*. Clearly Explained Series (Morgan Kaufmann, Los Altos, 2002)
22. T.A. Halpin, T. Morgan, *Information Modeling and Relational Databases*, 2nd edn. (Morgan Kaufmann, Los Altos, 2008)

Chapter 10
One-Attribute Dimensions

In the previous case studies, you might have seen dimensions with only one attribute. Is this dimension necessary? Can we delete this dimension from the star schema? This chapter addresses these questions.

Creating a star schema is achieved by transforming an operational database through some systematic steps of `create table`, `insert into`, `update`, `alter table` and other SQL commands. In other words, the data warehouse or the star schema takes an operational database as an input, performs these transformation steps and creates tables for the data warehouse. The data and attributes for the star schema are limited by what is given by the operational database. Because the operational databases used so far are simple databases and the information (or the attributes) in the operational databases is limited, we often ended up with dimensions with single attributes in the data warehouse. In the real world, additional information on each dimension can be obtained from other sources, outside the given operational databases, and consequently, dimensions with single attributes are rare, although this may occur occasionally. Therefore, in the previous case studies, although the dimensions have only one attribute, the dimensions are kept in the star schema, considering that in the real world, these dimensions would have many more attributes.

Why are attributes needed in the dimension? Even though the main information comes from the Fact Table, often when querying the Fact Table, further information about dimensions is needed in the report, and hence, additional attributes from the dimensions are needed by joining the required dimensions with the Fact Table. Therefore, it is handy to have additional attributes in the dimension to have more complete information about what the dimension is about.

However, in some cases, no additional attributes may be available for a dimension, which results in one-attribute dimensions. In this case, what should we do? Can we keep it, or should we transform it to another form? Is there any case where it is compulsory to keep the dimensions with single attributes?

© The Author(s), under exclusive license to Springer Nature Switzerland AG 2021
D. Taniar, W. Rahayu, *Data Warehousing and Analytics*, Data-Centric Systems
and Applications, https://doi.org/10.1007/978-3-030-81979-8_10

Some solutions to this case have been implicitly or explicitly explained in the previous chapters, as well as in the chapters ahead. So this chapter gives a comprehensive summary to dimensions with single attributes. The aim of this chapter is to become a one-stop shop to learn how to deal with one-attribute dimensions. Basically, there are two ways to deal with one-attribute dimensions: (*i*) Move it to the fact, or (*ii*) keep it in the dimension. These will be explored further in the next sections.

10.1   Move It to the Fact

10.1.1   Column-Based Solution in the Fact

In Chap. 7, there is an Olympic Games case study where the star schema has the Medal Type Dimension with only one attribute (see Fig. 10.1). The Medal Type attribute basically contains three records: Gold, Silver and Bronze; there are no other attributes in the Medal Type Dimension, simply because the Medal Type attribute itself is already self-explanatory. We could have additional attributes in the Medal Type Dimension, such as the Medal Description attribute to explain what each medal is for. But supposing that there are no other attributes in this dimension, this dimension becomes a one-attribute dimension.

A snapshot of the Fact Table is shown in Table 10.1; note the Medal Type column. The Medal Type Dimension table has only one attribute and three records, as shown in Table 10.2.

The Medal Type Dimension table seems unnecessary because Medal Type also exists in the Fact Table, and there is no need to reference to the Medal Type Dimension. Hence, the Medal Type Dimension can be removed from the star

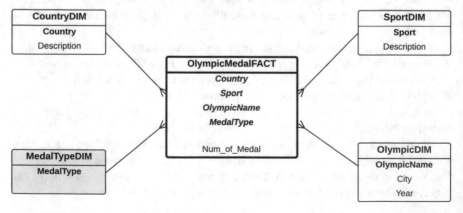

Fig. 10.1 An Olympic Games star schema with Medal Type Dimension as a one-attribute dimension

Table 10.1 Rio Olympic 2016 swimming medal tally

Country	Sport	Olympic name	Medal type	Num of Medals
USA	Swimming	Rio 2016	Gold	16
USA	Swimming	Rio 2016	Silver	8
USA	Swimming	Rio 2016	Bronze	9
Australia	Swimming	Rio 2016	Gold	3
Australia	Swimming	Rio 2016	Silver	4
Australia	Swimming	Rio 2016	Bronze	2
China	Swimming	Rio 2016	Gold	1
China	Swimming	Rio 2016	Silver	2
China	Swimming	Rio 2016	Bronze	3
Italy	Swimming	Rio 2016	Gold	1
Italy	Swimming	Rio 2016	Bronze	2

Table 10.2 Medal Type
Dimension

Medal type
Gold
Silver
Bronze

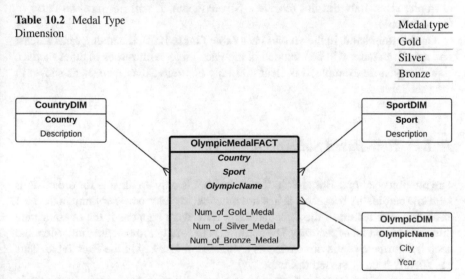

Fig. 10.2 An Olympic Games star schema with Pivoted Fact Table

schema, but the Medal Type in the fact is converted into three fact measures because
there are three types of medals. The fact measures now become Number of Gold
Medal, Number of Silver Medal and Number of Bronze Medal, instead of just
Number of Medal.

The new star schema is shown in Fig. 10.2. Moving the Medal Type to the fact
is possible because the Medal Type Dimension is a low-cardinality dimension, that
is, a dimension with only a small number of records, in this case only three records.
Hence, it is possible to expand the one fact measures to become three fact measures.
This fact measure structure is called *Column-Based Solution in the Fact* because we
are converting the dimension attribute into new columns (new fact measures) in the

Table 10.3   Rio Olympic 2016 swimming medal tally

Country	Sport	Olympic Name	Num of Gold	Num of Silver	Num of Bronze
USA	Swimming	Rio 2016	16	8	9
Australia	Swimming	Rio 2016	3	4	2
China	Swimming	Rio 2016	1	2	3
Italy	Swimming	Rio 2016	1	0	2

Fact Table. Table 10.3 shows the snapshot of the new Fact Tables with additional columns to represent the number of medals of different medal type.

Note the main difference between the two Fact Tables. The first Fact Table (Table 10.1) is created using an inner join between several tables in the operational database. As a result, only non-zero medals are included in the Fact Table. For example, since Italy did not receive a Silver medal, it will not have an entry in the Fact Table.

On the other hand, in the second Fact Table (Table 10.3), if a country has at least one medal, an entry will be included in the Fact Table, with zeroes in other medals. Using Italy as an example, it is clear that Italy has zero Silver medals, as shown in the Fact Table.

10.1.2   Row-Based Solution in the Fact

The previous Column-Based Solution in the Fact is only applicable for dimensions with low cardinality because it is just not practical to create a new column in the Fact Table for each record of the dimension. So now, what happens if the one-attribute dimension has a lot of records? The solution is to simply remove the dimension and keep the attribute of the dimension as a dimension-less key in the Fact Table. This has already been discussed in Chap. 9.

Using the same Olympic Games case study, the new star schema is shown in Fig. 10.3. The contents of the Fact Table are the same as in Table 10.1 shown previously. The main difference is that the Medal Type attribute does not reference any other table because there is no Medal Type Dimension anymore.

This solution is called the *Row-Based Solutions in the Fact* because each medal type is represented by records in the Fact Table, so it is Row Based, as opposed to Column Based.

The Olympic Games case study is perhaps not an exemplar case for the Row-Based solution because the Medal Type Dimension is a low-cardinality dimension. Now let's consider another case study. In Chap. 2 (as well as in Chap. 3), there is a Sales case study. The original star schema is shown in Fig. 10.4.

The Time Dimension contains only the Quarter attribute. Assuming that there are only four quarters (e.g. Q1, Q2, ..., Q4), regardless of the year, then this star schema may be used to compare the sales performance of, for example, Q1 (all years) and

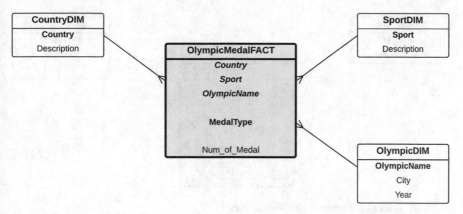

Fig. 10.3 An Olympic Games star schema with Medal Type as a Dimension-less Key

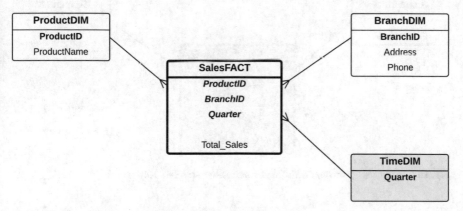

Fig. 10.4 A Sales star schema with Time as a one-attribute dimension

Q2 (all years), which might make sense in some circumstances if we would like to see the seasonal trend of the quarters. If this is the case, then it would be similar to the Olympic Games case study. Therefore, the Quarter could be shifted to the fact measures, resulting in a new star schema shown in Fig. 10.5.

However, if the Quarter also embeds the Year information, such as Q12020 for the first quarter in 2020 or Q22019 for the second quarter in 2019, then the Pivot Fact Table cannot be the solution because the number of possible records in the Time Dimension can be large if the period of time covered by this data warehouse is quite long (e.g. many years of data). Therefore, it won't be possible to have one fact measure for each quarter/year record in the Time Dimension because there will be a lot of fact measures that must be needed in the Fact Table. The solution is to have the Quarter attribute as a Dimension-less Key in the fact and to remove the Time Dimension from the star schema, as shown in Fig. 10.6. This representation is hence a Row-Based Solution because the Quarter is represented as records in the Fact

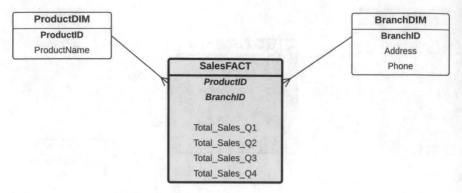

Fig. 10.5 A Sales star schema with Pivoted Fact

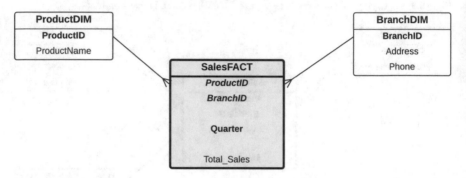

Fig. 10.6 A Sales star schema with Quarter as a Dimension-less Key

Table, rather than as columns. Using this Row-Based Solution, it is not restricted by the number of records in the one-attribute dimension.

In terms of the records in the Fact Table, actually, there is no difference between a star schema with a dimension-less key and the star schema with one-attribute dimension because the contents of the Fact Table are identical. The only difference is that the dimension-less key attribute (e.g. Quarter in the star schema in Fig. 10.6) has no references to the Time Dimension because the Time Dimension has been removed, whereas the Quarter in the star schema in Fig. 10.4 has a reference to the Time Dimension. But content-wise, the Fact Tables are identical.

If there is no difference between the Fact Tables in both star schemas, is it worth retaining the Time Dimension, even though it has only one attribute? Or should it be removed because it does not contribute anything to the Fact Table? Note early in Chap. 2, when introducing various notations for the star schema, the simplest notation does not include the attributes, only the names of the dimensions and facts. This is often used to show a high level of star schema design, as it shows the relationship between fact and dimensions, without going through the details.

Using this notation, if the Time Dimension is removed (due to having one-attribute only), the star schema has no notion of Time, as the attributes are not

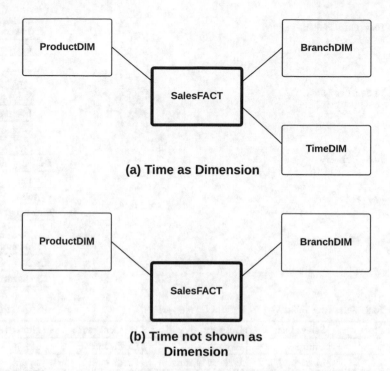

Fig. 10.7 With and without Time Dimension

shown. Figure 10.7 shows two level star schemas: one with a Time Dimension and the other without. If we only look at the level design, it is often more desirable to keep the one-attribute dimension, like the Time Dimension in this case. Hence, when the attributes are then included in the design, it is often desirable to keep the one-attribute dimensions to have a clearer design, considering that the Fact Table remains the same, whether it is with or without the one-attribute dimensions.

10.2   Keep It in the Dimension

10.2.1   Combine All One-Attribute Dimensions

If there is more than one dimension with a single attribute and these dimensions are not related to one other, it is often desirable to combine them into a Junk Dimension. Let's use the following Sales case study. It is used to analyse the number of sales based on several conditions, such as whether the sales were made in-store or online and whether a gift coupon was used. Also, other parameters were considered, such as delivery and payment modes. Hence, there are four dimensions: Gift Coupon, Sales Mode, Delivery and Payment Mode. These four tables are shown

Table 10.4 Gift Coupon

Gift coupon
Y
N

Table 10.5 Sales Mode

Sales mode
In-store
Online

Table 10.6 Delivery

Delivery mode
Normal
Express
Pick-up in store

Table 10.7 Payment Mode

Payment mode
Cash
Credit card

Table 10.8 Sales Fact Table

Gift coupon	Sales mode	Delivery	Payment mode	Num of sales
Y	Online	Express	Credit card	...
Y	In-store	Pick-up in store	Credit card	...
N	In-store	Pick-up in store	Cash	...
N	Online	Normal	Credit card	...
...	...	...	...	...

in Tables 10.4, 10.5, 10.6, and 10.7. Note that all of these four dimensions only have one attribute. Additionally, they are low cardinality as well (e.g. two to three records in each table). Surrogate keys could be added to these dimension tables; hence, each table will have two attributes (e.g. Gift CouponID and Gift Coupon Description in the Gift Coupon Table). But these two attributes can be simplified into one attribute, as currently shown in the tables. Therefore, these four dimension tables are one-attribute dimension tables.

The contents of the Fact Table are shown in Table 10.8.

Because these four dimensions are low cardinality, we can combine them into a Junk Dimension. Comparing the two star schemas, the first option is to have four one-attribute dimensions, and the second option is to have a Junk Dimension instead. See Fig. 10.8.

The contents of the Junk Dimension are shown in Table 10.9. By definition, a Junk Dimension is a Cartesian product of all the participating dimensions. However, if we examine this table more carefully, if the Payment Mode is Cash, the Sales Mode must be In-Store, and the Delivery Mode is simply Pick-up in store, simply because it is not possible to pay cash for an online sale. Additionally, if it is In-store sales, then there will be no need for delivery, whether it be normal delivery or

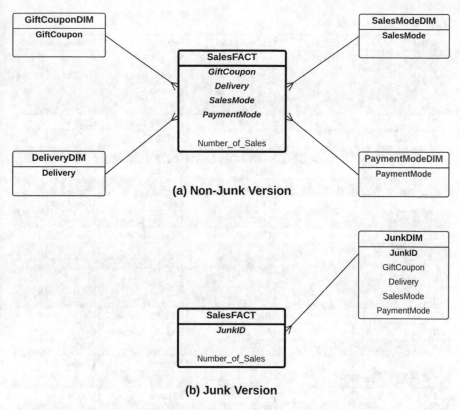

Fig. 10.8 Junk vs. Non-junk Dimension

express delivery. Hence, some of these records are greyed out in the Junk Dimension table. So this is not a pure Junk Dimension because there is an intrinsic relationship between the original dimensions. These greyed out records should be excluded from the junk table. The new junk table is then shown in Table 10.10. The JunkID is renumbered. Note the reduction in record numbers, from 24 to 10 only.

10.2.2   Combine with Other Normal Dimensions

It might be possible that the one-attribute dimension has a relationship with (or is related to) another normal dimension, and in this case, it is possible to combine the one-attribute dimension with the normal dimension.

In Chap. 5, there is a case study about Book Sales which we are going to discuss again here, in the context of one-attribute dimension. The star schema is shown in Fig. 10.9. The Category Dimension has only one attribute in this example. The other three dimensions are Store, Time and Book Dimensions.

Table 10.9 Junk Dimension

JunkID	Gift coupon	Sales mode	Delivery	Payment mode
1	Y	In-store	Normal	Cash
2	Y	In-store	Normal	Credit card
3	Y	In-store	Express	Cash
4	Y	In-store	Express	Credit card
5	Y	In-store	Pick-up in store	Cash
6	Y	In-store	Pick-up in store	Credit card
7	Y	Online	Normal	Cash
8	Y	Online	Normal	Credit card
9	Y	Online	Express	Cash
10	Y	Online	Express	Credit card
11	Y	Online	Pick-up in store	Cash
12	Y	Online	Pick-up in store	Credit card
13	N	In-store	Normal	Cash
14	N	In-store	Normal	Credit card
15	N	In-store	Express	Cash
16	N	In-store	Express	Credit card
17	N	In-store	Pick-up in store	Cash
18	N	In-store	Pick-up in store	Credit card
19	N	Online	Normal	Cash
20	N	Online	Normal	Credit card
21	N	Online	Express	Cash
22	N	Online	Express	Credit card
23	N	Online	Pick-up in Store	Cash
24	N	Online	Pick-up in Store	Credit card

Table 10.10 New Junk Dimension

JunkID	Gift coupon	Sales mode	Delivery	Payment mode
1	Y	In-store	Pick-up in store	Cash
2	Y	In-store	Pick-up in store	Credit card
3	Y	Online	Normal	Credit card
4	Y	Online	Express	Credit card
5	Y	Online	Pick-up in store	Credit card
6	N	In-store	Pick-up in store	Cash
7	N	In-store	Pick-up in store	Credit card
8	N	Online	Normal	Credit card
9	N	Online	Express	Credit card
10	N	Online	Pick-up in store	Credit card

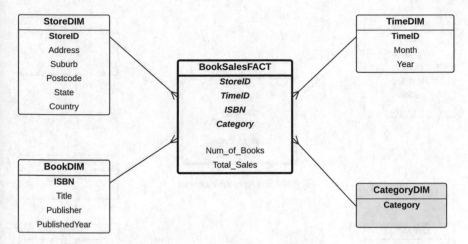

Fig. 10.9 Star schema for Book Sales

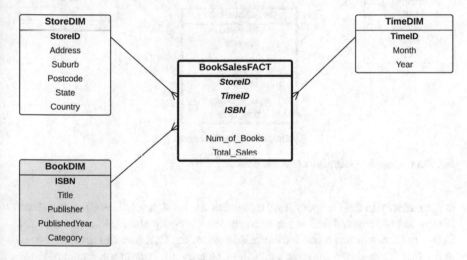

Fig. 10.10 Category inside the Book Dimension

Assuming that each book can only have one category, then Category is related to Book. In this case, we can simply move the category into the book. Therefore, the star schema will have only three dimensions with the Category Dimension removed (see Fig. 10.10). In other words, combining the one-attribute dimensions to normal dimensions will be possible if the one-attribute dimensions are related to the normal dimensions.

Looking at the star schema in Fig. 10.10, if we want to retain the Category Dimension as an independent dimension, it is possible to change the granularity of the star schema. In the star schema in Fig. 10.10, the granularity is determined by the Book. We could move it to a higher-level granularity (more general) by moving

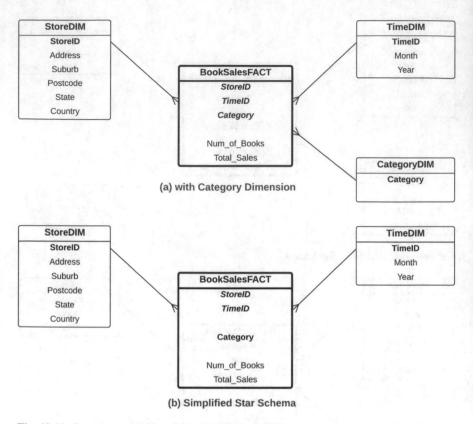

Fig. 10.11 Category granularity of the star schema

the granularity to the Category level instead of the Book level. Therefore, the Book Dimension is removed and is replaced by the Category Dimension. Because the Category Dimension is a one-attribute dimension, the Category can now be moved to the fact (Row-Based Solution), and the Category attribute in the fact can be the dimension-less key. Figure 10.11 shows the star schema with the Category as the granularity.

10.2.3   Determinant Dimension with One-Attribute Only

A Determinant Dimension which has a single attribute must be retained in the star schema. Let's use a Weather Data case study. The star schema is shown in Fig. 10.12. Star schema *a* is Level-0, containing one dimension called Weather Station. The fact has one dimension-less key, which is the Time (e.g. Weather Time). The fact measures are typical weather measurements, such as Temperature, Humidity, Wind Speed, Pressure and Rainfall. This Level-0 star schema means that each record is the weather data individually recorded at each weather station.

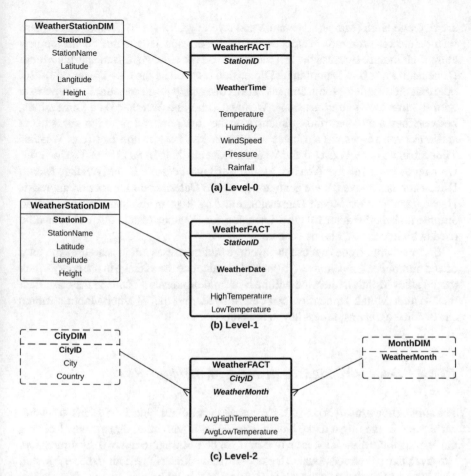

Fig. 10.12 Weather star schemas

Star schema *b* is Level-1, where the Weather Time in Level-0 is now converted into Weather Date. This means that weather records for the same day and from the same weather station are aggregated into one record in the Level-1 star schema. Hence, it is more general than Level-0. The fact measures are also aggregated, using maximum and minimum for high and low temperatures. The other fact measures from Level-0 are not included in Level-1, even though they could have been, for example, Max Humidity and Min Humidity. But usually, Maximum (High) Temperature and Minimum (Low) Temperature are the most common measures for daily weather.

Star schema *c* is even more general than star schema *b* because the Weather Date is now changed to Weather Month. The Weather Station Dimension is also made more general to become the City Dimension. This means the data from all the weather stations in the same city are aggregated by month. The fact measures are Average High (Max) Temperature and Average Low (Min) Temperature. These features are very common in monthly weather data, especially to find out the temperature each month. Because the fact measures use the average, this means the Month Dimension must be a Determinant Dimension (as well as the City Dimension). But note that the Weather Month Dimension is a one-attribute dimension. In comparison with the previous two dimensions, Weather Time and Weather Date Dimensions, because they are one-attribute dimensions, they are removed from the star schema and make the respective attributes dimension-less keys in the fact (e.g. Weather Time attribute in fact Level-0 and Weather Date attribute in fact Level-1). However, we cannot do it for the Weather Month attribute because if the Weather Month Dimension is removed, there is no notion of the Determinant Dimension anymore. Hence, the Weather Month Dimension must be kept in the star schema, also to enforce that the Weather Month Dimension (or Weather Month attribute) must be used in all queries to this star schema.

If star schema c does not use the average fact measures but instead uses the total of and number of temperatures (to be used to calculate the average in the query), then the Weather Month Dimension will be a normal dimension. If this is the case, then the Weather Month Dimension can be removed, and the Weather Month attribute can become a dimension-less key in the fact.

10.2.4   One-Attribute Dimension with Bridge

The above shows that it is possible to remove the one-attribute dimensions in several ways, such as moving it to the fact or combining it with other dimensions. Keeping the one-attribute dimension is an option if we do not want to remove the dimension. However, there is one case where the one-attribute dimension needs to be kept intact. This is when the dimension is connected to a bridge table.

Let's use the Book Sales case study again to illustrate this concept. Figure 10.13 shows a star schema where the Book Dimension is connected to the Author Dimension through a Bridge Table called Title Author Bridge. The reason for using the bridge table is because one book may have multiple authors and each author may have written several books. Therefore, analysing total sales must be based on books, not based on authors.

This star schema's granularity is at the Book granularity. If we would like to have a lower granularity than book but in this star schema, there is no category to go for, we need to "combine" several books into one group. One way to combine several books into one group of books is by looking at the authors, that is, to combine books if they are written by the same list of authors. Hence, instead of individual ISBN as in Fig. 10.13, we now use ISBN Group List (Fig. 10.14). The ISBN Group List can be created by using the `listagg` function which will be covered in a later chapter.

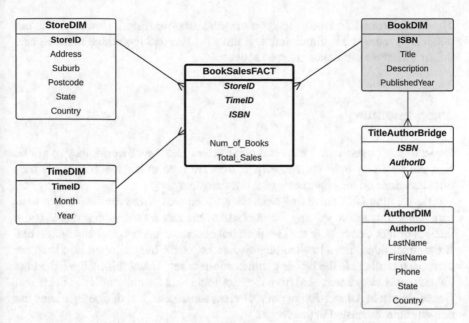

Fig. 10.13 Book Dimension with a Bridge Table

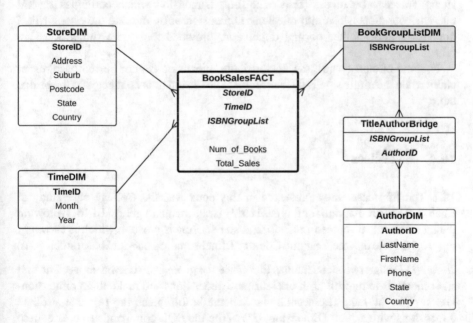

Fig. 10.14 Collection of books as the granularity

The Book Group List Dimension is a one-attribute dimension. However, it is not possible to remove this dimension as it links the fact and the bridge table. Hence, this one-attribute dimension must be kept as is.

10.3 Summary

One-attribute dimensions are special dimensions. In the real world, they might be rare, but they may exist occasionally. If they are used in a star schema, this may raise questions on the importance of this dimension.

One solution is to move this dimension to the fact. If the dimension is a low-cardinality dimension, shifting the attribute to the fact measures is possible if the number of fact measures in the fact does not become excessive and unmanageable. If the dimension is not a low-cardinality dimension, the only solution is to keep the dimension attribute in the fact as a dimension-less key. Using this method, the Fact Table is practically unchanged from the Fact Table with the one-attribute dimension. Hence, from a high level design point of view, sometimes, it is desirable to keep the one-attribute dimension in the design.

Another solution is to combine the one-attribute dimension with other dimensions. This can be in the form of a Junk Dimension which combines several one-attribute dimensions into one Junk Dimension or by moving the one-attribute dimension into another normal dimension, provided that these dimensions are related.

In some circumstances, the one-attribute dimension cannot be removed, especially when the dimension is a Determinant Dimension or is connected to the bridge table.

10.4 Exercises

10.1 The first case study presented in this book was the College case study in Chap. 2. The E/R diagram for this case study is shown again in Fig. 10.15. Following the requirements discussed in Chap. 2, a star schema is created, which is shown in Fig. 10.16. Note that the Year and Country Dimensions are one-attribute dimensions.

Tasks (*i*) Design two star schemas to replace the given star schema above. The first star schema is to remove both one-attribute dimensions and make them dimension-less keys in the fact. The second star schema is to replace the two one-attribute dimensions with a Junk Dimension. (*ii*) Write the SQL commands to create these two new star schemas.

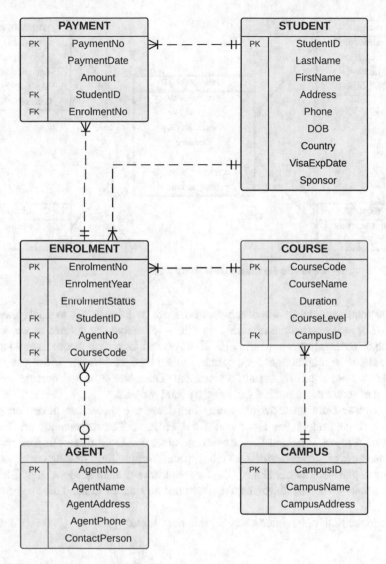

Fig. 10.15 An E/R diagram of the College case study

10.2 This exercise is based on one of the exercises in Chap. 2. The Information Technology Services (ITS) department of a University keeps track of projects on which they are working. The E/R diagram of the operational database is shown in Fig. 10.17.

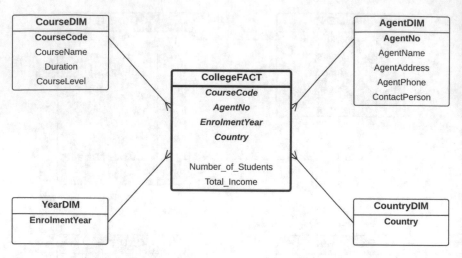

Fig. 10.16 A star schema for the College case study

The requirements for the data warehouse are as follows: (*i*) to show the total number of projects that are of a long duration—durations are defined as *short term* (less than 10 days), *medium term* (10–30 days) and *long term* (more than 30 days); (*ii*) to show the total budget for a certain project type; and (*iii*) to show an average budget cost per hour of a certain department. The average budget cost per hour is defined as total budget per hour divided by total projects.

It is clear from these requirements that there are three fact measures: Total Budget, Total Budget Per Hour and Total Projects. The dimensions are Project Duration, Project Type and Department. Assume that these three dimensions have only one attribute, such as duration (short term, medium term or long term), project type or department name. Hence, the fact will have dimension-less keys. The star schema will then only have the fact, without any dimensions. This is shown in Fig. 10.18.

Write the SQL commands to create this Fact Table.

10.5   Further Readings

This chapter specialises in a small topic of star schema design, particularly dimensions that have only one attribute (key attribute). Data design or entity modelling where the entity has only the key attribute is a general database design issue. Further discussions on data modelling and database design can be found in the following textbooks: [1–5] and [6].

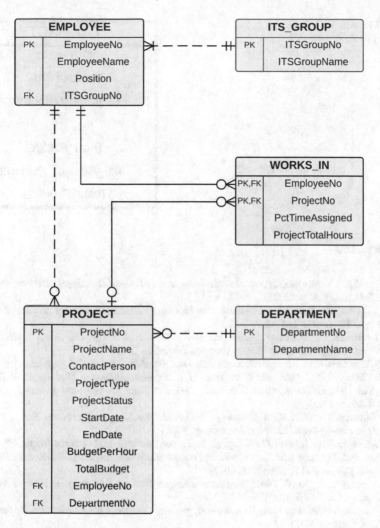

Fig. 10.17 An E/R diagram of the ITS Project case study

Star schema design involving various issues on dimensions, including dimensions with one attribute, is discussed in various data warehousing books, such as [7–16] and [17].

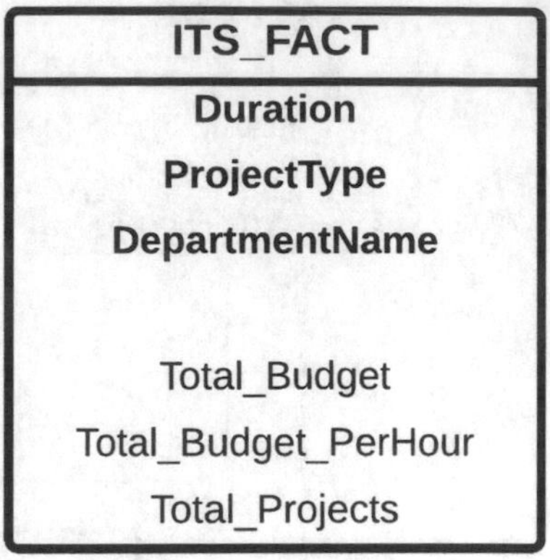

Fig. 10.18 The star schema
for the ITS Project case study

ITS_FACT
Duration
ProjectType
DepartmentName
Total_Budget
Total_Budget_PerHour
Total_Projects

References

1. T.A. Halpin, T. Morgan, *Information Modeling and Relational Databases*, 2nd edn. (Morgan Kaufmann, Los Altos, 2008)
2. J.L. Harrington, *Relational Database Design Clearly Explained*. Clearly Explained Series. (Morgan Kaufmann, Los Altos, 2002)
3. M.J. Hernandez, *Database Design for Mere Mortals: A Hands-On Guide to Relational Database Design* For Mere Mortals (Pearson Education, 2013)
4. N.S. Umanath, R.W. Scamell, *Data Modeling and Database Design* (Cengage Learning, 2014)
5. T.J. Teorey, S.S. Lightstone, T. Nadeau, H.V. Jagadish, *Database Modeling and Design: Logical Design*. The Morgan Kaufmann Series in Data Management Systems (Elsevier, Amsterdam, 2011)
6. G. Simsion, G. Witt, *Data Modeling Essentials*. The Morgan Kaufmann Series in Data Management Systems (Elsevier, Amsterdam, 2004)
7. C. Adamson, *Star Schema The Complete Reference* (McGraw-Hill Osborne Media, 2010)
8. R. Laberge, *The Data Warehouse Mentor: Practical Data Warehouse and Business Intelligence Insights* (Mcgraw-Hill, New York, 2011)
9. M. Golfarelli, S. Rizzi, *Data Warehouse Design: Modern Principles and Methodologies*. (Mcgraw-Hill, New York, 2009)
10. C. Adamson, *Mastering Data Warehouse Aggregates: Solutions for Star Schema Performance* (Wiley, London, 2012)
11. P. Ponniah, *Data Warehousing Fundamentals for IT Professionals* (Wiley, London, 2011)
12. R. Kimball, M. Ross, *The Data Warehouse Toolkit: The Definitive Guide to Dimensional Modeling* (Wiley, London, 2013)
13. R. Kimball, M. Ross, W. Thornthwaite, J. Mundy, B. Becker, *The Data Warehouse Lifecycle Toolkit* (Wiley, 2011)
14. W.H. Inmon, *Building the Data Warehouse*. ITPro Collection (Wiley, London, 2005)
15. M. Jarke, *Fundamentals of Data Warehouses*, 2nd edn. (Springer, Berlin, 2003)
16. E. Malinowski, E. Zimányi, *Advanced Data Warehouse Design: From Conventional to Spatial and Temporal Applications*. Data-Centric Systems and Applications (Springer, Berlin, 2008)
17. A. Vaisman, E. Zimányi, *Data Warehouse Systems: Design and Implementation*. Data-Centric Systems and Applications (Springer, Berlin, 2014)

Part IV
Multi-Fact and Multi-Input

Chapter 11
Multi-Fact Star Schemas

A data warehouse has four basic features: (i) Integrated, (ii) Subject Oriented, (iii) Time Variant, and (iv) Non-volatile. A *Subject-Oriented* data warehouse means that one star schema focuses on one subject only.

A subject refers to a topic of analysis. For example, a star schema might be built to analyse property sales. If we want to have a data warehouse that focuses on property rentals, it has to be a separate star schema because one star schema focuses on one subject only. In this example, property sales is one subject, and property rentals is another subject. The first star schema focuses on property sales, whereas the second star schema focuses on property rentals. These two star schemas should not be combined because they focus on different subjects. The input operational database for these star schemas might be one operational database.

This chapter focuses on *multi-fact* star schemas where a star schema has multiple Fact Tables. Figure 11.1 shows a star schema with two facts, and Dimension 3 and Dimension 4 are shared by the two facts. Actually, we can simply re-draw this as two separate star schemas as in Fig. 11.2 where each star schema has a fact and four dimensions. In other words, it does not matter whether we draw Dimension 3 and Dimension 4 in one star schema or in two star schemas. In practice, they are independent of the fact because in the implementation, the links between dimensions and the fact are implemented through value matching, like the Primary Key—Foreign Key value join. Hence, Dimension 3 and Dimension 4 and in fact all dimensions are independent to the facts.

In this chapter, we are going to learn how to create multi-fact star schemas. The first case study is a Book Sales case study which shows how multi-facts are used. When studying multi-fact star schemas, two important questions that often appear are what is subject oriented and whether multiple star schemas can be combined into one. These questions will also be discussed in this chapter, using the Private Taxi case study (for the subject-oriented question) and the Tutorials and the PhD supervision case studies (for combining multiple star schemas into one). Finally, this chapter will end with a case study where a multi-fact is caused by the different

© The Author(s), under exclusive license to Springer Nature Switzerland AG 2021
D. Taniar, W. Rahayu, *Data Warehousing and Analytics*, Data-Centric Systems
and Applications, https://doi.org/10.1007/978-3-030-81979-8_11

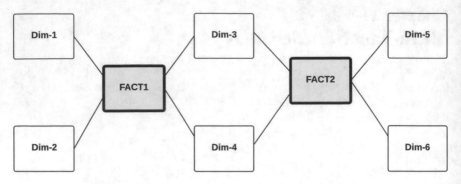

Fig. 11.1 A multi-fact star schema

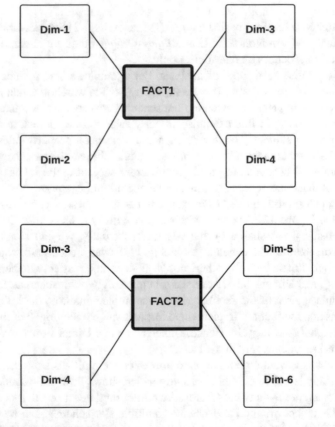

Fig. 11.2 Two separate star schemas

granularities of the dimensions. In other words, there are two main causes for a multi-fact: (*i*) different *Subject* and (*ii*) different *Granularity*.

11.1 Different Subject Multi-fact: The Book Sales Case Study

This case study is about a bookshop that has several stores and they sell books. The E/R diagram of an operational database is shown in Fig. 11.3.

The system stores information about books, including the authors, publishers, book categories and the reviews that each book has received. The "stars" attribute in the Review entity records the star rating for each review (e.g. five stars for excellent to one star for poor, etc.). One book may receive many reviews. For simplicity, it is assumed that a book will only have one category, as also shown in the E/R diagram.

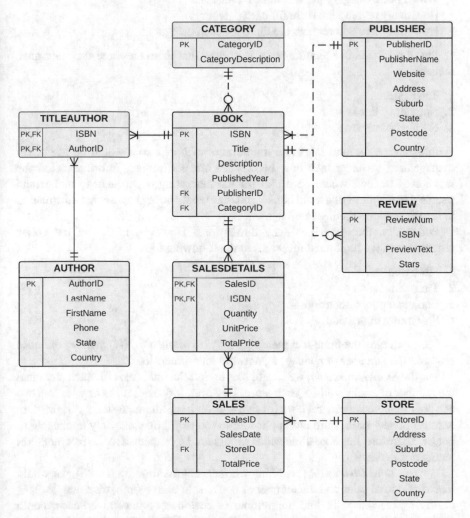

Fig. 11.3 An E/R diagram of the Bookshop System

The E/R diagram also includes entities related to the sales of books and the stores which sell the books. Each store has many sales transactions (i.e. the Sales entity), and each sales transaction may include several books (i.e. the Sales Details entity). The Total Price attribute in the Sales Details entity is Quantity multiplied by the Unit Price, whereas Total Price in the Sales entity is the total price of each sales transaction.

The requirements for the data warehouse are quite simple. The data warehouse must be able to answer at least the following questions:

- What are the total sales for each bookstore in a month?
- What is the number of books sold for each category?
- Which book category has the highest total sales?
- How many reviews are there for each category?
- How many five-star reviews are there for each category?

Based on the above requirements, it is clear that there are three fact measures, namely:

1. Total sales
2. Number of books sold
3. Number of reviews

Information on total sales and the number of books sold would be useful for management to understand how book sales are performing. Information on the number of reviews would be useful for the marketing department to understand people's perceptions of certain books, and they would be able to use this information to launch other marketing campaigns.

Potentially, there can be many dimensions. However, based on the above requirements, we limit the dimensions to the following four:

1. Store dimension
2. Time dimension
3. Book category dimension
4. Star rating dimension

Let's examine the three fact measures (e.g. (i) *total sales*, (ii) *number of books sold* and (iii) *number of reviews*) against the four dimensions.

For the *Store Dimension*, we would like to find the total sales for each store and the number of books sold for each store. However, it doesn't make any sense if we ask how many customer reviews (or five-star reviews) there are for a certain store because review ratings are not applicable to the store; they are only applicable to books. Therefore, the store dimensions should not be connected to the fact measure: number of reviews.

For the *Time Dimension* (assuming we measure the time by month), we would like to find total sales and the number of books sold each month, for example. This clearly makes sense. To find the number of customer reviews given each month seems to make sense too. However, if we inspect the E/R diagram closely, the month (or the date) is associated with the purchase of a book, not when a review is given by a customer. Consequently, the time dimension should not be connected to the fact measure: number of reviews.

It seems clear now that the first two fact measures, total sales and number of books sold, are related to the *Book Sales*, whereas the third fact measure, number of reviews, is not about book sales but is associated with *reviews of the books*. Therefore, these three fact measures are not within one common subject; in fact, they are two different subjects: *Book Sales* and *Book Reviews*.

Let's continue with the third dimension: *Book Category* (or simply *Category*). Finding the total sales and the number of books sold for each book category seems to be sensible. For example, how many fiction novels were sold and what is the total amount of sales for fiction novels? Finding how many five-star reviews there were for fiction novels is also reasonable because each book (or each novel) will receive reviews; therefore, calculating how many reviews for each particular book category is also reasonable. In other words, the book category dimension is applicable to all of the three fact measures: total sales, number of books sold and number of reviews.

The fourth dimension, the *Star Rating Dimension*, describes the meaning of each star rating, that is, from one star to five stars, as an example. Finding the total sales for books with five-star ratings seems to be fine. Because one book may receive many different kinds of star ratings, we need to aggregate these and come up with a one-star rating, such as 3.6 stars. Then this will be classified to 3.0 to 3.9 stars in the Star Rating Dimension. Hence, the Review Star Dimension is also applicable to all of the fact measures.

In summary, the first subject, "Book Sales", will be the first fact, with all four dimensions (e.g. Store, Time, Category and Star Rating dimensions). The fact measures are total sales and number of books sold. The second subject is "Reviews", with only two dimensions, Category and Star Rating dimensions. There is only one fact measure, which is the number of reviews. The star schema (refer to Fig. 11.4) will then have two fact entities, Book Sales Fact and Review Fact. Note that the Category and Star Rating dimensions are shared by the two facts. The same star schema can be drawn as having two separate star schemas, as shown in Fig. 11.5.

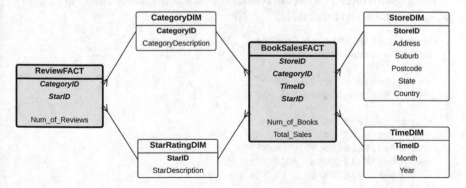

Fig. 11.4 A multi-fact star schema

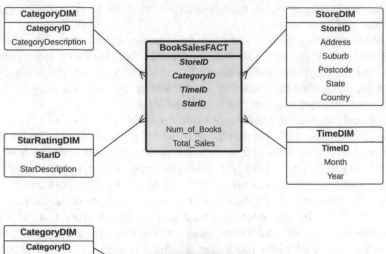

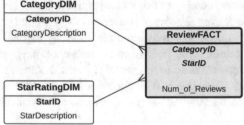

Fig. 11.5 Two star schemas

11.1.1 Implementation in SQL

The first step is to create the dimension tables. There are four dimensions in
the star schema. The SQL for creating the first three dimension tables—Category
Dimension, Store Dimension and Time Dimension—is quite straightforward; that
is, extract them directly from the operational database, namely from the Category
table, Store table and Sales table.

```
create table CategoryDim as
select * from Category;

create table StoreDim as
select * from Store;

create table TimeDim
as select distinct
  to_char(SalesDate, 'YYYYMM') as TimeID,
  to_char(SalesDate, 'MM') as Month,
  to_char(SalesDate, 'YYYY') as Year
from Sales;
```

The Star Rating Dimension can also be extracted from the operational database, particularly from the Review table which has the Stars attribute containing a star value from 0 to 5. However, in this exercise, we decided to create Star Rating Dimension manually in case there are books without any reviews. Books without reviews are actually not a problem for the Review Fact but will have a problem in the Book Sales Fact because every book must have an average rating. Hence, we need to create Star Rating Dimension manually. Note that creating this dimension manually will not require a TempFact because the table can be extracted from the operational database. After this dimension table is created, six records are inserted from zero stars to five stars.

```
create table StarRatingDim
(StarID number(1),
  StarDescription varchar2(15));

insert into StarRatingDim values (0, 'Unknown');
insert into StarRatingDim values (1, 'Poor');
insert into StarRatingDim values (2, 'Not Good');
insert into StarRatingDim values (3, 'Average');
insert into StarRatingDim values (4, 'Good');
insert into StarRatingDim values (5, 'Excellent');
```

After these four dimension tables are created, we can now create the first Fact Table, ReviewFact. Creating ReviewFact is quite straightforward, using the count function to count the number of reviews. The query joins tables Book and Review based on ISBN and then does an appropriate grouping for count.

```
create table ReviewFact as
select
    B.CategoryID,
    R.Stars as StarID,
    count(*) as Num_of_Reviews
from Book B, Review R
where B.ISBN=R.ISBN
group by B.CategoryID, R.Stars;
```

Creating the Book Sales Fact can be slightly tricky. The problem is because one book may receive many reviews. If we simply join all the necessary tables, including the tables Book and Review, there will be an incorrect counting of the number of books because books with multiple reviews will be counted as multiple instances of the book, and consequently, the book count will be incorrect. To solve this problem, each book should have only "one"-star rating instead of multiple star ratings. This can be easily achieved by calculating the average ratings of each book.

Another problem with book and reviews is that some books may not have any reviews. Hence, when we join Book and Review, we cannot use an inner join because if we do, books without reviews will be discarded. To solve this problem, an outer join must be used to preserve books without any reviews. For those books without reviews, a zero-star rating will be allocated.

In order to better understand this book-review process, we have divided the SQL into two steps: first is to do an outer join between Book and Review, and second is to do an average. The plus (+) sign at the end of the first SQL is used to indicate an outer join. The nvl function is used to handle a null value replacement. The rounding function in the second SQL is to get an integer average of the star ratings for each book.

```
create table TempBookWithStar as
select
  B.ISBN,
  B.CategoryID,
  nvl(R.Stars, 0) as Stars
from Book B, Review R
where B.ISBN = R.ISBN(+);

create table TempBookWithAvgStar as
select ISBN, CategoryID, round(avg(Stars)) as Avg_Stars
from TempBookWithStar
group by ISBN, CategoryID;
```

The above join in `create table TempBookWithStar` uses the (+) sign to indicate the *Left Outer Join*. Alternatively, we can use the ANSI syntax to create the TempBookWithStar table as follows:

```
create table TempBookWithStar as
select
  B.ISBN,
  B.CategoryID,
  nvl(R.Stars, 0) as Stars
from Book B left outer join Review R
  on B.ISBN = R.ISBN;
```

Once the TempBookWithAvgStar table is created, we can use this table instead of the original Book table when we create BookSalesFact. Creating BookSalesFact is quite straightforward by joining TempBookWithAvgStar, Sales and SalesDetails tables and summing the Quantity and TotalPrice attributes to get the two fact measures: Num_of_Books and Total_Sales.

```
create table BookSalesFact as
select
  T.CategoryID,
  to_char(S.SalesDate, 'YYYYMM') as TimeID,
  S.StoreID,
  T.Avg_Stars as StarID,
  sum(D.Quantity) as Num_of_Books,
  sum(D.TotalPrice) as Total_Sales
from
  TempBookWithAvgStar T, Sales S, SalesDetails D
where T.ISBN = D.ISBN
and S.SalesID = D.SalesID
group by
  T.CategoryID,
  to_char(S.SalesDate, 'YYYYMM'),
  S.StoreID,
  T.Avg_Stars;
```

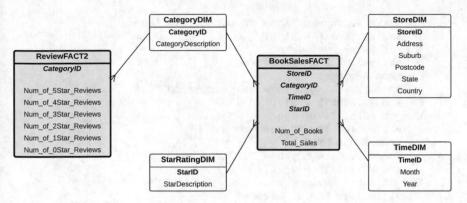

Fig. 11.6 Review star schema with Pivoted Fact

11.1.2   Multi-Fact with Pivot Table

Another possible option for the ReviewFact star schema is to have a Pivoted Fact Table and to remove the Star Rating Dimension from the Review Fact. As a result, the fact measures must incorporate all Star RatingIDs, ranging from 0 to 5. Figure 11.6 shows this star schema. Note that the Star Rating Dimension still exists but is not connected to Review Fact; rather it is connected to the Book Sales Fact only.

The SQL command to create the Review Fact2 Table is shown as follows. After the CategoryID attribute, six attributes to indicate Number of Reviews are added with an initial value of zero. After this table is created, the six Number of Reviews columns must be updated with the correct Number of Stars. We can use a nested query to check for each category and for each StarID. An nvl function must be used in case the subquery returns a null, indicating that there is no review for that particular StarID in that particular CategoryID.

```
create table ReviewFact2 as
select CategoryID,
  0 as Num_of_0Star_Reviews, 0 as Num_of_1Star_Reviews,
  0 as Num_of_2Star_Reviews, 0 as Num_of_3Star_Reviews,
  0 as Num_of_4Star_Reviews, 0 as Num_of_5Star_Reviews
from CategoryDim;

update ReviewFact2 F2
set
  Num_of_1Star_Reviews =
    nvl((select Num_of_Reviews
      from ReviewFact F1
      where F2.CategoryID = F1.CategoryID
      and F1.StarID = 1),0),
  Num_of_2Star_Reviews =
    nvl((select Num_of_Reviews
      from ReviewFact F1
```

```
      where F2.CategoryID = F1.CategoryID
      and F1.StarID = 2),0),
Num_of_3Star_Reviews =
  nvl((select Num_of_Reviews
    from ReviewFact F1
    where F2.CategoryID = F1.CategoryID
    and F1.StarID = 3),0),
Num_of_4Star_Reviews =
  nvl((select Num_of_Reviews
    from ReviewFact F1
    where F2.CategoryID = F1.CategoryID
    and F1.StarID = 4),0),
Num_of_5Star_Reviews =
  nvl((select Num_of_Reviews
    from ReviewFact F1
    where F2.CategoryID = F1.CategoryID
    and F1.StarID = 5),0);
```

11.2   Multi-Fact or Single Fact with Multiple Fact Measures: A Private Taxi Company Case Study

The concept of "Subject Oriented" can often be ambiguous. Consider the following case study. A private taxi company is a small business. The owner of this company owns five cars, which can be chartered like taxis. The business has many clients, and the number of clients grows due to word of mouth. When a client needs a service from this company (such as driving them from their home to the airport, or vice versa), they send a booking though an SMS or a social media chat (e.g. WhatsApp). Once the booking is confirmed, a driver will pick the clients up and send them to the destination. The charges are normally a pre-defined fixed price; for example, from an inner suburb of the city to the airport, the charge is $90. The charge might be slightly more expensive than a normal taxi, but for some reason, many clients prefer the service that this company offers.

The company has three full-time drivers, including the owner and five sessional or part-time drivers who may be called when the company needs drivers. Every time a client hires this taxi, the information (e.g. origin and destination, pick-up time and date, fares, driver, car, distance travelled, amount of petrol used, etc.) is recorded in their operational database.

On top of this operational database, the company also builds a data warehouse for reporting purposes. The data warehouse is a star schema with only three dimensions: Car Dimension, Driver Dimension and Week Dimension. The company would like to analyse their total income, total kilometres travelled and total fuel used. These then become the fact measures of the star schema.

The question is whether the three fact measures indicate that there are three different subjects: total income is about money, total kilometres is about distance, and total fuel is about volume. They all concern different measures. If we think they are different subjects, the star schema will end up as shown in Fig. 11.7.

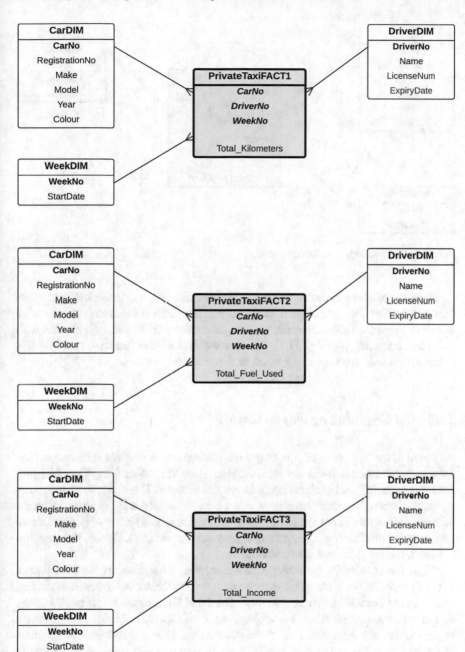

Fig. 11.7 Three facts, each with a different fact measure

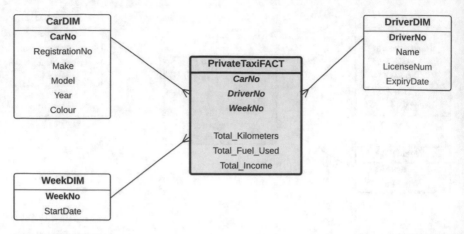

Fig. 11.8 One fact with three fact measures

The use of three facts is unreasonable because all the three use the same dimensions: Car, Driver and Week Dimensions. They indicate that the fact measures are all of the same subject. Hence, the star schema should be just a single fact with three fact measures (see Fig. 11.8). In other words, one fact can have multiple fact measures, as we have seen many times in the previous chapters.

11.3   To Combine or Not to Combine

As stated at the beginning of this chapter, a star schema is *Subject Oriented*, which means that it should focus on one subject. However, identifying that two star schemas have different subjects is not as easy as thought. Therefore, there are many situations where the difference in topics among multiple star schemas becomes unclear. When this becomes unclear, we will often ask whether it is possible to merge or to combine these star schemas into one star schema. Hence, the question is whether to combine or not to combine.

Often, but not always, the subject of a star schema is indicated by the dimensions. In the Private Taxi case study, the dimensions are Car, Driver and Week, and the fact measures are Total Kilometres, Total Fuel Used and Total Income. It is not about the unit of measure that identifies the subject, but the dimensions. In this case, all three fact measures are examined from three dimensions. Hence, the three star schemas focus on one subject, and they can be merged into one star schema, as shown in Fig. 11.8. So in short, if the star schemas have exactly the same dimensions (and the fact measures are different), the star schemas can simply be merged into one star schema, and the new star schema will take all the fact measures from the individual star schema. Merging multiple star schemas, in this case, can simply join the three Fact Tables.

```
create table PrivateTaxiFact as
select F1.CarNo, F1.WeekNo, F1.DriverNo,
  F1.Total_Kilometers, F2.Total_Fuel_Used, F3.Total_Income
from
  PrivateTaxiFact1 F1,
  PrivateTaxiFact2 F2,
  PrivateTaxiFact3 F3
where F1.CarNo = F2.CarNo
and F1.CarNo = F3.CarNo
and F1.WeekNo = F2.WeekNo
and F1.WeekNo = F3.WeekNo
and F1.DriverNo = F2.DriverNo
and F1.DriverNo = F3.DriverNo;
```

If different star schemas have the same dimensions and the same fact measures, it is very likely that each star schema focuses on a different time period. So although the star schema is identical, the contents are different. For example, suppose in the Private Taxi case study (see Fig. 11.7), the first star schema with Total Kilometres has two Fact Tables: PrivateTaxiFact1a and PrivateTaxiFact1b, each of which focuses on different time periods (e.g. the first Fact Table PrivateTaxiFact1a is based on WeekNo 1–26, whereas the second Fact Table PrivateTaxiFact1b is based on WeekNo 27–52). These two Fact Tables can be merged into one Fact Table because they focus on two different time periods. We can simply use the union command in SQL to merge the two Fact Tables. In this case, the two Fact Tables are union compatible, so uniting them is very straightforward.

```
create table PrivateTaxiFact1 as
select *
from PrivateTaxiFact1a
union
select *
from PrivateTaxiFact1b;
```

Now let's discuss the opposite, that is, when do we not to combine. The short answer to this is when the dimensions are not identical. If star schema-1 has dimensions A, B and C and star schema-2 has dimensions A, B, D and E, then it is clear that both star schemas are different, and consequently, they cannot be combined into one star schema. Therefore, it is somewhat convenient to use dimensions as a means to identify the subjects among star schemas, especially in the context of whether to combine or not to combine multiple star schemas.

However, when all but one dimension is different or has a slightly different context, then the question as to whether to combine or not to combine can sometimes be tricky. For example, if star schema-1 has dimensions A, B and C_1 and star schema-2 has dimensions A, B and C_2, where C_1 and C_2 can be the same but have different context or even when C_1 and C_2 are slightly different, then the question as to whether to combine or not to combine is not that quite straightforward. In this kind of scenarios, there are three possible options:

1. A Determinant Dimension Solution: the star schemas can be combined, but the combined star schema will have a Determinant Dimension.

2. A Non-determinant Dimension Solution: the star schemas can also be combined, and the result star schema does not need to have a Determinant Dimension.
3. Mutually Exclusive Star Schemas: the star schemas are all mutually exclusive; hence, combining them is quite straightforward.

These three options will be discussed in more detail with case studies in the following sections.

11.3.1   A Determinant Dimension Solution: Flight Charter Case Study

The Flight Charter case study is about a small company which hires small aircraft for touristic or expedition purposes. So it is similar to a taxi. The main difference is that in this Flight Charter case study, for a medium-sized aircraft, the flight has a pilot and a co-pilot. For a small-sized aircraft, there is only one pilot (without a co-pilot). The latter is similar to the Private Taxi case study where each taxi has one driver.

The company has 12 licensed pilots who can fly an aircraft as a pilot or as a co-pilot, whenever necessary. The company has eight aircraft of various sizes. Because the 12 pilots can also fly an aircraft as a co-pilot, in the database, the details of these 12 pilots are stored in a table called the Employee Table (in other words, there are no two separate tables to differentiate between a set of pilot records and another set of co-pilot records because pilot and co-pilot personnel come from the same pool, namely the Employee Table). The aircraft is stored in the Aircraft Table.

Suppose a star schema is already built as shown in Fig. 11.9. The star schema has three dimensions: Aircraft, Time (which Month) and Pilot. The fact measures

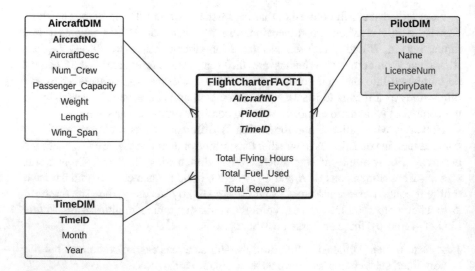

Fig. 11.9 Flight Charter pilot star schema

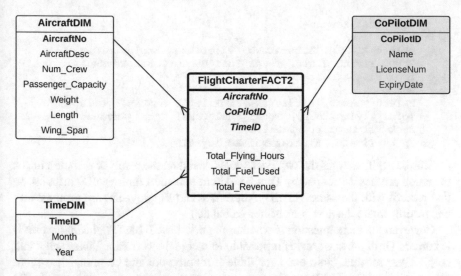

Fig. 11.10 Flight Charter co-pilot star schema

are Total Flying Hours, Total Fuel Used and Total Revenue. Using this star schema, we can calculate the number of flying hours for each pilot in a particular month, for example, or calculate the total fuel used in a particular aircraft in a particular month, or simply calculate the total revenue in a particular month.

However, this star schema (refer to Fig. 11.9) focuses on pilot only. If a pilot flew an aircraft as a co-pilot (with someone else as the pilot), this star schema doesn't capture it. In order to analyse the co-pilot, suppose there is another star schema for co-pilot (see Fig. 11.10) which has the same dimensions and the same fact measures as in the pilot star schema. The only difference is the actual contents of the Fact Table. The co-pilot star schema contains the total flying hours, total fuel used and total revenue for each employee (as a co-pilot), in each month and flying each aircraft.

Since the two star schemas are identical, in terms of the dimensions and the fact measures, is it possible to merge both star schemas into one star schema so that we can analyse pilot and co-pilot at the same time? Technically, to merge two Fact Tables, we can use a union command in SQL, something like:

```
select * from FlightCharterFact1
union
select * from FlightCharterFact2
```

We can put this union command in the create table as select command as follows:

```
create table FlightCharterFact3 as
select TimeID, AircraftNo, EmployeeID,
    sum(Total_Flying_Hours),
    sum(Total_Fuel_Used),
```

```
      sum(Total_Revenue)
from (
      select TimeID, AircraftNo, PilotID as EmployeeID,
      Total_Flying_Hours, Total_Fuel_Used, Total_Revenue
      from FlightCharterFact1
      union
      select TimeID, AircraftNo, CoPilotID as EmployeeID,
      Total_Flying_Hours, Total_Fuel_Used, Total_Revenue
      from FlightCharterFact2)
group by TimeID, AircraftNo, EmployeeID;
```

This basically merges the two Fact Tables, and then the result of this merging or the union process is grouped by TimeID, AircraftNo and EmployeeID attributes, so that records with the same TimeID, AircraftNo and EmployeeID are grouped into one record, where the new sum is then calculated.

However, the main question is whether the new Fact Table (FlightCharterFact3) is correct. Or in other words, is it possible to merge the two Fact Tables (pilot and co-pilot star schemas) into one Fact Table (for both pilot and co-pilot)? If we join the two original Fact Tables as shown in the above SQL command, an EmployeeID (either as a pilot or a co-pilot) flying the same aircraft at the same TimeID (e.g. a certain month) will be merged and has one new Total Flying Hours. For example, Employee 101 has 500 flying hours as a pilot flying aircraft A_1 on January and 250 flying hours as a co-pilot flying the same aircraft on January and will have 750 total flying hours, which is correct. However, for the flights where this Employee flew as a co-pilot, someone else was the pilot. This means that the 250 h that Employee 101 has a co-pilot will also contribute to other Employee who was the pilot. This will result in the double counting of the number of hours, not from each individual perspective, but from the Month perspective, as well as from the Aircraft perspective. Therefore, the combined star schema will be incorrect.

The combined star schema will only be correct if the Pilot/Co-pilot Dimension is a *Determinant Dimension*; this means that the query must always use the Pilot/Co-pilot Dimension. The combined star schema is shown in Fig. 11.11. In this combined dimension, we use the Pilot Dimension to reflect the combined dimension, and it is a Determinant Dimension. Using this star schema, any queries to the star schema must use the Pilot Dimension, whether it is asking for the Total Flying Hours or Total Fuel Used or Total Revenue, and it will give the correct answer for the queried pilot whether this pilot was the pilot or the co-pilot for the flights.

However, if we need to query without using the Pilot Dimension, then we need to go back to the original star schemas, whether it is the pilot star schema or the co-pilot star schema. Therefore, it is not a matter of simply merging the original two star schemas, and once the combined star schema is created, the original two star schemas can be discarded. This will not be the case. So when we need to query the Aircraft and/or the Month, we only need to use the original pilot star schema because not all flights have co-pilots but must have pilots. So the complete data is in the pilot star schema. But if we want to query the co-pilot (and possibly with Aircraft and/or Month), then we need to use the co-pilot star schema. Nevertheless, we cannot discard the two original star schemas, even when the combined star schema has been created.

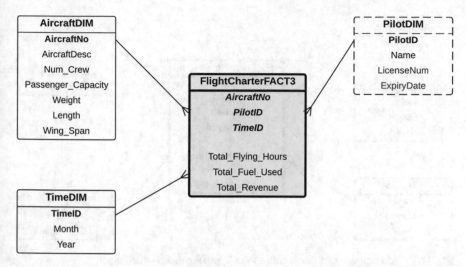

Fig. 11.11 A combined star schema with a Determinant Dimension—Solution 1

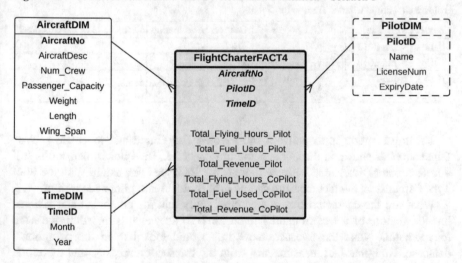

Fig. 11.12 A combined star schema with a Pivoted Fact Table—Solution 2

Another possible solution for the combined star schema is to use a *Pivoted Fact Table* as shown in the star schema in Fig. 11.12. The Pilot Dimension is still a Determinant Dimension, but the fact measures are now explicitly elaborated for pilots and co-pilots. This star schema is more complete than the star schema in Fig. 11.12 because of the breakdown of the fact measures. This is a similar case to the Private Taxi case study where the original three star schemas have a different fact measure and they are then merged. So a join operation is used to combine the two star schemas.

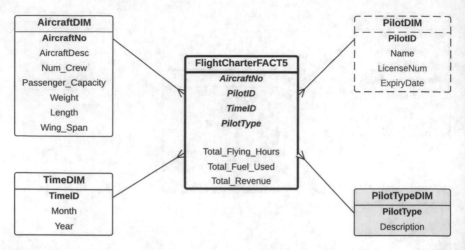

Fig. 11.13 A combined star schema with a Pilot Type Dimension—Solution 3

Table 11.1 Flight Charter Transaction Table

TripID	Trip date	Pilot	Co-pilot	Aircraft no	Destination	Distance	...	...	Customer
1012	4-Jan-2018	Jeremy	Peter	A8876	...	...			...
1013	5-Jan-2018	Jeremy		AB133	...	...			...
1014	6-Jan-2018	Peter	Daniel	A8876	...	...			...
...	...	...	...	...	...	...			...

The third option is to pull out the Pivoted Fact Table and to create a Type Dimension, as shown in the star schema in Fig. 11.13. The Pilot Dimension is still a Determinant Dimension, but now there is a Pilot Type Dimension. With the Pilot Type Dimension, the fact measure can be broken down into pilot and co-pilot.

What are the characteristics of this case study that the combined star schema has the feature of a Determinant Dimension? To answer this question, we need to understand what the transaction records would look like in the operational database. Note that fact measures are from the transaction records, and therefore, understanding the transaction records is crucial in designing star schemas.

In this Flight Charter case study, a transaction record is basically a charter of a flight. Every time a customer charters a flight, a transaction record is created in the operational database. This transaction record records the details of the flight, including who is the pilot and, if applicable, who is the co-pilot. Table 11.1 shows a snapshot of the Flight Charter Transaction Table. Each record contains the details of each transaction, including two columns: Pilot and Co-pilot (within each transaction record). As shown in this example, not all trips have a co-pilot (but every trip must have a pilot). Additionally, pilots and co-pilots are drawn from the same pool. For example, Peter was a co-pilot on one trip but was a pilot on another trip. So the transaction records have the feature of "one pool (of personnel or objects) on the same transaction record". The one pool is the pilot/co-pilot pool, and both pilot and co-pilot appear on one transaction.

For the "one pool (of personnel or objects) on the same transaction record" cases, there are three possible solutions, as shown in this Flight Charter case study.

1. A Determinant Dimension to represent the one pool of personnel or objects. For example: the Pilot Dimension. This is the basic solution which will also be used in the other two solutions below.
2. A Pivoted Fact Table to break down the fact measures. For example: Total Flying Hours for Pilot and for Co-pilot.
3. Instead of the Pivoted Fact Table, a Type Dimension to break down the fact measure. For example: Pilot Type and Co-pilot Type.

Because the combined star schema features a Determinant Dimension, the queries to the star schema will be rather limited because every query must use the Determinant Dimension. Therefore, the two original star schemas must still be kept to answer the other queries which do not involve the Determinant Dimension.

11.3.2 A Non-determinant Dimension Solution: Bachelor/Master Final Projects Case Study

The Bachelor/Master Final Projects case study involves the final projects that every Bachelor and Master's student must complete as their capstone projects. The project may be done in groups or individually, and each project must have one Manager, who is a Lecturer (i.e. a Faculty Member, such as a Lecturer or a Professor). A Faculty Member may supervise many projects at both Bachelor or Master levels.

The star schema for this case study focuses on the Manager. We would like to analyse a number of projects (and a number of students) for each Faculty Member who acts as a Manager of a project. The other dimensions of the star schema are Research Area and Year Dimensions. Some projects may focus on Software Engineering, whereas others may be on AI or Networking, or any other research areas. Each project is of 1-year duration, say, from March (beginning of the first semester) till November (end of the second semester).

Two star schemas have been created, one focusing on Bachelor Final Projects and the other focusing on Master Final Projects. The two star schemas are shown in Fig. 11.14.

Before discussing how to combine these two star schemas into one star schema, understanding what the transaction records look like is crucial. In this case study, assume there are two transaction tables, one to record Bachelor Final Projects and the other to record Master Final Projects. These two transaction tables are shown in Tables 11.2 and 11.3.

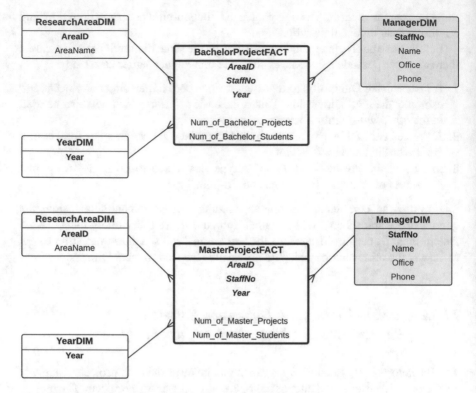

Fig. 11.14   Bachelor and Master Final Projects star schemas

Table 11.2   Bachelor Final Projects

ProjectNo	Project name	Description	Research area	Year	Students	Manager
1025	Big Data	...	Databases	2018	...	Dan Tan
1026	Blockchain	...	Security	2018	...	Lee Yungli
1027	MobileApp	...	AppDevelop	2018	...	Armit Armito
...	...	...	...	...	...	...

Table 11.3   Master Final Projects

ProjectNo	Project name	Description	Research area	Year	Students	Manager
6221	Mobile BI	...	BI	2018	...	James Chung
6222	CryptoCurr	...	Security	2018	...	Lee Yungli
6223	Map Query	...	Databases	2018	...	Dan Tan
...	...	...	...	...	...	...

This case study is known as "one pool (of personnel or objects) but in different transaction records", which is different from the Flight Charter case study where the one pool is on the same transaction records. Even when the transaction table is combined into one transaction table (see Table 11.4), only one manager exists

Table 11.4 Bachelor/Master Final Projects

ProjectNo	Project name	Description	Project level	Research area	Year	Students	Manager
1025	Big Data	...	Bachelor	Databases	2018	...	Dan Tan
1026	Blockchain	...	Bachelor	Security	2018	...	Lee Yungli
1027	MobileApp	...	Bachelor	AppDevelop	2018	...	Armit Armito
...	...	...	...	...	...	...	...
6221	Mobile BI	...	Master	BI	2018	...	James Chung
6222	CryptoCurr	...	Master	Security	2018	...	Lee Yungli
6223	Map Query	...	Master	Databases	2018	...	Dan Tan
...	...	...	...	...	...	...	...

per project. So it is one pool in different transaction records. The pool, in this case study, is the Manager because the Manager is from the pool of Faculty Members (e.g. Lecturers or Professors).

There are three possible solutions for the combined star schema. These three solutions are almost identical to the solutions for the Flight Charter case study (one pool, same transaction records). The only difference is that the Manager Dimension in the combined star schema is not a Determinant Dimension. It is a normal dimension. Hence, the three possible solutions are:

1. A Normal Dimension to represent the one pool of personnel or objects, for example, the Manager Dimension. This is the basic solution which will also be used in the other two solutions below.
2. A Pivoted Fact Table to break down the fact measures, for example, the Number of Projects for Bachelor degrees and for Master degrees.
3. Instead of the Pivoted Fact Table, a Type Dimension to break down the fact measure, for example, the Bachelor Type and Master Type.

Figure 11.15 shows the combined star schema, which looks the same as the two original star schemas. The new star schema basically combines the records from both original Fact Tables and recalculates the fact measures. This is done by unioning the two original Fact Tables and then recalculating the fact measures by using sum and group by, as shown previously in the Flight Charter case study.

The second solution is shown in Fig. 11.16 where the fact measures are broken down into Bachelor and Master Levels. This Fact Table is created by a simple join between the two original Fact Tables.

The third solution is to pull out the Pivoted Fact Table and create a Type Dimension. Figure 11.17 shows the third star schema.

Comparing these three solutions, the first solution as shown in Fig. 11.15 is more general than the other two because the Fact Table does not indicate the level of project (e.g. Bachelor or Master projects). The second and the third solution, as shown in Figs. 11.16 and 11.17, have the same level of details because the project level (e.g. Bachelor or Master projects) is indicated. The only difference is that in

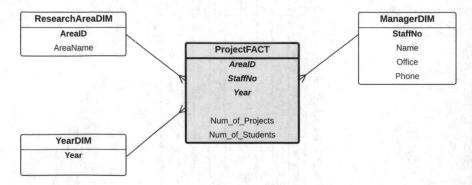

Fig. 11.15 A combined star schema—the basic solution: Solution 1

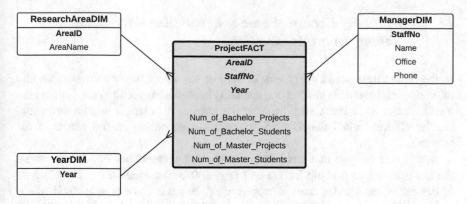

Fig. 11.16 A combined star schema—Solution 2 with Pivoted Fact

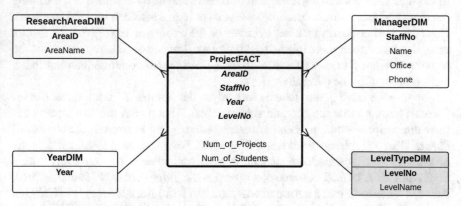

Fig. 11.17 A combined star schema—Solution 3 with Type Dimension

the second solution, the project level is indicated by the fact measures (e.g. Number of Bachelor Projects and Number of Master Projects), as opposed to an additional dimension (e.g. Level Type Dimension) in the third solution, where the project level is indicated by this dimension.

Unlike the Flight Charter case study which has a Determinant Dimension, the Bachelor/Master Final Projects case study does not have a Determinant Dimension. Consequently, once a combined star schema is created, there is no need to keep the original star schemas anymore. We can simply query the combined star schema to get any information we want.

11.3.3   *Mutually Exclusive Star Schemas: Lecturer/Tutor Taking Tutorials Case Study*

In the university system, every unit or subject has two components, lectures and tutorials (or laboratories). A lecture, normally held in a big lecture hall, is delivered by a Lecturer (or a Professor). So after the lecture, the class is broken down into smaller classes, called tutorials or laboratories (depending on the nature of the subject).

A unit or a subject that has a large number of students has one lecture as all the students fit into one big lecture hall (e.g. 200–300 students). But each tutorial or laboratory has a much smaller capacity (e.g. 20 students or even less). Hence, a unit or a subject that has 200 students has one lecture per week and has maybe up to ten tutorials (or laboratories). Each tutorial is taught by a Tutor (or a Teaching Assistant), which is often a postgraduate student (e.g. a Master's or a PhD student). However, it is common that the Lecturer or the Professor who gives the lecture must also take one or more tutorials/laboratories. Some exceptions apply so that the lecturer may not take any of the tutorials, but it is more common than not that a lecturer must take one tutorial.

Suppose we have a star schema to analyse the number of tutorials or classes for each tutor, unit or subject, and semester year. This means the star schema has three dimensions—tutor, unit and semester—and the fact measure is the number of classes. The star schema is shown in Fig. 11.18. Note that each tutor, although he or she may be a postgraduate student, is a staff member in the university with a Staff Number. A typical example of a record in the Tutor Fact is "FIT3003, 2/2018, Adam, 4", meaning that a tutor named Adam had four tutorials for unit FIT3003 in semester 2/2018. Another record could be something like: "FIT2094, 2/2018, Adam, 2", meaning that Adam took another two tutorials in a different unit (FIT2094) in the same semester.

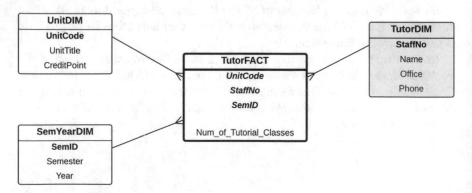

Fig. 11.18 Tutor star schema

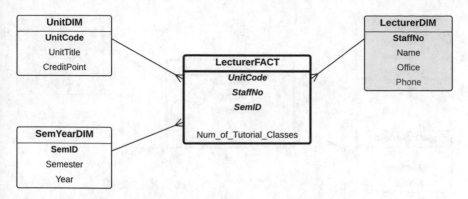

Fig. 11.19 Lecturer star schema

We can also have another star schema for the lecturer because many times, we would like to find out the number of classes (tutorials) for each lecturer, unit and semester. Therefore, the lecturer star schema has also three dimensions—Lecturer, Unit and Semester—and one fact measure: Number of Classes. The star schema for the lecturer is shown in Fig. 11.19. An example of a record in the Lecturer Fact is "FIT3003, 2/2018, Robert, 1", meaning that lecturer or Professor Robert took one tutorial for FIT3003 in semester 2/2018. Another example could be "FIT5137, 2/2018, Robert, 1", which indicates that in that semester, Professor Robert also took one tutorial for another unit, which is FIT5137.

This Tutorial case study is different from the previous two case studies. In this case study, there are "two pools" of staff: one pool for Tutors (i.e. Teaching Assistants who are normally postgraduate students) and another pool for Faculty Members (e.g. Lecturers, Professors). The two previous case studies have "one pool" of personnel (e.g. Pilots or Managers who are Faculty Members).

The two pools also exist in different transaction records in this Tutorial case study because for each tutorial, there is only one Tutor who might be a Tutor (Teaching Assistant) or a Faculty Member (Lecturer or Professor). Therefore, this Tutorial case study has the feature of "two pools (of personnel or objects) in different transaction records".

The combined star schema is shown in Fig. 11.20. The combined star schema also looks identical to the original star schemas. The Staff Dimension contains both Tutors and Faculty Members, and it only includes attributes which are identical to both pools, such as Staff No, Name, Office and Phone.

It will not make any sense if there are two separate dimensions, one for Tutor and the other for Lecturer (see Fig. 11.21), because there are two different pools, and consequently, the fact measure will not have any values. That's why only one dimension is needed, which is the Staff Dimension, which combines the two pools.

The Fact Table of the new star schema combines the records from the two original Fact Tables, without the need to recalculate the fact measures; they are simply union-ed. Hence, we only use the union command in SQL to merge the two

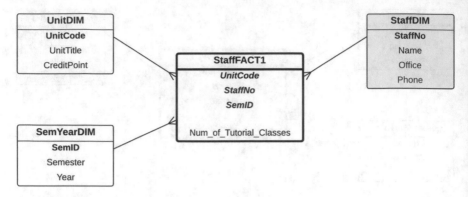

Fig. 11.20 A combined star schema—Tutor and Lecturer

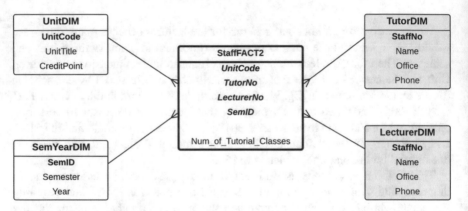

Fig. 11.21 A combined star schema with Tutor and Lecturer Dimensions—doesn't make sense

original Fact Tables, without any `group by`. The reason for this is because the original two Fact Tables are *mutually exclusive*; hence, the title of this subsection is Mutually Exclusive Star Schemas.

The combined star schema also does not have a Determinant Dimension. This means that once the combined star schema is created, there is no need to keep the original two star schemas any longer. Any queries can be directed to the combined star schema.

If the Flight Charter case study and the Bachelor/Master Final Projects case study have two other solutions, Pivoted Fact Table and Type Dimension, can the Mutually Exclusive Star Schema cases have these two other solutions? Figure 11.22 shows a second solution with a Pivoted Fact Table. However, this solution does not make any sense because for a Tutor, there is no Number of Tutorial Classes by Lecturer. Also, for a Lecturer, there is no Number of Tutorial Classes by Tutor. Hence, a Pivoted Fact Table solution is not applicable for Mutually Exclusive Star Schema cases.

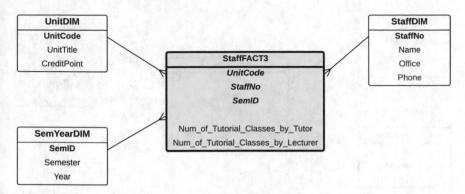

Fig. 11.22 A combined star schema with Pivoted Fact Table—this doesn't make sense

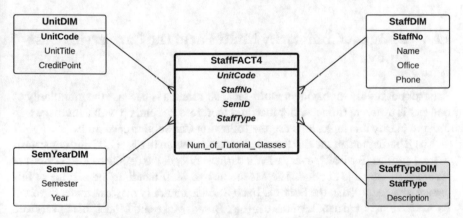

Fig. 11.23 A combined star schema with a Type Dimension—this doesn't make sense

Figure 11.23 shows a third option, that is, a Type Dimension, instead of a Pivoted Fact Table. This star schema does not make too much sense either because there is no combination of a Lecturer (from the Staff Dimension) and a Tutor Type (from the Type Dimension).

If we need to incorporate a Type, it will be better to have a Type inside the Staff Dimension, as shown in Fig. 11.24. This Type attribute is additional information for each staff record in the Staff Dimension.

As a conclusion, for Mutually Exclusive Star Schemas, the combined star schema has only one solution, which is very much similar to the original star schemas. The Pivoted Fact Table or the Type Dimension solutions are not applicable.

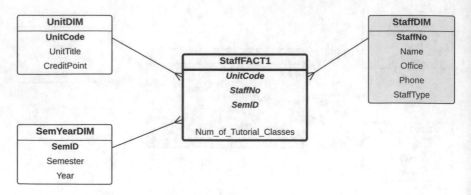

Fig. 11.24   A combined star schema with a Type attribute

11.4   Different Granularity Multi-Fact: The Car Service Case Study

Another possibility for having a multi-fact star schema is because the granularity of one fact is different from another fact, and therefore, we end up with a multi-fact. In order to clarify this topic, let's use the following Car Service case study.

The E/R diagram of the Car Service System is shown in Fig. 11.25 and consists of several entities. Basically, every time a car is serviced, the details of the service are recorded in the table Service. The Mechanic and Staff handling the service of this car are also recorded in the Service. Each Service naturally may use many different Parts. Obviously, the details of the car (e.g. Brand, Make and Model) being serviced are recorded, together with the Customer details.

A simple star schema could have two fact measures: Total Service Cost and Number of Services. We could choose a number of dimensions, such as Time (Month), Car Brand, Mechanic and Service Dimensions. If we would like to include the Part Dimension, a bridge table is needed between Service and Part Dimensions. This star schema is shown in Fig. 11.26.

Having a bridge table between Service and Part Dimensions is clearly understood because each Part may be used in many different Services, and vice versa. The cardinality relationship is an *m-m* between Service and Part in the E/R diagram.

However, we could actually count the Number of Services from the Part granularity if the Part Dimension is a Determinant Dimension. In other words, it is possible to skip the Service and use the Part directly. This means that we would count the Number of Services for any Parts (and possibly together with other dimensions, such as Time, Car Brand and Mechanic Dimensions). The use of Part in the query is critical because each part may be used by different services. By not using Part in the query, we would double the count of Number of Services. Hence, Part Dimension is a Determinant Dimension. The star schema is then shown in Fig. 11.27.

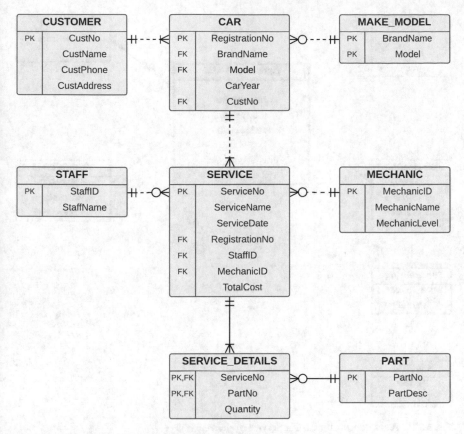

Fig. 11.25 An E/R diagram for the Car Service Systems

A couple of things can be noted from this star schema. Firstly, the granularity of the two star schemas is different. The star schema with the Service Dimension (Fig. 11.26) is at a Service granularity. Therefore, the counting for Number of Services is obvious, where each service contributes one count to the Number of Services.

On the other hand, the star schema with the Part Dimension (Fig. 11.27) is at a Part granularity; it is more detailed and has a higher granularity than that of the Service Dimension. So we need two star schemas as shown above, one for the Service level and the other for the Part level. Both star schemas are valid; that is, they show the correct Number of Services. However, for the star schema at the Part level of granularity, because each part may contribute to many services, the Part Dimension must be a Determinant Dimension. If not, the Number of Services would be double counted in many cases.

Secondly, for the star schema with the Part level of granularity, it does not make sense to have Total Service Cost as the fact measure because each part may contribute to different services and Total Service Cost is associated with the Service,

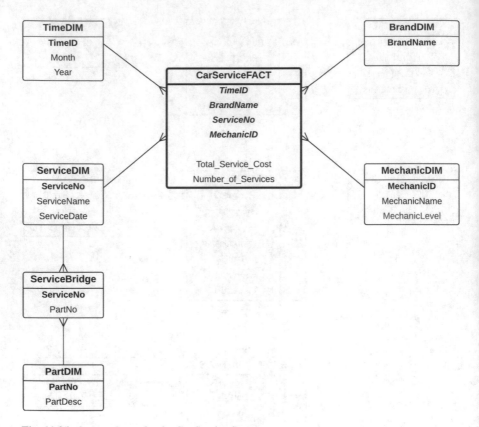

Fig. 11.26 A star schema for the Car Service Systems

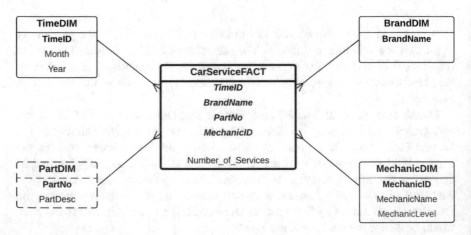

Fig. 11.27 A different level of granularity of the Car Service star schema

not with the Part. Consequently, it may not be possible to determine the cost of each part in each service, at least through the given E/R diagram, because the cost of each part is not stated, only the Total Cost of the Service. If the E/R diagram records the cost of each part in each service, it would then make sense to have Total Service Cost fact measure in the star schema with the Part level of granularity.

Nevertheless, this case study shows that the multi-fact star schema is created due to the different granularities of the star schema.

11.5   Summary

A star schema with multi-fact or multiple star schemas are needed because each Fact focuses on a subject or on a different granularity. A subject and a granularity are often determined by the dimensions. Therefore, star schemas with different dimensions can be said to have different subjects. When multiple star schemas have different subjects (or different dimensions), they cannot be merged into one star schema.

However, in the case where the dimensions are identical or similar in nature, a question is raised as to whether or not to combine these star schemas. There are three general cases that fall into this category:

1. *One pool in the same transaction records*. There are three solutions for the combined star schemas: (i) the new star schema has a Determinant Dimension; (ii) the new star schema also has a Pivoted Fact Table; (iii) the new star schema replaces the Pivoted Fact Table with a Type Dimension. The fact measures need to be recalculated.
2. *One pool in different transaction records*. The three solutions are similar to the "one pool in the same transaction records", but the combined star schema does not have a Determinant Dimension. The fact measures need to be recalculated too.
3. *Two pools in different transaction records*. There is only one solution, which is the first solution like the "one pool in different transaction records". The fact measures do not need to be recalculated.

11.6   Exercises

11.1 Swappers is a new online house swap service where users can advertise their houses which are available to swap with others for holiday purposes. To participate, users need to register as a member, and currently, the membership is free.

The website provides feedback from existing users. When users leave feedback, they can divide this into two categories. The first is their satisfaction level with the destination accommodation, and the second is their level of satisfaction with how their own houses have been taken care of during the Swappers program.

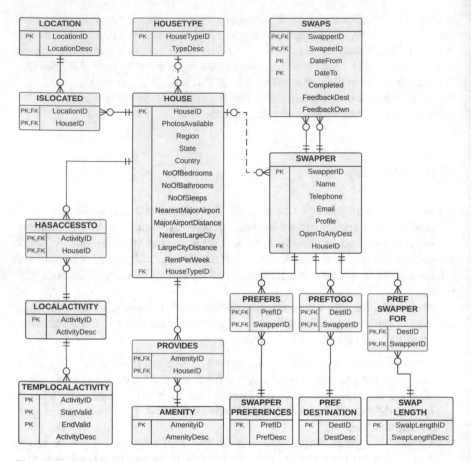

Fig. 11.28 Holiday house swappers E/R diagram

Currently, the operational database system consists of several tables. The E/R diagram is shown in Fig. 11.28. A holiday house is located in a certain location, has access to local activities and has amenities. A house is owned by a person who has his/her own preferences to go to certain destinations and for a period of time. Feedback is also recorded in the swaps table.

In the data warehouse, suppose we would like to measure five things: (*i*) total completed swaps, (*ii*) total failed swaps, (*iii*) total satisfaction rates, (*iv*) total rental savings and (*v*) total number of swappers. Information on the total completed swaps and the total failed swaps would be useful for management to understand whether the swapping program is successful or not. Total satisfaction rates are also used to measure the success of swaps. Total rental savings is used to understand the impact of financial saving provided by this program. Total number of swappers is used to analyse the preferences of swappers based on their preferences.

Table 11.5 PhD students

No	PhD student	Research area	Starting year	Supervisors
1	Adam Smith	AI	2017	Wayne Ong (main) Jim Carr (co)
2	Bernie Tan	Networking	2018	Jeremy Theo (main)
3	Grace Park	Security	2018	Kazumi Tanaka (main) Xu Yang (co) Yijie Lee (co)
...	...	...	...	...

Since the operational database is big, it has many tables and attributes; potentially, there could be more than ten dimensions. Some potential dimensions are:

1. Season or month
2. Length of stay
3. Destination
4. House type
5. Distance from city or from airport
6. Local activities
7. Amenities
8. Natural location
9. Preferred length of stay
10. Preferred destination
11. Preferred season
12. And others

These are all valid dimensions. Some dimensions can also be very complex, involving bridge tables. However, for simplicity, we choose only a few basic dimensions: (*i*) season, (*ii*) length of swap, (*iii*) destination, (*iv*) suburb (or state) of the house and (*v*) member profile.

Tasks: Design a star schema with multi-fact using the above-stated fact measures and dimensions. Also, identify the "subject" of each fact.

11.2 This case study is about PhD student supervision. Each PhD student may have one or more supervisors, that is a main supervisor and one or more co-supervisors. Table 11.5 shows a list of PhD students in the Faculty of Information Technology. Note that the table is not necessarily in a normalised form. The research area shown in the table is the main research area of the thesis. The supervisor is a faculty member (e.g. Lecturer, Professor) who is eligible to supervise a PhD student. Any eligible supervisors may supervise PhD students in various capacities as either a main supervisor or a co-supervisor.

A star schema has been created to analyse the main supervisors. This includes the number of students which each supervisor supervises as the main supervisor. This star schema for main supervisors is particularly important due to the important role a main supervisor plays in the supervision of PhD students. For simplicity, the

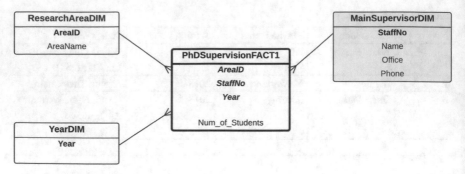

Fig. 11.29 PhD student supervision (main supervisor)

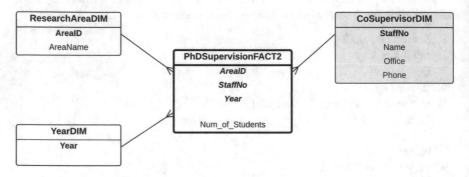

Fig. 11.30 PhD student supervision (co-supervisor)

dimensions included in the star schema are Research Area, Year and Supervisor. The star schema is shown in Fig. 11.29.

The main supervisor star schema does not include supervisors in the co-supervision role. Hence, another star schema is created to cater for co-supervisors. The co-supervisor star schema is shown in Fig. 11.30. The dimensions and fact of the co-supervisor star schema are identical to those of the main supervisor star schema. The Co-supervisor Dimension draws from the same pool as the Main Supervisor Dimension, that is, all eligible faculty members (e.g. Lecturers, Professors) who supervise PhD students in some capacity. The Fact Table would, however, have different contents; the Number of Students in the Main Supervisor star schema can be drilled down by the Main Supervisor, Research Area and Year Dimensions, whereas the same fact measure attribute in the co-supervisor star schema is related to co-supervisors.

It would be more beneficial if the two star schemas are combined into one star schema, if at all possible, so that the new star schema covers both types of supervisors. Using the new star schema, we can directly query the number of students that a particular Professor has, whether as a main supervisor or as a co-supervisor.

Questions

1. Design a new star schema that combines these two star schemas. Discuss three different options for designing this combined star schema. Also discuss whether the fact measure in the new star schema needs to be recalculated from the original two star schemas or whether it is sufficient to only "union" or "combine" the original fact measure values.
2. Discuss the differences between these three options.
3. Discuss whether using the new star schema is sufficient and the original two star schemas can now be discarded.

11.3 Faculty of Engineering has a different practice from other faculties in the university in relation to how they conduct their tutorials. It is quite common that a tutorial class in the Faculty of Engineering is very large (90–120 students per class) as opposed to small tutorial classes (e.g. 15–20 students) in other faculties. Therefore, a tutorial class in the Faculty of Engineering is usually run by several Teaching Assistants (TA), also known as Tutors. Tutors are usually postgraduate students (e.g. PhD or Master's students) and are part-time or sessional. They are paid by the hour for the number of hours they conduct the class.

In a typical tutorial in the Faculty of Engineering, there are three to four tutors for each tutorial. One of the tutors is called the "Lead Tutor", whereas the rest are "Tutors". Two star schemas, one for Lead Tutors and the other for Tutors, are created as shown in Fig. 11.31. For simplicity, there are only three dimensions: Unit Dimension, which lists all available units; Semester Dimension; and Lead Tutor/Tutor Dimension. The fact measure is Number of Tutorials.

Tasks

1. Design a new star schema that combines these two star schemas. Also, discuss three possible solutions.
2. Discuss if the new fact measure could simply "aggregate" the values from the previous star schemas.
3. Discuss if the original two star schemas can be discarded once the new star schema that combines the two has been created.

11.4 A cleaning company sends their cleaners to client houses or offices. They have domestic clients (e.g. domestic households), as well as company offices. This cleaning company employs two types of cleaners: Full-Time cleaners and Part-Time cleaners. It is a common practice in this cleaning company for them to send cleaners in pairs, both of whom could be Full-Time cleaners, Part-Time cleaners or a combination. Every time the company receives an order from a client, it schedules two cleaners to do the job.

Two star schemas have been created (see Fig. 11.32). Both star schemas have three dimensions: Job Type Dimension (e.g. domestic or office), Time Dimension (e.g. month) and Full-Time/Part-Time Cleaner Dimension. For Full-Time cleaners (or Full-Time employees), they have their annual salaries, whereas Part-Timers

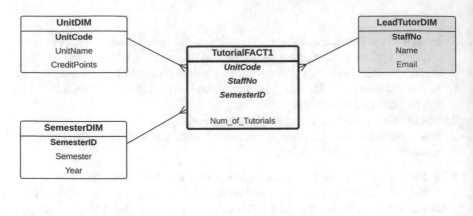

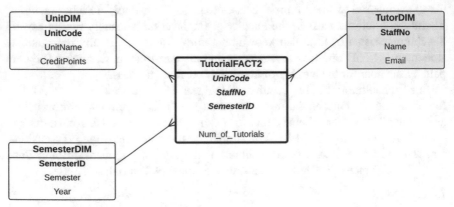

Fig. 11.31   Lead Tutor and Tutor star schemas

have an hourly rate. The face measures are Number of Jobs, Total Hours and Total Income.

Tasks: Combine these two star schemas. Show three possible options for the newly combined star schema.

Discussions: This Cleaner case study has two pools of personnel: Full-Time cleaners and Part-Time cleaners, possibly with many different attributes between these two types (e.g. Full-time cleaners have annual salary, whereas Part-Time cleaners do not have annual salary but have an hourly rate). But the work itself does not differentiate between Full-Time and Part-Time cleaners.

In contrast, in the previous two exercises (e.g. PhD Supervision case study and Faculty of Engineering Tutorial case study), there is only one pool of personnel. In the PhD Supervision case study, the pool of personnel is the faculty member which contains eligible supervisors (e.g. Lecturers, Professors, etc.). But in the work itself, there are two different roles: main supervisors and co-supervisors.

The Faculty of Engineering case study is quite similar to the PhD Supervision case study, which only has one pool of personnel, namely Teaching Assistants (TA)

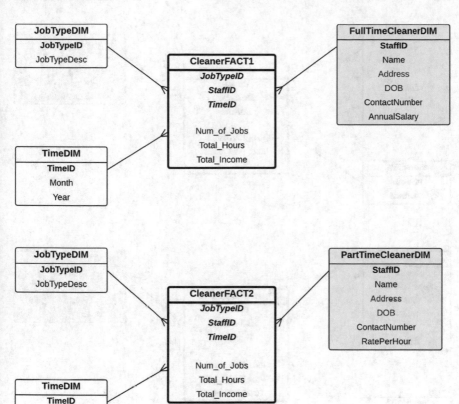

Fig. 11.32 Cleaner star schemas

or Tutors who are usually postgraduate students, but in the job, there are two roles: Lead Tutor and Tutor.

In terms of the combined star schemas, would there be any differences in relation to the approach and result between the Cleaner case study and the two previous case studies (e.g. PhD Supervision case study and Tutorial case study)?

11.5 A taxi company employs two types of drivers: Full-Time Drivers and Part-Time Drivers. Full-Time Drivers drive on a certain number of days, following certain shifts adopted by the company. Part-Time Drivers are more flexible and drive on a fewer number of days. The driving schedule is usually fixed 1 week ahead. If there is a deviation to the schedule, the affected drivers need to organise a swap among themselves.

The company has developed two star schemas: one star schema for Full-Time Drivers and another star schema for Part-Time Drivers. The star schemas are shown in Fig. 11.33. Both star schemas have three dimensions: Car Dimension,

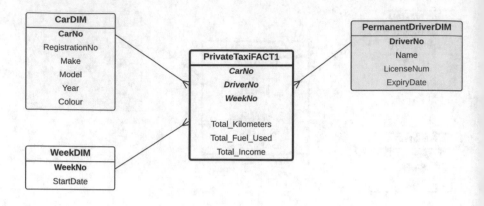

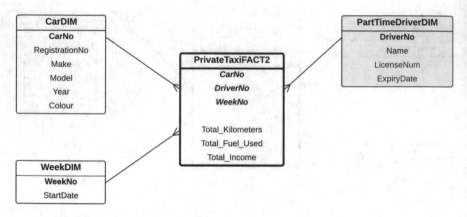

Fig. 11.33   Taxi driver star schemas

Week Dimension and Driver Dimension. There are three fact measures in both star schemas: Total Kilometres, Total Fuel Used and Total Income.

Tasks and Discussions

1. Design a new star schema that incorporates both Full-Time and Part-Time drivers. If there is more than one possible answer to combine these two star schemas, discuss each of them.
2. Are the fact measures in the new star schemas merely a union of the fact measures from the two previous star schemas or should the values of the new fact measures be recalculated?
3. Why doesn't it make sense for the new star schema to use a Determinant Dimension, or a Type Dimension and a Pivoted Fact?
4. In this case study, there are two pools of personnel: Full-Time and Part-Time drivers. The previous exercise on the Cleaner case study has also two pools of personnel: Full-Time and Part-Time cleaners. However, their combined star schemas are different. What are the differences and why are they different?

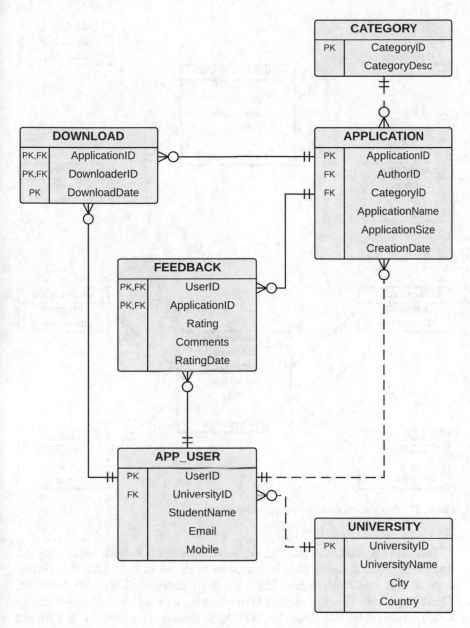

Fig. 11.34 E/R diagram of Apps Downloads

11.6 In Chap. 3, there is a case study on Apps Creation and Download. The E/R diagram to keep track of Apps downloads is shown in Fig. 11.34. Basically, in this case study, the App User, who is a member of a university, may create and download apps. This E/R diagram maintains this.

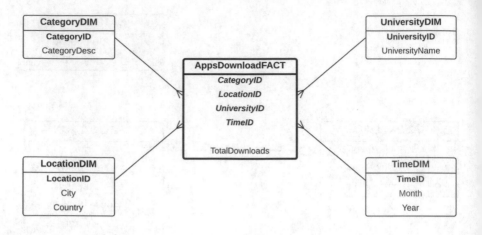

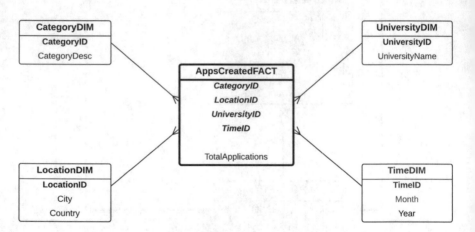

Fig. 11.35 Two star schemas for the Apps case study

Two star schemas, one focusing on Total Downloads and the other on Total Applications (Total Apps), have been created as shown in Fig. 11.35. The dimensions of both star schemas are identical, namely Category, University, Location, Time Dimensions. Hence, using these star schemas, we are able to query something like "Retrieve Total Downloads or Total Apps for a certain category in a certain month" (using the Category and Time Dimensions).

Both Fact Tables are created using the following SQL commands:

```
create table AppsDownloadFact as
select
  DownloadMonth, LocationID,
  CategoryID, UniversityID,
  count(*) as TotalDownloads
```

```
from TempFact
group by
   DownloadMonth, LocationID,
   CategoryID, UniversityID;

create table TotalAppsFact as
select
   CreationMonth, LocationID,
   CategoryID, UniversityID,
   count(distinct ApplicationID) as TotalApps
from TempFact
group by
   CreationMonth, LocationID,
   CategoryID, UniversityID;
```

Technically, the difference is rather small. In the first Fact Table, it uses the Download Month attribute, whereas in the second Fact Table, it uses the Creation Month attribute, both from the respective Temp Fact Tables. Another difference is that the first Fact Table uses the count(*) function, whereas the second uses the count(distinct) function.

Since both star schemas are on the same topic or subject and the dimensions are also identical, it becomes possible to merge both star schemas into one with two fact measures. The new star schema is shown in Fig. 11.36.

Write the SQL command to merge the two Fact Tables into one new Fact Table. Note that in some months, no apps might be created, but there are downloads. Therefore, for that particular month, Total Apps must be shown as zero. The opposite is also applicable, as in some months, there were no downloads but some apps were created. Both original Fact Tables do not store zeroes in the fact measures, but in the merged Fact Table, and if any of the aforementioned cases occurred, zeroes must be recorded in one of the fact measures.

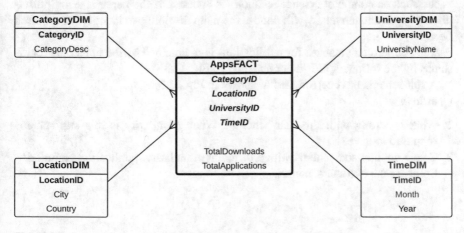

Fig. 11.36 A merged star schema for the Apps case study

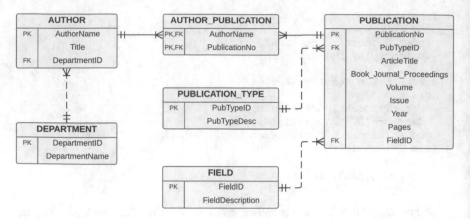

Fig. 11.37 Publication E/R diagram

11.7 It is common that professors, lecturers and all academics working in a university conduct research and publish their research in journals, conference proceedings or edited books. This practice is commonly known as "publish or perish", which implies that they must publish in order to keep their jobs.

The university keeps track of all their academics' (e.g. professors, researchers, lecturers) publications. When a publication is entered into the system, the system will generate a publication number as an identifier. Other details such as author names, article titles, journal volume, issue, year and page numbers are also recorded. It is common that an article (a paper) has several authors. There are several publication types. Each paper is categorised into one publication type, whether it is a journal paper or a conference paper or a book chapter.

For classification purposes, the university allocates a research field for each paper, such as computer science, economics and law. If the paper is a multidisciplinary paper, the university will choose the main discipline to identify the research field.

The operational database for publications is maintained by the university and is shown by the following E/R diagram (refer to Fig. 11.37).

A star schema has been created as shown in Fig. 11.38.

Questions

1. What is wrong with this star schema? What granularity is this star schema supposed to have?
2. Create another star schema with a different granularity: publication granularity. Compare the two star schemas.

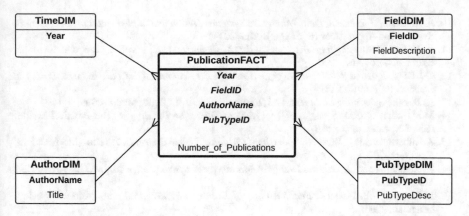

Fig. 11.38 Publication star schema

11.7 Further Readings

Multi-fact star schema is one of the design aspects in data warehouse design, emphasising the concept of *Subject* of data warehousing. Other concepts of data warehousing, such as Non-volatile, Time Variant and Integrated, can be found in various data warehousing books, such as [1–9] and [10].

More complete descriptions on SQL functions (e.g. nvl) and outer join can be found in SQL books, such as [11–13] and [14].

Combining Fact Tables uses the union-compatible concept in the relational theory. More complete information about relational theory can be found in the following books: [15–20] and [21].

References

1. C. Adamson, *Star Schema The Complete Reference* (McGraw-Hill Osborne Media, 2010)
2. R. Laberge, *The Data Warehouse Mentor: Practical Data Warehouse and Business Intelligence Insights* (McGraw-Hill, New York, 2011)
3. M. Golfarelli, S. Rizzi, *Data Warehouse Design: Modern Principles and Methodologies* (McGraw-Hill, New York, 2009)
4. C. Adamson, *Mastering Data Warehouse Aggregates: Solutions for Star Schema Performance* (Wiley, London, 2012)
5. R. Kimball, M. Ross, *The Data Warehouse Toolkit: The Definitive Guide to Dimensional Modeling* (Wiley, London, 2013)
6. R. Kimball, M. Ross, W. Thornthwaite, J. Mundy, B. Becker, *The Data Warehouse Lifecycle Toolkit* (Wiley, London, 2011)
7. W.H. Inmon, *Building the Data Warehouse*. ITPro Collection (Wiley, London, 2005)
8. M. Jarke, *Fundamentals of Data Warehouses*, 2nd edn. (Springer, Berlin, 2003)
9. E. Malinowski, E. Zimányi, *Advanced Data Warehouse Design: From Conventional to Spatial and Temporal Applications*. Data-Centric Systems and Applications (Springer, Berlin, 2008)

10. A. Vaisman, E. Zimányi, *Data Warehouse Systems: Design and Implementation*. Data-Centric Systems and Applications (Springer, Berlin, 2014)
11. J. Melton, *Understanding the New SQL: A Complete Guide*, vol. I, 2nd edn. (Morgan Kaufmann, Los Altos, 2000)
12. C.J. Date, *SQL and Relational Theory - How to Write Accurate SQL Code*, 2nd edn. Theory in Practice. (O'Reilly, 2012)
13. A. Beaulieu, *Learning SQL: Master SQL Fundamentals* (O'Reilly Media, 2009)
14. M.J. Donahoo, G.D. Speegle, *SQL: Practical Guide for Developers*. The Practical Guides (Elsevier, Amsterdam, 2010)
15. A. Silberschatz, H.F. Korth, S. Sudarshan, *Database System Concepts*, 7th edn. (McGraw-Hill, New York, 2020)
16. R. Ramakrishnan, J. Gehrke, *Database Management Systems*, 3rd edn. (McGraw-Hill, New York, 2003)
17. R. Elmasri, S.B. Navathe, *Fundamentals of Database Systems*, 3rd edn. (Addison-Wesley-Longman, 2000)
18. C.J. Date, *An introduction to Database Systems*, 7th edn. (Addison-Wesley-Longman, 2000)
19. E.F. Codd, *The Relational Model for Database Management, Version 2* (Addison-Wesley, Reading, 1990)
20. H. Darwen, *An Introduction to Relational Database Theory* (Ventus Publishing, Erie, 2009)
21. C.J. Date, *The New Relational Database Dictionary: Terms, Concepts, and Examples* (O'Reilly Media, 2015)

Chapter 12
Slicing a Fact

Slicing a fact is often thought of as an easy topic and therefore overlooked. This chapter discusses comprehensively how to slice facts. Slicing a fact is to slice the Fact Table into two or more Fact Tables which has the effect of creating multi-fact star schemas. So, the topic of slicing a fact is almost the opposite of a multi-fact, which was discussed in the previous chapter. In Chap. 11, one of the issues was whether or not to combine a multi-fact into one fact. In contrast, this chapter focuses on how to slice a Fact Table into multiple Fact Tables, creating a multi-fact star schema.

In principle, there are two slicing methods: (*i*) Vertical Slice and (*ii*) Horizontal Slice. However, this chapter will also discuss slicing the fact where the star schema features a Determinant Dimension. The details, together with some examples, will be discussed in the next sections.

12.1 Vertical Slice

A Vertical Slice is known as *Vertical Data Partitioning* in Distributed Databases, whereby a table is vertically partitioned by dividing the attributes of a table into multiple tables. Since the Fact Table in a star schema is a table, a vertical slice is to partition the Fact Table vertically by assigning different fact measures into the new Fact Tables. Consequently, one Fact Table is sliced or partitioned into multiple Fact Tables, creating a multi-fact star schema.

Let's use the Private Taxi Company case study which was used in Chap. 11. This case study presents a star schema of a Private Taxi Company which has a number of taxis for private charter. The star schema has three fact measures, Total Kilometres, Total Fuel Used and Total Income, and three dimensions, Car Dimension, Week Dimension and Driver Dimension. The star schema is shown in Fig. 12.1.

© The Author(s), under exclusive license to Springer Nature Switzerland AG 2021
D. Taniar, W. Rahayu, *Data Warehousing and Analytics*, Data-Centric Systems
and Applications, https://doi.org/10.1007/978-3-030-81979-8_12

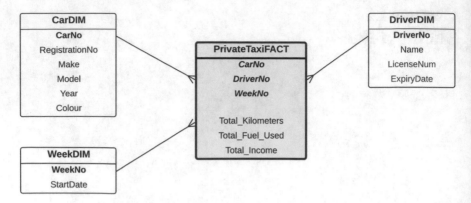

Fig. 12.1 A Private Taxi Company star schema

Remember that a data warehouse is subject-oriented, meaning that each star schema has one subject only. Chapter 11 has already discussed the vagueness of the concept of subject-oriented, as many times, it is difficult to state whether a star schema has only one subject or topic or potentially has more than one subject.

Assuming that in this case study where the fact has three different kinds of fact measures, one about distance travelled, one about fuel consumption and the other about income, if we want to split the Fact Table into three Fact Tables, we need to vertically slice the Fact Table. In this example, the three fact measures will be separated into three Fact Tables. As a result, three star schemas are created, each star schema having one fact measure. The three star schemas are shown in Fig. 12.2. Note that the dimensions are unchanged and so are the dimension identifiers in the Fact Table.

The SQL commands to vertically partition a Fact Table consist of multiple `create table` statements, as shown as follows. It is as simple as choosing the fact measures for the new Fact Table.

```
create table PrivateTaxiFact1 as
select CarNo, DriverNo, WeekNo, Total_Kilometers
from PrivateTaxiFact;

create table PrivateTaxiFact2 as
select CarNo, DriverNo, WeekNo, Total_Fuel_Used
from PrivateTaxiFact;

create table PrivateTaxiFact3 as
select CarNo, DriverNo, WeekNo, Total_Income
from PrivateTaxiFact;
```

Note that the dimension tables and records in the dimension are not changed. The only difference is that the Fact Table is now partitioned into multiple Fact Tables.

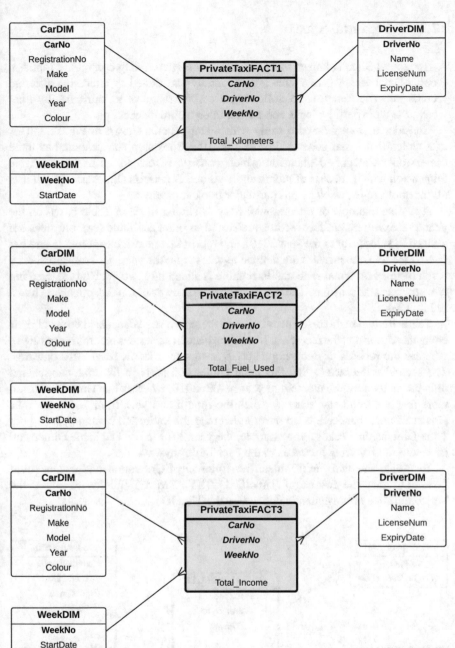

Fig. 12.2 Slicing the fact into three star schemas

12.2   Horizontal Slice

A Horizontal Slice is known as *Horizontal Data Partitioning* commonly found in Parallel Databases, where a table is horizontally partitioned by allocating different records into a different table, following a certain range of an attribute. In other words, the table partitioning is actually an inter-record distribution.

Consider a Bookshop case study with a star schema shown in Fig. 12.3. This is a typical sales star schema. In this case, the Bookshop star schema has three dimensions: Category Dimension which consists of categories of books, Time Dimension which consists of month and year and Bookshop Dimension which lists all the bookshops owned by this particular book company.

A typical example of a Horizontal Slice is slicing the Fact Table based on the year. For example, old fact records are stored in one Fact Table (e.g. old sales are defined as sales prior to the year 2010), and recent fact records are stored in another Fact Table (e.g. recent sales are defined as sales from the year 2010 onwards). As a result of this horizontal slice, the Fact Table is sliced into two Fact Tables, creating two separate star schemas. The two star schemas, however, look identical, as shown in Fig. 12.4.

There are two main differences between these two star schemas in terms of their contents. One is that the records in Bookshop Fact 2a have sales records before 2010, whereas the records in Bookshop Fact 2b have sales records from 2010 onwards. The second difference is that the Time Dimension table in the first star schema only needs to store the month and year before 2010, whereas the Time Dimension table in the second star schema stores the month and year from 2010 onwards. However, note that there is no harm in keeping the entire month and year in both Time Dimension, because, for example, the year 2011 in the first Time Dimension table will not have any instances in the Fact Table anyway.

To create the two Fact Tables, the following SQL commands can be used (assuming that the format of TimeID is "YYYYMM"). In this example, the horizontal slice is conveniently defined by the TimeID.

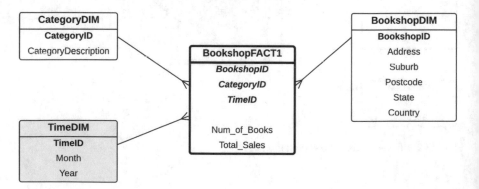

Fig. 12.3 A Bookshop star schema

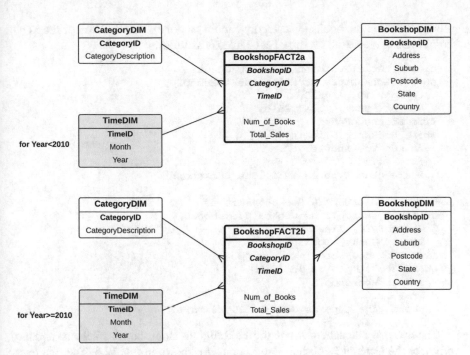

Fig. 12.4 Two bookshop star schemas sliced by year

```
create table BookshopFact2a as
select *
from BookshopFact1
where TimeID < 201001;

create table BookshopFact2b as
select *
from BookshopFact1
where TimeID >= 201001;
```

Now let's assume that the partitioning is not based on TimeID but on the Bookshop Dimension. For example, this company has two main businesses, the first one is the bookshop entity specialising in selling religious books and accessories, and the second one is a bookshop entity selling general and popular books. Let's use the star schema in Fig. 12.3 as the original star schema. The Horizontal Slice of the fact is hence based on the BookshopID attribute, not the TimeID attribute. Hence, we need to know which bookshops fall into the first category (i.e. religious bookshops) and which ones fall into the second category (i.e. general bookshops). Since there is no Type attribute or Description attribute in the Bookshop Dimension, checking must be done against information from the operational database (note that if the bookshop category is not kept in the operational database, horizontal slicing based on the bookshop category cannot be done). The skeleton of the SQL commands to create the two new Fact Tables is as follows:

```
create table ReligiousBookshopFact as
select BookshopID as ReligiousBookshopID,
  CategoryID, TimeID,
  Num_of_Books, Total_Sales
from BookshopFact1
where BookshopID is in (
  select BookshopID
  from ...
  where <the type is Religious Bookshop>);

create table GeneralBookshopFact as
select BookshopID as GeneralBookshopID,
  CategoryID, TimeID,
  Num_of_Books, Total_Sales
from BookshopFact1
where BookshopID is in (
  select BookshopID
  from ...
  where <the type is General Bookshop>);
```

The two new star schemas, one for the Religious Bookshop star schema and the other for the General Bookshop star schema, are shown in Fig. 12.5. The first Fact

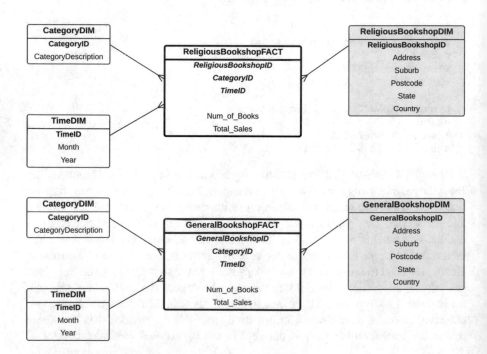

Fig. 12.5 Two bookshop star schemas sliced by bookshop type

Table contains book sales from the religious bookshops, whereas the second Fact Table contains book sales from the general bookshops.

For the Bookshop Dimension itself, we can split it into two new dimensions (e.g. Religious Bookshop Dimension and General Bookshop Dimension) using the following SQL commands:

```
create table ReligiousBookshopDim as
select BookshopID as ReligiousBookshopID,
   Address, Suburb,
   Postcode, State, Country
from BookshopDim
where BookshopID is in (
   select BookshopID
   from ...
   where <the type is Religious Bookshop>);

create table GeneralBookshopDim as
select BookshopID as GeneralBookshopID,
   Address, Suburb,
   Postcode, State, Country
from BookshopDim
where BookshopID is in (
   select BookshopID
   from ...
   where <the type is General Bookshop>);
```

If the original Bookshop Dimension has a Type Category (e.g. BookshopType attribute) as shown in Fig. 12.6, then creating the two new Fact Tables and the two new Bookshop Dimension tables becomes more straightforward, without the need to check the bookshop category in the operational database.

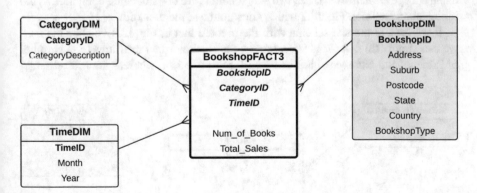

Fig. 12.6 The original bookshop star schema with Bookshop Type attribute in Bookshop Dimension

12.3   Vertical or Horizontal Slice?

A Vertical Slice and a Horizontal Slice as described in the previous sections seem to be quite straightforward and easy to digest and implement. However, in some cases, this might not be the case. This section will showcase an example of these.

Using the Final Year Bachelor and Master Project case study presented in Chap. 11, the star schema is shown in Fig. 12.7. This case study is about the final capstone projects that students must take when finishing their respective degrees, either at a Bachelor or a Master's level. Each project is supervised by a faculty member (e.g. lecturer or professor). The faculty member who supervises a project is called the Manager of the project. Some faculty members become managers of both Bachelor and Master's projects, whereas others may only become managers of either Bachelor or Master's projects, or neither. The star schema shown in Fig. 12.7 basically includes two fact measures, Number of Projects and Number of Students, with three dimensions, Research Area, Year and Manager. Using this star schema, we can easily retrieve the number of projects that each faculty member supervises or manages, including other details such as research areas and year.

The Multi-Fact chapter also shows two other possible star schemas to capture this Bachelor-Master's Final Year Project case study. The first one, as shown in Fig. 12.8, includes the breakdown of the level (e.g. Bachelor or Master's) in the Fact; hence, the fact is a pivoted fact. This star schema has more detailed information compared to the previous star schema that does not include a breakdown of the level.

The second one, shown in Fig. 12.9, includes a new dimension called the Project Level Dimension, which indicates whether it is a Bachelor or a Master's project. This star schema has the same level of detail as the pivoted fact star schema because the breakdown of the level is given in the star schema, either using a pivoted fact or using a Type Dimension. These two star schemas are more detailed compared to the initial star schema in Fig. 12.7 which does not have the breakdown.

If we look at the star schema with the pivoted fact in Fig. 12.8, to slice the fact, we can simply do a *Vertical Slice*, that is, the Bachelor fact measures (e.g. Number of Projects and Number of Students) will go to one fact, whereas the Master's fact

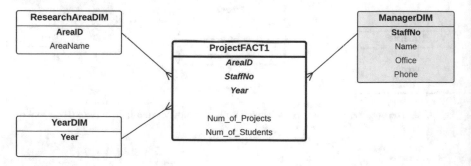

Fig. 12.7 Bachelor/Master Project

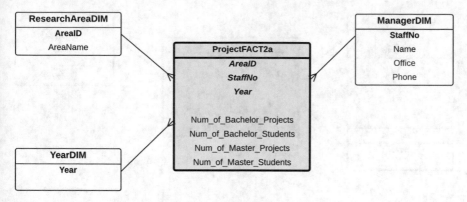

Fig. 12.8 Bachelor/Master's Project with Pivoted Fact

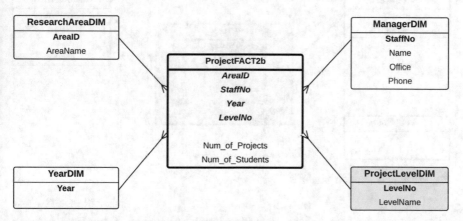

Fig. 12.9 Bachelor/Master's Project with Project Level Dimension

measures (e.g. Number of Projects and Number of Students) will go to another fact. So, this is a simple Vertical Slice of the fact.

However, if we look at the star schema with a Type Dimension as shown in Fig. 12.9, to slice the fact, we can simply do a *Horizontal Slice*, where the fact records belonging to the Bachelor Level will go to one fact, and the fact records belonging to the Master's Level will go to another fact. So, this is a simple Horizontal Slice of the fact.

After slicing the original fact, the two new star schemas are shown in Fig. 12.10. The fact measures are unchanged and so are the three dimensions. The Fact Table will have different records, depending on whether it is Bachelor or Master's Project. Note that when we do a Horizontal Slice of the star schema in Fig. 12.9, in the sliced version in Fig. 12.10, there is no need to keep the Project Level Dimension, because the project level is already reflected in the respective star schemas.

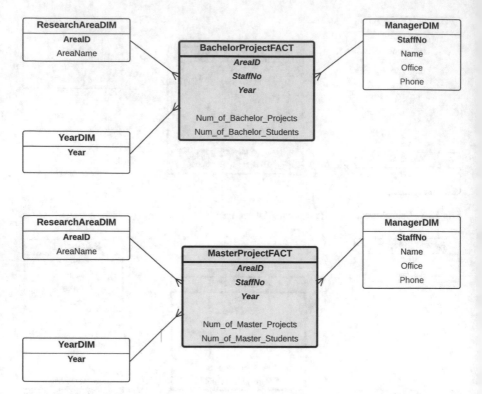

Fig. 12.10 Split into two facts

So, slicing the fact in this case study is done either using a Vertical Slice or a Horizontal Slice, depending on whether the fact is a pivoted fact or the star schema has a Project Level Dimension. However, if the star schema that is given is only the original star schema as shown in Fig. 12.7, there is no way to slice the fact, because there is no information as to whether the project is a Bachelor or a Master's project. Slicing the fact based on project level, in this case study, is only possible if the star schema has information about the project level, which is actually provided by the other star schemas.

As a conclusion, if there is no existing star schema that includes more detailed information about the slicing attribute, it is not possible to slice the fact. We cannot slice the fact easily if the fact is too general. In this case study, the intended slice is not based on Staff, but based on the Project, which is not shown in the dimension nor in the fact. Hence, slicing the fact of the star schema shown in Fig. 12.7 based on the project is not possible.

12.4   Determinant Dimension

How do we slice a star schema based on a Determinant Dimension? Let's use the
Flight Charter case study as an example. The original star schema is shown in
Fig. 12.11. The star schema has three fact measures, namely Total Flying Hours,
Total Fuel Used and Total Revenue. There are three dimensions, which are Aircraft,
Time and Pilot Dimensions. In this case, the Pilot Dimension is a *Determinant
Dimension*. This means that to query the star schema, the Pilot Dimension must
be used either in the filtering phase of the query or in the results of the query. An
example of a Pilot Dimension used in the filtering phase of the query is to retrieve
all the three fact measure values for a certain pilot in a certain month. In this case,
the pilot is used to filter the query results. An example of Pilot Dimension used in
the query results is to show the Total Flying Hours of all Pilots. In this case, the
query results will include Pilot information (at least PilotID).

Why is the Pilot Dimension a Determinant Dimension in this case study? As
discussed in the previous chapter, the Pilot Dimension contains the records of
qualified pilots. These pilots on certain flights may act as a Co-Pilot. Because there
is only one pool of qualified pilots, there is no need to separate them into two pools,
pilots and Co-Pilots. Both are simply put into the Pilot Dimension. Hence, to query
this star schema (refer to Fig. 12.11), the information about the pilot must be used,
especially when retrieving Total Flying Hours, for example. One particular "pilot"
with certain Total Flying Hours is actually composed of Total Flying Hours when
this pilot was a pilot, as well as when this pilot was a Co-Pilot. If the Pilot Dimension
was not a Determinant Dimension, the fact measure would be wrong, because for a
certain month, for example, the Total Flying Hours (without any pilot information

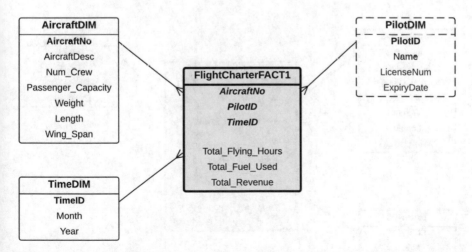

Fig. 12.11 Flight Charter with a Determinant Dimension

in the query) will be double counting. This is why the Pilot Dimension has not been used in the query; hence, the Pilot Dimension is a Determinant Dimension.

This Flight Charter star schema (Fig. 12.11) is very similar to the Bachelor/Master's Project star schema (Fig. 12.7) discussed in the previous section. The Bachelor/Master's Project star schema also consists of three dimensions, Research Area, Year and Manager, and the Flight Charter star schema also contains three dimensions, Aircraft, Time (or Year) and Pilot (which is personnel, like the Manager Dimension in the Bachelor/Master's Project star schema). The main difference is that the Pilot Dimension is a Determinant Dimension, whereas the Manager Dimension is not. The reason why the Manager Dimension is not a Determinant Dimension is that we can query the Bachelor/Master's Project star schema without using the Manager Dimension either in the filtering phase or in the query results phase. For example, to retrieve Number of Projects run last year, there won't be any double counting because there is only one Manager per project. This is different from the Flight Charter star schema. Each flight may have one or two pilots (the second pilot is a Co-Pilot). Hence, when we retrieve Total Flying Hours last year, for example, there will be double counting, especially by those flights that have two pilots.

Similar to the Bachelor/Master's Project star schema (Fig. 12.7), it is not possible to slice the fact of the Flight Charter star schema (Fig. 12.11) without any breakdown of the fact measure. We need star schemas with a lower level of detail. In the Bachelor/Master's Project star schema, there are two additional star schemas, one with the Pivoted Fact (Fig. 12.8) and the other with the Type Dimension, which is the Project Level Dimension (Fig. 12.9). For the Flight Charter case study, there are also two additional star schemas, one with the Pivoted Fact (Fig. 12.12) and the other with a Type Dimension, called the Pilot Type Dimension (Fig. 12.13).

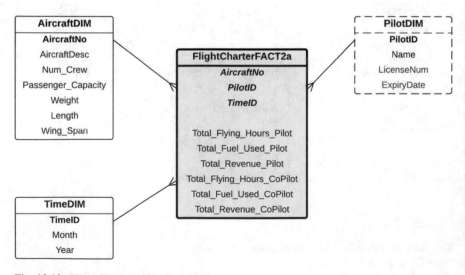

Fig. 12.12 Flight Charter with a Pivoted Fact

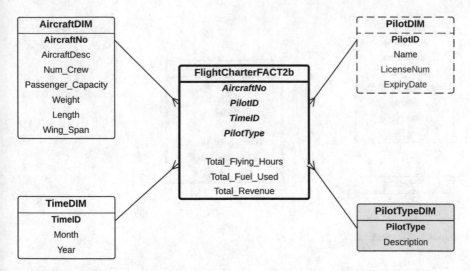

Fig. 12.13 Flight Charter with a Type attribute

Based on these two additional star schemas, it is clear that to slice the Pivoted Fact star schema (Fig. 12.12), we use a Vertical Slice, where the fact measure attributes in the Fact Table are split into two facts. The fact measures related to Pilot will go to the new Pilot star schema, whereas the fact measures related to Co-Pilot will go to the new Co-Pilot star schema. The star schema with the Pilot Type Dimension will be done using a Horizontal Slice, where the records in the Fact Table are distributed to two new Fact Tables based on the Pilot Type attribute value. The two new star schemas, after slicing, are shown in Fig. 12.14.

Note that the Pilot Dimension and Co-Pilot Dimension in the two new star schemas are no longer Determinant Dimensions. The reason for this is the fact measure value in the new star schema is solely based on one personnel type, and hence, there won't be any double counting, even when the Pilot Dimension (or the Co-Pilot Dimension) is not used in querying the star schema.

12.5 Summary

This chapter studies in more detail the concept of the slicing fact, using a number of case studies. There are two main methods: (*i*) Vertical Slice and (*ii*) Horizontal Slice.

A Vertical Slice is vertical data partitioning as in Distributed Databases, where each fact measure attribute is placed into a new star schema. A Horizontal Slice is horizontal data partitioning as in Parallel Databases, where each record of the Fact Table is placed in or distributed to a new Fact Table.

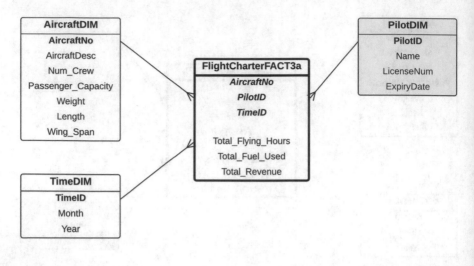

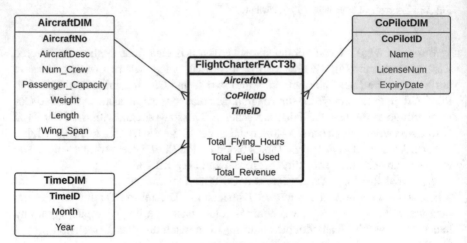

Fig. 12.14 Split into two facts

If the star schema is too general, where the partitioning attribute or the partitioning information is not present in the star schema, there is no way to slice the Fact Table. Therefore, it is critical for the star schema to contain the necessary partitioning information, such as Pivoted Fact, Type attribute or determinant dimension.

If the star schema has a Pivoted Fact, then it is a Vertical Slice. If the star schema has a Type Dimension, then it is a Horizontal Slice. Note though that if a Determinant Dimension is used in the slicing, the Determinant Dimension becomes a normal dimension in the new sliced star schemas.

12.6 Exercises

12.1 A tutorial is a small class (either in a lab setting or in a small classroom setting) and is taught by either a Teaching Assistant (known as a tutor who is usually a Master's or a PhD student) or a Professor (i.e. lecturer of the subject). A star schema for the Tutorial case study is shown in Fig. 12.15. It consists of three dimensions, namely Unit, Semester Year and Staff Dimensions. In the Staff Dimension, there is an attribute called Staff Type to indicate whether the staff member is a Teaching Assistant (tutor) or a Professor (lecturer). There is only one fact measure, that is, the Number of Tutorial Classes. In other words, this star schema keeps track of the number of tutorial classes that each staff member (either a Teaching Assistant or a Professor) teaches.

Because there are only two categories of staff, a Teaching Assistant who is a postgraduate student (e.g. a Master's or PhD student) and a Professor/Lecturer who is in charge of the subject (or unit), they are mutually exclusive. Therefore, it is sensible to slice the fact into two star schemas, one for the Teaching Assistant and the other for the Professor. Your task is to design these two new star schemas.

Suppose that the Staff Dimension does not have a Staff Type attribute as shown in Fig. 12.16, discuss how the two new star schemas (one for Teaching Assistant and the other for Professor) are created.

12.2 A PhD student is usually supervised (or advised) by one main supervisor and potentially one or more co-supervisors. The star schema shown in Fig. 12.17 keeps track of the number of students that each supervisor (i.e. Professor) supervises. The star schema contains three dimensions: Research Area, Year and Supervisor. The Supervisor Dimension in this case is a Determinant Dimension. The main reason for this is that each student may have multiple supervisors, which affects the calculation of the Number of Students fact measure.

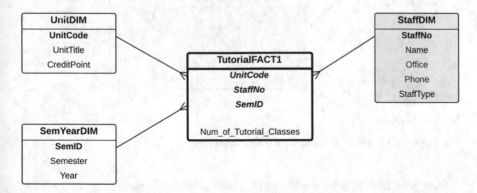

Fig. 12.15 Tutorial star schema with Staff Type attribute

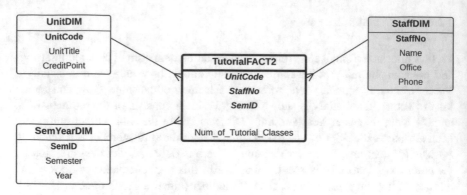

Fig. 12.16 Tutorial star schema without Staff Type attribute

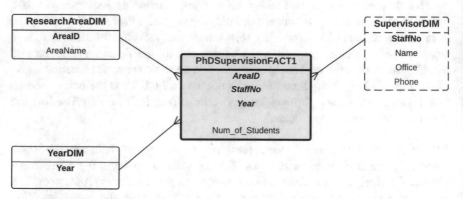

Fig. 12.17 PhD Supervision star schema with a Determinant Dimension

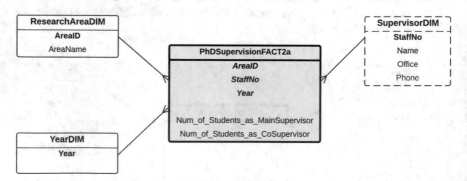

Fig. 12.18 PhD Supervision star schema with Pivoted Fact

The lower level of details of the PhD Supervision star schema in Fig. 12.17 exists in two forms, one is with a Pivoted Fact (see the star schema in Fig. 12.18) and the other is with the Supervisor Type Dimension (see the star schema in Fig. 12.19).

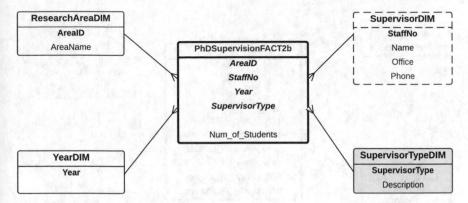

Fig. 12.19 PhD Supervision star schema with Supervisor Type Dimension

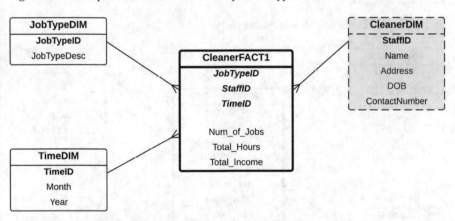

Fig. 12.20 Cleaner star schema with a Determinant Dimension

Your task is to slice the PhD Supervision Fact into two star schemas, one for Main Supervisor and the other for the Co-Supervisor.

12.3 A cleaning company employs Full-Time and Part-Time cleaners to be assigned to domestic or commercial (offices) cleaning jobs. Each job may need one or two (or even three) cleaners. These cleaners may be a mixture of Full-Time and Part-Time cleaners or all of one type (e.g. all Full-Time cleaners or all Part-Time cleaners). From the company's point of view, the difference between Full-Time and Part-Time cleaners is only administrative. For example, a Full-Time cleaner, apart from working Full-Time, is entitled to other benefits, such as superannuation, paid annual leave, etc. On the other hand, a Part-Time cleaner is paid hourly based on the work performed, without any other benefits.

The star schema shown in Fig. 12.20 contains three dimensions: Job Type, Time and Cleaner Dimensions. The Job Type Dimension indicates various job types, such as domestic cleaning, general office cleaning, etc. The Cleaner Dimension is a

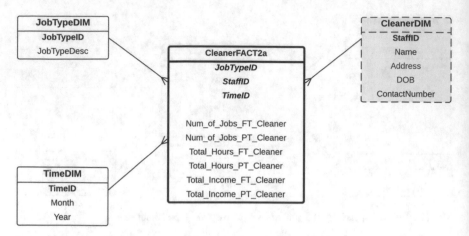

Fig. 12.21 Cleaner star schema with Pivoted Fact

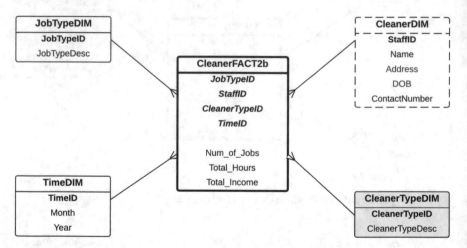

Fig. 12.22 Cleaner star schema with Cleaner Type Dimension

Determinant Dimension because the calculation of the fact measures (e.g. Number of Jobs, Total Hours and Total Income) is drawn from both Full-Time and Part-Time cleaners.

This Cleaner star schema (Fig. 12.20) has a lower level of details as shown in the following two star schemas. Figure 12.21 shows the breakdown of the fact measures, through the Pivoted Fact. Figure 12.22 shows an addition of a Cleaner Type Dimension to indicate whether a cleaner is a Full-Time cleaner or a Part-Time cleaner.

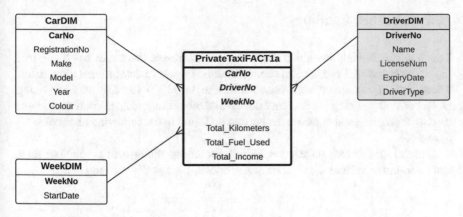

Fig. 12.23 Taxi Driver star schema with Driver Type attribute

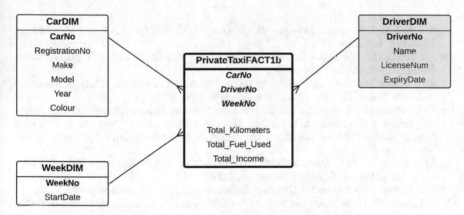

Fig. 12.24 Taxi Driver star schema without Driver Type attribute

Your task is to slice the fact by different types of cleaners (e.g. Full-Time and Part-Time cleaners). Hence, create two new star schemas, one for Full-Time cleaners and the other for Part-Time cleaners.

12.4 A Taxi company also employs two kinds of drivers, Full-Time drivers and Part-Time drivers. The star schema in Fig. 12.23 shows three dimensions: Car, Week and Driver Dimensions. The Driver Dimension also includes the Driver Type attribute to indicate if the driver is a Full-Time or a Part-Time driver.

Your task is to slice the fact by creating two new star schemas, one focusing on Full-Time drivers and the other on Part-Time drivers, as indicated by the Driver Type attribute in the Driver Dimension. To make the exercise more challenging, suppose the star schema does not have a Driver Type attribute in the Driver Dimension, as shown in Fig. 12.24. Discuss how the fact is sliced into two star schemas.

12.7   Further Readings

This book is the only book that discusses and addresses the issues of slicing fact, including horizontal and vertical slicing, and the impact to determinant dimensions. These are all illustrated with case studies to highlight the use of fact slicing. However, entity slicing is a database design and modelling issue. Further discussions on data modelling and database design can be found in the following textbooks: [1–5] and [6].

General discussions on fact modelling, including slicing and combining facts, can be found in various data warehousing books, such as [7–16] and [17].

References

1. T.A. Halpin, T. Morgan, *Information Modeling and Relational Databases*, 2nd edn. (Morgan Kaufmann, Los Altos, 2008)
2. J.L. Harrington, *Relational Database Design Clearly Explained*. Clearly Explained Series (Morgan Kaufmann, Los Altos, 2002)
3. M.J. Hernandez, *Database Design for Mere Mortals: A Hands-On Guide to Relational Database Design*. For Mere Mortals (Pearson Education, 2013)
4. N.S. Umanath, R.W. Scamell, *Data Modeling and Database Design* (Cengage Learning, 2014)
5. T.J. Teorey, S.S. Lightstone, T. Nadeau, H.V. Jagadish, *Database Modeling and Design: Logical Design*. The Morgan Kaufmann Series in Data Management Systems (Elsevier, Amsterdam, 2011)
6. G. Simsion, G. Witt, *Data Modeling Essentials*. The Morgan Kaufmann Series in Data Management Systems (Elsevier, Amsterdam, 2004)
7. C. Adamson, *Star Schema The Complete Reference* (McGraw-Hill Osborne Media, 2010)
8. R. Laberge, *The Data Warehouse Mentor: Practical Data Warehouse and Business Intelligence Insights* (McGraw-Hill, New York, 2011)
9. M. Golfarelli, S. Rizzi, *Data Warehouse Design: Modern Principles and Methodologies* (McGraw-Hill, New York, 2009)
10. C. Adamson, *Mastering Data Warehouse Aggregates: Solutions for Star Schema Performance* (Wiley, New York, 2012)
11. P. Ponniahm, *Data Warehousing Fundamentals for IT Professionals* (Wiley, London, 2011)
12. R. Kimball, M. Ross, *The Data Warehouse Toolkit: The Definitive Guide to Dimensional Modeling* (Wiley, London, 2013)
13. R. Kimball, M. Ross, W. Thornthwaite, J. Mundy, B. Becker, *The Data Warehouse Lifecycle Toolkit* (Wiley, London, 2011)
14. W.H. Inmon, *Building the Data Warehouse*. ITPro Collection (Wiley, London, 2005)
15. M. Jarke, *Fundamentals of Data Warehouses*, 2nd edn. (Springer, Berlin, 2003)
16. E. Malinowski, E. Zimányi, *Advanced Data Warehouse Design: From Conventional to Spatial and Temporal Applications*. Data-Centric Systems and Applications (Springer, Berlin, 2008)
17. A. Vaisman, E. Zimányi, *Data Warehouse Systems: Design and Implementation*. Data-Centric Systems and Applications (Springer, Berlin, 2014)

Chapter 13
Multi-Input Operational Databases

As mentioned in the first chapter of this book, creating a data warehouse using star schema modelling is basically a transformation process from an operational database, through procedures known as ETL (Extract-Transform-Load), whereby the operational database is extracted, transformed, cleaned, aggregated, etc., and as a result, a data warehouse is created. Hence, the operational database is an input to the transformation process to create a data warehouse. In all the previous chapters, the input to this transformation process is one operational database, which is represented by an E/R diagram. So, taking an E/R diagram as an input, the transformation process will produce a star schema (or multiple star schemas) as the output.

This chapter focuses on case studies where there is not one input operational database, but multiple; hence, the title of this chapter is "Multi-Input Operational Databases". There are two main methods to be discussed in this chapter, *Vertical Stacking* and *Horizontal Stacking*, each discussed with a case study. The first case study is a University Student Clubs case study, where a university has many student clubs, ranging from hobby types of clubs, to sport and cultural clubs, to more academic types of clubs. This is a case study for Vertical Stacking.

The second case study, which illustrates Horizontal Stacking, is a typical Bookstore case study which covers in-store sales as well as online sales.

13.1 Vertical Stacking: University Student Clubs Case Study

A university typically has many student clubs. The aim of these clubs is to offer students extracurricular activities by bringing together students who share the same interests. Some clubs focus on sports, for example, tennis clubs, basketball clubs, etc. Others focus on religion, such as Christian clubs, Muslim clubs, etc., or language or country-based clubs, such as Malaysian clubs, Taiwanese clubs, etc.

© The Author(s), under exclusive license to Springer Nature Switzerland AG 2021   329
D. Taniar, W. Rahayu, *Data Warehousing and Analytics*, Data-Centric Systems
and Applications, https://doi.org/10.1007/978-3-030-81979-8_13

All clubs receive a small amount of annual funding from the university. The clubs of course have to be self-sufficient by attracting other income, such as membership fees, sponsorships, etc. Because the university provides the clubs with a small amount of funding every year, the university wants to examine and analyse how the funds are spent; and therefore, a data warehouse for the clubs is needed.

Because each club maintains its own operational database, the data warehouse required by the university will take the operational databases from all the clubs as the input and will produce an integrated data warehouse. In other words, the input is Multi-Input Operational Databases, and the output is an integrated star schema. Having one star schema to analyse all clubs would be more convenient for university central management rather than having one star schema for each student club. Note that it is common that there are more than 50 student clubs in most universities; hence, it would not be convenient for the central university management to analyse 50 star schemas.

For simplicity, in this chapter, we only focus on three clubs: The first, a Student Orchestra Club, is a kind of hobby club which students and faculty members who play classical music instruments may join. The second, a Business and Commerce Club, is a more academic type of club, focusing on academic activities to support business and commerce students. The third, a Japanese Club, is a social club where Japanese and other students can practise their Japanese language skills.

13.1.1   Student Orchestra Club

The Philharmonic Society is a student university orchestra. They rehearse once a week and have public concerts a few times a year. Each member plays a musical instrument. They also have a professional conductor from outside the university. When they perform concerts, they sell tickets. Their operational database contains information about their members, concert activities and expenditure. The E/R diagram of this University Orchestra Club is shown in Fig. 13.1.

Each member must play a classical instrument (e.g. violin, trumpet, etc.) with a sufficient level of proficiency, and this club does not impose any membership fees on their members. The club's main event is the Concert, to which they sell tickets (e.g. full-price and concession tickets). The typical expenditure for a concert is the hosting costs (booking cost, advertising, ticket printing, etc.) and the conductor's fee. So, a simple equation for total expenditure and total income is as follows:

$$Total\ Expenditure = Total\ (Hosting\ Cost + Conductor\ Fee) \qquad (13.1)$$

$$Total\ Income = Total\ Ticket\ Sales \qquad (13.2)$$

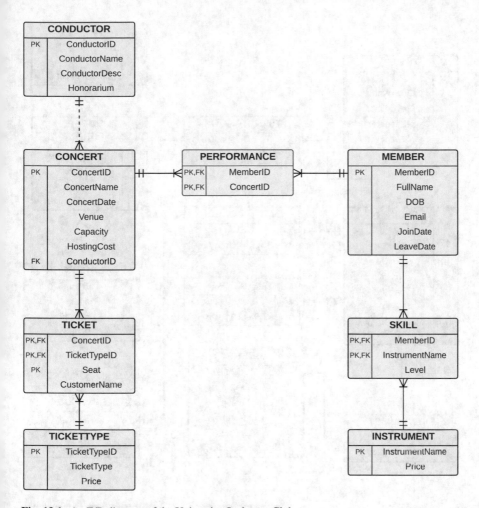

Fig. 13.1 An E/R diagram of the University Orchestra Club

13.1.2   Business and Commerce Students' Society

The Business and Commerce Students' Society is a student club for business and commerce students. Because this club is more academically oriented, their activities are academic, including events relating to job orientation, seminars, etc. Several companies sponsor this club. Their operational database (refer to the E/R diagram shown in Fig. 13.2) contains information about their members, events and seminars, as well as expenditure, which may include financial support from sponsors.

This club has a membership fee. Each membership card is valid for 1 year and needs to be renewed every year. The club holds three types of events: Professional Events, General Events and Social Events. A member will receive a fixed discount

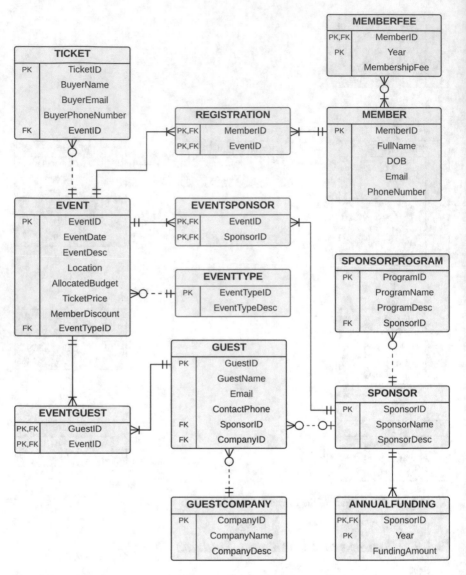

Fig. 13.2 An E/R diagram of the Business and Commerce Students' Society

to attend an event, whereas a non-member has to pay the full price for a ticket to an event.

The club receives an annual funding from sponsors, which, in turn, sponsors may advertise many programmes via the club website (e.g. internship programmes), and the membership fees from student members. Financial support from sponsor and membership fees are all spent on the events, as an allocated expenditure. Each event generates income through ticket sales. Club members receive a discount when purchasing a ticket event, whereas non-members need to pay the full amount for the ticket.

The total expenditure and total income for this club can be calculated as follows:

$$Total\ Expenditure = Total\ (Allocated\ Expenditure) \tag{13.3}$$

$$Total\ Income = Total\ (Ticket\ Price \times Number\ of\ Tickets+$$

$$Ticket\ Price \times (1 - Discount) \times Number\ of\ Member\ Registration) \tag{13.4}$$

13.1.3 Japanese Club

The Japanese Club is a cultural club for those who are interested in Japanese culture. This may include students who study Japanese at university or students who can speak Japanese and would like to mingle with Japanese native speakers or other students who speak Japanese. Their events are more cultural and social. Their operational database contains information about their members, events and expenditure.

This club does not have a membership fee. It has only one type of event: Culture/Social Event. The club receives funding for each event but does not receive annual funding from sponsors. Total expenditure and income can be calculated as follows:

$$Total\ Expenditure = Total\ (Event\ Funding\ Amount) \tag{13.5}$$

$$Total\ Income = Total\ (Member\ Ticket\ Count \times Member\ Price+$$

$$NonMember\ Ticket\ Count \times NonMember\ Price) \tag{13.6}$$

Because the club is only a small organisation, it maintains its database in an Excel file. A snapshot of the Excel file is shown in Fig. 13.3.

Using a standard database tool, such as Oracle SQL Developer, you can easily import/export a database from/to Excel files. Based on the Excel file, the equivalent E/R diagram is shown in Fig. 13.4.

	A	B	C	D	E	F
1	EVENTID	EVENTTIME	EVENTDESC	EVENTLOCATION	MEMBERPRICE	NONMEMBERPRICE
2	1	01-FEB-14	Okonomiyaki BBQ	Back bar,67 green st, windsor, vic	40	45
3	2	05-MAR-14	Trivia Night	Imperial 522 Chapel St South Yarra VIC 3141	30	32
4	3	06-APR-13	Onigiri Night	The Vic 281 Victoria St Abbotsford VIC 3067	15	20
5	4	09-MAY-13	Origami Night	Little Red Pocket Cocktail Bar 422 Little Collins St Melbourne VIC 3000	32	36
6	5	11-JUN-12	Karaoke	Melbourne Karaoke Lounge 465 Spencer St West Melbourne VIC 3003	26	32

EVENTFUNDING MJCSPONSOR SOCIALEVENT MEMBERTICKET NONMEMBERTICKET MJCMEMBER +

Fig. 13.3 An Excel file maintained by the Japanese Club

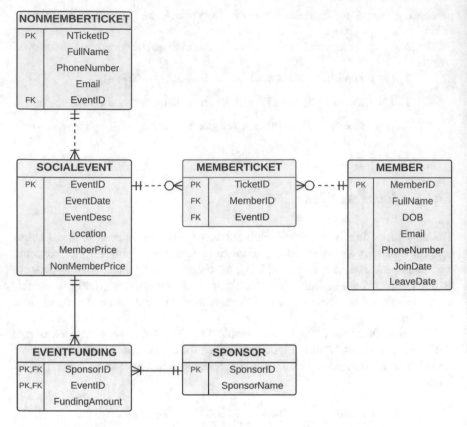

Fig. 13.4 An E/R diagram of the Japanese Club

13.1.4   Building an Integrated Data Warehouse

The aim of this study is to build an integrated data warehouse where there are three different input operational databases. The diagram in Fig. 13.5 shows that there are three input operational databases which create one data warehouse.

The information stored in each of these operational databases are members, activities and financial expenditure. Although they are different operational databases, they have the same theme, which is activities and expenditure. So, it becomes possible to create one data warehouse which combines information from these different operational databases.

The diagram in Fig. 13.6 shows a star schema which combines the three operational databases. For simplicity, the star schema has only two dimensions: Club and Year Dimensions. The star schema focuses on Total Expenditure and Total Income.

Fig. 13.5 An integrated data warehouse

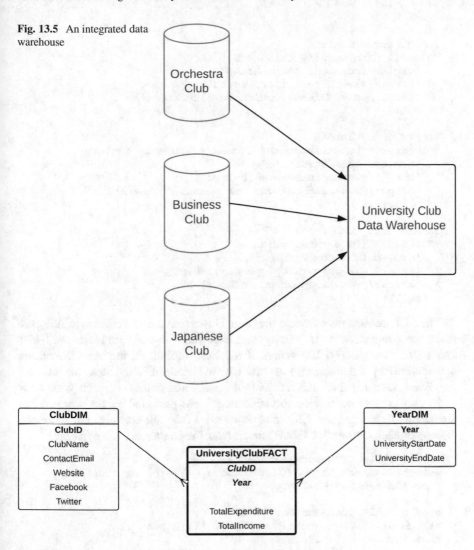

Fig. 13.6 An integrated data warehouse for the Student Club case study

The first step is to create the two dimensions: Club Dimension and Year Dimension. The SQL commands to create the Club Dimension are as follows. It has three records to store the details of the three clubs.

```
create table ClubDim (
   ClubID        integer,
   ClubName      varchar2(100),
   ContactEmail  varchar2(100),
   Website       varchar2(100),
   Facebook      varchar2(100),
   Twitter       varchar2(100)
);
```

```
insert into ClubDim
values(1, 'University Orchestra Club',
    'gro.sbulchsanom@cinomrahlihp',
    'http://www.universityorchestra.org',
    'http://www.facebook.com/universityorchestra',
    null);

insert into ClubDim
values(2, 'The Business and Commerce Students Society',
    'gro.sbulchsanom@sscb',
    'http://www.businesscommerce.net',
    'https://www.facebook.com/businessandcommerceclub',
    null);

insert into ClubDim
values(3, 'The Japanese Club',
    'gro.sbulchsanom@esenapaj',
    'http://www.universityjapaneseclub.org/',
    'https://www.facebook.com/nihonclub',
    null);
```

The SQL commands to create the Year Dimension are as follows. In this case study, we assume that the University Start Date and University End Date are 1-Mar and 1-Dec, respectively. The values of the Year attribute in the Year Dimension are obtained by the respective events. By unioning these values, we have all the Years. Because there are three input operational databases, each with a set of its own tables, we assume the table names are preceded by the *Club Name*: Orchestra Club, Business Club and Japanese Club, respectively. For example, "OrchestraClub.Concert" is Table Concert from the Orchestra Club database.

```
create table OrchestraYearDim as
select distinct to_char(ConcertDate, 'YYYY') as Year
from "OrchestraClub.Concert";

create table BusinessYearDim as
select distinct to_char(EventDate, 'YYYY') as Year
from "BusinessClub.Event";

create table JapaneseYearDim as
select distinct to_char(EventDate, 'YYYY') as Year
from "JapaneseClub.SocialEvent";

create table YearDim as
select distinct *
from (
  select * from OrchestraYearDim
  union
  select * from BusinessYearDim
  union
  select * from JapaneseYearDim
);
```

```
alter table YearDim add (
  UniversityStartDate date,
  UniversityEndDate    date
);

update YearDim
set UniversityStartDate =
  to_date('01-Mar-'||Year, 'DD-MM-YYYY');

update YearDim
set UniversityEndDate =
  to_date('01-Dec-'||Year, 'DD-MM-YYYY');
```

The next step is to process the first operational database, which is the Orchestra Database. Since the data warehouse focuses on expenditure and income, we need to aggregate several attributes from the Orchestra operational database, such as Hosting Cost, Honorarium for the Conductor, as well as the income obtained from the Ticket Sales.

To obtain these values, we need to join four tables: Concert, Conductor, Ticket and Ticket Type, using the following SQL command:

```
create table OrchestraTempFact1 as
select T.ConcertID,
  to_char(CT.ConcertDate, 'YYYY') as Year,
  CT.HostingCost, CR.Honorarium,
  TT.Price
from "OrchestraClub.Ticket" T,
  "OrchestraClub.TicketType" TT,
  "OrchestraClub.Concert" CT,
  "OrchestraClub.Conductor" CR
where
  T.TicketTypeID = TT.TicketTypeID and
  CT.ConcertID = T.ConcertID and
  CT.ConductorID = CR.ConductorID;
```

Table Concert in the Orchestra Club database has a higher level of aggregation than Table Ticket because one concert has many tickets. The above Orchestra TempFact1 is at the Ticket level, meaning that the income is for each ticket. Hence, we need to aggregate concert tickets to calculate the total income, using the following SQL command:

```
create table OrchestraTempFact2 as
select ConcertID, Year,
  HostingCost, Honorarium,
  sum(Price) as Income
from OrchestraTempFact1
group by ConcertID, Year, HostingCost, Honorarium
order by ConcertID, Year, HostingCost, Honorarium;
```

Before the Fact Table is created, a ClubID must be added:

```
alter table OrchestraTempFact2
add (ClubID integer);
```

```
update OrchestraTempFact2
set ClubID = 1;
```

Finally, the Fact Table for the Orchestra star schema is created by aggregating the Hosting Cost, Honorarium and Income, using the following SQL command:

```
create table OrchestraFact as
select ClubID, Year,
  sum(HostingCost + Honorarium) as TotalExpenditure,
  sum(Income) as TotalIncome
from OrchestraTempFact2
group by ClubID, Year;
```

A similar series of steps must be applied to process the Business and Commerce Club Database. For this club, the income source is mainly from ticket sales. Members who attend an event must purchase a ticket, but they receive a member discount. Non-members are able to attend an event but must pay the full price. The first process is to count the number of tickets bought by members and non-members using the following SQL commands. Registration Count and Ticket Count indicate the number of tickets bought by members and non-members, respectively.

```
--Number member tickets for each event
create table EventRegistration as
select E.EventID, count(*) as RegistrationCount
from
  "BusinessClub.Event" E,
  "BusinessClub.Registration" R
where E.EventID = R.EventID
group by E.EventID;

--Number non-member tickets for each event
create table EventTicket as
select E.EventId, count(*) as TicketCount
from "BusinessClub.Event" E,
  "BusinessClub.Ticket" T
where  E.EventID = T.EventID
group by E.EventID;
```

For each event, we will have the number of tickets bought by members and non-members. We also need to get the Ticket Price, Allocated Expenditure and Member Discount (percentage) when creating a TempFact. The ClubID must also be added to the TempFact.

```
create table BusinessCommerceTempFact as
select
  E.EventID,
  to_char(E.EventDate, 'YYYY') as Year,
  E.TicketPrice,
  E.AllocatedBudget,
  E.MemberDiscount,
  T.TicketCount,
  R.RegistrationCount
```

```
from "BusinessClub.Event" E,
  EventTicket T,
  EventRegistration R
where E.EventID = T.EventID
and E.EventID = R.EventID;

alter table BusinessCommerceTempFact
add (ClubID integer);

update BusinessCommerceTempFact
set ClubID = 2;
```

Finally, the Business Commerce Fact Table is created by aggregating the required attributes from the TempFact to calculate the Total Expenditure and Total Income.

```
create table BusinessCommerceFact as
select ClubID, Year,
  sum(AllocatedExpenditure) as TotalExpenditure,
  sum(RegistrationCount * (1-MemberDiscount) * TicketPrice +
    TicketCount * TicketPrice) as TotalIncome
from BusinessCommerceTempFact
group by ClubID, Year;
```

The third and last operational database to be processed in this case study is the Japanese Club operational database. Their events also have two types of tickets, one for members and the other for non-members. So, first we need to count the number of tickets bought by members and by non-members (indicated by attributes Member Ticket Count and Non-Member Ticket Count, respectively):

```
--Number of member tickets for each event
create table EventMemberTicket as
select E.EventID, count(*) as MemberTicketCount
from "JapaneseClub.SocialEvent" E,
  "JapaneseClub.MemberTicket" M
where E.EventID = M.EventID
group by E.EventID;

--Number of non-member tickets for each event
create table EventNonMemberTicket as
select E.EventID, count(*) as NonMemberTicketCount
from "JapaneseClub.SocialEvent" E,
  "JapaneseClub.NonMemberTicket" M
where E.EventID = M.EventID
group by E.EventID;
```

Once the ticket counts are calculated, we can then create a TempFact, which includes the other required data, such as Funding Amount, Member Price (ticket price for members) and Non-Member Price (ticket price for non-members).

```
create table JapaneseTempFact as
select
  E.EventID,
  to_char(E.EventDate, 'YYYY') as Year,
  F.FundingAmount,
```

```
       M.MemberTicketCount,
       N.NonMemberTicketCount,
       E.MemberPrice,
       E.NonMemberPrice
   from "JapaneseClub.SocialEvent" E,
       EventNonMemberTicket N,
       EventMemberTicket M,
       "JapaneseClub.EventFunding" F
   where E.EventID = N.EventID
   and E.EventID = M.EventID
   and E.EventID = F.EventID;

   alter table JapaneseTempFact
   add (ClubID integer);

   update JapaneseTempFact
   set ClubID = 3;
```

Finally, the Fact Table is created by aggregating the Funding Amount, as well as calculating the Total Income using the information about ticket count and ticket price for both members and non-members.

```
   create table JapaneseFact as
   select ClubID, Year,
       sum(FundingAmount) as TotalExpenditure,
       sum(MemberPrice * MemberTicketCount +
           NonMemberPrice * NonMemberTicketCount)
           as TotalIncome
   from JapaneseTempFact
   group by ClubID, Year;
```

Once each club has its own star schema, we can then simply combine the three star schemas using the union command in SQL:

```
   create table MonashClubFact as
   select * from OrchestraFact
   union
   select * from BusinessCommerceFact
   union
   select * from JapaneseFact;
```

As a conclusion, creating an integrated star schema in this case study simply involves combining the individual star schemas. Combining the individual star schemas from each club is possible because the dimensions and the fact measures are identical. It uses only two dimensions, Club and Year, and it has two fact measures, Total Expenditure and Total Income. Therefore, when combining the individual star schemas into one integrated star schema, we can simply use union and the sum aggregate function.

The method to combine the initial multiple star schema is a *Union*. The union operator is like stacking multiple tables on top of each other. Consequently, it is like *Vertical Stacking*, where all the tables are stacked vertically and combined into one table.

13.2   Horizontal Stacking: Real-Estate Property Case Study

This second case study illustrates how an integrated data warehouse (or star schema) is created from multiple input operational databases. This is a case study of a real-estate company which sells properties (e.g. houses, apartments, townhouses, units, etc.). This company has a number of operational databases, each for different purposes.

The first operational database is the *Inspection Database*. In this database, all properties registered for sale are recorded. The E/R diagram of this Inspection Database is shown in Fig. 13.7. The main table is Table Property, which records all the properties for sale. This includes the details of the properties, like address and all the facilities (e.g. car spaces). Each property falls into a category, such as house, apartment, townhouse, unit, etc., which is indicated by the Property Type. Some properties may also have a range of expected prices, which is particularly useful when the property is advertised for an auction. The date the property is first listed is used to calculate a range of statistical analyses, such as how many properties are newly listed in a certain month.

Properties are also advertised in various ways, such as sales websites, newspapers, magazines, etc. If the property is for auction, the details of the auction, such as the date, are also recorded.

Each property is normally open for inspection, and in many cases, there are inspection schedules. At every inspection, the details of the visitors, usually their name and contact number, are recorded. Note that the cardinality between Property and Inspection, and between Inspection and Visitor, is both $1 - m$. This means that if the same visitor visits the same property on different inspection dates, this will be recorded twice. The main reason for this is that the real-estate agent does not maintain a client master list. Table Visitor simply logs all the visitors who attended the inspection.

Finally, each property is usually assigned to two staff from the branch which handles the property.

The second operational database that this real-estate agency maintains is related to auction, and hence it is called an *Auction Database*. This database, as the name states, records the properties that have been auctioned and their auction results, that is, whether the property was sold at auction, sold after auction or not sold. Some properties which are not sold at auction may be sold to the highest bidder, after negotiation with the owner of the property, which is referred to as "sold after auction". If the negotiation fails, then the property's auction result is "not sold".

The E/R diagram of the Auction Database is shown in Fig. 13.8. Note that in this Auction Database, it has its own Table Property, which may not necessarily have the same attributes as the Table Property in the Inspection Database. For example, the Listed Date attribute is not included in Table Property in the Auction Database. Table Auction in this Auction Database has more information compared to Table Auction in the Inspection Database. The reason for this is simple: Table Auction

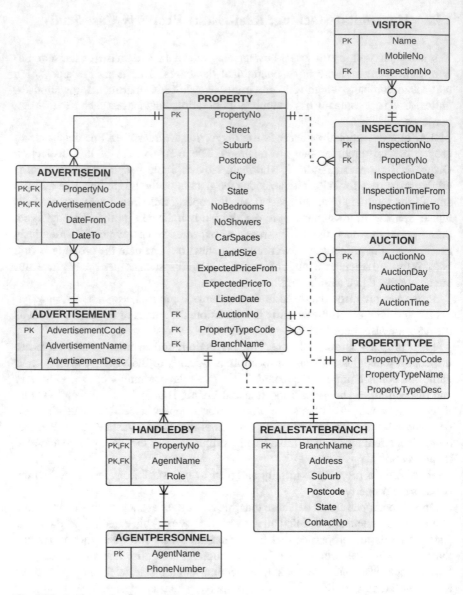

Fig. 13.7 An Inspection Property Database

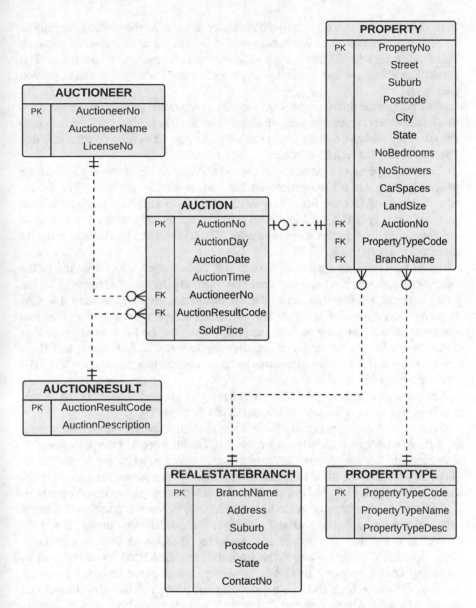

Fig. 13.8 An Auction Database

in the Inspection Database is only an indicator as to when the auction is going to be held, whereas Table Auction in the Auction Database contains more information about the auction activity itself, such as the name of the auctioneer and the auction result. The auctioneer and auction result are of course irrelevant in the Inspection Database.

Also note that in this database, only records on properties for auction are recorded in Table Property. Properties that are not for auction (for private sale, for instance) are not recorded here, although these private sale properties are recorded in Table Property in the Inspection Database.

The third operational database is the *Sold Property Database*. This database keeps the records of all properties that are sold, either through auction or private sale. This means that the listed properties are not yet sold, nor are they entered in Table Property in this database. Also note that not all properties are going to be auctioned. Many properties are private sales. The E/R diagram for the Sold Property Database is shown in Fig. 13.9.

In the Sold Property Database, there is also Table Property. However, unlike the other two databases, this table does not have auction information. However, all the sold properties, whether they are sold at auction or not, are listed here. In this Sold Property Database, there is Table Owner, which does not exist in the other two databases. If there are joint owners, the joint owners will be listed as one record in Table Owner. Hence, the cardinality relationship between Table Property and Table Owner is $1 - 1$. If the same owner owns another property, this information will exist as different records in Table Owner.

Other tables related to the selling and buying of properties, such as the Buyer, the Bank and the Lawyer, are maintained in this database.

As a summary, the three databases in this real-estate company are all independent of each other because each of them focuses on different aspects of the properties that they manage. The Inspection Database contains information on the inspection of properties for sale, the Auction Database contains information on the auction itself, and the Property Sold Database contains information on properties which have sold.

This real-estate company would like to build an integrated data warehouse which takes all three operational databases as input for the data warehouse. There are several analyses that they would like to make. The first is *Auction Clearance Rate*. Auction Clearance Rate is an important measure used by many people, including banks, property investors, government and other financial agencies. Auction Clearance Rate shows the percentage of properties being auctioned that are actually sold. A high Auction Clearance Rate indicates a high auction success rate. This often indicates that the property market is getting hot, meaning that there are many bidders pushing up property prices which is known as a property boom era. On the other hand, during a property crash or property slump era, there are not many successful auctions, which indicates that property prices are falling. So, Auction Clearance Rate is Total Successful Auctions divided by Total Properties in Auction. These two fact measures will be included in the integrated star schema.

The other important indicator in the property market is Number of New Listings. Number of New Listings indicates how many properties are newly listed every

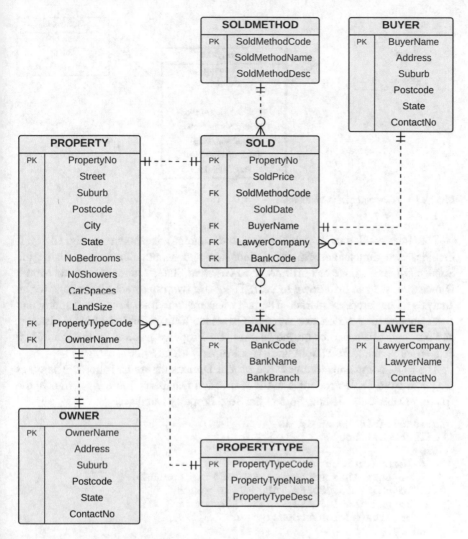

Fig. 13.9 A Property Sold Database

month. During a property boom era, this indicator usually shows an increase, meaning that there are many people selling their houses. One of the reasons for this is because property prices are increasing so property owners will make a profit by selling in a boom era. On the other hand, during a slow time when the property market is weak, many people hold onto their properties and do not sell; hence, the number of newly listed properties declines. The last indicator for this integrated data warehouse is the price. For simplicity, we only need to keep track of the mean price of the property prices. Hence, we need Total Properties Sold and Total Sold Price.

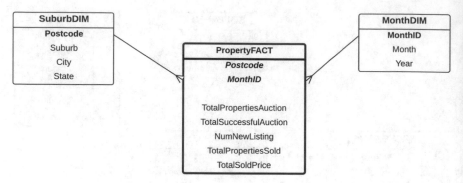

Fig. 13.10 A Property Data Warehouse

The integrated data warehouse (or star schema) is shown in Fig. 13.10. It includes two dimensions: Suburb Dimension and Month Dimension. Using the Suburb Dimension, we can drill down to City and State, whereas using the Month Dimension, we can drill down to year. These are two important dimensions for an analysis of the property market. The fact measures are Total Property Auction and Total Successful Auction (for Auction Clearance Rate), Number of New Listings, and Number of Properties on the Market (for Property Auction Ratio), and Total Properties Sold and Total Sold Price (for Property Mean Price).

The SQL commands to create the Month Dimension are as follows. The dates are taken from Table Properties in the Inspection Database, Table Properties in the Auction Database and Table Sold in the Sold Property Database.

```
create table MonthDim as
select distinct *
from (
  select distinct
    to_char (ListedDate, 'YYYYMM') as MonthID,
    to_char(ListedDate, 'Mon') as Month,
    to_char(ListedDate, 'YYYY') as Year
  from "Inspection.Property"
  union
  select distinct
    to_char (AuctionDate, 'YYYYMM') as MonthID,
    to_char(AuctionDate, 'Mon') as Month,
    to_char(AuctionDate, 'YYYY') as Year
  from "Auction.Auction"
  union
  select distinct
    to_char (SoldDate, 'YYYYMM') as MonthID,
    to_char(SoldDate, 'Mon') as Month,
    to_char(SoldDate, 'YYYY') as Year
  from "SoldProperty.Sold"
);
```

The SQL commands to create the Suburb Dimension are as follows. All are taken from the respective Table Properties in the three operational databases.

```
create table SuburbDim as
select distinct *
from (
  select distinct Postcode, Suburb, City, State
  from "Inspection.Property"
  union
  select distinct Postcode, Suburb, City, State
  from "Auction.Property"
  union
  select distinct Postcode, Suburb, City, State
  from "SoldProperty.Property"
);
```

The first fact measure is Auction Clearance Rate, which is Total Properties Auction and Total Successful Auction. These two attributes are from the Auction Database. The SQL commands to create a TempFact are as follows:

```
create table PropertyTempFact1 as
select
  P.Postcode,
  to_char(A.AuctionDate, 'YYYYMM') as MonthID
  count(*) as TotalPropertiesAuction,
  0 as TotalSuccessfulAuction,
  0 as NumNewListing,
  0 as TotalPropertiesSold,
  0 as TotalSoldPrice
from "Auction.Auction" A, "Auction.Property" P
where A.AuctionNo = P.AuctionNo
group by
  P.Postcode,
  to_char(A.AuctionDate, 'YYYYMM');

create table PropertyTempFact2 as
select
  P.Postcode,
  to_char(A.AuctionDate, 'YYYYMM') as MonthID
  0 as TotalPropertiesAuction,
  count(*) as TotalSuccessfulAuction,
  0 as NumNewListing,
  0 as TotalPropertiesSold,
  0 as TotalSoldPrice
from "Auction.Auction" A, "Auction.Property" P
where A.AuctionNo = P.AuctionNo
and A.AuctionResultCode = 'SA'
group by
  P.Postcode,
  to_char(A.AuctionDate, 'YYYYMM');
```

13 Multi-Input Operational Databases

The next fact measure to be implemented is Number of New Listings. It uses the Inspection Databases. Basically, it counts the number of records in Table Property by grouping the Listed Date.

```
create table PropertyTempFact3 as
select
  P.Postcode,
  to_char(P.ListedDate, 'YYYYMM') as MonthID,
  0 as TotalPropertiesAuction,
  0 as TotalSuccessfulAuction,
  count(*) count as NumNewListing,
  0 as TotalPropertiesSold,
  0 as TotalSoldPrice
from "Inspection.Property" P
group by
  P.Postcode, to_char(P.ListedDate, 'YYYYMM');
```

Finally, the last two fact measures, Total Properties Sold and Total Sold Price, are calculated using the Sold Property Database. The SQL commands are listed here:

```
create table PropertyTempFact4 as
select
  P.Postcode,
  to_char(S.SoldDate, 'YYYYMM') as MonthID,
  0 as TotalPropertiesAuction,
  0 as TotalSuccessfulAuction,
  0 count as NumNewListing,
  count(*) as TotalPropertiesSold,
  sum(S.SoldPrice) as TotalSoldPrice
from "SoldProperty.Property" P, "SoldProperty.Sold" S
where P.PropertyNo = S.PropertyNo
group by
  P.Postcode, to_char(S.SoldDate, 'YYYYMM');
```

Once these four Property TempFact Tables are created, the final Fact Table is created by combining them. When the four Property TempFact Tables are unioned, the results need to be aggregated by summing all the fact measure attributes.

```
create table PropertyFact as
select
  Postcode, MonthID,
  sum(TotalPropertiesAuction),
  sum(TotalSuccessfulAuction),
  sum(NumNewListing),
  sum(TotalPropertiesSold),
  sum(TotalSoldPrice)
from (
  select
    Postcode, MonthID,
    TotalPropertiesAuction,
    TotalSuccessfulAuction,
    NumNewListing,
    TotalPropertiesSold,
    TotalSoldPrice
```

```
    from PropertyTempFact1
    union
    select
      Postcode, MonthID,
      TotalPropertiesAuction,
      TotalSuccessfulAuction,
      NumNewListing,
      TotalPropertiesSold,
      TotalSoldPrice
    from PropertyTempFact2
    union
    select
      Postcode, MonthID,
      TotalPropertiesAuction,
      TotalSuccessfulAuction,
      NumNewListing,
      TotalPropertiesSold,
      TotalSoldPrice
    from PropertyTempFact3
    union
    select
      Postcode, MonthID,
      TotalPropertiesAuction,
      TotalSuccessfulAuction,
      NumNewListing,
      TotalPropertiesSold,
      TotalSoldPrice
    from PropertyTempFact4)
  group by Postcode, MonthID;
```

The method to combine multiple star schemas is *Horizontal Stacking*, where each original Fact Table has its own fact measures and these fact measures are different from those of other Fact Tables. We push all the fact measures from their original Fact Table into the new Fact Table.

13.3 Summary

This chapter discusses how a star schema is created using multiple input operational databases. Two case studies are studied in this chapter: the University Student Club case study and the Real-Estate Property case study.

In the University Student Club case study, each operational database creates its own star schema. Then these star schemas are combined using a `union` command in SQL. Basically, each star schema is independent of each other, and hence combining them is simply putting all the records from the individual Fact Table into one combined Fact Table.

In the Real-Estate Property case study, it is slightly different. Each star schema has its own fact measure. Therefore, when creating the integrated star schema, all the fact measures must be included in the Fact Table. So, if the University Student

Club is a *Vertical Stacking*, where multiple Fact Tables are stacked vertically into one Fact Table, the Real-Estate Property case study is a *Horizontal Stacking*, where multiple tables are stacked horizontally into one Fact Table.

13.4   Exercises

13.1 This exercise extends the Real-Estate Property case study described in the previous section. This case study has three input operational databases, Inspection Database, Auction Database and Sold Property Database. The respective E/R diagrams are shown in Figs. 13.7, 13.8, and 13.9. The integrated star schema is shown in Fig. 13.10 and includes five fact measures, Total Properties Auction, Total Successful Auctions, Number of New Listings, Total Properties Sold and Total Sold Price.

Now we would like to revise the fact by including more fact measures. The first fact measure to be added is Number of Properties on the Market. This is particularly useful to determine the ratio between properties for auction and total properties. This is known as Property Auction Rate. Property Auction Rate often indicates whether the property market is going up or down. When the property market is good, many sellers prefer an auction because they may achieve a higher selling price. On the contrary, when the property market is weak, many sellers opt for a private sale because the auction results are not encouraging.

The second fact measure to be added to the fact is Total Days on the Market. This fact measure will be used to calculate the Average Days of Properties on the Market by dividing Total Days on the Market with Number of Properties on the Market. When the property market is strong, properties tend to sell quickly; hence, the Average Days of Properties on the Market will be low. The opposite is also true; that is, when the property market is not strong, it is harder to sell properties and it takes a longer time.

The third and fourth new fact measures to be added are related to the Inspection. Number of Properties for Inspection and Number of People for Inspection will be used to calculate the average number of visitors for each open for inspection. This indicator is very useful to show the demand for a property. Obviously, the more people who inspected a property, the more likely the property will be sold for a higher price.

The new star schema is shown in Fig. 13.11. The task is now to write the SQL commands to implement this new star schema.

13.2 A bookshop offers two kinds of sales: in-shop sales and online sales. Each of these maintains its own database system. The In-Shop Sales E/R diagram is shown in Fig. 13.12. In this database, the details of the books are stored in various tables, namely Book, Author, Title Author, Publisher and Category. Information on the sales are stored in tables: Sales and Sales Details. Because this information relates to In-Shop Sales, the Employee who performed the sales is recorded. The Employee

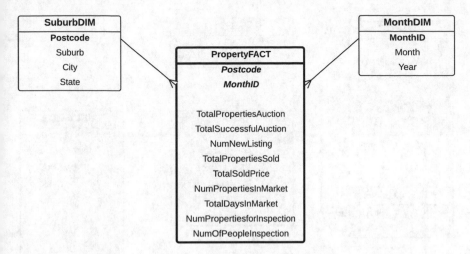

Fig. 13.11 An Extended Property star schema with new fact measures

is also attached to a particular Branch, and of course, each sale is performed in a branch. Note that this kind of circular relationship of entities does not follow the Boyce-Codd Normal Form (BCNF).

The Online Sales E/R diagram is shown in Fig. 13.13. The details on the books are the same as those in the In-Shop Sales E/R diagram, namely Book, Author, Title Author, Publisher and Category. However, the Sales are simpler. It has only two entities: Online Sales and Sales Details. However, it has one additional entity, Delivery, which records the delivery address. Information about the Employee who performed the sale is not recorded.

An integrated star schema is requested by this bookshop. The star schema will take both In-Store Sales and Online Sales databases. The star schema needs to include Total Books Sold and Total Price. The star schema is shown in Fig. 13.14. Your task is to implement this star schema using the SQL commands.

13.5 Further Readings

The issues of Multi-Input Operational Databases have not been explicitly discussed in other resources. This book is the most comprehensive book on the topic. This covers vertical and horizontal stacking. The topic of data linkage in general has been an active research area. Some of the books on the topic of data linkage or record matching are [1, 2] and [3].

General star schema design methods, which may touch on the use of multiple operational databases as the source for creating a data warehouse, can be found in various data warehousing books, such as [4–13] and [14].

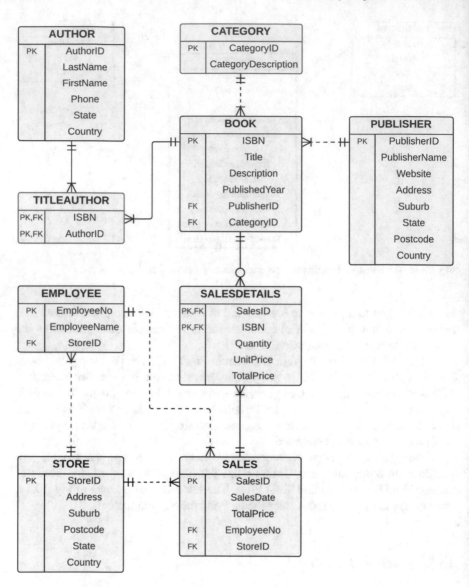

Fig. 13.12 Bookshop case study: In-Shop Sales

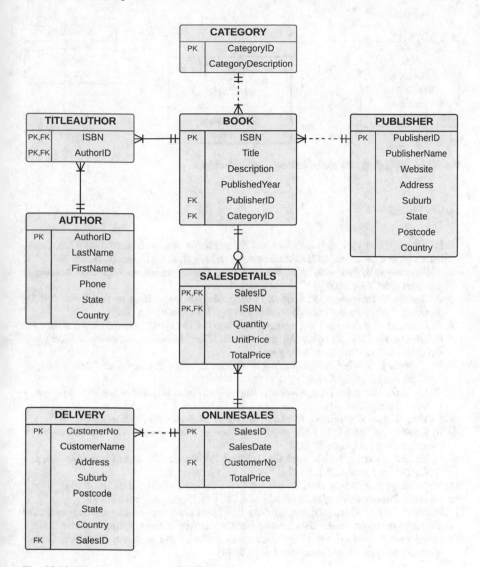

Fig. 13.13 Bookshop case study: Online Sales

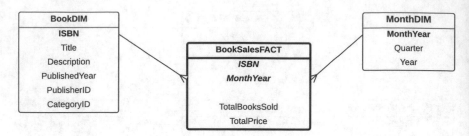

Fig. 13.14 Bookshop case study: an integrated star schema

References

1. P. Christen, *Data Matching: Concepts and Techniques for Record Linkage, Entity Resolution, and Duplicate Detection*. Data-Centric Systems and Applications (Springer, Berlin, 2012)
2. T.N. Herzog, F.J. Scheuren, W.E. Winkler, *Data Quality and Record Linkage Techniques* (Springer New York, 2007)
3. P. Christen, T. Ranbaduge, R. Schnell, *Linking Sensitive Data: Methods and Techniques for Practical Privacy-Preserving Information Sharing* (Springer, Berlin, 2020)
4. C. Adamson, *Star Schema The Complete Reference* (McGraw-Hill Osborne Media, 2010)
5. R. Laberge, *The Data Warehouse Mentor: Practical Data Warehouse and Business Intelligence Insights* (McGraw-Hill, New York, 2011)
6. M. Golfarelli, S. Rizzi, *Data Warehouse Design: Modern Principles and Methodologies* (McGraw-Hill, New York, 2009)
7. C. Adamson, *Mastering Data Warehouse Aggregates: Solutions for Star Schema Performance* (Wiley, London, 2012)
8. P. Ponniah, *Data Warehousing Fundamentals for IT Professionals* (Wiley, London, 2011)
9. R. Kimball, M. Ross, *The Data Warehouse Toolkit: The Definitive Guide to Dimensional Modeling* (Wiley, London, 2013)
10. R. Kimball, M. Ross, W. Thornthwaite, J. Mundy, B. Becker, *The Data Warehouse Lifecycle Toolkit* (Wiley, London, 2011)
11. W.H. Inmon, *Building the Data Warehouse*. ITPro Collection (Wiley, London, 2005)
12. M. Jarke, *Fundamentals of Data Warehouses*, 2nd edn. (Springer, Berlin, 2003)
13. E. Malinowski, E. Zimányi, *Advanced Data Warehouse Design: From Conventional to Spatial and Temporal Applications*. Data-Centric Systems and Applications (Springer, Berlin, 2008)
14. A. Vaisman, E. Zimányi, *Data Warehouse Systems: Design and Implementation*. Data-Centric Systems and Applications (Springer, Berlin, 2014)

Part V
Data Warehousing Granularity and Evolution

Chapter 14
Data Warehousing Granularity and Levels of Aggregation

The central concept of data warehousing is *Aggregation*. In the previous chapters, we have been focusing on fact measures as the aggregated values. Since fact measures are the focus of star schemas, data warehousing is then about aggregation and aggregate values.

Digging deeper into the nature of aggregate values, aggregate values have different levels of granularity. For example, Total Sales per Year has lower granularity than Total Sales per Quarter; or Number of Logins in the lab per Semester has a different level of focus (or granularity) than Number of Logins in the lab per Month. Consequently, there is a notion of levels of granularity, or level of detail, and in the context of data warehousing, we use the term *levels of aggregation*. This chapter focuses on levels of aggregation, which is basically the foundation of *data warehousing granularity*.

14.1 Levels of Aggregation

Let's use the Computer Lab Usage case study to illustrate the concept of *levels of aggregation* and *data warehousing granularity*. Figure 14.1 shows a simple operational database that keeps track of the usage of computers in the labs.

This computer lab usage operational database has only four tables. Student Table is a typical table that stores student details, including the Degree in which the student is enrolled and the Study Type (e.g. Full-Time, Part-Time, Online, etc.). The Degree Code attribute is obviously a Foreign Key to the Degree Table that stores a list of degrees offered by the university (e.g. Bachelor of Computer Science, Master of Information Technology, etc.). The Computer Table stores the details of each computer in the labs, including Computer Model, Operating Systems, etc. The transaction table is the Lab Activities table which keeps the details of every student who logged in to each computer. Each time a student logs in, a Login No is created

© The Author(s), under exclusive license to Springer Nature Switzerland AG 2021
D. Taniar, W. Rahayu, *Data Warehousing and Analytics*, Data-Centric Systems
and Applications, https://doi.org/10.1007/978-3-030-81979-8_14

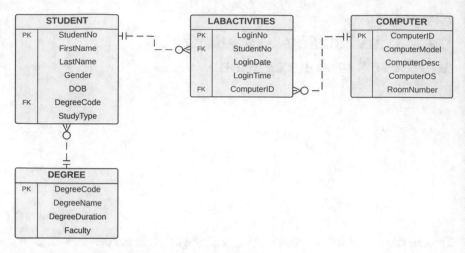

Fig. 14.1 An E/R diagram of the Computer Lab Usage

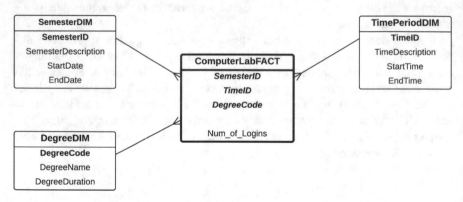

Fig. 14.2 The Computer Lab Usage star schema

and other details are stored, including Student No, Login Date and Time as well as ComputerID.

A star schema is created to analyse the usage of computers in the labs. For simplicity, only one fact measure is included in the star schema, that is, Num of Logins. The dimensions are Semester Dimension (e.g. semester 1, semester 2), Time Period Dimension (e.g. morning, afternoon, night) and Degree Dimension. The star schema is shown in Fig. 14.2. Using this simple star schema, we can query the number of logins per semester, per time period and per degree. Because we count the number of logins, the same student who logins several times will be counted as many times as the student logs in.

Data warehousing is basically precomputed aggregate values. The fact measure, Num of Logins, is basically a precomputed value by counting the number of records in the transaction table, namely the Lab Activities table, grouped by Semester, Time

Period and Degree. With this precomputed aggregate value, querying to the star schema can directly obtain the values (i.e. number of logins), by retrieving the fact measure, by specifying the desired dimensions, whether it be Semester, Time Period and/or Degree. With this precomputed aggregate value, the retrieval of the Num of Logins can be done more efficiently. This is why a data warehouse is preferred by management in relation to the decision-making process, simply because the required information is already precomputed. To be more specific, a data warehouse is built through a transformation process from the operational database by extracting, cleaning and aggregating the original data in the operational databases. This is why decision-making does not normally query operational databases because decision-making queries usually focus on aggregate values and it will be more convenient if the required information is readily available in the data warehouse because the data warehouse is already cleaned, transformed, aggregated, etc.

Because decision support queries usually focus on aggregate values, the Fact Tables often contain highly aggregated values, such as Num of Logins, grouped by Semester, Time Period and Degree. When management finds interesting data in the Fact Table, they often want to drill down to find more detailed information. For example, in the Computer Lab Usage case study, if the number of logins at night in semester 1 is rather high, we might want to see the breakdown, such as the number of logins per hour at night, etc. The main question is how to do this. Unfortunately, because the Fact Table does not store this information (e.g. the hour by hour breakdown), the obvious answer is to drill down into the operational database. However, this is not a preferable practice because the operational database might not be easily accessible (note that a data warehouse might be built from several operational databases and some operational databases are not even a database system, such as an Excel spreadsheet, etc.). Additionally, the data from the operational databases have been cleaned, transformed, etc. when building the data warehouse. Therefore, accessing the operational database for drilling down purposes is not desirable.

To address this issue, the concept of *data warehousing granularity* is introduced. A data warehousing granularity consists of a number of granularity layers of a star schema. The highest granularity star schema which contains the most detailed data is known as **Level-0**. **Level-1** is built on top of Level-0 star schemas, and Level-1 star schemas have a lower granularity of the fact measure (e.g. less detail data as the data is already aggregated). Subsequently, **Level-2** star schemas are built on top of Level-1 star schemas and have an even lower granularity of the fact measures. Figure 14.3 shows data warehousing granularity consisting of several levels of star schemas in various levels of aggregations.

So, Level-0 has the highest level of granularity where no aggregation exists. All domain tables are incorporated as dimension tables, and there is no grouping in the dimension attributes. Basically, Level-0 means no aggregation. Since Level-0 has no aggregation, Level-0 star schema is almost identical to the E/R diagram but with a different structure. Level-0 star schema is structured in a star schema model, whereas the E/R diagram is not. But in essence, both Level-0 star schema and the

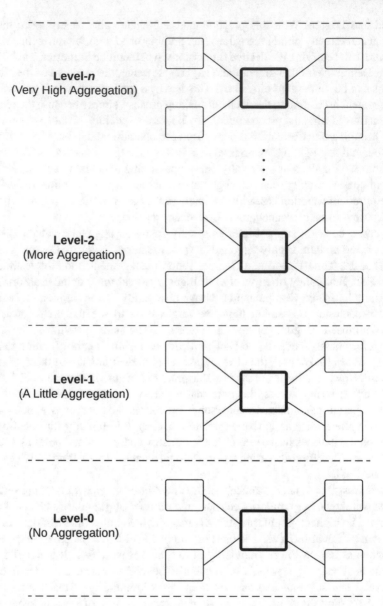

Fig. 14.3 Data warehousing granularity

corresponding E/R diagram convey the same information at the same granularity level.

The lower the level of granularity, the higher the level of aggregation. Because star schemas and data warehousing are about aggregation and aggregate values, we will use the term **level of aggregation** to refer to Level-0, Level-1, Level-2, etc. (instead of level of granularity). The higher the level of aggregation, the higher the level number (e.g. Level-2 has more aggregation than Level-1, Level-3 has more aggregation than Level-2, etc.). There is no particular rule about determining the level of aggregation (with the exception that Level-0 always means no aggregation). The numbering is a sequence number. Therefore, given a star schema, we can only determine whether it is Level-0 or not Level-0. If it is not Level-0, then it depends on how many non-Level-0 there are below this star schema and this will determine the position of this star schema in the data warehousing architecture.

Now let's return to the Computer Lab Usage star schema as shown in Fig. 14.2. This star schema is definitely not a Level-0 star schema because the fact measure is an aggregate value. When creating the Fact Table, a count function is used in the SQL and a group by SemesterID, TimeID and Degree Code attributes. This star schema must be a higher level of aggregation (e.g. Level-1 or Level-2 or higher). This is the reason why we are not able to drill down into the Num of Logins when using this star schema. To be able to drill down to a lower level of aggregation, we need to query the lower level of aggregation of this star schema. Therefore, we must create another star schema with a lower level of aggregation.

There are two ways to lower the level of aggregations of a star schema:

1. *Add a new dimension*. When we add a new dimension, each value in the fact measure will literally be broken down more details on each record of the new dimension.
2. *Replace an existing dimension with a higher granularity dimension*. The values of the fact measures will also be broken down more details because the fact measure has a lower detail dimension.

Using the star schema in Fig. 14.2, to lower this star schema, we could add a new dimension called Student Dimension (see a new star schema in Fig. 14.4). For simplicity, the star schema in Fig. 14.2 is **Level-2**, and the new star schema in Fig. 14.4 is **Level-1**. This Level-1 star schema is created by adding other information available from the tables in the operational database. We can add the student information, and this will subsequently lower the level of aggregation of the Fact Table. This Level-1 star schema is the drill down of the Num of Logins fact measure, which provides the details of who (which students) contributed to this Num of Logins fact measure. In other words, the Fact Table now contains the StudentID. Note that the lower the level of aggregation, the more data we will have in the Fact Table.

To lower the Level-1 star schema, in this example, we use the second method for lowering the aggregation level, that is, by changing the level of granularity of the existing dimensions. The Semester Dimension is now changed to the highest level of granularity, namely Login Date Dimension. This means that the level of aggregation

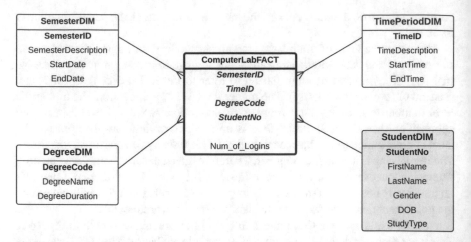

Fig. 14.4 The Computer Lab Usage star schema—a lower level of aggregation

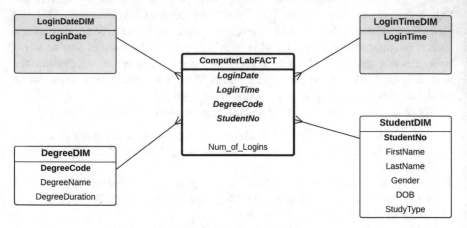

Fig. 14.5 The Computer Lab Usage star schema—Level-0

of the Num of Logins has now dropped to the actual login date, which is the lowest possible level of detail. Note that Level-0 must use the lowest dimension level, and the Login Date Dimension is the lowest from the Semester Dimension perspective.

Another dimension that needs to be lowered is the Time Period Dimension. At the moment, the Time Period Dimension indicates either morning, afternoon or night only (e.g. TimeID 1, 2, 3 represents morning, afternoon, night, respectively). This is a high level of abstraction. To lower this dimension, a Login Time Dimension is used which represents the actual login time for each login.

Level-0 star schema is shown in Fig. 14.5. In this example, Level-0 star schema is the drilling down of the semester and time period which should now contain the actual login date (instead of semester) and the actual login time (instead of time period). This is the level where we do not have any grouping of dimension attributes, and hence Level-0.

In summary, Level-0 star schema provides the most detailed information about the data warehouse. The upper levels provide some levels of aggregated information which have a higher level of aggregation. There is no particular guideline on how many levels we should have in the data warehousing architecture. It can be less or more than three, depending on how many levels of aggregation are needed. Additionally, there is no particular rule on what kind of aggregation is needed on a particular level (except that Level-0 must not have any aggregation). The only rule is that level $(x + 1)$ star schema is more aggregated than level x star schema. Moreover, there could be more than one star schema in each level. For example, Level-0 has one star schema, but Level-1 might have five star schemas, and Level-2 might have ten star schemas.

The upper-level star schema is often implemented as a *dashboard*, where users are able to interact with the dashboard which provides users with highly aggregated information for decision-making. In the implementation, Level-0 star schema (dimensions and Fact Tables) is implemented by the `create table` statements in SQL, whereas the upper-level star schemas are often implemented as a view (implemented by the `create view` statements in SQL).

Now let's compare the three levels of the Computer Lab Usage star schemas, by exploring the differences in their respective Fact Tables. Sample data in the Fact Table of the Level-2 star schema are shown in Table 14.1. Note that $S1$ is Semester 1, and TimeID 3 indicates an evening time period.

Suppose that the management retrieves this data from the Fact Table and is interested in knowing further details about Num of Logins of 1500 which is the number of students in lab in Semester 1, evening, doing a Bachelor of IT. In other words, the management would like to drill down further to find out more information about these 1500 logins. We cannot do this in Level-2 because the Level-2 Fact Table only has data about SemesterID, TimeID and DegreeCode. In order to drill down further, we need a lower level of aggregation of star schema (e.g. Level-1 star schema).

So, a sample data for Level-1 star schema is as follows (see Table 14.2). It basically breaks down the 1500 logins (semester 1, evening, Bachelor of IT) into a number of login figures based on Student No. The total of logins of semester 1, evening, Bachelor of IT must be equal to 1500.

Now suppose the management would like to drill down into the 120 logins (see the highlighted cell in Table 14.2). In particular, management would like to see the actual date and time of login. Then we must go to Level-0 star schema. Level-0 Fact Table is shown in Table 14.3.

Table 14.1 Level-2 Fact Table

SemesterID	TimeID	DegreeCode	Num of logins
S1	3	BIT	**1500**
S1	3	BEng	1250
S1	3	BSc	788
S1	3	MBA	980
...	...	...	...

Table 14.2 Level-1 Fact Table

SemesterID	TimeID	DegreeCode	StudentNo	Num of logins
S1	3	BIT	21002	**120**
S1	3	BIT	21023	90
S1	3	BIT	21025	55
S1	3	BIT	21066	37
S1	3	BIT	...	...
S1	3	BIT	...	...
...	...	...	...	...

Table 14.3 Level-0 Fact Table

Login date	Login time	DegreeCode	StudentNo	Num of logins
4-Apr	19:00	BIT	21002	1
6-Apr	21:20	BIT	21002	1
8-Apr	02:30	BIT	21002	1
3-May	18:55	BIT	21002	1
7-May	19:30	BIT	21002	1
8-May	19:45	BIT	21002	1
...	...	...	...	...

In this example, S1 is broken down into the actual dates that represent Semester 1, and TimeID 3 (night) is broken down into the actual login time that represents the time period night. It is natural that the fact measure in Level-0 star schema contains 1s only, and this shows no aggregation because there is only one login per student at each particular date and time. The total number of logins for this particular student in this example (in Semester 1 at night) must be equal to 120 as indicated by the Level-1 Fact Table.

These three Fact Tables illustrate how the values of the fact measure (e.g. Num of Logins in this example) are broken down into more detail in the lower levels of aggregations. Because the fact measure is a count, the fact measure in Level-0 contains all 1s.

As previously mentioned, there could be more than one star schema at each level of aggregation. For example, we could have a new level in between Level-0 and Level-1 in the Computer Lab Usage case study. The current Level-1 has SemesterID (S1 and S2) and TimeID (1 = morning, 2 = afternoon, 3 = night), whereas Level-0 has the actual LoginDate and the actual LoginTime. We could have an intermediate level where the Semester is broken down into months rather than directly into the actual LoginDate. Similarly, TimeID (morning, afternoon, night) can be broken down into hours rather than directly into the actual LoginTime. Hence, we could have two star schemas in the new Level-1 (note that the old Level-1 becomes Level-2 and the old Level-2 becomes Level-3), as shown in Tables 14.4–14.8.

Note that there are two star schemas in the new Level-1, Level-1a and Level-1b. Level-1a is a lower version of Level-2 because SemesterID in Level-2 is lower than MonthID in Level-1a. In Level-2, the data on the student who has 120 logins

Table 14.4 The new Level-3 Fact Table

SemesterID	TimeID	DegreeCode	Num of logins
S1	3	BIT	**1500**
S1	3	BEng	1250
S1	3	BSc	788
S1	3	MBA	980
...	...	...	...

Table 14.5 The new Level-2 Fact Table

SemesterID	TimeID	DegreeCode	StudentNo	Num of logins
S1	3	BIT	21002	**120**
S1	3	BIT	21023	90
S1	3	BIT	21025	55
S1	3	BIT	21066	37
S1	3	BIT	...	...
S1	3	BIT	...	...
...	...	...	...	...

Table 14.6 The new Level-1a Fact Table

MonthID	TimeID	DegreeCode	StudentNo	Num of logins
Jan	3	BIT	21002	**10**
Feb	3	BIT	21002	**5**
Mar	3	BIT	21002	**7**
Apr	3	BIT	21002	**10**
May	3	BIT	21002	**25**
Jun	3	BIT	21002	**20**
Jul	3	BIT	21002	**34**
...	...	...	...	...

Table 14.7 The new Level-1b Fact Table

SemesterID	HourID	DegreeCode	StudentNo	Num of logins
S1	6:00-6:59pm	BIT	21002	**25**
S1	7:00-7:59pm	BIT	21002	**10**
S1	8:00-8:59pm	BIT	21002	**5**
S1	9:00-9:59pm	BIT	21002	**5**
S1	10:00-10:59pm	BIT	21002	**3**
...	...	...	...	...

is broken down into each month in Level-1a. If the duration of Semester 1 is from 1-Jan to 14-Jul and Semester 2 is from 15-Jul to 31-Dec, then the total logins for this student from Jan to Jul will not necessarily be equal to 120 because the month of July covers two semesters.

Table 14.8 Level-0 Fact Table

Login date	Login time	DegreeCode	StudentNo	Num of logins
4-Apr	19:00	BIT	21002	1
6-Apr	21:20	BIT	21002	1
8-Apr	02:30	BIT	21002	1
3-May	18:55	BIT	21002	1
7-May	19:30	BIT	21002	1
8-May	19:45	BIT	21002	1
...	...	...	...	...

Although the month of July covers two semesters, the fact measure (Num of Logins) will not be double counted, because the raw records will be assigned correctly to the month, without double allocation. Consequently, the Month Dimension does not need to be a Determinant Dimension.

The new Level-1b does not lower down to Semester; rather, it lowers down to TimeID, instead. So, instead of TimeID of morning, afternoon, night, in Level-1b, it uses HourID in which the TimeID is broken down into an hourly base but the SemesterID is kept. The number of logins for this student in Level-1b must be equal to 120, as indicated in Level-2.

Comparing Level-1a and Level-1b, it cannot be said that one level is higher than the other; they are simply not comparable, because MonthID in Level-1a is lower than SemesterID in Level-1b, but TimeID in Level-1a is not lower than HourID in Level-1b. Hence, they are not comparable. But what we are certain of is that both star schemas (Level-1a and Level-1b) are lower than Level-2 and higher than Level-0.

Level-0 Fact Table is not changed by the introduction of the new Level-1a and Level-1b star schemas.

14.2 Facts Without Fact Measures

Looking at the Level-0 star schema of the Computer Lab Usage case study, as shown in Fig. 14.5, we can see that the fact measure, which is Num of Logins, has all 1s in Table 14.8. This means that there is only 1 login for a particular student (who is enrolled in a particular degree) at a certain date and time. Because it is Level-0, this means there is no aggregation, and as a result, each Num of Logins must be equal to 1.

If the fact measure only has the value of 1, then the fact measure may be excluded from the fact. Consequently, the Level-0 star schema of the Computer Lab Usage case study has no fact measure (see the new star schema in Fig. 14.6). This is then known as **facts without fact measures**. Facts without fact measures are only possible if the fact measure is a count, such as Num of Logins in this case. If the

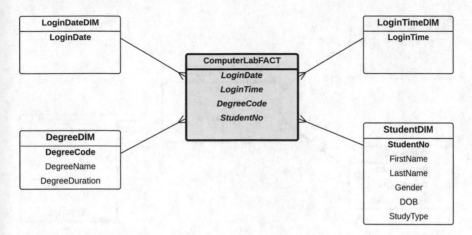

Fig. 14.6 Facts without fact measures

fact measure is not a count, but some other aggregate functions such as sum (e.g. Total Sales), then Level-0 star schema must still include the fact measure (e.g. the amount of sales of each sales).

When designing Level-0 star schema, people often choose the ID as the smallest unit, such as StudentID as the fact measure. The reason is that later in Level-1, when the granularity is reduced, the StudentID will be replaced by an aggregate function (e.g. count), and therefore the fact measure will become Number of Students. However, this kind of design for Level-0 using StudentID as the fact measure is not desirable, because StudentID is linked to a dimension, namely Student Dimension. Fact measure should not be linked to any dimension. So, in this case, it is more desirable to use Fact without Fact Measure, as described above.

The star schema in Fig. 14.6 can be made more compact by combining LoginDate and LoginTime Dimensions into one dimension, called Lab Activities Dimension, because login date and time attributes are part of the Lab Activities entity in the E/R diagram. The new star schema is shown in Fig. 14.7. It does not change the level of granularity because there is only one login for each LoginID.

The most complete Level-0 star schema for the Computer Lab Usage case study is shown in Fig. 14.8. It has four dimensions, including the new Computer Dimension. A comparison of the E/R diagram of the Computer Lab Usage case study (see Fig. 14.1) shows that the star schema in Fig. 14.8 contains all the information from the E/R diagram, but it is structured differently in the star schema. A star schema is basically an *n*-ary relationship between all dimensions with the fact, and the fact actually represents the transactions in the E/R diagram.

When designing a data warehouse, we can start designing from a higher level of aggregation, as shown in this book. Lowering down the level of aggregation can be done either by adding a new dimension to the star schema or by changing the level of granularity of an existing dimension. However, there is a second method for designing a data warehouse, that is, by starting from Level-0 by converting an

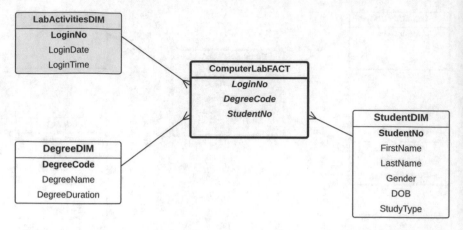

Fig. 14.7 Facts without fact measures—combining LoginDate and LoginTime into LabActivities

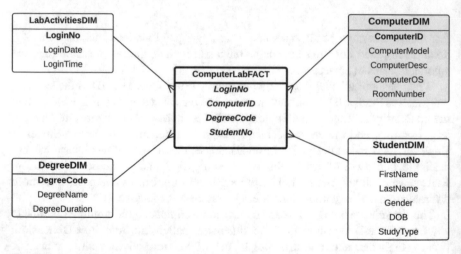

Fig. 14.8 Fact without Fact Measure—most complete star schema

E/R diagram into a star schema structure, which is an *n*-ary relationship between all dimensions and the fact. Then to move up a level, the opposite method is applied, that is, changing the level of granularity of a dimension or removing an existing dimension.

14.3   Star Schemas with No Aggregation

When designing a data warehouse, it is common to begin with a star schema at a high level of aggregation. If we started from Level-0, the data in the data warehouse is basically identical to the data in the operational database, only with a different way of structuring it. The reason why it is the same is because Level-0 star schema has no aggregation. If we started designing a star schema from Level-0, the question as to why we need a star schema in the first place is often asked. This is why designing a star schema often starts from a high level of aggregation, where it is clear that the fact measures are basically aggregated values.

Then to lower the level of aggregation of a star schema, two options are usually available: (*i*) add a new dimension to the star schema, or (*ii*) change the level of granularity of an existing dimension to a higher granularity (or more detailed dimension). If we keep lowering the level of aggregation of a star schema, it will ultimately reach the lowest level of aggregation, known as **Level-0** of aggregation, which is no aggregation in the star schema. Level-0 star schema has all the fact measures not aggregated.

This section examines some complexities and pitfalls of identifying whether or not a star schema is on Level-0. Consider the following Purchase Order case study as an example. A typical E/R diagram of the Purchase Order operational database is shown in Fig. 14.9. A Customer may have several Purchase Orders (POs). The Customer entity contains typical customer details; the Purchase Order entity contains OrderID, Order Date, Pay Method (e.g. credit card, cash, etc.), Ordering Method (e.g. Online Sales, Direct Sales, etc.) and a reference to CustID.

Each Purchase Order may contain several Order Lines, and each Order Line is an Item with a specified quantity of order. The relationship between Purchase Order and Item is a *m-m* through the Order Line entity. The Order Line entity records the Order Price and Quantity of each Item being Purchase Ordered. Each item contains ItemID, QOH (Quantity On Hand), ProductID, Size and Colour. Finally, the Product entity which consists of ProductID, Current Price, Description and Category may

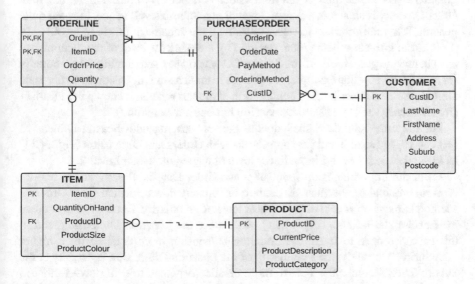

Fig. 14.9 Purchase Order E/R diagram

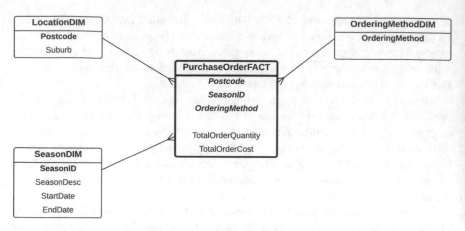

Fig. 14.10   Purchase Order star schema—high level of aggregation

have several Items. In other words, Product is an abstraction and Item is a physical item, which is why Product-Item cardinality is 1-*m*.

A simple star schema captures two fact measures, Total Order Quantity and Total Order Cost (both can be calculated by summing the Order Price and Quantity attributes from the Order Line table). For simplicity, the star schema includes three dimensions, Location Dimension (which is the Suburb and Postcode of Customer), Season and Ordering Method (e.g. Online Sales, Direct Sales, etc.). The star schema is shown in Fig. 14.10 (for simplicity, the Ordering Method Dimension includes one attribute only). This star schema is at a high level of aggregation, because the granularity (or the details) of the dimensions is rather low. For example, the Total Order Quantity is shown at a Season level or at a Suburb level; hence, they are quite general. It is quite common to start designing a star schema at this high level.

To make this star schema more detailed by lowering the level of aggregation, we can change the granularity of the Season Dimension to make it a higher granularity (or more detailed). For example, we change from Season Dimension to Order Date Dimension. This lower level of aggregation of the star schema is shown in Fig. 14.11 (for simplicity, the Order Date Dimension has one attribute only).

For the time being, we shall denote the star schema with Season Dimension (Fig. 14.10) as **Level-2** and the star schema with Order Date Dimension (Fig. 14.11) as **Level-1**, where Level-1 has a lower level of aggregation than Level-2.

From the time dimension point of view, Order Date is already at the highest level of granularity because it cannot be broken down further. The Ordering Method Dimension is also at the highest level of granularity. Hence, to lower down the Level-1 star schema (Fig. 14.11), we choose the Location Dimension, and in this case, we need to make the Location Dimension more detailed (or a higher granularity). Currently, the granularity of the Location Dimension is actually at the Suburb level. To make the Suburb more detailed, we need to drill down further to the addresses within each suburb. Assuming the address of each customer is unique

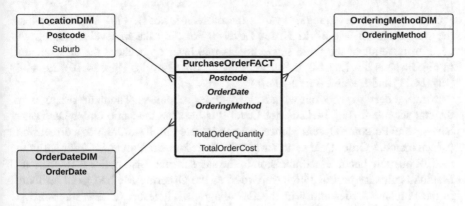

Fig. 14.11 Purchase Order star schema—with Order Date Dimension

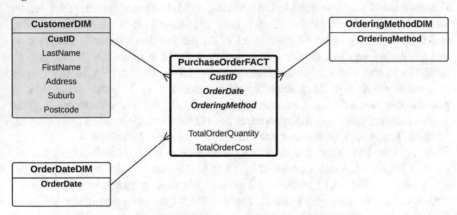

Fig. 14.12 Purchase Order star schema—with Customer Dimension

(hence, each customer is one household), to make the Location Dimension (Suburb) a higher granularity, we can use Customer as a dimension, and hence, Customer Dimension. The new star schema is shown in Fig. 14.12, which is lower than the Level-1 star schema in Fig. 14.11.

As a result, there are three star schemas at different levels of aggregation: **Level-2** (Fig. 14.10) with Season Dimension, **Level-1** (Fig. 14.11) with Order Date Dimension and **Level-0** (Fig. 14.12) with Customer Dimension. The method used to lower the level of aggregation is by changing the level of granularity of a dimension.

Is the star schema in Fig. 14.12 at the lowest possible level (aka Level-0)? Note that Level-0 is the level where there is no aggregation. In other words, in Level-0, it is not possible to break down the fact measures further as there are no aggregate values. Looking at the Purchase Order E/R diagram, as shown earlier in Fig. 14.9, Order Date in Purchase Order is not a candidate key, meaning that it is possible that there is more than one Purchase Order on the same date. When there is more than one Purchase Order per day and the grouping is based on day, this means that

Purchase Order is an aggregated value; therefore, it is not Level-0. This means that the star schema in Fig. 14.12 is **not Level-0**. For the sake of level numbers, we need to move all of the three star schemas one level up because the star schema in Fig. 14.12 is not Level-0; therefore, they become **Level-3** (Fig. 14.10), **Level-2** (Fig. 14.11) and **Level-1** (Fig. 14.12).

Now we need to work out what a Level-0 star schema is. The main reason why the star schema in Fig. 14.12 is not Level-0 is because the same Order Date may have several Purchase Orders. Hence, we need to use Purchase Order as a dimension (which includes Order Date as an attribute). Therefore, the Purchase Order entity in the E/R diagram becomes a dimension in the star schema. Also notice that Ordering Method is already part of Purchase Order, as the Ordering Method is an attribute in the Purchase Order entity in the E/R diagram. Therefore, a new star schema should have a Purchase Order Dimension which basically combines the Order Date Dimension with the Ordering Method Dimension. In the Purchase Order Dimension, where OrderID is the identifier, there is only one OrderID for each Purchase Order, and therefore the number of Purchase Orders per Purchase Order is always one, and it is not aggregated. This new star schema, with the Purchase Order Dimension, as shown in Fig. 14.13, is lower than the Level-1 star schema (Fig. 14.12). Notice that when we change the granularity of a dimension to be more detailed, it is possible that two existing dimensions are captured in the new dimension (e.g. the new Purchase Order Dimension captures both Order Date Dimension and Ordering Method Dimension), so the new star schema has one less dimension.

So now we have four star schemas: **Level-3** (Fig. 14.10), **Level-2** (Fig. 14.11), **Level-1** (Fig. 14.12) and **Level-0** (Fig. 14.13). The main question is whether the star schema in Fig. 14.13 is really a Level-0? We need to check whether the fact measures are still aggregated values. Level-0 should not have aggregated values in the fact measures. However, note that the two dimensions (Customer Dimension and Purchase Order Dimension) are already their lowest level and it is not possible to break them down further. But having the lowest level dimensions does not guarantee that the fact measure will not be an aggregated value. To check whether the fact measures are aggregated or not, we need to go back to the E/R diagram.

Fact measures in the star schema always focus on transactions. In this case study, the transaction happens when a Purchase Order is raised or when a purchase is

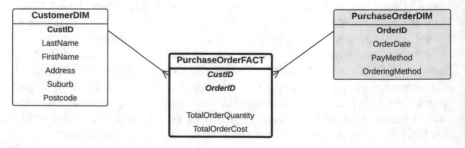

Fig. 14.13 Purchase Order star schema—with Purchase Order Dimension

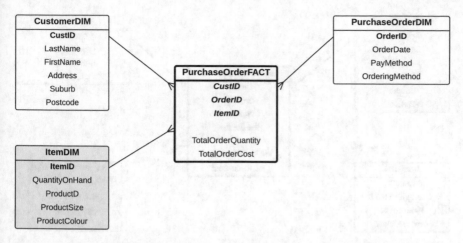

Fig. 14.14 Purchase Order star schema—with Item Dimension

ordered. The fact measures of this star schema are about calculating the quantity
and the cost for each purchase order. Because one Purchase Order can have several
Items, this means the lowest level of transaction is about each item in the Order
Line—each individual record in the Order Line table. A transaction usually occurs
in a *m-m* relationship, which in this case is the relationship between Purchase Order
and Item entities, which is basically the Order Line entity.

The star schema in Fig. 14.13 is **not Level-0** because the fact measures do not
capture both sides of the *m-m* relationship between Purchase Order and Item entities
in the E/R diagram; rather, it captures one side only, which is from the Purchase
Order side and not from the Item side. As a result, the fact measures are still
aggregated as one Purchase Order may contain several Order Lines. To include the
Item, we need an Item Dimension in the star schema. The new star schema with
Item Dimension is shown in Fig. 14.14. This is a true **Level-0** star schema. The final
line-up becomes:

- **Level-4** (Fig. 14.10)
- **Level-3** (Fig. 14.11)
- **Level-2** (Fig. 14.12)
- **Level-1** (Fig. 14.13)
- **Level-0** (Fig. 14.14)

As a conclusion, the Level-0 star schema should focus on the transaction level of
the E/R diagram, which is denoted by the *m-m* relationship. If we need to include all
possible dimensions in the star schema, we can add Product Dimension to the star
schema, as shown in Fig. 14.15. Note that this star schema does not have a lower
level of aggregation than previous star schema shown in Fig. 14.14 as both have the
same level of detail.

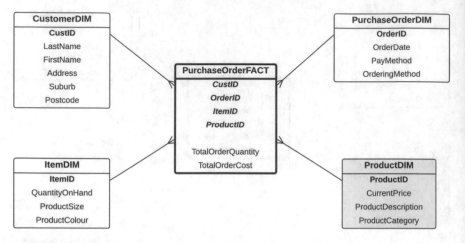

Fig. 14.15 Purchase Order star schema—with Product Dimension

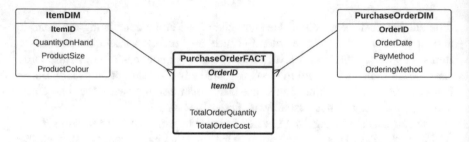

Fig. 14.16 Purchase Order star schema—the minimum requirement

The reason for this is that both star schemas already focus on the transaction level of the operational database, which is the record level of the Order Line table. One Item has one product, and if the star schema is already at the Item level, having Product Dimension will not lower the detail. This is also the same with the Customer because the star schema is at the Purchase Order level and one Purchase Order has only one Customer. Therefore, having a Customer Dimension does not lower the star schema either.

The bottom line is that the Level-0 star schema must have the Purchase Order Dimension and Item Dimension as its core. Therefore, the star schema with these two dimensions, as shown in Fig. 14.16, is Level-0. So there are four possible Level-0 star schemas, the minimal star schema with Purchase Order Dimension and Item Dimension (Fig. 14.16), the most complete star schema with Customer Dimension and Product Dimension (Fig. 14.15) or with either Customer Dimension (Fig. 14.14) or Product Dimension only.

14.4   Understanding the Relationship Between Transactions and Fact Measures

The central point of understanding the Level-0 star schema is by understanding the concept of transactions in the operational databases and where transactions are recorded in the E/R diagram.

In the Computer Lab Usage case study, the transaction is captured in the Lab Activities entity, which is a *m-m* relationship between Student and Computer entities. The fact measure of the star schema is Num of Logins, which is a `count` of records in the Lab Activities table grouped by the dimensions (Student, Degree, Login Date and Login Time).

Using the Purchase Order case study, the transaction is captured in the *m-m* relationship between Purchase Order entity and Item entity through Order Line entity. The fact measures of the star schema in the Purchase Order case study are Total Order Quantity and Total Order Cost (see the star schema in Fig. 14.14). Total Order Quantity is a `sum` of the Quantity attribute in the Order Line entity, whereas the Total Order Cost is also a `sum` of (Order Price × Quantity). Both attributes are also in the Order Line entity.

Suppose that in the Purchase Order case study, the fact measure is Num of Purchase Orders instead of Total Order Quantity and Total Order Cost. The star schema is shown in Fig. 14.17. The transaction focus which is the basis for this star schema is now different from the previous star schema in Fig. 14.14.

In the new star schema with Num of Purchase Orders as the fact measure, the focus of the transaction is no longer on the Order Line entity, but on the Purchase Order entity because the counting of the Num of Purchase Orders is in the Purchase Order entity and not in the Order Line entity. The implication is not seen in the star schema at this level (Fig. 14.17) as the star schema looks similar to the star schema in Fig. 14.9, as both have three dimensions, Location Dimension,

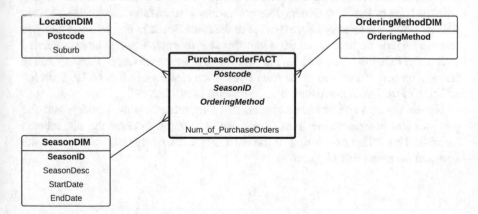

Fig. 14.17 Purchase Order star schema—with Number of Purchase Orders as the fact measure

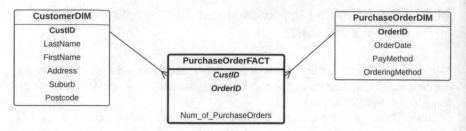

Fig. 14.18 Purchase Order star schema—Level-0

Season Dimension and Ordering Method Dimension. But when we lower down the level of aggregation, by changing the Season Dimension and the Ordering Method Dimension into the Purchase Order Dimension, and the Location Dimension into the Customer Dimension, as shown in Fig. 14.18, this star schema is now already in Level-0. In contrast, the same star schema with Customer and Purchase Order Dimensions, shown in Fig. 14.13, is not Level-0 as previously discussed. Where is the difference?

The difference is in the focus of the fact measure in the transactions. The focus of the previous star schema with Total Order Quantity and Total Order Cost is in the Order Line entity in the E/R diagram. Order Line entity is a *m-m* relationship between Purchase Order entity and Item entity. On the other hand, the star schema with the Num of Purchase Orders has a focus on the Purchase Order entity, not on the Order Line entity. So, the focus of the transaction is different. The granularity of the focus of the transaction is different. This is why the star schema in Fig. 14.18 is Level-0, whereas the star schema in Fig. 14.13 is not.

The difference is highlighted even more when we add a new dimension called Item Dimension to the star schema. In the star schema with Total Order Quantity and Total Order Cost, the star schema needs to have Item Dimension to make the star schema Level-0 (see the star schema in Fig. 14.16). In contrast, the star schema with the Num of Purchase Orders does not have to have an Item Dimension, because the star schema is already Level-0 (refer to the star schema in Fig. 14.18). But if we want to include an Item Dimension into this star schema, a bridge table is needed to connect to the Item Dimension because for each Purchase Order, there are many Items. The new star schema with Item Dimension is shown in Fig. 14.19. With the addition of the Item Dimension, the star schema is still Level-0.

Notice the main difference between this star schema with a bridge and the previous star schema where Item Dimension is required to make the star schema Level-0. This difference is due to the different focus of the fact measure in the transactions in the E/R diagram.

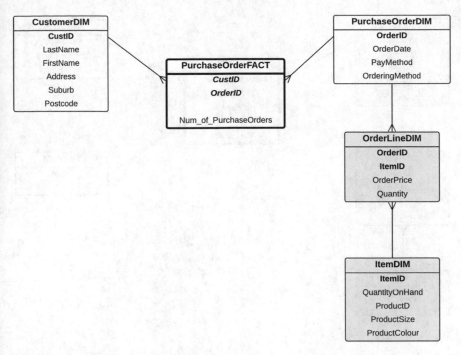

Fig. 14.19 Purchase Order star schema—Level-0 with a Bridge to Item Dimension

14.5 Levels of Aggregations, Hierarchy and Multi-Fact

Levels of aggregation are often discussed in the context of Dimension Hierarchy as well as Multi-Fact. In order to understand these, let's have a look at the Sales star schema, as shown in Fig. 14.20. This star schema is a simple Sales star schema with a hierarchy. The hierarchy is Branch, City and Country Dimensions. There are two other dimensions: Product and Time (Year). The Sales Fact has one fact measure, which is Total Sales.

The Sales Fact shows that Total Sales is actually based on the Branch granularity. The hierarchy is merely an implementation of a normalisation concept in the Relational Database Management Systems (RDBMS), to minimise repetition and to avoid anomalies. The hierarchy is not meant for the drilling down or rolling up of the fact measure.

The hierarchy shows the level of granularity. The hierarchy should start from the most detailed, which in this case is Branch, and go down to the more general, which is City and then Country. If these dimensions in the hierarchy are split into three star schemas, then there will be three Fact Tables. The three star schemas are shown in Fig. 14.21. The first star schema uses the Branch Dimension; therefore, Total Sales is based on the Branch granularity. To differentiate this from the other

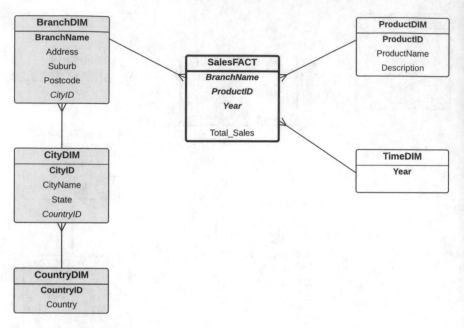

Fig. 14.20 Sales star schema

two star schemas, the Fact Table is called the Branch Sales Fact with Branch Total Sales as the fact measure.

The second star schema uses the City Dimension instead of the Branch. Therefore, the Total Sales fact measure is based on the City granularity. The Fact Table is renamed to City Sales Fact with City Total Sales as the fact measure. Note that the City star schema is more general than the Branch star schema.

The third star schema uses the Country Dimension, which is the most general. The Fact Table becomes the Country Sales Fact with Country Total Sales as the fact measure.

These three star schemas (i.e. Branch star schema, City star schema and Country star schema) focus on the same fact measure, which is Total Sales. The difference is that they have a different level of granularity or level of aggregation. The Branch star schema is the most detailed, whereas the Country star schema is the most general. Hence, there are three levels in this data warehousing architecture, Level-1 for the Branch star schema, Level-2 for the City star schema and Level-3 for the Country star schema.

These three levels can be joined into a multi-fact star schema through the hierarchy, as shown in Fig. 14.22. The hierarchy is maintained among the dimensions from a different level of granularity (e.g. Branch, City and Country Dimensions), and the other dimensions (e.g. Product and Time) are shared by the three Fact Tables.

So, in short, there is a correlation between the concept of data warehousing granularity (or level of granularity), dimension hierarchy and multi-fact. The

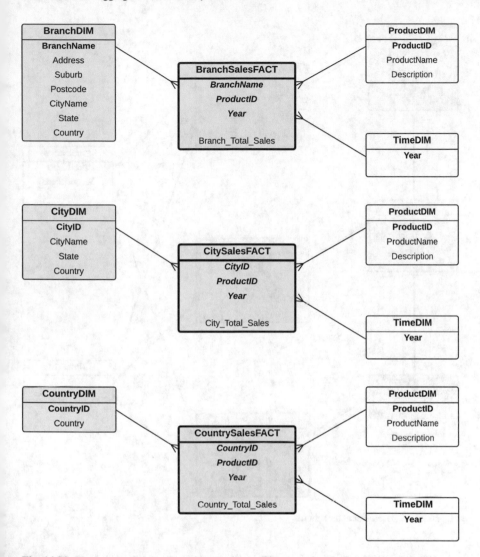

Fig. 14.21 Three star schemas (Branch star schema, City star schema, Country star schema)

different levels of star schema depict the multi-fact and the dimensions in different levels of the star schema can form a hierarchy. As a matter of fact, hierarchy is often used in conjunction with the level of granularity in the data warehousing architecture.

The star schema will become more complex when multiple hierarchies exist. Suppose the Time Dimension has a hierarchy, such as Quarter and then Year, as shown in the star schema in Fig. 14.23.

Using the hierarchy, multi-fact and level of granularity concepts described above, we can have a star schema consisting of two facts, one for Quarterly Sales and the

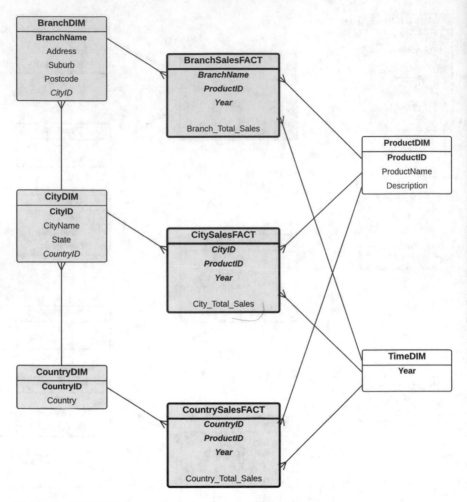

Fig. 14.22 A Multi-Fact star schema

other for Yearly Sales. The hierarchy between Quarter and Year is maintained in the following star schema (refer to Fig. 14.24). So, this is another star schema with Time hierarchy.

If we combine the previous star schema (with the Branch hierarchy) and this star schema (with the Time hierarchy), we will end up with six Fact Tables, as shown in the following star schema (Fig. 14.25) (for simplicity, the Product Dimension is not shown in the picture). This multi-fact, multi-hierarchy star schema is often called a *fact constellation*.

Using this combined star schema, drilling down and rolling up the Total Sales can be done by querying the right level of the star schema; as there are six options to choose from, namely Quarterly and Yearly Branch, City and Country Facts, each with a different granularity of Total Sales.

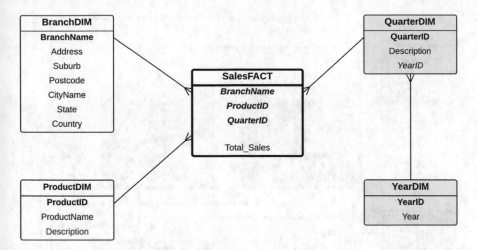

Fig. 14.23 Another hierarchy (Quarter, Year)

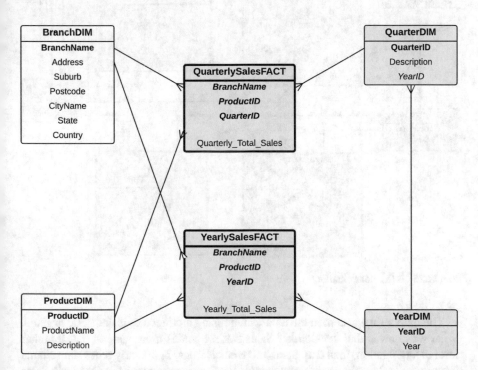

Fig. 14.24 A multi-fact star schema with Quarter and Year hierarchy

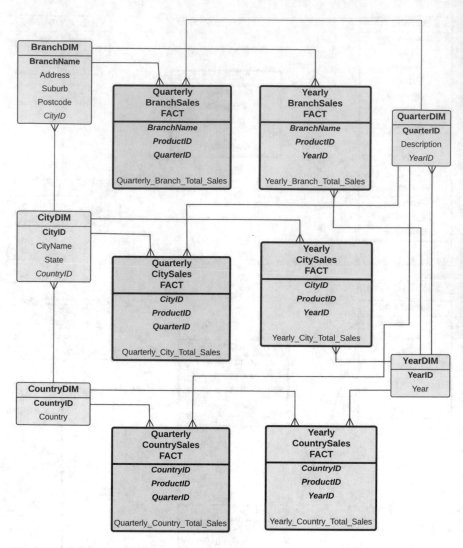

Fig. 14.25 A fact constellation

The numbering system in the data warehousing architecture is not static or fixed. What we know is that the Country Sales star schema is more general (has a higher level of aggregation) than City Sales and Branch Sales. The Yearly Sales star schema is more general and hence has a higher level of aggregation than the Quarterly Sales star schema. The following six levels of aggregation for the above star schema are not correct:

- Level-6: Yearly Country Sales
- Level-5: Quarterly Country Sales

- Level-4: Yearly City Sales
- Level-3: Quarterly City Sales
- Level-2: Yearly Branch Sales
- Level-1: Quarterly Branch Sales

Take the Quarterly Country Sales and the Yearly City Sales as an example. It is incorrect to say that the Quarterly Country Sales is more general than the Yearly City Sales because they are not formed by one hierarchy. This is the reason why the above level numbering is incorrect.

When comparing two star schemas, we only need to know whether one star schema is more general than the other or whether the two star schemas cannot be compared. In the above example, Quarterly Country Sales and Yearly City Sales are not comparable, whereas Yearly Country Sales is definitely more general than Quarterly Country Sales. Therefore, we should have two partitions or groups of levels of granularity, one based on time and the other based on location.

The relationship between the six star schemas is actually depicted in Fig. 14.26. It can be seen that there are four different "paths" of the level of aggregation. One path is from Country Yearly being the most general to City Yearly less general, then to Branch Yearly and finally to Branch Quarterly being the most detailed.

The second path of the level of aggregation is also from Country Yearly being the most general to Country Quarterly less general, then to City Quarterly and to Branch Quarterly which is the most detailed. The third path of the level of aggregation is Country Yearly, City Yearly, City Quarterly and Branch Quarterly, and the fourth path is Country Yearly, Country Quarterly, Branch Yearly and Branch Quarterly.

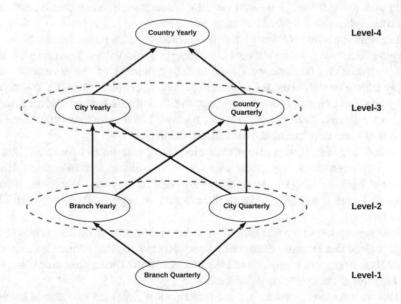

Fig. 14.26 A simpler fact constellation

All of the four paths have the same most general star schema, which is Country Yearly, and the most detailed star schema, which is Branch Quarterly. Each of the four paths of the level of aggregation guarantees that the level of aggregation within each path is accurate. For example, in the first path, City Yearly is more general than Branch Yearly.

It is clear from these four paths of the level of aggregation that Country Yearly being the most general is at the top level of the data warehousing architecture (e.g. Level-4 in this example) and the next level down (e.g. Level-3) is City Yearly and Country Quarterly. Level-2 contains Branch Yearly and City Quarterly. Level-1 which is the most detailed is Branch Quarterly.

14.6 Summary

This chapter introduces the concept of levels of aggregation. A series of star schemas actually form a hierarchy of levels, whereby Level-0 aggregation is the star schema without aggregation in the fact measures. The higher the level of aggregation, the more aggregation in the fact measure. This hierarchy of the level of aggregation is called *data warehousing granularity*.

It is important to understand the concept of data warehousing architecture because a data warehouse is built primarily for drilling down to find some interesting data for business decisions. Some users (such as top managers) may focus on high levels of aggregation as the data is already highly aggregated, whereas other users might want to dig further down into the data; hence, more detailed data is needed—a data warehouse with a lower level of aggregation.

It is common when we design a data warehouse that we start from a high level of aggregation where fact measures contain aggregated values. Lowering the level of aggregation can be done by changing the granularity of the dimension or by simply adding new dimensions. Designing a data warehouse from the lowest level of aggregation (Level-0) is often counterproductive because the operational database itself is the lowest level; hence, creating a Level-0 data warehouse can be seen as unnecessary and duplicating the operational database.

Level-0 data warehouses also often include facts without fact measures because there is no need for a numerical value to represent the fact measures. This is permitted because Level-0 contains no aggregation anyway. However, when it moves to Level-1, the fact measure is necessary to capture the aggregated value of the fact.

Determining whether a star schema is in Level-0 or not can be tricky. Not having an aggregated fact measure does not always mean that the star schema is in Level-0. Hence, it is important to understand the concept of the transaction recorded in the E/R diagram of the operational database.

Data warehousing granularity that contains star schemas of various levels of aggregation can be seen as multi-fact star schemas formed in a global hierarchy, which is also known as *fact constellation*. Hence, having a global overview of all star schemas in the fact constellation is important, especially in data investigation during business decision-making.

14.7 Exercises

14.1 In the Computer Lab Usage case study explained earlier in this chapter, the E/R diagram of the operational database is shown in Fig. 14.1. It has four tables: Computer, Lab Activities, Student and Degree. The three levels of star schemas are shown in Fig. 14.2 for Level-2, Fig. 14.4 for Level-1 and Fig. 14.5 for Level-0.

Tasks: Create these three star schemas, including their Fact Tables and all of the dimension tables, using SQL statements.

14.2 In the Purchase Order case study in this chapter, the E/R diagram of the operational database is shown in Fig. 14.9. This Purchase Order operational database has five tables: Customer, Purchase Order, Order Line, Item and Product. Five levels of star schema have been designed. The Level-4 star schema (Fig. 14.10) consists of the Location Dimension, Season Dimension and Ordering Method Dimension. The Level-3 star schema (Fig. 14.11) replaces the Season Dimension with the Order Date Dimension. The Level-2 star schema (Fig. 14.12) replaces the Location Dimension with the Customer Dimension, as the location is part of each customer. The Level-1 star schema (Fig. 14.13) combines the Order Date Dimension and Ordering Method Dimension to become the Purchase Order Dimension, because the order date and ordering method attributes are both in the Purchase Order entity. Finally, the Item Dimension is added to Level-0 (Fig. 14.14).

Tasks: Create these five star schemas, including their Fact Tables and all of the dimension tables, using SQL statements. As there are two more Level-0 star schemas (see Figs. 14.15 and 14.16), create these two Level-0 star schemas as well.

14.3 Note that the Level-0 star schemas represent the transactions in the operational database. A transaction is normally denoted by a m-m cardinality relationship. In the Computer Lab Usage case study, as shown in the E/R diagram in Fig. 14.1, the transactions are represented by the m-m relationship between Student and Computer entities, through the Lab Activity entity. In the Product Order case study (the E/R diagram is shown in Fig. 14.9), the transaction is a m-m relationship between Purchase Order and Item entities, through the Order Line entity. Both Level-0 star schemas of these two case studies focus on these particular entities and entity relationships.

In the Purchase Order case study, the Level-0 star schema includes two main dimensions: Purchase Order and Item Dimensions (refer to the Level-0 star schema in Fig. 14.16), because these two dimensions are basically the two entities in the E/R diagram which are the basis of the purchase order transactions. It may include Customer and Product Dimensions, but these are optional.

On the other hand, in the Computer Lab Usage case study, the Level-0 star schema includes Student Dimension, Degree Dimension, Login Date Dimension

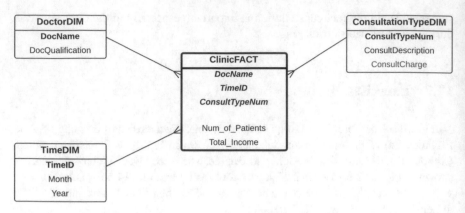

Fig. 14.27   Clinic Star Schema Version 1

and Login Time Dimension. The transaction in the Computer Lab Usage case study is a *m-m* relationship between Student and Computer entities. The Computer Dimension is not used in the Level-0 star schema (see Fig. 14.5). Why is Computer Dimension not used in this Level-0 star schema, and yet the star schema in Fig. 14.5 is Level-0? In contrast, the Item Dimension must be used in the Level-0 star schema in the Purchase Order case study. What is the difference between these two case studies to result in the Level-0 star schema solutions being different?

14.4   A medical clinic employs four general practitioners (doctors): Dr Adele, Dr Ben, Dr Kate and Dr Chris. Some of these doctors do not practise every day. For example, Dr Adele practises on Mondays and Wednesdays only, whereas Dr Ben is there only on Thursdays. When a patient comes to the clinic and has a consultation with a doctor, the patient pays a certain consultation fee, depending on the type of consultation the patient had. For example, a general consultation fee (code 113) is $37.50. Because of the nature of medical practice, there are more than 100 different codes for different types of consultations.

The clinic maintains an operational database that records every payment for each consultation by every doctor. A data warehouse is needed for reporting purposes. There are two versions of star schema for this clinic. Figures 14.27 and 14.28 show the two star schemas.

Questions: Which schema (Star Schema Version 1 or Star Schema Version 2) has a higher level of aggregation, or are they of the same level of aggregation? State your answer, and explain your reasons as well.

14.5   A tourist bus company hires buses to groups of people for tourism purposes. Their customers include primary and secondary schools, various organisations, groups of private foreign tourists, etc. The company has many buses of various sizes (e.g. 45-seater buses, 24-seater buses or minibuses and 12-seater cars). When clients hire a bus, the driver is included in the hiring package as well.

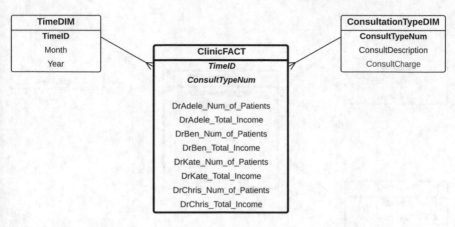

Fig. 14.28 Clinic Star Schema Version 2

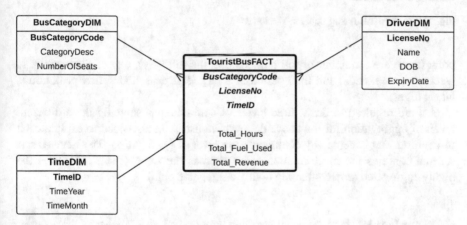

Fig. 14.29 Tourist Bus Star Schema Version 1

Two star schemas have been developed. Figures 14.29 and 14.30 show the two star schemas.

Questions: Which schema (Star Schema Version 1 or Star Schema Version 2) has a higher level of aggregation, or are they of the same level of aggregation? State your answer and explain your reasons as well.

14.6 There is a toll way (or toll road) in a metropolitan city (such as CityLink or EastLink in Melbourne, or any similar toll roads in other major cities in the world). This toll road has a number of toll points where motorists are electronically charged for travelling on the toll road. Every time a motorist passes through this tool point, the registration number of the vehicle, vehicle type (e.g. car, bus, truck, etc.), amount paid and time are recorded in the operational database.

A data warehouse needs to be built to analyse the revenue from the toll payments. The management would like to drill down into this revenue based on the toll

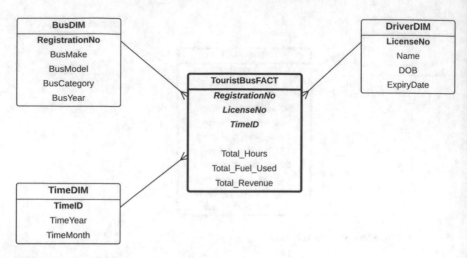

Fig. 14.30 Tourist Bus Star Schema Version 2

point (there are a number of toll points along the toll way), day of the week (e.g. weekdays, weekends) and time period of a day (e.g. peak hours, non-peak hours, late at night).

You are required to draw three levels of star schemas showing three different levels of aggregation for the above data warehouse. You also need to explain each of the three star schemas by contrasting the level of aggregation. The Level-0 star schema contains the most detailed data, whereas the Level-2 star schema is the highly aggregated (e.g. containing highly aggregated data).

Tasks

- Draw a Level-2 star schema and explain why it is a Level-2 schema.
- Draw a Level-1 star schema and explain why it is a Level-1 schema. You may want to add a new dimension called vehicle (e.g. cars, trucks, buses, etc.). You also need to explain the difference between Level-1 and Level-2 schemas.
- Draw a Level-0 star schema and explain why it is a Level-0 schema. You also need to explain the difference between Level-1 and Level-0 schemas.

14.8   Further Readings

The concept of granularity is crucial in understanding in-depth star schema design, without which an incorrect perception of the design will come to surface quite easily. This book is the only book that discusses the concept of granularity in data warehousing, including star schema design using higher level and lower level of aggregation. More general star schema design and modelling has been covered in various data warehousing books, such as [1–10] and [11]. Data modelling books, such as [12–16] and [17], would also give readers a broader perspective on differences between various levels of aggregation in data modelling.

The concept of data granularity has been used in various applications. For example, in the digital image processing area, images or pictures can be stored in multiple pixel sizes [18]. The sharper the picture, the denser the number of pixels. For the same images, it can be stored in different levels of granularity. The concept of granularity is also used in digital video processing and audio processing, where video and audio files can be stored in various levels of details [19, 20].

Data produced by sensors in the form of data streams, which is raw data, may be processed into lower granularity, by aggregating data into coarser-grained data [21–23].

Granularity, particularly time granularity, has been discussed in the context of databases and data mining [24], as well as in distributed database transactions and real-time database logical design [25, 26]. Granularity in data modelling is discussed in [27].

References

1. C. Adamson, *Star Schema The Complete Reference* (McGraw-Hill Osborne Media, 2010)
2. R. Laberge, *The Data Warehouse Mentor: Practical Data Warehouse and Business Intelligence Insights* (McGraw-Hill, New York, 2011)
3. M. Golfarelli, S. Rizzi, *Data Warehouse Design: Modern Principles and Methodologies* (McGraw-Hill, New York, 2009)
4. C. Adamson, *Mastering Data Warehouse Aggregates: Solutions for Star Schema Performance* (Wiley, London, 2012)
5. P. Ponniah, *Data Warehousing Fundamentals for IT Professionals* (Wiley, London, 2011)
6. R. Kimball, M. Ross, *The Data Warehouse Toolkit: The Definitive Guide to Dimensional Modeling* (Wiley, London, 2013)
7. R. Kimball, M. Ross, W. Thornthwaite, J. Mundy, B. Becker, *The Data Warehouse Lifecycle Toolkit* (Wiley, London, 2011)
8. W.H. Inmon, *Building the Data Warehouse*. ITPro Collection (Wiley, London, 2005)
9. M. Jarke, *Fundamentals of Data Warehouses*, 2nd edn. (Springer, Berlin, 2003)
10. E. Malinowski, E. Zimányi, *Advanced Data Warehouse Design: From Conventional to Spatial and Temporal Applications*. Data-Centric Systems and Applications (Springer, Berlin, 2008)
11. A. Vaisman, E. Zimányi, *Data Warehouse Systems: Design and Implementation*. Data-Centric Systems and Applications (Springer, Berlin, 2014)
12. T.A. Halpin, T. Morgan, *Information Modeling and Relational Databases*, 2nd edn. (Morgan Kaufmann, Los Altos, 2008)
13. J.L. Harrington, *Relational Database Design Clearly Explained*. Clearly Explained Series (Morgan Kaufmann, Los Altos, 2002)
14. M.J. Hernandez, *Database Design for Mere Mortals: A Hands-On Guide to Relational Database Design*. For Mere Mortals (Pearson Education, 2013)
15. N.S. Umanath, R.W. Scamell, *Data Modeling and Database Design* (Cengage Learning, 2014)
16. T.J. Teorey, S.S. Lightstone, T. Nadeau, H.V. Jagadish, *Database Modeling and Design: Logical Design*. The Morgan Kaufmann Series in Data Management Systems (Elsevier, Amsterdam, 2011)
17. G. Simsion, G. Witt, *Data Modeling Essentials*. The Morgan Kaufmann Series in Data Management Systems (Elsevier, Amsterdam, 2004)

18. B. Jähne, *Digital Image Processing*. Engineering Online Library (Springer, Berlin, 2005)
19. M. Parker, S. Dhanani, *Digital Video Processing for Engineers: A Foundation for Embedded Systems Design* (Elsevier, Amsterdam, 2012)
20. M.G. Christensen, *Introduction to Audio Processing* (Springer, Berlin, 2019)
21. P. Sangat, M. Indrawan-Santiago, D. Taniar, Sensor data management in the cloud: data storage, data ingestion, and data retrieval. Concurr. Comput. Pract. Exp. **30**(1), e4354 (2018)
22. L. Golab, M.T. Ozsu, *Data Stream Management*. Synthesis Lectures on Data Management (Morgan & Claypool Publishers, 2010)
23. H.C.M. Andrade, B. Gedik, D.S. Turaga, *Fundamentals of Stream Processing: Application Design, Systems, and Analytics* (Cambridge University Press, Cambridge, 2014)
24. C. Bettini, S. Jajodia, S. Wang, *Time Granularities in Databases, Data Mining, and Temporal Reasoning* (Springer, Berlin, 2000)
25. M.T. Özsu, P. Valduriez, *Principles of Distributed Database Systems*. (Springer, New York, 2011)
26. T.J. Teorey, *Database Modeling and Design*. The Morgan Kaufmann Series in Data Management Systems (Elsevier, Cambridge, 1999)
27. G. Powell, *Database Modeling Step by Step* (CRC Press, Boca Raton, 2020)

Chapter 15
Designing Lowest-Level Star Schemas

So far, this book has explained that designing a star schema starts from a higher level of aggregation. The main reason for this is that Level-0 star schemas are often very similar, if not identical, to the operational database represented by the E/R diagram. Therefore, from a pedagogical point of view, it is often difficult to understand why a data warehouse is needed if we can get everything from the operational database.

A higher-level star schema represents the precomputed values of fact measures. With these, data retrieval from the star schema may well suit decision-making, which is often based on a higher level of aggregation as it depicts a global picture of the business operation. Therefore, a Level-0 star schema is often seen as unnecessary. Additionally, going up or down the level of aggregation can be done quite easily by changing the granularity of the dimensions and/or adding/removing dimensions from the star schema.

However, in some cases, designing a Level-0 star schema is needed. This chapter focuses on when and how to design a Level-0 star schema. The first case study is the median house price and the second case study is exploring the use of some of the statistical functions that require the star schemas to be at Level-0.

15.1 Median House Price

An operational database which stores the sales records of properties (e.g. houses, apartments, units) already exists. The E/R diagram is shown in Fig. 15.1. The database is very simple, as it contains only four tables. The main table is the Property table which keeps the details of each property, with Property No as its Primary Key. The transaction table is the Sold table, which stores the sold price and sold date. For simplicity of design, the relationship between Property and Sold is 1-1. There are two other minor tables: Sold Method (e.g. auction, private sale) and Property Type (e.g. house, apartment, unit) tables.

© The Author(s), under exclusive license to Springer Nature Switzerland AG 2021 391
D. Taniar, W. Rahayu, *Data Warehousing and Analytics*, Data-Centric Systems
and Applications, https://doi.org/10.1007/978-3-030-81979-8_15

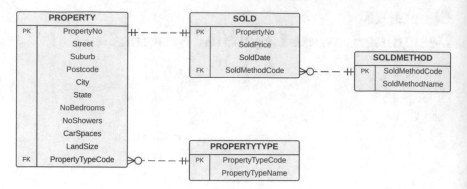

Fig. 15.1 Sold Properties E/R diagram

One of the main indicators in the housing market is *median price*, which is the "middle" price. If the houses are sorted based on the Sold Price, the Sold Price of the house in the middle of this list is the median price. This means we need to keep each individual Sold Price as it is not possible to aggregate the houses, either based on Suburb, or Month, or any other parameters. Each individual house is needed in the data warehouse, because without this, the median price cannot be determined. Consequently, the star schema must be at the lowest level, which is Level-0. Assume we would like to have four dimensions, such as Suburb, Month, Property Type and Sold Method, and the query to the data warehouse is something like: "what is the median price of houses sold in 2020 in a certain suburb?" This kind of question is often asked in the property market which may determine the performance of properties in a certain place.

If we examine, particularly, the Suburb and Month Dimensions, these dimensions are not in the highest granularity; rather, they are already coarse-grained because month is more general than dates and suburb is more general than streets, for example. Because determining the median price needs each individual property, we need to go to the lowest level of these two dimensions. Instead of Suburb, we need to go down to the actual address of the property (note that just the street is not detailed enough because there is a possibility that there is more than one house on the same street in this database). Also, for the Month, we need to go down to the actual sold date, as this is the most detailed and cannot be broken down further.

Therefore, the star schema's four dimensions are Address, Sold Date, Property Type and Sold Method Dimensions. Note that the Property Type and Sold Method Dimensions are already the most detailed or at the highest granularity. The fact measure is Sold Price. Note that the fact measure is not an aggregated value, it is an individual value, that is, a Level-0 fact measure (e.g. no aggregation). The star schema is shown in Fig. 15.2. Note that the Address Dimension uses Property No as the identifier as it is unique to each property.

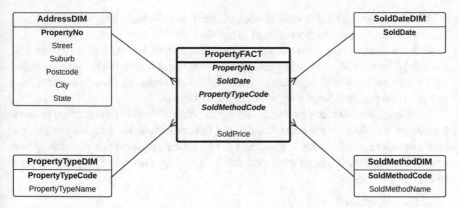

Fig. 15.2 Level-0 star schema

Property Type and Sold Method Dimensions are identical to the respective tables/entities in the E/R diagram. The Address Dimension is also taken from the Property entity, with only selected attributes, and the Sold Date Dimension contains the Sold Date attribute taken from the Sold entity. The SQL commands to create the Level-0 star schema are as follows:

```
create table AddressDim as
select
   PropertyNo,
   Street,
   Suburb,
   Postcode,
   City,
   State
from Property;

create table SoldDateDim as
select distinct SoldDate
from Sold;

create table PropertyTypeDim as
select * from PropertyType;

create table SoldMethodDim as
select * from SoldMethod;

create table PropertyFact as
select
   P.PropertyNo,
   S.SoldDate,
   P.PropertyTypeCode,
   S.SoldMethodCode,
   S.SoldPrice
from Property P, Sold S
where P.PropertyNo = S.PropertyNo;
```

The Property Fact Table only needs data from the Property and Sold tables, and there is no aggregate function used in the SQL.

If we examine this Level-0 star schema further, it looks similar to the E/R diagram. The only difference is perhaps the view; the star schema is centred around the fact, so the view is clear, whereas the E/R diagram shows the binary relationships between two entities. So, the focus is slightly different.

With this Level-0 star schema, to answer the query "retrieve the median price of a house sold in 2020 in the Kew suburb", the SQL is as follows. In this example, the median aggregate function is used. But not all DBMS support this function. As an exercise at the end of this chapter, you will be asked to write an SQL query without using this function.

```
select median(T.SoldPrice)
from
   (select
      A.Suburb,
      P.PropertyTypeName,
      to_char(F.SoldDate, 'YYYY'),
      F.SoldPrice
   from PropertyFact F, AddressDim A, PropertyTypeDim P
   where F.PropertyNo = A.PropertyNo
   and F.PropertyTypeCode = P.PropertyTypeCode
   and A.Suburb = 'Kew'
   and P.PropertyTypeName = 'House'
   and to_char(F.SoldDate, 'YYYY') = '2020'
   order by F.SoldPrice) T;
```

If we would like to create a higher-level star schema from this Level-0 star schema, we could change the granularity of the two dimensions, Address and Sold Date Dimensions, to more general dimensions. So, instead of the Address Dimension, we change it to the Suburb Dimension, and instead of the Sold Date Dimension, we change it to the Sold Month Dimension. By changing these two dimensions, the star schema becomes more general. However, the Sold Price fact measure needs to change as well because when the properties are grouped together, the individual Sold Price disappears.

In the new star schema with a high level of aggregation, the individual Sold Price needs to be aggregated. So, instead of answering the median query, the query for the higher level of star schema could be the mean query (or the average query). In order to answer the mean query, the star schema must have the Total Sold Price and Number of Properties; these two become the new fact measure of the higher-level star schema, as shown in Fig. 15.3. Note that to answer the mean or the average query, the star schema does not need to be in Level-0.

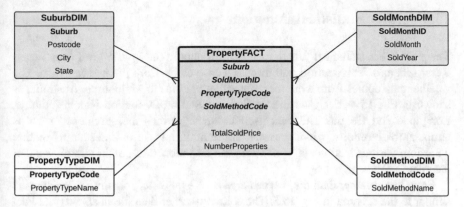

Fig. 15.3 Higher level of aggregation star schema

The SQL commands to create this new star schema are as follows. Note that we do not need to recreate the Property Type Dimension and Sold Method Dimension tables. They can be reused from the Level-0 star schema.

```
create table SuburbDim as
select distinct
   Suburb,
   Postcode,
   City,
   State
from Property;

create table SoldMonthDim as
select distinct
   to_char(SoldDate, 'MMYYYY') as SoldMonthID,
   to_char(SoldDate, 'Mon') as SoldMonth,
   to_char(SoldDate, 'YYYY') as SoldYear
from Sold;

create table PropertyFactLevel1 as
select
   P.Suburb,
   to_char(S.SoldDate, 'MMYYYY') as SoldMonthID,
   P.PropertyTypeCode,
   S.SoldMethodCode,
   sum(S.SoldPrice) as TotalSoldPrice,
   count(*) as NumberProperties
from Property P, Sold S
where P.PropertyNo = S.PropertyNo
group by
   P.Suburb,
   to_char(S.SoldDate, 'MMYYYY'),
   P.PropertyTypeCode,
   S.SoldMethodCode;
```

15.2   Other Statistical Functions

There are other statistical functions that need data at the lowest level (e.g. Level-0 star schema). This section will discuss a case study from the *weather* data. The weather data comes from a number of weather stations in Melbourne, Australia, as shown in Fig. 15.4. Each weather station has StationID, Station Name, Latitude, Longitude and Height. The data itself contains various measurements, such as temperature, humidity, wind, pressure and rainfall, all with a timestamp. Weather data from various stations in Melbourne can be unioned to obtain a large weather data file.

From this weather data file, we can create a Level-0 star schema, which appears similar to the schema in Fig. 15.5. The source weather data file is already the Fact Table of the data warehouse. However, for simplicity, only a few fact measures are selected, namely, Temperature, Humidity, Wind Speed, Wind Gust, Pressure and Rainfall.

The star schema has three chosen dimensions: Weather Station, Time and Wind Direction Dimensions. Weather Station and Time Dimensions are quite obvious because the weather data measurements are recorded based on the time at a particular weather station. However, the Wind Direction Dimension is not that obvious. If we look at the source weather data, there is important information about wind, particularly wind speed, as well as wind direction. Wind Speed is a fact measure because it is a numerical value, which later can be aggregated in a higher level of star schema. However, Wind Direction is not; hence, it is not suitable to become a fact measure of a data warehouse. Additionally, it is not possible to aggregate Wind Direction in a higher level of data warehouse. A solution for this is to make Wind Direction a dimension. But this violates the concept of the *two-*

Latest Weather Observations for Melbourne Airport

IDV60901

Issued at 7:41 pm EDT Thursday 14 November 2019 (issued every 10 minutes, with the page automatically refreshed every 10 minutes)

Station Details ID: 086282 Name: MELBOURNE AIRPORT Lat: -37.67 Lon: 144.83 Height: 113.4 m

Data from the previous 72 hours. | See also: Recent months at Melbourne Airport

Date/Time EDT	Temp °C	App Temp °C	Dew Point °C	Rel Hum %	Delta-T °C	Wind Dir	Wind Spd km/h	Wind Gust km/h	Wind Spd kts	Wind Gust kts	Press QNH hPa	Press MSL hPa	Rain since 9am mm
14/07:30pm	19.1	15.2	8.6	50	5.5	SW	19	28	10	15	1011.8	1011.6	0.0
14/07:00pm	20.3	15.8	8.5	46	6.2	WSW	22	30	12	16	1011.8	1011.6	0.0
14/06:30pm	20.5	15.4	7.8	44	6.6	SW	24	33	13	18	1011.9	1011.7	0.0
14/06:00pm	20.6	15.8	7.4	42	6.8	W	22	30	12	16	1011.9	1011.7	0.0
14/05:30pm	21.0	16.8	7.5	41	7.0	WSW	19	22	10	12	1012.0	1011.8	0.0
14/05:00pm	21.9	17.7	7.5	39	7.5	WNW	19	32	10	17	1012.0	1011.8	0.0
14/04:30pm	21.3	16.5	7.2	40	7.3	WNW	22	37	12	20	1012.3	1012.1	0.0
14/04:00pm	21.3	17.5	7.4	40	7.2	WNW	17	30	9	16	1012.3	1012.1	0.0
14/03:30pm	20.2	16.8	7.6	44	6.5	WNW	15	20	8	11	1012.9	1012.7	0.0
14/03:00pm	20.5	16.1	7.5	43	6.7	W	20	30	11	16	1013.4	1013.2	0.0

Fig. 15.4 Sample data from Melbourne Airport Weather Station

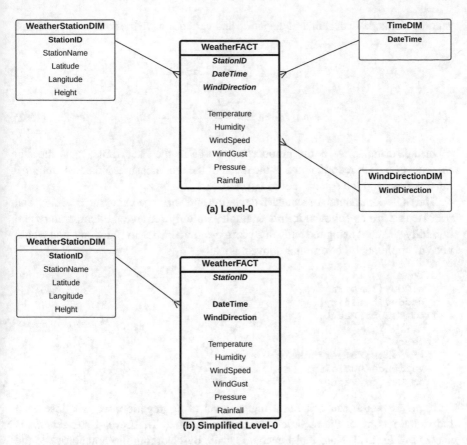

Fig. 15.5 Level-0 star schema for the weather data

column table methodology introduced earlier to validate a star schema, since the Wind Direction Dimension is only related to one fact measure, Wind Speed, and not to other fact measures.

Since two dimensions, Time and Wind Direction Dimensions, are one-attribute dimensions, the star schema is simplified by making Data Time and Wind Direction dimensionless keys in the Fact Table. Figure 15.5 shows the original and simplified Level-0 star schema for the weather data. It is easily noticeable that the Fact Table is almost *identical* to the source weather data (the Fact Table only selected a few important fact measures from the source weather data even though it could have selected all), with only an additional Weather Station Dimension which stores the details of the weather station.

With this Level-0 star schema, we could perform more statistical queries about the weather, such as *Mean*, *Standard Deviation* and *Variance*. The mean (or average) does not require the Fact Table to be at Level-0, as already discussed in the earlier chapters. But for the standard deviation and variance, we need to get each individual record. Mathematical, standard deviation and variance have the following formulas:

$$stddev = \sqrt{\frac{\sum_{i=1}^{N}(x_i - \bar{x})^2}{N}} \tag{15.1}$$

$$variance = \frac{\sum_{i=1}^{N}(x_i - \bar{x})^2}{N} \tag{15.2}$$

In the formulas, N is the number of records in the Fact Table, x_i is the fact measure of the ith record, and $\bar{x}$ is the mean of the fact measure. Note that *standard deviation* is simply a square root of *variance*.

The SQL commands for standard deviation and variance using the `stddev` and `variance` are as follows. In this example, we only retrieve the temperature and humidity. Without using the built-in `stddev` and `variance` functions, you would need to implement the formulas above manually.

```
select
  stddev(Temperature),
  stddev(Humidity)
from WeatherFact0;

select
  variance(Temperature),
  variance(Humidity)
from WeatherFact0;
```

If we make the star schema a higher level of aggregation, we will lose each individual record, as the records are being aggregated. The Level-1 star schema is shown in Fig. 15.6. Note that Level-1 is made by changing the granularity of the Time Dimension in Level-0 to be the Date Dimension in Level-1. Consequently, the fact measures need to change. In this example, Level-1 has three fact measures: High Temperature, Low Temperature and Highest Wind Gust. High and Low Temperature indicate the maximum and minimum temperature recorded for each day, whereas Highest Wind Gust is the maximum value of Wind Gust for each day. These maximum and minimum values are aggregated values from the original records. Therefore, it will only have one value for each day instead of one value for each recording, which is, in this case, every 30 min. For the clarity of the star schema of Level-1, the Weather Date Dimension and Wind Direction Dimension are kept as one-attribute dimensions. They could be changed to dimensionless keys in the fact if these two dimensions are removed from the star schema. Obviously, using the Level-1 star schema, we cannot query the standard deviation and variance, as they require the original records and they cannot work with already aggregated records. However, calculating the mean can be done in a higher level of aggregation star schema if we have the total of a fact measure and the number of records of a fact measure. However, in this example, the Level-1 star schema does not have these two fact measures. Hence, calculating the mean (e.g. of temperature) still needs to use a Level-0 star schema.

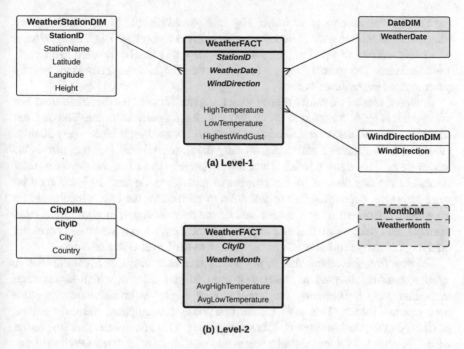

Fig. 15.6 Higher level of aggregate star schemas for the weather data

Figure 15.6 also shows the Level-2 star schema where the Weather Date Dimension is now changed to the Weather Month Dimension in order to make it more general. Usually, monthly weather information only involves Average High and Low Temperatures. In this case, the average is used; hence, the Weather Month Dimension becomes a Determinant Dimension, and because it is a Determinant Dimension, it has to be kept as a dimension. The Determinant Dimension cannot be removed from the star schema.

15.3 Querying Level-0 or a Higher-Level Star Schema

When we want to query the data warehouse, sometimes we need to determine which level of star schema will be able to answer the query. If the query requires the original individual records, obviously we need to query the Level-0 star schema. Previous sections discuss examples of median queries as well as standard deviation and variance queries. These require the original individual records; hence, the query needs to be directed to the Level-0 star schema.

If we need to query something else, and it is already provided by the fact measure, then it is also obvious to which star schema we should send our queries. For example, if we want to find the High Temperature on a particular date, then the Level-1 star schema as shown in Fig. 15.6 would be able to answer the query immediately. Or if we want to find the Temperature (i.e. Average High Temperature) of a particular city in a particular month, then the query can be directed to a Level-1 star schema. The primary reason for this is that the fact measures have already precomputed the required values.

However, what if we want to query something that is not in the fact measures? For example, it is quite common to query the rainfall on a particular date. The original record has rainfall data every 30 min as the weather recording is taken every 30 min. However, in the Level-1 star schema (daily data), the rainfall fact measure is no longer there. This daily rainfall is not the "average" rainfall of the day (e.g. total rainfall of the day divided by the number of recordings on that day), but may be something else, depending on the definition of rainfall for the day, which could be maximum, or something else determined by the domain expert or regulator. In this example, this daily rainfall is not captured in the Level-1 star schema. Therefore, querying the daily rainfall needs to be directed to the Level-0 star schema.

So, querying something that is not in the fact measure of a higher-level star schema must be directed to the lowest level of star schema which stores each individual record. Basically, fact measures of higher-level star schemas store precomputed values. This also means that these precomputed values must be predefined at the design stage of data warehousing. This also means that we predict what precomputed values might be useful and used quite often later. Obviously, not all aggregated values can be predicted in advance. If this is the case, querying these values must be directed to the lowest level of star schema. On the other hand, if the desired fact measure is already there in the higher-level star schema, querying them will be direct and easy.

15.4 Summary

Designing a star schema from Level-0 is sometimes necessary due to some expected queries using statistical functions requiring data to be presented in its original form. This chapter illustrates the use of some of these basic statistical functions, including median, standard deviation and variance.

Level-0 star schemas are very similar to the E/R diagram, but with a different focus. Star schemas focus on fact measures, whereas the E/R diagram focuses on the relationships between entities. The Level-0 fact measure is non-aggregated. This is very different from fact measures of higher-level star schemas where the fact measures must be aggregated.

Examining higher-level star schemas further, we can see that they are actually precomputed fact measures. Using these precomputed fact measures, querying high-level star schemas can be simplified, compared with retrieving high-level information from Level-0 star schemas. Therefore, precomputed fact measures in high-level star schemas are useful, simple and fast.

Making a star schema a higher level can be done easily by changing the granularity of dimension. This will make the new star schema more general with a higher level of aggregation. Nevertheless, from the pedagogical point of view, sometimes it is easier to learn a star schema by designing the star schema from a higher level of aggregation. Making the star schema more detailed can be done by changing the granularity of the dimension to be more specific as well as adding new dimensions. This method was covered in the previous chapter and will further be explored in the next chapter.

15.5 Exercises

15.1 There are many possible ways to write an SQL query to find the median value. Using the Median House Price case study above, write an SQL query to retrieve the median house price sold in January 2020 in the suburb called Kew without using the median function.

15.2 Again, using the Median House Price case study, write the SQL commands to retrieve the *Mean*, *Standard Deviation* and *Variance* house prices sold in 2020 (e.g. the Property Type is House, and the houses involved in these calculations are located anywhere in the database).

15.6 Further Readings

This book is the only book that discusses thoroughly the distinction between star schema design using higher level and lower level of aggregation. This book particularly focuses on Level-0 which is a star schema without any aggregation. Understanding this concept is crucial in learning how to design star schemas.

General star schema design can be found in various data warehousing books, such as [1–10] and [11].

Data modelling books, such as [12–16] and [17], would also give readers a broader perspective on differences between various levels of aggregation in data modelling.

References

1. C. Adamson, *Star Schema The Complete Reference* (McGraw-Hill Osborne Media, 2010)
2. R. Laberge, *The Data Warehouse Mentor: Practical Data Warehouse and Business Intelligence Insights* (McGraw-Hill, New York, 2011)
3. M. Golfarelli, S. Rizzi, *Data Warehouse Design: Modern Principles and Methodologies* (McGraw-Hill, New York, 2009)

4. C. Adamson, *Mastering Data Warehouse Aggregates: Solutions for Star Schema Performance* (Wiley, London, 2012)
5. P. Ponniah, *Data Warehousing Fundamentals for IT Professionals* (Wiley, London, 2011)
6. R. Kimball, M. Ross, *The Data Warehouse Toolkit: The Definitive Guide to Dimensional Modeling* (Wiley, London, 2013)
7. R. Kimball, M. Ross, W. Thornthwaite, J. Mundy, B. Becker, *The Data Warehouse Lifecycle Toolkit* (Wiley, London, 2011)
8. W.H. Inmon, *Building the Data Warehouse*. ITPro Collection (Wiley, London, 2005)
9. M. Jarke, *Fundamentals of Data Warehouses*, 2nd edn. (Springer, Berlin, 2003)
10. E. Malinowski, E. Zimányi, *Advanced Data Warehouse Design: From Conventional to Spatial and Temporal Applications*. Data-Centric Systems and Applications (Springer, Berlin, 2008)
11. A. Vaisman, E. Zimányi, *Data Warehouse Systems: Design and Implementation*. Data-Centric Systems and Applications (Springer, Berlin, 2014)
12. T.A. Halpin, T. Morgan, *Information Modeling and Relational Databases*, 2nd edn. (Morgan Kaufmann, Los Altos, 2008)
13. J.L. Harrington, *Relational Database Design Clearly Explained*. Clearly Explained Series (Morgan Kaufmann, Los Altos, 2002)
14. M.J. Hernandez, *Database Design for Mere Mortals: A Hands-On Guide to Relational Database Design*. For Mere Mortals (Pearson Education, 2013)
15. N.S. Umanath, R.W. Scamell, *Data Modeling and Database Design* (Cengage Learning, 2014)
16. T.J. Teorey, S.S. Lightstone, T. Nadeau, H.V. Jagadish, *Database Modeling and Design: Logical Design*. The Morgan Kaufmann Series in Data Management Systems (Elsevier, Amsterdam, 2011)
17. G. Simsion, G. Witt, *Data Modeling Essentials*. The Morgan Kaufmann Series in Data Management Systems (Elsevier, Amsterdam, 2004)

Chapter 16
Levels of Aggregation:
Adding and Removing Dimensions

From the previous chapter, we learned that lowering the level of aggregation of a star schema can be done by adding new dimensions to the star schema. Naturally, when a new dimension is added, in which a new attribute is added to the Fact Table, the level of details of the fact measure will become more detailed; hence, the level of aggregation will be lower.

The opposite is also true, in that, when a dimension is removed from a star schema, the level of aggregation of the star schema becomes higher, resulting in a more general star schema.

These are general rules about adding or removing dimensions to lower or raise the level of aggregation. However, there are some exceptions, that is, when adding a new dimension, the level of aggregation of the star schema does not become lower. We are also going to learn some complexities when a dimension is removed from a star schema. Hence, the aim of this chapter is to understand the impact of the level of aggregation of a star schema when we add or remove a dimension from the star schema.

16.1 Adding New Dimensions

We must be careful when adding a star schema with a new dimension. In particular, there are two things: (i) in some cases, adding a new dimension may not lower down the level of aggregation, and (ii) adding a new dimension may double count values in the fact measure, and hence the fact measure will be incorrect.

16.1.1   Adding New Dimensions Does Not Lower
Down the Level of Aggregation

Consider the Olympic star schema (refer to Fig. 16.1), consisting of three dimensions: Country Dimension, Sport Dimension and Olympic Dimension. For simplicity, in each dimension, only the identifier attribute is listed and none of the non-identifier attributes are included. The Country Dimension maintains a list of countries which have participated in the Olympic Games. The Olympic Dimension lists all the Olympic names, including the city and the year (e.g. the London Olympics was in London in 2012, and the Rio Olympics was in Rio, Brazil, in 2016). The Sport Dimension lists all the sport types in the Olympics, such as Swimming, Athletics, Volleyball, etc. Each sport is a category of the sport, not the actual sport event. So, for example, Swimming is a sport type instead of the actual event, such as "100 m Butterfly Men", or "4 × 100 m Freestyle Relay Women". As another example, Volleyball is a sport type, whereas the events are "Volleyball Men" or "Volleyball Women". Hence, Sport Dimension has a high level of aggregation.

Now we are going to add a new dimension called **Medal Dimension**, which has only three records, Gold, Silver and Bronze. The fact measure is reduced from three to one, from Number of Gold Medals, Number of Silver Medals and Number of Bronze Medals, to just Number of Medals. The new star schema with four dimensions is shown in Fig. 16.2.

Note that usually when a dimension is added to a star schema, the level of aggregation of the new star schema is naturally lower than the star schema before the new dimension is added. *Does this statement hold for this case study?* Does the new star schema with four dimensions have a lower level of aggregation than the previous star schema with only three dimensions? To answer this question, we need to examine the data (or the records) in the Fact Table.

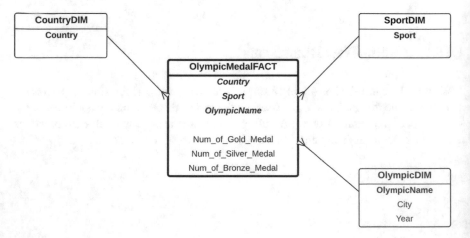

Fig. 16.1 The Olympic Star Schema Version 1: with three dimensions

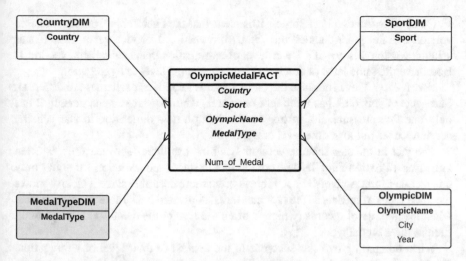

Fig. 16.2 The Olympic Star Schema Version 2: with four dimensions

Table 16.1 The Olympic Games Fact Table (star schema version 1)

Country	Sport	Olympic name	Num of Gold	Num of Silver	Num of Bronze
USA	Swimming	London 2012	16	9	6
China	Swimming	London 2012	5	1	4
Australia	Swimming	London 2012	1	6	3

Table 16.2 The Olympic Games Fact Table (star schema version 2)

Country	Sport	Olympic name	Medal type	Num of medal
USA	Swimming	London 2012	Gold	16
USA	Swimming	London 2012	Silver	9
USA	Swimming	London 2012	Bronze	6
China	Swimming	London 2012	Gold	5
China	Swimming	London 2012	Silver	1
China	Swimming	London 2012	Bronze	4
Australia	Swimming	London 2012	Gold	1
Australia	Swimming	London 2012	Silver	6
Australia	Swimming	London 2012	Bronze	3

The Fact Table for star schema version 1 (without Medal Dimension) has six attributes: three from the dimensions and the other three for the fact measures. The contents of the Fact Table are shown in Table 16.1. For simplicity, we only show three countries which participated in the London 2012 Olympic Games. The Fact Table for star schema version 2 (with Medal Dimension) consists of only five attributes, four from the dimensions but only one fact measure (see Table 16.2).

Comparing the two Fact Tables, it is clear that the Fact Table from star schema version 2 is not more detailed than that from version 1 or vice versa. Hence, star schema version 2 is not of a lower level of aggregation than star schema version 1. Both have the same level of detail and hence the same level of aggregation.

Why is this? This is because the fact measures in both star schemas are different. Star schema version 1 has three fact measures, whereas star schema version 2 has only one fact measure. So, in this case, adding a new dimension to star schema version 1 does not lower the level of aggregation.

The fact in the star schema version 1, which has three fact measures, is often known as a **Pivoted Fact Table** from the fact in star schema version 1 that has only one fact measure. A pivoted Fact Table is often more desirable since it clearly shows the three types of medals as the fact measures. A pivoted Fact Table is only possible when the number of records (which is often the *type* or the *category*) is very small, such as three medal types.

When comparing two star schemas in the context of their level of aggregation, the fact measures must be identical. They cannot be different as in this example. If they are different, the level of aggregation between the two star schemas is not comparable.

16.1.2   Adding New Dimensions May Result in a Double Counting in the Fact Measure

Before discussing the double counting problem, let's lower the level of aggregation of star schema version 2 in the previous section by changing the level of granularity of Sport Dimension. We now replace Sport Dimension (which is basically the sport category) with **SportEvent Dimension**, as shown in Fig. 16.3.

In SportEvent Dimension, we keep the actual sport event, not the sport category. For example, instead of "Swimming", we now have "100 m Butterfly Men", "4 × 100 m Freestyle Relay Women", etc. Naturally, this star schema has a lower level of aggregation compared to the previous star schema.

Now let's examine the records in the Fact Table. In comparing and contrasting the Fact Tables of the two star schemas, let's look at the content of both Fact Tables (Fig. 16.2 is now a Level-1 star schema and Fig. 16.3 is a Level-0 star schema). For simplicity, we focus on the Australian records only (refer to Tables 16.3 and 16.4).

By changing the level of granularity from Sport to Sport Event, the level of granularity of the fact changes. Instead of having only one record for swimming with a bronze medal, now we have the breakdown, which is the three bronze medal records. The Fact Table with Sport Event naturally has a higher level of granularity (or a lower level of aggregation) compared to the Fact Table with Sport, because the Fact Table with Sport Event has 1s in the number of medals (i.e. the fact measure). This star schema is a Level-0 star schema, which means there is no aggregation.

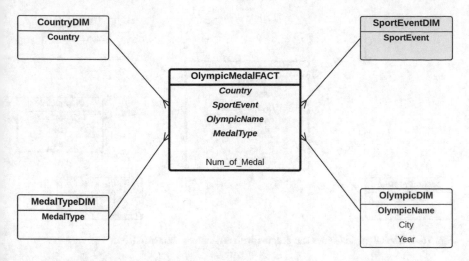

Fig. 16.3 The Olympic star schema: with SportEventDim

Table 16.3 The Olympic Games Fact Table (Level-1 star schema)

Country	Sport	Olympic name	Medal type	Num of medals
Australia	Swimming	London 2012	Gold	1
Australia	Swimming	London 2012	Silver	6
Australia	Swimming	London 2012	Bronze	3

Table 16.4 The Olympic Games Fact Table (Level-0 star schema)

Country	Sport event	Olympic name	Medal type	Num of medal
Australia	4 × 100 m Freestyle Relay Women	London 2012	Gold	1
Australia	4 × 100 m Medley Relay Women	London 2012	Silver	1
Australia	4 × 200 m Freestyle Relay Women	London 2012	Silver	1
Australia	100 m Breaststroke Men	London 2012	Silver	1
Australia	100 m Freestyle Men	London 2012	Silver	1
Australia	200 m Individual Medley Women	London 2012	Silver	1
Australia	100 m Backstroke Women	London 2012	Silver	1
Australia	4 × 100 m Medley Relay Men	London 2012	Bronze	1
Australia	100 m Butterfly Women	London 2012	Bronze	1
Australia	200 m Freestyle Women	London 2012	Bronze	1

What will happen if we decide to add a new dimension called **Athlete Dimension**. The rationale behind this is very simple, that is, to drill down to each winning athlete. The new star schema is then shown in Fig. 16.4.

A snapshot of the Fact Table, focusing on the Australian records, is shown in Table 16.5. By looking at the Fact Table, the last record in this example shows that *Bronte Barratt* was the athlete who received a bronze medal in the 200 m Freestyle

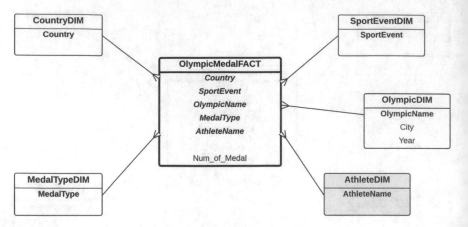

Fig. 16.4 The Olympic Games star schema with an Athlete Dimension

Women. So, drilling down to the winning athlete seems to be correct. From the Fact Table, it is also correct to see that *Bronte Barratt* received two medals: one bronze (200 m Freestyle Women) and one silver (4 × 200 m Freestyle Relay Women).

However, this Fact Table is incorrect because if we query the number of gold medals that Australia received in swimming in the London Olympics 2012, it will return four instead of one. Therefore, adding a new dimension, Athlete Dimension, will result in an incorrect fact measure (e.g. number of medals). The reason for this is because it is double counting the medals.

Double counting the number of medals occurs due to a different focus or a different subject of the star schema. One focus is from the Country point of view, and another focus is from the Athlete point of view. The two cannot be mixed and combined into one star schema.

The solution is to have two star schemas, one star schema which focuses on country and the other which focuses on athlete. The multi-fact star schema is shown in Fig. 16.5.

The Fact Tables are shown in Tables 16.6 and 16.7. If we examine these carefully, we can see that table AthleteMedalFact (see Table 16.7) is also incorrect because when we query "How many gold medals for 4 × 100 Freestyle Relay Women in London 2012 Olympic Games", the answer is four, which is incorrect. The query on the AthleteMedalFact will be correct if the Athlete Dimension is *always* used in any data retrieval on the AthleteMedalFact. Hence, Athlete Dimension must be a *Determinant Dimension*.

The correct multi-fact star schema is shown in Fig. 16.6 (note that we can still keep Country Dimension in the CountryMedalFact schema because Athlete Dimension is a Determinant Dimension).

Table 16.5 Athlete Fact Table

Country	Sport Event	Athlete	Olympic name	Medal type	Num of medal
Australia	4 × 100 m Freestyle Relay Women	Alicia Coutts	London 2012	Gold	1
Australia	4 × 100 m Freestyle Relay Women	Cate Campbell	London 2012	Gold	1
Australia	4 × 100 m Freestyle Relay Women	Brittany Elmslie	London 2012	Gold	1
Australia	4 × 100 m Freestyle Relay Women	Melanie Schlanger	London 2012	Gold	1
Australia	4 × 100 m Medley Relay Women	Emily Seebohm	London 2012	Silver	1
Australia	4 × 100 m Medley Relay Women	Leisel Jones	London 2012	Silver	1
Australia	4 × 100 m Medley Relay Women	Alicia Coutts	London 2012	Silver	1
Australia	4 × 100 m Medley Relay Women	Melanie Schlanger	London 2012	Silver	1
Australia	4 × 200 m Freestyle Relay Women	Bronte Barratt	London 2012	Silver	1
Australia	4 × 200 m Freestyle Relay Women	Melanie Schlanger	London 2012	Silver	1
Australia	4 × 200 m Freestyle Relay Women	Kylie Palmer	London 2012	Silver	1
Australia	4 × 200 m Freestyle Relay Women	Alicia Coutts	London 2012	Silver	1
Australia	100 m Breaststroke Men	Christian Sprenger	London 2012	Silver	1
Australia	100 m Freestyle Men	James Magnussen	London 2012	Silver	1
Australia	200 m Individual Medley Women	Alicia Coutts	London 2012	Silver	1
Australia	100 m Backstroke Women	Emily Seebohm	London 2012	Silver	1
Australia	4 × 100 m Freestyle Relay Men	Hayden Stoeckel	London 2012	Bronze	1
Australia	4 × 100 m Freestyle Relay Men	Christian Sprenger	London 2012	Bronze	1
Australia	4 × 100 m Freestyle Relay Men	Matt Targett	London 2012	Bronze	1
Australia	4 × 100 m Freestyle Relay Men	James Magnussen	London 2012	Bronze	1
Australia	100 m Butterfly Women	Alicia Coutts	London 2012	Bronze	1
Australia	200 m Freestyle Women	Bronte Barratt	London 2012	Bronze	1

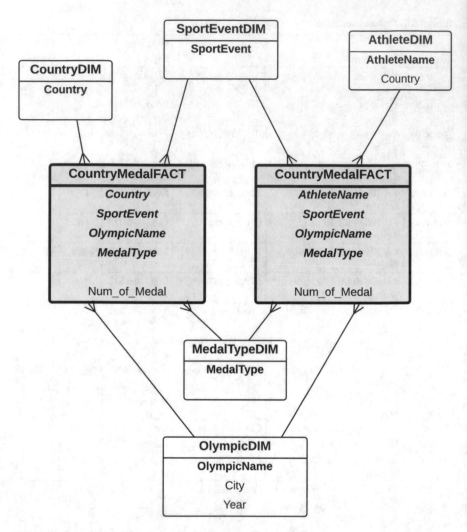

Fig. 16.5 A multi-fact star schema for the Olympic Games, which is still incorrect

16.1.3   The Final Star Schemas

We can avoid a multi-fact schema by having a different level of granularity for the Fact Table with Athlete Dimension. The levels of aggregation for the Olympic Games case study are shown in Figs. 16.7, 16.8, and 16.9.

The corresponding Fact Tables are previously shown in Tables 16.2, 16.6, and 16.7. Note that there are two Level-0s. Level-0a in Table 16.6 shows that the fact measure (e.g. Num_of_Medal) contains 1s (also shown in the Fig. 16.8 star schema). Adding the Athlete Dimension in Fig. 16.9 does not lower down the level of aggregation because neither of the star schemas (e.g. Figs. 16.8 and 16.9) have

Table 16.6 Country Medal Fact

Country	Sport event	Olympic name	Medal type	Num of medal
Australia	4 × 100 m Freestyle Relay Women	London 2012	Gold	1
Australia	4 × 100 m Medley Relay Women	London 2012	Silver	1
Australia	4 × 200 m Freestyle Relay Women	London 2012	Silver	1
Australia	100 m Breaststroke Men	London 2012	Silver	1
Australia	100 m Freestyle Men	London 2012	Silver	1
Australia	200 m Individual Medley Women	London 2012	Silver	1
Australia	100 m Backstroke Women	London 2012	Silver	1
Australia	4 × 100 m Medley Relay Men	London 2012	Bronze	1
Australia	100 m Butterfly Women	London 2012	Bronze	1
Australia	200 m Freestyle Women	London 2012	Bronze	1

any aggregation, and by definition, both are Level-0. Hence, they become Level-0a and Level-0b.

A multiple Level-0 star schema was illustrated in the previous example (see the Purchase Order case study). The thinnest Level-0 star schema has two dimensions, Purchase Order Dimension and Item Dimension. We can have additional two dimensions, Product Dimension and Customer Dimension, without lowering the star schema. The main reason for this is that the star schemas with Purchase Order and Item Dimensions are already at the lowest level of detail as they represent the transactions in the operational database. Each Purchase Order has only one Customer, and each Item has only one Product. Therefore, adding Customer and Product Dimensions does not break down the values of the fact measures.

It is also the same case with the Olympic case study. The star schema in Fig. 16.8 is already at the lowest level because the fact measure contains 1s. Each medal has one Athlete only; hence, adding the Athlete Dimension will not lower down the star schema.

The most important thing about the Level-0 star schema is that each record in the Fact Table in the Level-0 star schema represents a single record in the transaction table in the operational database. If the fact measure is a count, the fact measure Level-0 will contain 1s. If the fact measure is a sum, the fact measure in Level-0 is the original value of the transaction record. Once the star schema reaches Level-0, adding new dimensions will not break down the fact measures and hence will not lower the level of the star schema.

16.1.4   Summary

In general, adding a new dimension will result in a lower level of aggregation. However, one must take into account the following issues:

1. Adding a new dimension will not lower the level of aggregation, especially when the fact measure has changed. For example, the fact measure changes from three

Table 16.7 Athlete Medal Fact

Athlete	Sport event	Olympic name	Medal type	Num of medal
Alicia Coutts	4 × 100 m Freestyle Relay Women	London 2012	Gold	1
Cate Campbell	4 × 100 m Freestyle Relay Women	London 2012	Gold	1
Brittany Elmslie	4 × 100 m Freestyle Relay Women	London 2012	Gold	1
Melanie Schlanger	4 × 100 m Freestyle Relay Women	London 2012	Gold	1
Emily Seebohm	4 × 100 m Medley Relay Women	London 2012	Silver	1
Leisel Jones	4 × 100 m Medley Relay Women	London 2012	Silver	1
Alicia Coutts	4 × 100 m Medley Relay Women	London 2012	Silver	1
Melanie Schlanger	4 × 100 m Medley Relay Women	London 2012	Silver	1
Bronte Barratt	4 × 200 m Freestyle Relay Women	London 2012	Silver	1
Melanie Schlanger	4 × 200 m Freestyle Relay Women	London 2012	Silver	1
Kylie Palmer	4 × 200 m Freestyle Relay Women	London 2012	Silver	1
Alicia Coutts	4 × 200 m Freestyle Relay Women	London 2012	Silver	1
Christian Sprenger	100 m Breaststroke Men	London 2012	Silver	1
James Magnussen	100 m Freestyle Men	London 2012	Silver	1
Alicia Coutts	200 m Individual Medley Women	London 2012	Silver	1
Emily Seebohm	100 m Backstroke Women	London 2012	Silver	1
Hayden Stoeckel	4 × 100 m Freestyle Relay Men	London 2012	Bronze	1
Christian Sprenger	4 × 100 m Freestyle Relay Men	London 2012	Bronze	1
Matt Targett	4 × 100 m Freestyle Relay Men	London 2012	Bronze	1
James Magnussen	4 × 100 m Freestyle Relay Men	London 2012	Bronze	1
Alicia Coutts	100 m Butterfly Women	London 2012	Bronze	1
Bronte Barratt	200 m Freestyle Women	London 2012	Bronze	1

fact measures (Number of Gold Medals, Number of Silver Medals and Number of Bronze Medals) to one fact measure (Number of Medals). Adding a new Medal Dimension to the star schema will not change the level of granularity of the star schema when the fact measures have also changed.

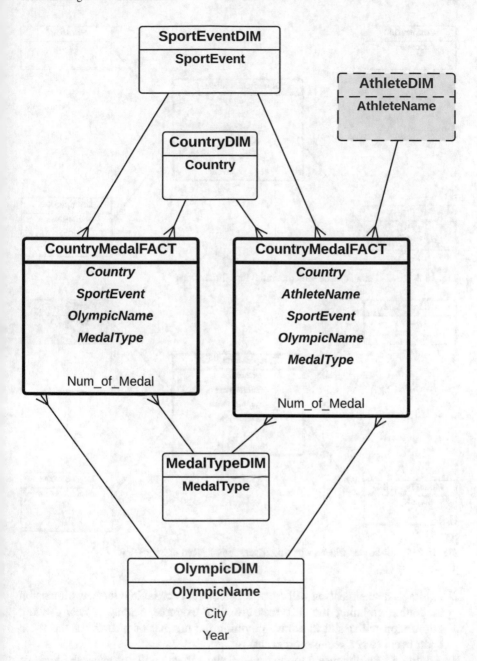

Fig. 16.6 Athlete Dimension is a Determinant Dimension (the correct multi-fact star schema)

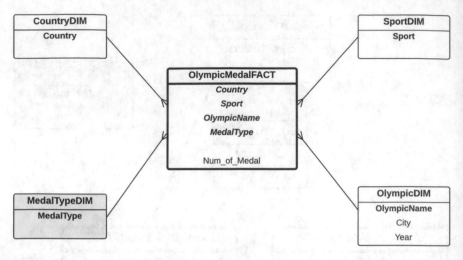

Fig. 16.7 Level-1 star schema with the initial four dimensions

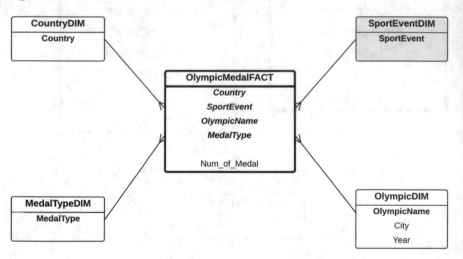

Fig. 16.8 Level-0a star schema with SportEventDim, instead of SportDim

2. Adding a new dimension will create an incorrect Fact Table if the new dimension is double counting the fact measure. For example, adding a new Athlete Dimension will result in double counting the number of medals for the same sport event (e.g. a group sport event).
3. Adding a new dimension to the Level-0 star schema will not lower the level of aggregation of the new star schema. For example, adding Athlete Dimension to Level-0 star schema will not lower down the star schema, because it is already the lowest level.

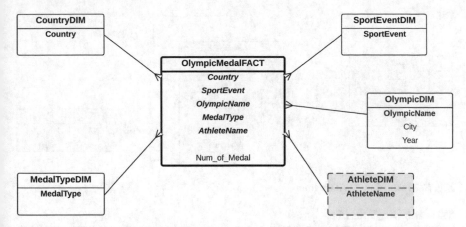

Fig. 16.9 Level-0b star schema with AthleteDim as a Determinant Dimension

16.2   Removing Dimensions

There are two important lessons from the previous section on "Adding New Dimensions" and its implications for the level of aggregation: (*i*) a double counting of the fact measure problem when adding a new dimension which is a Determinant Dimension and (*ii*) the breaking down of 1s in the fact measure into a lower level of aggregation.

In this section, we are going to examine these two elements, namely (*i*) the impact of removing dimensions to level of aggregation, especially in regard to removing a Determinant Dimension, and (*ii*) aggregating values of the fact measure into a higher level of aggregation. Hence, in this section, we are going learn about the impact of Determinant Dimensions on the level of aggregation.

16.2.1   An Employee Case Study

Consider the E/R diagram as shown in Fig. 16.10 which contains employee, department and job entities. Each employee has a department and a jobID. The sample records are shown in Tables 16.8, 16.9, and 16.10. For simplicity, the details of some attributes are not displayed. However, pay particular attention to the Hire Date attribute which indicates when each employee commenced employment.

The star schema as shown in Fig. 16.11 captures the number of employees for each job department and time (month). The calculation of the fact measure, namely Number of Employees, is not straightforward. Employee 101 started her job in November 2016, and she was the only employee to be recruited that month. It was not until February 2017 that Employee 102 started their job. This means that in

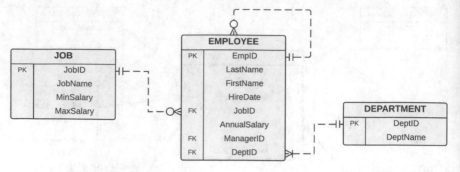

Fig. 16.10 Employee E/R diagram

Table 16.8 Employee table

EmpID	Last name	First name	Hire date	JobID	Annual salary	ManagerID	DeptID
101	Koh	Katie	1-Nov-2016	SRep			D01
102	Li	Liam	1-Feb-2017	SRep			D01
103	Mao	Mary	1-May-2017	SRep			D01
104	Qi	Queeny	1-May-2017	Acc			D02
...	...	...	...	...			...

Table 16.9 Department table

DeptID	Department name
D01	Cosmetic
D02	Accounting
...	...

Table 16.10 Job table

JobID	Job name	Min salary	Max salary
SRep	Sales Representative		
Acc	Accounting		
...	...		

February 2017, there were two employees (employees 101 and 102). Therefore, the
TimeID is not only the Hire Date but also the months after the hire date.

Let's assume that the data warehouse captures only for the duration from
November 2016 to June 2017 (see the Time Dimension shown in Table 16.11). The
Fact Table is shown in Table 16.12.

The SQL command to create the Employee Fact Table is as follows. Assume that
the Time Dimension table has been created, which consists of records showing each
month between November 2016 and June 2017. When creating the Employee Fact
Table, we need to join the Employee table and the Time Dimension table and the
join condition is Hire Date $\leq$ TimeID. This means that the months after the hire date
will capture the information that the employee is still working. Consequently, when
counting the number of employees in any month after the hire date, this employee
will be counted.

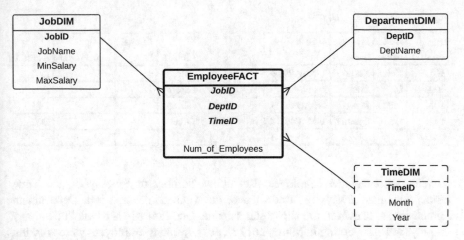

Fig. 16.11 Employee star schema

Table 16.11 Time
Dimension

TimeID	Month	Year
201611	Nov	2016
201612	Dec	2016
201701	Jan	2017
201702	Feb	2017
201703	Mar	2017
201704	Apr	2017
201705	May	2017
201706	Jun	2017

Table 16.12 Employee Fact

JobID	DeptID	TimeID	Num of employees
SRep	D01	201611	1
SRep	D01	201612	1
SRep	D01	201701	1
SRep	D01	201702	2
SRep	D01	201703	2
SRep	D01	201704	2
SRep	D01	201705	3
SRep	D01	201706	3
Acc	D02	201705	1
Acc	D02	201706	1

```
create table EmployeeFact as
select E.JobID, E.DeptID, T.TimeID,
   count(*) as Num_of_Employees
from Employee E, TimeDim T
where to_char(E.HireDate, 'YYYYMM') <= T.TimeID
group by JobID, DeptID, TimeID;
```

Table 16.13 Employee table

EmpID	Last name	First name	Hire Date	Cease Date	JobID	Annual Salary	ManagerID	DeptID
101	Koh	Katie	1-Nov-2016	31-Mar-2017	SRep			D01
102	Li	Liam	1-Feb-2017	null	SRep			D01
103	Mao	Mary	1-May-2017	null	SRep			D01
104	Qi	Queeny	1-May-2017	null	Acc			D02
...	...	...	...	...	...			...

When querying the Employee Fact about Number of Employees, the Time Dimension must always be used; hence, the Time Dimension is a Determinant Dimension, as shown in the above star schema. One example is to ask "how many employees were working in March 2017?", or "how many employees were working in June 2017?". The answers are 2 and 4, respectively. The SQL for the second query is as follows:

```
select TimeID, sum(Num_of_Employees)
from EmployeeFact
where TimeID = '201706'
group by TimeID;
```

The problem will be more complex if there is an attribute Cease Date for each employee record (see the sample data in Table 16.13). For example, Employee 101 quits her job at the end of March 2017. The number of employees must be counted month by month. In this case, there were two employees in Feb and Mar 2017, but in Apr 2017, there was only one employee (not two) because one employee quit her job.

The SQL to create the EmployeeFact Table is as follows. Note that it needs to incorporate the CeaseDate attribute in the join condition.

```
create table EmployeeFact as
select E.JobID, E.DeptID, T.TimeID,
    count(*) as Num_of_Employees
from Employee E, TimeDim T
where to_char(E.HireDate, 'YYYYMM') <= T.TimeID
and to_char(E.CeaseDate, 'YYYYMM') >= T.TimeID
group by JobID, DeptID, TimeID;
```

16.2.2   Removing a Determinant Dimension

In the previous lesson, we learned that to make a star schema more detailed (i.e. lowering the level of aggregation), you can add a dimension to the star schema. The opposite is also true. When we remove a dimension from a star schema, the level of aggregation of the star schema is higher and the star schema becomes more general.

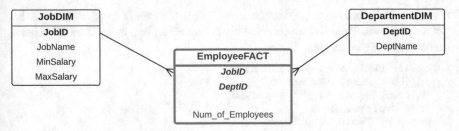

Fig. 16.12 Employee star schema with two dimensions

Table 16.14 Employee
Fact—incorrect

JobID	DeptID	Num of employees
SRep	D01	15
SRep	D01	2

Table 16.15 Employee
Fact—correct

JobID	DeptID	Num of employees
SRep	D01	3
SRep	D01	1

Suppose we remove the Time Dimension from the above star schema (refer to Fig. 16.11). The new star schema is shown in Fig. 16.12. The new star schema now has only two dimensions, Job and Department Dimensions.

When moving the level of aggregation up or down, usually the value in the fact measure is broken down (when we lower the level of aggregation) or is aggregated or summed up (when we increase the level of aggregation). For example, in the case of Num of Employees, when we remove the Time Dimension, we expect that the Num of Employees of the same TimeID will simply be aggregated or summed up. However, when we do this, we will get the incorrect number of employees, as shown in Table 16.14. There are 15 employees employed in a Sales Rep job in Department D01. Therefore, simply summing up the records from the lower level of aggregation is not correct in this case study. The correct Employee Fact should be as shown in Table 16.15. There are 3 employees (not 15) employed in a Sales Rep job in Department D01.

The main question is why doesn't simply summing up the fact measure when we move to a higher level of aggregation work in this case? The answer is because the dimension that we remove is a "Determinant Dimension". Therefore, removing a Determinant Dimension implies that we need to recalculate the fact measure as simply aggregating up or summing up will not produce the correct results.

The SQL command to produce the new Employee Fact is as follows. Note that joining with the Time Dimension is removed because there is no Time Dimension in this level of star schema.

```
create table EmployeeFact as
select E.JobID, E.DeptID,
    count(*) as Num_of_Employees
from Employee E
group by JobID, DeptID;
```

However, if we remove the Time Dimension, as in the second star schema, the calculation of the Number of Employees can be difficult if the star schema does not maintain the history of employees. Therefore, this higher level of aggregation star schema (the second star schema above) will not accurately reflect the status of the number of employees as the Time Dimension is missing.

16.2.3 Summary

In general, removing a dimension will result in a higher level of aggregation (less granularity or more general). However, one must take into account the following issues:

1. Removing a Determinant Dimension may affect the calculation of the fact measure when increasing the level of aggregation.
2. Aggregating or summing up the fact measure values to a higher level may be incorrect. Hence, the fact measure must be recalculated at each level.

16.3 Exercises

16.1 Given the Fact Table in Table 16.2, write the SQL commands to convert this table to the table in Table 16.1.

16.2 Looking at the Country Fact Table in Table 16.6, write the SQL command (using Group By Rollup) to produce the report as shown in Table 16.16.

16.3 Given the Athlete Fact Table in Table 16.7, write the SQL commands to convert this table to the Country Fact Table shown in Table 16.6.

16.4 Further Readings

This book is the only book that discusses thoroughly the distinction between star schema design using higher level and lower level of aggregation. Understanding this

Table 16.16 Country Medal Report

Olympic name	Country	Sport event	Medal type	Num of medal
London 2012	Australia	4 × 100 m Freestyle Relay Women	Gold	1
London 2012	Australia	All Sports Events	Gold	1
London 2012	Australia	4 × 100 m Medley Relay Women	Silver	1
London 2012	Australia	4 × 200 m Freestyle Relay Women	Silver	1
London 2012	Australia	100 m Breaststroke Men	Silver	1
London 2012	Australia	100 m Freestyle Men	Silver	1
London 2012	Australia	200 m Individual Medley Women	Silver	1
London 2012	Australia	100 m Backstroke Women	Silver	1
London 2012	Australia	All Sports Events	Silver	6
London 2012	Australia	4 × 100 m Medley Relay Men	Bronze	1
London 2012	Australia	100 m Butterfly Women	Bronze	1
London 2012	Australia	200 m Freestyle Women	Bronze	1
London 2012	Australia	All Sports Events	Bronze	3
London 2012	Australia	All Sports Events	All Medals	10

concept is crucial in learning how to design star schemas, and this chapter focuses on the impact of adding and removing dimensions to the granularity level of a star schema.

General star schema design can be found in various data warehousing books, such as [1–11]. Data modelling books, such as [12–17], would also give readers a broader perspective on the differences between various levels of aggregation in data modelling.

References

1. C. Adamson, *Star Schema The Complete Reference* (McGraw-Hill Osborne Media, New York, 2010)
2. R. Laberge, *The Data Warehouse Mentor: Practical Data Warehouse and Business Intelligence Insights* (McGraw-Hill, New York, 2011)
3. M. Golfarelli, S. Rizzi, *Data Warehouse Design: Modern Principles and Methodologies* (McGraw-Hill, New York, 2009)
4. C. Adamson, *Mastering Data Warehouse Aggregates: Solutions for Star Schema Performance* (Wiley, New York, 2012)
5. P. Ponniah, *Data Warehousing Fundamentals for IT Professionals* (Wiley, New York, 2011)
6. R. Kimball, M. Ross, *The Data Warehouse Toolkit: The Definitive Guide to Dimensional Modeling* (Wiley, New York, 2013)
7. R. Kimball, M. Ross, W. Thornthwaite, J. Mundy, B. Becker, *The Data Warehouse Lifecycle Toolkit* (Wiley, New York, 2011)
8. W.H. Inmon, *Building the Data Warehouse*. ITPro Collection (Wiley, New York, 2005)
9. M. Jarke, *Fundamentals of Data Warehouses*, 2nd edn. (Springer, Berlin, 2003)
10. E. Malinowski, E. Zimányi, Advanced data warehouse design: from conventional to spatial and temporal applications, in *Data-Centric Systems and Applications* (Springer, Berlin, 2008)

11. A. Vaisman, E. Zimányi, Data warehouse systems: design and implementation, in *Data-Centric Systems and Applications* (Springer, Berlin, 2014)
12. T.A. Halpin, T. Morgan, *Information Modeling and Relational Databases*, 2nd edn. (Morgan Kaufmann, Los Altos, 2008)
13. J.L. Harrington, Relational database design clearly explained, in *Clearly Explained Series* (Morgan Kaufmann Publishers, New York, 2002)
14. M.J. Hernandez, *Database Design for Mere Mortals: A Hands-On Guide to Relational Database Design*. For Mere Mortals. (Pearson Education, London, 2013)
15. N.S. Umanath, R.W. Scamell, *Data Modeling and Database Design* (Cengage Learning, Boston, 2014)
16. T.J. Teorey, S.S. Lightstone, T. Nadeau, H.V. Jagadish, Database modeling and design: logical design, in *The Morgan Kaufmann Series in Data Management Systems* (Elsevier, Amsterdam, 2011)
17. G. Simsion, G. Witt, Data modeling essentials, in *The Morgan Kaufmann Series in Data Management Systems* (Elsevier, Amsterdam, 2004)

Chapter 17
Levels of Aggregation and Bridge Tables

The levels of aggregation discussed so far are in the context of normal dimensions which are directly linked to the Fact. This chapter discusses the impact of the levels of aggregation when bridge tables exist in the star schema. The main question to be answered is how a dimension connected to a bridge table can be made more general, making the level of aggregation of the star schema higher.

17.1 Bridge Table: Truck Delivery Case Study

Let's revisit the Truck Delivery case study, which has a bridge table connecting the Trip and the Store Dimensions. The E/R diagram of the Truck Delivery operational database system is shown in Fig. 17.1.

The sample data is shown in Tables 17.1, 17.2, 17.3, and 17.4 (we focus only on the tables that will be used in the star schema, i.e. the Trip table, Truck table, Destination table and Store table).

The star schema of the Truck Delivery system is shown in Fig. 17.2. The Trip Dimension is linked to the Store Dimension through a Bridge Table.

The Trip Dimension has a WeightFactor and a StoreGroupList (ListAgg) attributes. The Trip Dimension table is shown in Table 17.5. The StoreGroupList attribute contains a list of stores for each trip, whereas the WeightFactor indicates the proportion of each store's contribution to the Total Delivery Cost in the Fact. The Bridge Table and Store Dimension table are shown in the Destination and Store tables in Tables 17.3 and 17.4. We can easily see that the StoreGroupList attribute values in the Trip Dimension table come from the Destination table, which lists all the stores for each trip.

The granularity of the Fact Table is determined by the Trip. In other words, the granularity is on a "Trip" level. Each trip will have an entry in the Fact Table. The Fact Table is shown in Table 17.6.

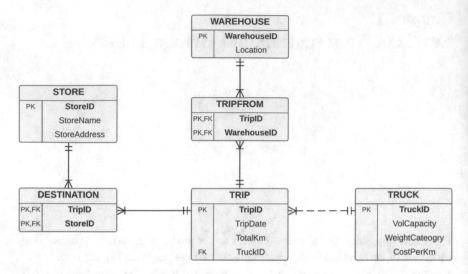

Fig. 17.1 Truck Delivery E/R diagram

Table 17.1 Trip table

TripID	Date	TotalKm	TruckID
Trip1	14-Apr-2018	370	Truck1
Trip2	14-Apr-2018	250	Truck2
Trip3	15-Apr-2018	375	Truck1
Trip4	16-Jul-2018	340	Truck1
Trip5	16-Jul-2018	175	Truck2
...	...	...	...

Table 17.2 Truck table

TruckID	VolCapacity	WeightCategory	CostPerKm
Truck1	250	Medium	$1.20
Truck2	300	Medium	$1.50
...	...	...	...

In order to understand this Fact Table, let's look at the Trip table in the operational database, which is shown in Table 17.1. There are five trips (namely Trip1 to Trip5). Trip1 to Trip3 were done in April (in the Autumn season), whereas Trip4 and Trip5 were done in July (in the Winter season). Because the granularity of the Fact Table is based on the Trip, the Fact Table has five records, the same number of records as in the Trip table.

Table 17.3 Destination table

TripID	StoreID
Trip1	M1
Trip1	M2
Trip1	M4
Trip1	M3
Trip1	M8
Trip2	M4
Trip2	M1
Trip2	M2
Trip3	M1
Trip3	M2
Trip3	M3
Trip3	M4
Trip3	M8
Trip4	M1
Trip4	M3
Trip4	M4
Trip4	M2
Trip4	M8
Trip5	M1
Trip5	M2
...	...

Table 17.4 Store table

StoreID	StoreName	Address
M1	MyStore City	Melbourne
M2	MyStore Chaddy	Chadstone
M3	MyStore HiPoint	High Point
M4	MyStore Westfield	Doncaster
M5	MyStore North	Northland
M6	MyStore South	Southland
M7	MyStore East	Eastland
M8	MyStore Knox	Knox City
...	...	...

17.1.1   Combining Trips: TripGroupList

One method to increase the level of aggregation of a star schema is by changing the granularity of an existing dimension by making the dimension more general or a lower granularity. Since the Truck Delivery star schema has the Trip as the granularity, to increase the level of aggregation of this star schema, we should not use the Trip as the level of granularity, because the Trip level granularity is rather high. In other words, the level of aggregation is low because it is done at a Trip level. Increasing the level of aggregation means that some trips need to be combined.

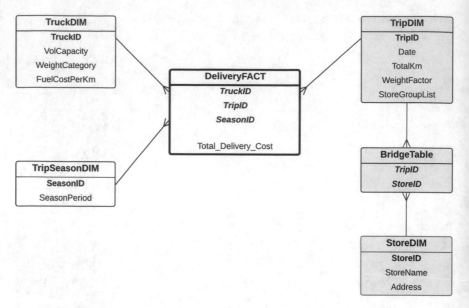

Fig. 17.2 Truck Delivery star schema

Table 17.5 Trip Dimension table

TripID	Date	TotalKm	WeightFactor	StoreGroupList
Trip1	14-Apr-2018	370	0.20	M1_M2_M3_M4_M8
Trip2	14-Apr-2018	250	0.33	M1_M2_M4
Trip3	15-Apr-2018	375	0.20	M1_M2_M3_M4_M8
Trip4	16-Jul-2018	340	0.20	M1_M2_M3_M4_M8
Trip5	16-Jul-2018	175	0.50	M1_M2
...	...	...	...	...

Table 17.6 Fact Table

TruckID	SeasonID	TripID	Total Delivery Cost
Truck1	Autumn	Trip1	444.00
Truck2	Autumn	Trip2	375.00
Truck1	Autumn	Trip3	450.00
Truck1	Winter	Trip4	408.00
Truck2	Winter	Trip5	262.50

Consequently, the level of aggregation is based on "multiple trips", not based on "single trips". The question is how to combine several trips into one record. What is the base for combining multiple trips into one?

Remember that each trip may deliver goods to many stores. For example, Trip1 delivers to five stores, namely M1, M2, M3, M4 and M8, as shown by the StoreGroupList attribute in the Trip Dimension table (refer to Table 17.5). Also notice that Trip3 delivers to the same five stores, also by Truck1 and in Autumn.

Table 17.7 TripGroupList Dimension table

TripGroupListID	TotalKm	WeightFactor	StoreGroupList
Trip1_Trip3	745	0.20	M1_M2_M3_M4_M8
Trip2	250	0.33	M1_M2_M4
Trip4	340	0.20	M1_M2_M3_M4_M8
Trip5	175	0.50	M1_M2
...	...	...	...

These two trips (e.g. Trip1 and Trip3) should be combined into one record, making the level of granularity lower (e.g. more general). Therefore, instead of storing Trip1 and Trip3 separately, we combine them into one record in the dimension table and make the TripGroupListID the identifier of the new dimension. In this example, we can simply concatenate both trips to become a new trip identifier (e.g. Trip1_Trip3). We call this new dimension table TripGroupListDim as shown in Table 17.7.

The TripGroupList Dimension replaces the Trip Dimension table with TripGroupListID as the identifier. This new dimension consists of three other attributes, which are TotalKm, WeightFactor and StoreGroupList. The TripDate attribute, originally in the Trip Dimension table, is no longer applicable here because one record potentially combines several trips on different dates (but in the same season, as indicated by the Season Dimension). The WeightFactor and StoreGroupList attributes are the same as those in the original Trip Dimension because those trips which are now combined share the same list of stores and hence have the same weight factor. The TotalKm is simply a sum of each individual trip.

The new star schema that uses Trip Group List as the base on the level of aggregation is shown in Fig. 17.3. So, if originally we have, for example, 100 trips, after combining several trips based on their destinations, we potentially reduce the number of records to much less than 100, thereby increasing the level of aggregation. Since the identifier of the TripGroupList Dimension, namely TripGroupListID, is new, this dimension identifier is now in the Fact Table as well as in the Bridge Table to replace the original TripID attribute.

The new Fact Table is shown in Table 17.8. Note that the third column of the Fact Table is TripGroupListID, which links to the TripGroupList Dimension table as shown in Table 17.7. Also compared with the original Fact Table shown earlier in Table 17.6, the new Fact Table in Table 17.7 has a higher level of aggregation because Trip1 and Trip3 are combined into one record. Also notice that the new Total Delivery Cost for these two trips is the sum of the original individual Total Delivery Cost.

The SQL commands to implement this new star schema can be challenging because we need to combine trips. To create the new dimension TripGroupList Dimension table, we need to join the Trip and Destination tables. In addition to calculating the WeightFactor and StoreGroupList, the TripDate attribute needs to be kept as well, because combining trips must have the same Season, Truck as well as StoreGroupList. So, it is essential, for the time being, to include TripDate.

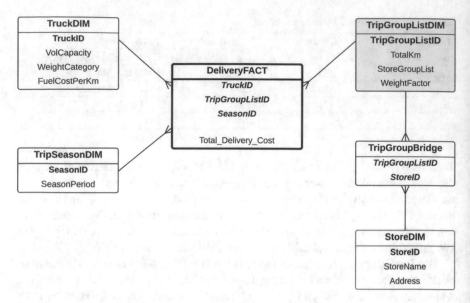

Fig. 17.3 A new star schema with TripGroupList Dimension

Table 17.8 Fact Table

TruckID	SeasonID	TripGroupListID	Total Delivery Cost
Truck1	Autumn	Trip1_Trip3	894.00
Truck2	Autumn	Trip2	375.00
Truck1	Winter	Trip4	408.00
Truck2	Winter	Trip5	262.50

After TripDate is converted into SeasonID, we can then combine trips based on SeasonID, WeightFactor and StoreGroupList. At the same time, the TotalKm must be recalculated. The SQL commands to create the TripGroupList Dimension table are as follows:

```
create table TripGroupListDimTemp as
select T.TripID, T.TripDate, T.TotalKm,
  1.0/count(D.StoreID) as WeightFactor,
  listagg (D.StoreID, '_') within group
    (order by D.StoreID) as StoreGroupList
from Trip T, Destination D
where T.TripID = D.TripID
group by T.TripID, T.TripDate, T.TotalKm;

alter table TripGroupListDimTemp
add (SeasonID varchar2(10));

update TripGroupListDimTemp
set SeasonID = 'Summer'
where to_char(TripDate, 'MM') in ('12', '01', '02');
```

```
update TripGroupListDimTemp
set SeasonID = 'Autumn'
where to_char(TripDate, 'MM') in ('03', '04', '05');

update TripGroupListDimTemp
set SeasonID = 'Winter'
where to_char(TripDate, 'MM') in ('06', '07', '08');

update TripGroupListDimTemp
set SeasonID = 'Spring'
where to_char(TripDate, 'MM') in ('09', '10', '11');

create table TripGroupListDim as
select
    listagg(TripID, '_') within group (order by TripID)
      as TripGroupListID,
    sum(TotalKm) as TotalKm,
    WeightFactor, StoreGroupList
from TripGroupListDimTemp
group by SeasonID, WeightFactor, StoreGroupList;
```

As can be seen from the star schema in Fig. 17.3, the Bridge Table connects TripGroupList Dimension and Store Dimension. It contains two attributes, Trip-GroupListID and StoreID. This Bridge Table is shown in Table 17.9.Note that Trip1 and Trip3 are now combined. They both share the same five stores. But also notice that Trip4, although having the same five stores, is not combined with Trip1 and Trip3. The reason for this is that Trip4 is in a different Season from Trip1 and Trip3. Notice group by SeasonID, WeightFactor,

Table 17.9 The New Bridge Table

TripGroupListID	StoreID
Trip1_Trip3	M1
Trip1_Trip3	M2
Trip1_Trip3	M3
Trip1_Trip3	M4
Trip1_Trip3	M8
Trip2	M4
Trip2	M1
Trip2	M2
Trip4	M1
Trip4	M2
Trip4	M3
Trip4	M4
Trip4	M8
Trip5	M1
Trip5	M2
...	...

StoreGroupList when creating the TripGroupList Dimension table. Hence,
only the trips sharing the same season (as well as the WeightFactor and StoreGroupList) will be combined. This is why Trip4 is not combined with Trip1 and Trip3.

The SQL to create the Bridge Table is as follows. It needs to join the TripGroupList Dimension and the Destination table, where the TripID in the Destination
appears in the TripGroupList Dimension.

```
create table TripGroupBridge as
select distinct TripGroupListID, StoreID
from    TripGroupListDim T, Destination D
where   T.TripGroupListID LIKE ('%'||D.TripID||'%');
```

Creating the TempFact and final Fact Tables is quite standard, except the
listagg and sum when creating the Fact Table.

```
create table TruckTempFact2 as
select
   tk.TruckID,
   tp.TripID,
   tp.TripDate,
   (tk.CostPerKm * tp.TotalKm) as ShipmentCost,
   StoreGroupList
from Truck tk, Trip tp, TripGroupListDim tg
where tk.TruckID = tp.TruckID
and tg.TripGroupListID like ('%'||tp.TripID||'%');

alter table TruckTempFact2
add (SeasonID varchar2(10));

update TruckTempFact2
set SeasonID = 'Summer'
where to_char(TripDate, 'MM') in ('12', '01', '02');

update TruckTempFact2
set SeasonID = 'Autumn'
where to_char(TripDate, 'MM') in ('03', '04', '05');

update TruckTempFact2
set SeasonID = 'Winter'
where to_char(TripDate, 'MM') in ('06', '07', '08');

update TruckTempFact2
set SeasonID = 'Spring'
where to_char(TripDate, 'MM') in ('09', '10', '11');

create table TruckFact3 as
select
  TruckID, SeasonID,
  listagg(TripID, '_') within group (order by TripID)
    as TripGroupListID,
  sum(ShipmentCost) as TotalCost
from TruckTempFact2
group by TruckID, SeasonID, StoreGroupList;
```

17.1.2   Combining Trips: StoreGroupList

In the previous chapter, combining trips is done by concatenating trips sharing the same SeasonID, WeightFactor and StoreGroupList. For example, Trip1 is combined with Trip3 to become Trip1_Trip3. Hence, TripID becomes TripGroupListID.

A second method to combine trips is to use the StoreGroupList. The trips with the same StoreGroupList are combined into one record. For example, the trips sharing the same StoreGroupList, such as M1_M2_M3_M4_M8, will have only one record in the StoreGroupList Dimension. The new star schema is shown in Fig. 17.4, and the StoreGroupList Dimension table is shown in Table 17.10.

The main difference between the StoreGroupList Dimension and the TripGroupList Dimension is that in the StoreGroupList Dimension, there is no TripID or TripGroupListID. Also, TotalKm is not included. Another difference is that in StoreGroupList, trips with the same StoreGroupList will be combined into one, that is, Trip1, Trip3 and Trip4 with the same StoreGroupList are combined into one record (although the TripID is not shown). This is not possible in the TripGroupList method because TripGroupListID appears in the Fact. If Trip1, Trip3 and Trip4 are combined in one TripGroupListID, then in the Fact Table, this will conflict

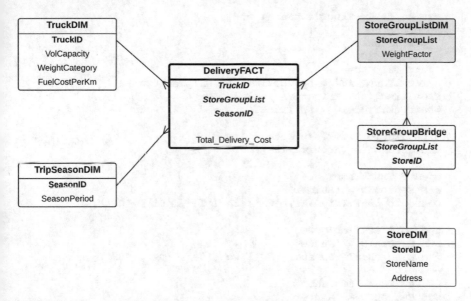

Fig. 17.4 A new star schema with StoreGroupList Dimension

Table 17.10 StoreGroupList Dimension table

StoreGroupList	WeightFactor
M1_M2_M3_M4_M8	0.20
M1_M2_M4	0.33
M1_M2	0.50
...	...

Table 17.11 Fact Table

TruckID	SeasonID	StoreGroupList	Total Delivery Cost
Truck1	Autumn	**M1_M2_M3_M4_M8**	894.00
Truck2	Autumn	M1_M2_M4	375.00
Truck1	Winter	**M1_M2_M3_M4_M8**	408.00
Truck2	Winter	M1_M2	262.50

with other dimensions, especially the Season Dimension because the three trips were not made in the same Season. However, with the StoreGroupList approach, it is possible to combine the trips with the same StoreGroupList, although they have different Season and different Truck because the Fact has StoreGroupList, not TripGroupListID.

The new Fact Table is shown in Table 17.11. Note that StoreGroupList M1_M2_M3_M4_M8 is repeated twice, one for Truck1 Autumn, which is basically Trip1 and Trip3, and the other for Truck1 Winter, which is Trip4. Imagine that in the TripGroupList approach, as discussed in the previous section, if Trip1, Trip3 and Trip4 are combined into one TripGroupListID, the Fact Table will be incorrect.

The SQL codes to create the TempFact and the new Fact are as follows:

```
create table TruckTempFact4 as
select
    tk.TruckID,
    tp.TripID,
    tp.TripDate,
    (tk.CostPerKM * tp.TotalKm) As ShipmentCost
from Truck tk, Trip tp
where tk.TruckID = tp.TruckID;

alter table TruckTempFact4
add (SeasonID varchar2(10));

update TruckTempFact4
set SeasonID = 'Summer'
where to_char(TripDate, 'MM') in ('12', '01', '02');

update TruckTempFact4
set SeasonID = 'Autumn'
where to_char(TripDate, 'MM') in ('03', '04', '05');

update TruckTempFact4
set SeasonID = 'Winter'
where to_char(TripDate, 'MM') in ('06', '07', '08');

update TruckTempFact4
set SeasonID = 'Spring'
where to_char(TripDate, 'MM') in ('09', '10', '11');

alter table TruckTempFact4
add (StoreGroupList varchar(100));
```

```
update TruckTempFact4 T
set StoreGroupList = (
  select listagg(D.StoreId, '_') within group
          (order by D.TripId) as StoreGroupList
  from Destination D
  where T.TripId = D.TripId
);

create table TruckFact4 as
select
  TruckId, SeasonId, StoreGroupList,
  sum(ShipmentCost) as TotalCost
from TruckTempFact4
where StoreGroupList is not null
group by TruckId, SeasonId, StoreGroupList;
```

The Bridge Table must also use StoreGroupList instead of TripGroupListID. It is shown in Table 17.12.

The SQL codes to create the Bridge Table are shown as follows:

```
create table StoreGroupBridge as
select distinct StoreGroupList, StoreID
from   TripGroupListDimTemp S, Destination D
where  S.StoreGroupList like ('%'||D.StoreID||'%');
```

This is similar to creating the Bridge Table in the TripGroupList method, that is, join the new dimension with the Destination table. But in this case, we check if the StoreID is part of the StoreGroupList, whereas in the Bridge Table using the TripGroupList method, it checks if the TripID is part of the TripGroupListID.

The Bridge Table in Table 17.12 is rather meaningless as the information is all redundant and unnecessary. We can actually bypass the Bridge Table and directly connect the StoreGroupList Dimension table with the Store Dimension table, as shown in the new star schema in Fig. 17.5. The relationship between the StoreGroupList Dimension and the Store Dimension is not purely a conventional primary key—foreign key relationship because StoreGroupList is a list of stores, whereas

Table 17.12 The New Bridge Table

StoreGroupList	StoreID
M1_M2_M3_M4_M8	M1
M1_M2_M3_M4_M8	M2
M1_M2_M3_M4_M8	M3
M1_M2_M3_M4_M8	M4
M1_M2_M3_M4_M8	M8
M1_M2_M4	M1
M1_M2_M4	M2
M1_M2_M4	M4
M1_M2	M1
M1_M2	M2
...	...

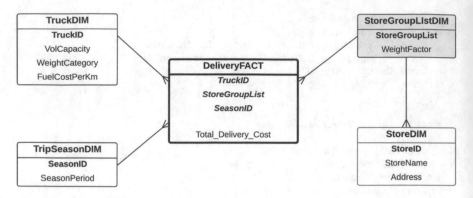

Fig. 17.5 The final star schema without the Bridge

StoreID is one store. So, when we join the StoreGroupList Dimension table and the Store Dimension table, it is not an equi-join like `where StoreGroupList = StoreID`, but is a non-equi-join using `where StoreGroupList like ('%'||StoreID||'%')`.

On the other hand, with the TripGroupList approach as shown in the star schema previously in Fig. 17.3 and the Bridge Table in Table 17.9, it is not possible to remove the Bridge Table, because the Bridge Table uses the TripGroupListID attribute, not StoreGroupList.

17.1.3   Summary

There are two options to combine the trips into one group of trips. The first method is by combining trips that share the same Season, Truck, WeightFactor and StoreGroupList. This is the TripGroupList method. The TripGroupListID, which is the identifier for the TripGroupList Dimension, is a concatenation of individual trips which are now combined. The second method is by combining StoreGroupList that shares the same stores. This is the StoreGroupList method. The StoreGroupList is now used instead of TripGroupListID. Because several trips are combined into one dimension record, the level of aggregation of the star schema is increased and the star schema becomes more general.

17.2   Bridge Table: Product Sales Case Study

The Truck Delivery case study mentioned in the above section shows two methods of combining trips (Trip Dimension) to become a higher level of aggregation: (*i*) combining trips using TripGroupList, which is a concatenation of the combined

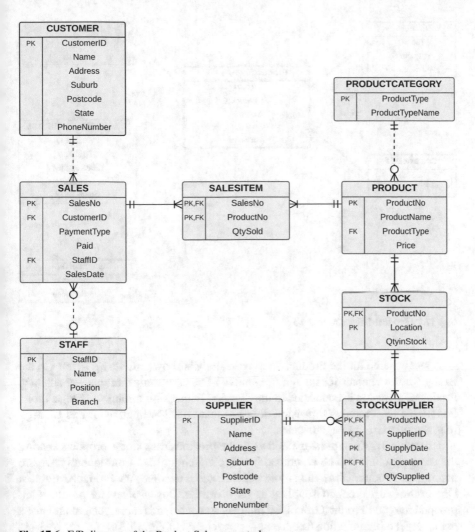

Fig. 17.6 E/R diagram of the Product Sales case study

trips, and (*ii*) combining trips using StoreGroupList using the list of stores of the combined trips.

This section shows that there is a third method to make a dimension that is connected to a bridge table less granular (making the star schema more general or having a higher level of aggregation). Let's use the Product Sales case study (this case study, together with the Truck Delivery case study, was also used in the Bridge Tables chapter). The E/R diagram of the Product Sales case study is shown in Fig. 17.6. Pay attention to the ProductCategory Entity, in which one Product Category may have several Products.

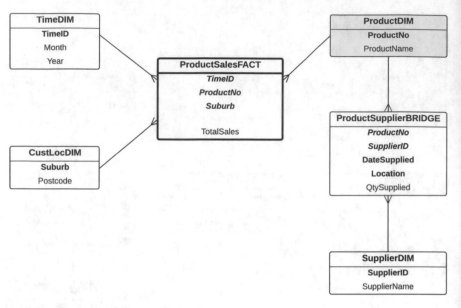

Fig. 17.7 Product Sales star schema

A star schema for the Product Sales case study is shown in Fig. 17.7 (refer to the Bridge Tables chapter for the requirements of this case study). In this star schema, Product Dimension is connected to the Supplier Dimension through a Bridge Table. For simplicity, Product Dimension does not feature the WeightFactor and ListAgg (e.g. SupplierGroupList) attributes.

Using the TripGroupList method as in the previous case study, products sharing the same suppliers can be combined in a ProductGroupList. Consequently, we are making the Product Dimension more general, and it becomes the ProductGroupList Dimension with ProductGroupList as its identifier. Because several products are grouped into one ProductGroupList, the new star schema is more general and has a higher level of aggregation (see the new star schema in Fig. 17.8).

Also notice that the Fact and the Bridge now have ProductGroupList instead of ProductNo. Because the Product Dimension does not have the WeightFactor and SupplierGroupList (e.g. ListAgg) attributes, the ProductGroupList Dimension has only one attribute, which is a combined ProductNo called ProductGroupList. Although the ProductGroupList Dimension has only one attribute, we might think that we can eliminate this dimension and make the Bridge directly connect to the Fact through the ProductGroupList attribute, as both of them share the same attribute. However, this is not possible because the cardinality relationship between the Fact and ProductGroupList Dimension is m-1 and ProductGroupList Dimension to the Bridge is 1-m. Otherwise, it would be a m-m between the Fact and the Bridge, which is not allowed in the design.

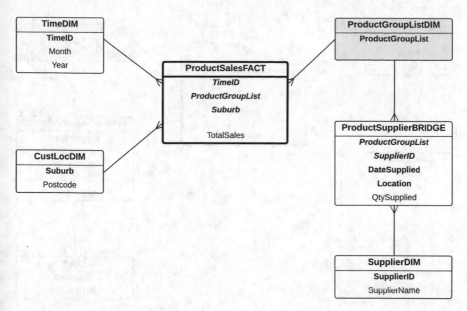

Fig. 17.8 ProductGroupList Dimension

Because the original Product Dimension does not have a SupplierGroupList attribute (e.g. the ListAgg attribute), we cannot apply the second method, like the StoreGroupList method in the Truck Delivery case study. So, the ProductGroupList method is the method that can be used to merge several products together, making the granularity lower or more general. However, if we think more deeply about this ProductGroupList approach, although it is correct that several products share the same supplier list, this is rather meaningless because having this ProductGroupList, each of which is a list of products with the same suppliers, will not have any meaning for the Total Sales. We may not need to find the total sales of a ProductGroupList, that is, finding the total sales of products with the same group of suppliers. Finding the total sales of a particular product, like the original star schema in Fig. 17.7, makes sense, but finding the total sales of ProductGroupList may not be.

So why does TripGroupList make sense but not ProductGroupList? One explanation is that in the Truck Delivery case study, the Total Delivery Cost is based on trips, although logically, it should have been based on stores. This is only because data on the delivery cost per score is not recorded in the operational database. So that is a logical error in the operational database due to the way the E/R diagram is designed. On the contrary, in the Product Sales case study, the Total Sales is already correctly attributed to Product and not to Supplier. Subsequently, grouping Products into ProductGroupList, which has suppliers as the basis, does not make any sense in the context of Total Sales.

Furthermore, there is no reason why grouping products based on suppliers makes sense. Each product has a product category (i.e. one product category may

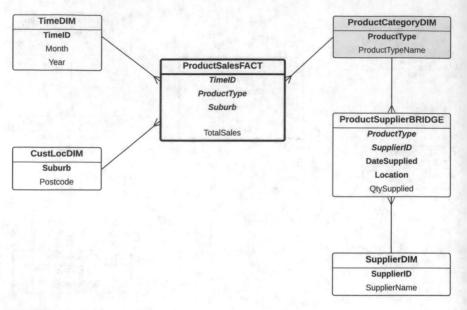

Fig. 17.9 ProductCategory Dimension

have several products). If we want to make the product more general, we should
have used Product Category. Using Product Category instead of Product not only
makes it more general or with a lower granularity, it also makes sense in terms
of Total Sales as we can query the Total Sales for a particular Product Category.
Therefore, to make Product more general, we should use Product Category instead
of ProductGroupList. The revised star schema is shown in Fig. 17.9, where the
ProductCategory Dimension replaces the ProductGroupList Dimension.

By using the ProductCategory Dimension where the ProductType attribute is the
identifier of this dimension, the Fact and Bridge now use the ProductType attribute.
The new star schema is able to answer queries such as finding the Total Sales of
a particular Product Category and later drill down into which suppliers supplied
that Product Category. However, the relationship between Product Category and
Supplier through the Bridge may not be as strong as that between Product and
Supplier in the original star schema. It is not necessary to use the Bridge and
Supplier Dimension at all in the star schema with the ProductCategory Dimension.
Hence, it makes more sense to have a simpler star schema with ProductCategory
Dimension and without the Bridge and Supplier Dimension. This simple star schema
is shown in Fig. 17.10.

Figure 17.11 shows the two levels of the Product Sales star schemas. The lower
level contains the Product Dimension with the Bridge, whereas the upper level
replaces the Product Dimension with the ProductCategory Dimension but without
the Bridge.

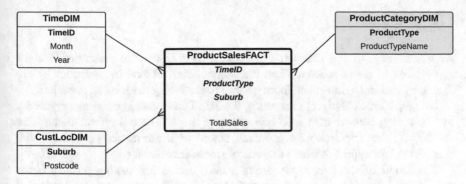

Fig. 17.10 ProductCategory Dimension without the Bridge

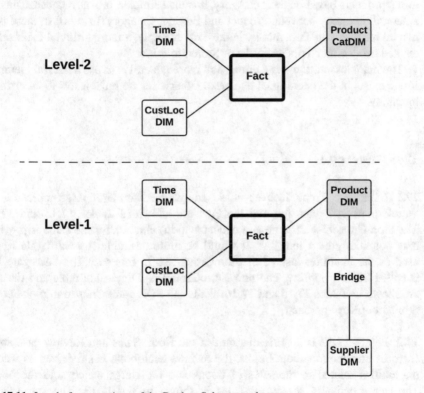

Fig. 17.11 Level of aggregations of the Product Sales star schemas

17.3   Summary

As a summary, in this chapter, we have learned how to increase the level of aggregation of a star schema when the original star schema has a Bridge Table, that is, by making a dimension more general (or lowering the level of granularity).

Using the Truck Delivery case study, the Trip Dimension is made more general, by combining several trips into one group of trips. There are two methods to do this: (*i*) the TripGroupList approach where trips are combined and (*ii*) the StoreGroupList approach where the lists of stores are combined.

The Product Sales case study offers a third option for making a star schema more general by replacing the Product Dimension with a more general dimension in the E/R diagram, namely Product Category. Not only does it make more sense to group products based on their category, it is also simpler. It is also because there is a natural hierarchy between Product and Product Category. In contrast, there is no natural hierarchy in Trip, that is, in the Trip example, a more artificial hierarchy is created using either TripGroupList or StoreGroupList.

The main lesson to make a dimension more general is to use a natural hierarchy if one exists in the operational database. Otherwise, we must resort to an artificial hierarchy.

17.4   Exercises

17.1 Using the Truck Delivery case study, the StoreGroupList approach, the StoreGroupList Dimension and the Fact are shown in Tables 17.13 and 17.14. The StoreGroupList attribute is a concatenation of stores for the same trip, which may potentially be a long list. It would be more desirable if a Surrogate Key is used as the identifier instead of StoreGroupList. In this case, the Surrogate Key is called StoreGroupKey. The new StoreGroupList Dimension table and the Fact are shown in Tables 17.15 and 17.16. Write the SQL commands to implement this StoreGroupKey approach.

17.2 Figure 17.12 is an E/R diagram of the Book Sales and Review case study. It stores information about books, the reviews each book has received as well as the entities related to the sales of books and the stores which sold the books. The star schema of the Book Sales is shown in Fig. 17.13. The star schema

Table 17.13 StoreGroupList Dimension table

StoreGroupList	WeightFactor
M1_M2_M3_M4_M8	0.20
M1_M2_M4	0.33
M1_M2	0.50
...	...

Table 17.14 Fact Table

TruckID	SeasonID	StoreGroupList	Total Delivery Cost
Truck1	Autumn	M1_M2_M3_M4_M8	894.00
Truck2	Autumn	M1_M2_M4	375.00
Truck1	Winter	M1_M2_M3_M4_M8	408.00
Truck2	Winter	M1_M2	262.50

Table 17.15 StoreGroupList Dimension table with StoreGroupKey as the Surrogate Key

StoreGroupKey	StoreGroupList	WeightFactor
1	M1_M2_M3_M4_M8	0.20
2	M1_M2_M4	0.33
3	M1_M2	0.50
...	...	...

Table 17.16 The New Fact Table with StoreGroupKey

TruckID	SeasonID	StoreGroupKey	Total Delivery Cost
Truck1	Autumn	1	894.00
Truck2	Autumn	2	375.00
Truck1	Winter	1	408.00
Truck2	Winter	3	262.50

has four dimensions: Store Dimension, Time Dimension, Category Dimension and Book Dimension. The Book Dimension has a Bridge Table connecting to the Author Dimension. In the Book Dimension, there are also WeightFactor and AuthorGroupList attributes.

If we want to make the star schema more general, we need to combine several books into one. There are two methods to do this:

1. The first method is by grouping books written by the same authors into one group of books. The star schema for this method is shown in Fig. 17.14. The Book Dimension is now replaced with the BookGroupList Dimension where ISBNGroupList becomes the new identifier, which is basically a concatenation of the ISBN of the books with the same authors.
2. The second method is by using the Category Dimension, to replace the original Book Dimension. The star schema of this method is shown in Fig. 17.15.

Compare and contrast these two approaches. It is also fair to assume that only a few authors are really prolific.

17.3 Figure 17.16 shows an E/R diagram of a Cable Television case study. The operational database stores information such as TV Programs, Channels and Plan, as well as Customer and Contract. The star schema of this Cable Television case study is shown in Fig. 17.17. The UsagePlan Dimension is connected to the Paid Channel Dimension through the Bridge. The WeightFactor and PaidChannelList attributes are also used in the UsagePlan Dimension.

Redesign the star schema to make it more general by combining several Plans into one group of Plans. Explain the two approaches, and design a star schema for

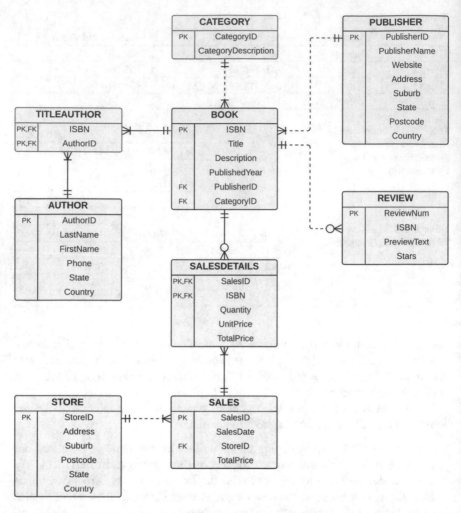

Fig. 17.12 Book Sales and Review E/R diagram

each approach. As a hint, make use of the information from the Paid Channel List, which is a `listagg` attribute.

17.4 The *Rural University of Victoria* has a number of campuses in several cities and towns in the state of Victoria. Each campus has several departments. Staff from one campus may need to travel to a different campus, for example, to teach, to attend administrative meetings or to undertake collaborative research with staff on another campus. To facilitate these travels, some departments have a fleet of cars which staff may borrow to travel between campuses in different cities. The campus also has a fleet of cars which staff may borrow if all the department cars are booked. These cars need a regular service as well as non-regular maintenance, including replacing

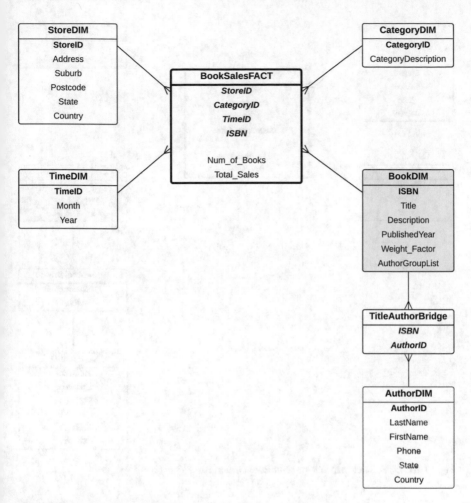

Fig. 17.13 Book Sales star schema

tyres and fixing other faults outside the regular service period. The E/R diagram of the operational database is shown in Fig. 17.18.

The star schema of this case study is shown in Fig. 17.19. The Car Dimension is connected to the User Dimension through the Bridge. The Car Dimension has also the WeightFactor and UserGroupList attributes.

We would like to redesign the star schema to make it more general, that is, by grouping several cars into one group of cars. Explain the four approaches, and design a star schema for each approach.

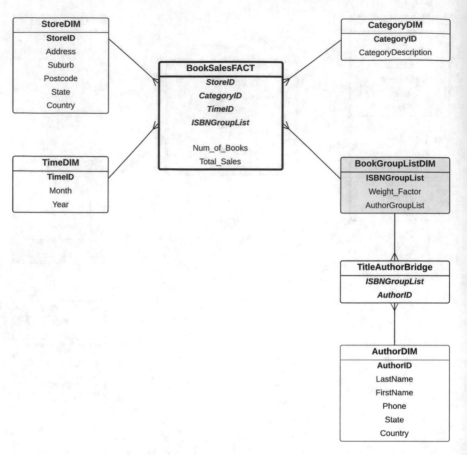

Fig. 17.14 Book Sales star schema with BookGroupList

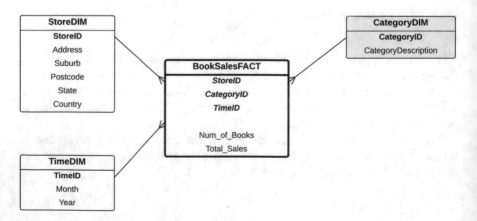

Fig. 17.15 Book Sales star schema with Category

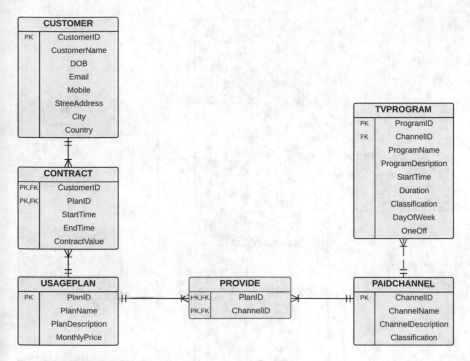

Fig. 17.16 Cable TV E/R diagram

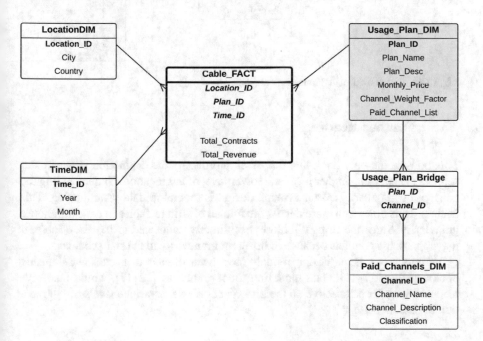

Fig. 17.17 Cable TV star schema with Bridge

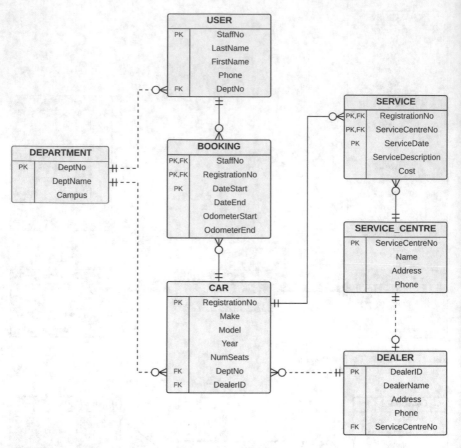

Fig. 17.18 Car Booking and Service E/R diagram

17.5 Further Readings

This book is the only book that discusses thoroughly the distinction between star schema design using higher level and lower level of aggregation. Understanding the concept of granularity in star schema design is crucial in data warehousing. This chapter drills down into more details the impact of Bridge Tables in data warehouse granularity. Since the Bridge Table is not directly connected to the Fact Table, it may raise some conflicts because of different granularity in the star schemas.

Various star schema design models have been discussed in data warehousing books, such as [1–11]. Data modelling books, such as [12–17], would also give readers a broader perspective on the differences between various levels of aggregation in data modelling.

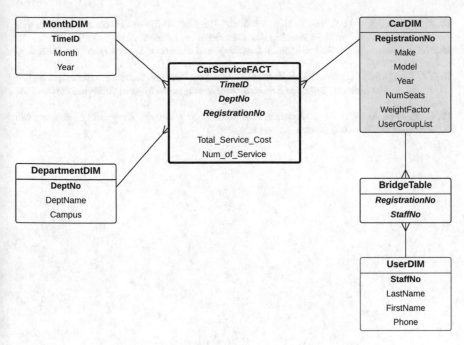

Fig. 17.19 Car Booking and Service star schema

References

1. C. Adamson, *Star Schema The Complete Reference* (McGraw-Hill Osborne Media, New York, 2010)
2. R. Laberge, *The Data Warehouse Mentor: Practical Data Warehouse and Business Intelligence Insights* (McGraw-Hill, New York, 2011)
3. M. Golfarelli, S. Rizzi, *Data Warehouse Design: Modern Principles and Methodologies* (McGraw-Hill, New York, 2009)
4. C. Adamson, *Mastering Data Warehouse Aggregates: Solutions for Star Schema Performance* (Wiley, New York, 2012)
5. P. Ponniah, *Data Warehousing Fundamentals for IT Professionals* (Wiley, New York, 2011)
6. R. Kimball, M. Ross, *The Data Warehouse Toolkit: The Definitive Guide to Dimensional Modeling* (Wiley, New York, 2013)
7. R. Kimball, M. Ross, W. Thornthwaite, J. Mundy, B. Becker, *The Data Warehouse Lifecycle Toolkit* (Wiley, New York, 2011)
8. W.H. Inmon, *Building the Data Warehouse* ITPro Collection (Wiley, New York, 2005)
9. M. Jarke, *Fundamentals of Data Warehouses*, 2nd edn. (Springer, Berlin, 2003)
10. E. Malinowski, E. Zimányi, Advanced Data Warehouse Design: From Conventional to Spatial and Temporal Applications. *Data-Centric Systems and Applications* (Springer, Berlin, 2008)
11. A. Vaisman, E. Zimányi, Data warehouse systems: design and implementation *Data-Centric Systems and Applications* (Springer, Berlin, 2014)
12. T.A. Halpin, T. Morgan, *Information Modeling and Relational Databases*, 2nd edn. (Morgan Kaufmann, Los Altos, 2008)
13. J.L. Harrington, Relational database design clearly explained, in *Clearly Explained Series* (Morgan Kaufmann, Los Altos, 2002)

14. M.J. Hernandez, *Database Design for Mere Mortals: A Hands-On Guide to Relational Database Design*. For Mere Mortals (Pearson Education, London, 2013)
15. N.S. Umanath, R.W. Scamell, *Data Modeling and Database Design* (Cengage Learning, Boston, 2014)
16. T.J. Teorey, S.S. Lightstone, T. Nadeau, H.V. Jagadish, Database modeling and design: logical design, in *The Morgan Kaufmann Series in Data Management Systems* (Elsevier, Amsterdam, 2011)
17. G. Simsion, G. Witt, Data modeling essentials, in *The Morgan Kaufmann Series in Data Management Systems* (Elsevier, Amsterdam, 2004)

Chapter 18
Active Data Warehousing

An *Active Data Warehousing* is where the data warehouse is immediately updated when the operational database is updated. The concept looks simple and attractive. However, there are a lot of complexities involved. This chapter will discuss the complexities thoroughly, covering (*i*) incremental updates, (*ii*) data warehousing schema evolution and (*iii*) operational database evolution. Firstly, the differences between Passive and Active Data Warehousing will be described.

18.1 Passive vs. Active Data Warehousing

A *Passive Data Warehouse* is a data warehouse that, once built, will remain unchanged. All of the data warehouses studied in the previous chapters are passive data warehouses. These are traditional data warehouses where a data warehouse contains star schemas that are subject-oriented, and the purpose of the data warehouse is decided at the design state. They are generally small and targeted star schemas. For example, a Product Sales star schema is for the sole purpose of analysing product sales and nothing else. A Computer Lab Activities star schema is for the sole purpose of analysing the use of computer labs by students. A Hospital Admission star schema is for the sole purpose of analysing patient admissions to hospitals. These are subject-oriented data warehouse, and the star schema is built for one specific purpose.

Although a data warehouse may contain a multi-fact star schema, it is still considered subject-oriented. Even in a multi-fact star schema, it only covers one subject, which is further divided into sub-subjects. For example, in the Domestic Cleaning Company case study, they employ cleaners to clean their clients' houses and offices. The star schema has two facts: one for Full-Time cleaners and the other for Part-Time cleaners, as they have different requirements, such as the number of hours they commit to the job, different payment rates and other conditions. Although

© The Author(s), under exclusive license to Springer Nature Switzerland AG 2021   449
D. Taniar, W. Rahayu, *Data Warehousing and Analytics*, Data-Centric Systems
and Applications, https://doi.org/10.1007/978-3-030-81979-8_18

they are separated into two facts, they have the same subject, which is recording their cleaning job activities (e.g. number of hours, number of cleaning jobs, payment, etc.).

Additionally, they are built based on historical data. For example, the Product Sales star schema only covers sales for a particular period, such as from 2010 to 2015; or the Hospital Admission star schema only covers patient admissions for a specific year, e.g. 2018. Once the data warehouse is built, there won't be any updates to the data warehouse. This is because the data warehouse is built based on historical data and for a specific time range. Hence, the data warehouse is not only subject-oriented but also non-volatile and time-variant. Non-volatile indicates that the data in the data warehouse will never change, whereas time-variant means that it is for a specific range of time, which is a past time period.

Data warehousing also provides various levels of granularity to support different levels of decision-making, where the upper levels precompute the fact measures from the lower levels. With the availability of multi-levels of star schema granularity, decision-makers can focus directly on a specific level of granularity of star schemas for more efficient data retrieval and subsequently more effective data analysis.

As Passive Data Warehousing is subject-oriented, targeted for a specific purpose, time-variant and non-volatile, the data warehouse is seen as a complete, self-contained and standalone package, once built. Decision-makers can then look into this package and analyse the data in it. As these packages are portable and the information in it is for a specific purpose, these packages are also known as *Data Marts*. In other words, the data is already in a self-contained package which is then saleable.

An *Active Data Warehousing* on the other hand is dynamic. When the operational database is updated, the data warehouse is immediately updated; hence, the data warehouse becomes "active". The data warehouse is always up-to-date, and it is no longer purely historical. Updates are also escalated to all levels of granularity of the star schema. The data warehouse is assumed to have direct access to the operational database, which is normally within the same system and environment. Figure 18.1 shows the architecture of Active Data Warehousing. The raw operational databases have been transformed into an integrated operational database, and the tables are directly accessible from (and available to) the data warehouse. Consequently, create view might be used in various levels of the star schema, which directly reflects any changes in the underlying tables, as views are virtual tables based on the underlying physical tables. Any changes in the underlying physical tables will automatically be reflected in the virtual tables. Additionally, certain database triggers can be used to automatically insert, update or delete the records in the data warehouse when an event (e.g. insert, update or delete) occurs in the underlying physical tables.

Designing an Active Data Warehousing adopts a bottom-up approach, where the star schema is built from Level-0, and the upper levels (e.g. Level-1, Level-2, etc.) are built on top of the immediate lower levels. Passive data warehouses may be built using the same approach, but as seen in the previous chapters, it is often easier to

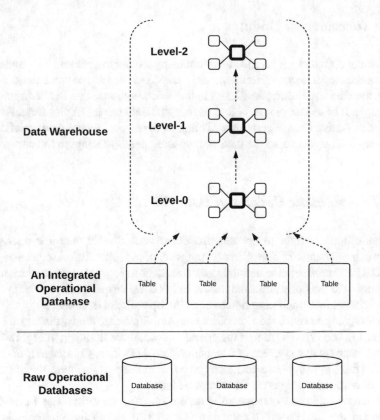

Fig. 18.1 Active Data Warehousing

build a data warehouse top-down, that is, to start from a reasonable upper level (e.g. Level-2) and move down level by level to the very bottom level (e.g. Level-0).

As this chapter focuses on Active Data Warehousing, we would like to examine what happens when Level-0 is updated in real time and what the impact is on the upper levels. This means that in Active Data Warehousing, updates are done automatically. This can be performed using virtual tables and database triggers to automatically execute the updates whenever needed.

Although most data warehousing is passive, some applications may benefit from Active Data Warehousing. This includes some monitoring sensors or real-time systems, where monitoring is done in real time using the current data, and hence, Active Data Warehousing is needed. Other systems that require continuous monitoring include air traffic control, the stock market, certain medical monitoring systems, etc.

18.2   Incremental Updates

Incremental updates in Active Data Warehousing are not to automatically update the data warehouse; there are other elements to consider, such as the "recentness" of the data in the data warehouse, as all data in the data warehouse need to be removed as they may not be as relevant as before. Hence, the data has an Expiry Date. Another element in Active Data Warehousing is that as time goes by, the data warehousing rules may have changed, so the data warehousing needs to adapt to the new rules.

18.2.1   Automatic Updates of Data Warehouse

Automatic updates refer to updates that occur immediately once the operational database has changed. The primary purpose of an operational database (or a transactional database) is to maintain new transaction records, as recent transactions keep coming to the operational databases. In technical terms, it is about the insertion of new records into the transaction table in the operational database.

We are going to reuse the Computer Lab Activities case study, discussed earlier in Chap. 14. The E/R diagram of the operational database is shown in Fig. 18.2. It is a simple database that consists of four tables: Lab Activities, Computer, Student and Degree. The data warehouse will focus on a fact measure that counts the number of usage (or number of logins) in the computer labs.

The data warehouse consists of three levels of granularity. The Level-0 star schema (or the lowest level), shown in Fig. 18.3, does not have any aggregation in the fact measure. The level of aggregation increases as the level of the star

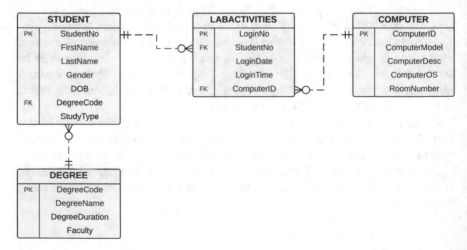

Fig. 18.2 E/R diagram for the Computer Lab Activities case study

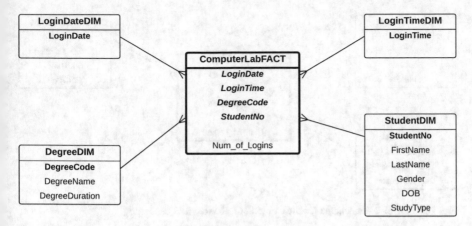

Fig. 18.3 Level-0 star schema for the Computer Lab Activities case study

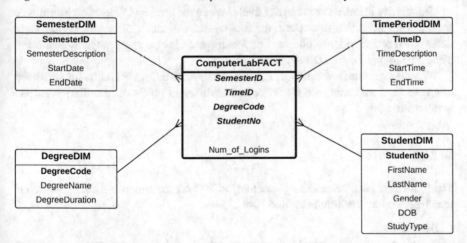

Fig. 18.4 Level-1 star schema for the Computer Lab Activities case study

schema increases. Figures 18.4 and 18.5 show the Level-1 and Level-2 star schemas, respectively.

The Level-0 star schema has four dimensions: Login Date, Login Time, Student and Degree Dimensions. These are the highest levels of granularity of data without any aggregation. Each record in the Fact Table refers to one login of one student at any particular date and time.

When the Login Date and Login Time are generalised to become Semester and Time Period Dimensions, the fact measure Number of Logins becomes aggregated. Hence, the granularity of the data is reduced. The star schema becomes Level-1. Note that the granularity is still determined by each individual student, but each student may log in multiple times during the semester, which is why the fact measure is now aggregated.

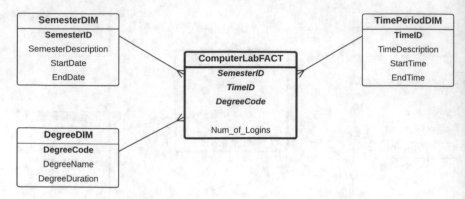

Fig. 18.5 Level-2 star schema for the Computer Lab Activities case study

Granularity is further reduced at Level-2 when the Student Dimension is removed from the star schema. As a result, the fact measure Number of Logins is no longer determined by individual students. The fact measure is now aggregated by Semester, Time Period and Degree Dimensions.

We will examine how the SQL commands in Active Data Warehousing are used to create (and maintain) the star schemas at different levels of granularity in data warehousing.

18.2.1.1   Level-0

The first step in star schema implementation is to create dimension tables, which is then followed by the TempFact and Fact Tables.

- **Dimensions**

  A common way to create dimension tables is by importing the necessary attributes and records from the relevant tables in the operational database. In this simple case study, only three tables from the operational database are used in the data warehouse, namely, Lab Activities, Student and Degree. Table Computer is not yet used in the data warehouse. So, for simplicity, any change to the Computer table will not affect the data warehouse.

```
create table LoginDateDim as
select distinct(LoginDate)
from LabActivities;

create table LoginTimeDim as
select distinct(LoginTime)
from LabActivities;

create table DegreeDim as
select DegreeCode, DegreeName, DegreeDuration
from Degree;
```

```
create table StudentDim as
select StudentNo, FirstName, LastName, Gender, StudyType
from Student;
```

Assuming that the operational database that contains the tables is within the same system, when new records are being added to the tables in the operational database, the dimension tables of the data warehouse must be updated automatically. Therefore, `create view` can be used to create the dimension tables, instead of `create table`.

```
create or replace view LoginDateDim as
select distinct(LoginDate)
from LabActivities;

create or replace view LoginTimeDim as
select distinct(LoginTime)
from LabActivities;

create or replace view DegreeDim as
select DegreeCode, DegreeName, DegreeDuration
from Degree;

create or replace view StudentDim as
select StudentNo, FirstName, LastName, Gender, StudyType
from Student;
```

If we would like to keep the dimension table as a table rather than a view (or a virtual table), we can use the `create table` command, but then we must have a database trigger that triggers an insertion of a new record. Take Student Dimension as an example. The trigger will trigger an insertion into the Student Dimension every time there is a new record inserted into the Student table in the operational database.

```
create table StudentDim as
select StudentNo, FirstName, LastName, Gender, StudyType
from Student;

create or replace trigger InsertNewStudent
  after insert on Student
  for each row
begin
  insert into StudentDim
  values (:new.StudentNo, :new.FirstName, :new.LastName,
    :new.Gender, :new.StudyType);
  commit;
end InsertNewStudent;
```

The advantage of using create view is that the update in the underlying table will be reflected immediately in the virtual table, so there is no need to have a separate trigger program that triggers an insertion to the dimension table.

For simplicity, we deal with the insertion of new records to the operational database only, that is, when new students log in to the Computer Lab Activities table or when there is a new degree program or when new students enrol. Update and Delete will be dealt with separately.

- **Fact**

The Fact Table is usually created in this way:

```
create table ComputerLabFactLevel0 as
select distinct
  LoginDate, LoginTime,
  S.DegreeCode, L.StudentNo,
  count(L.StudentNo) as Num_of_Logins
from Student S, LabActivities L
where S.StudentNo = L.StudentNo
group by
  LoginDate, LoginTime,
  S.DegreeCode, L.StudentNo;
```

It is not possible to use create view to create the Fact Table, because the Fact Table needs an additional attribute which is not in the operational database. This attribute is the fact measure, which is the Number of Logins attribute. If we use fact without fact measure, we could simply use create view to create the Fact Table, as no additional attributes are required in the Fact Table. But with additional attributes, we need to use create table and then use a database trigger to automatically add a record to the fact when there is a new record inserted into the transaction table (e.g. Lab Activities table in the operational database).

```
create or replace trigger UpdateFactLevel0
  after insert on LabActivities
  for each row
  declare
    Degree Student.DegreeCode%type;
begin
  select DegreeCode into Degree
  from Student
  where StudentNo = :new.StudentNo;

  insert into ComputerLabFactLevel0
  values (:new.LoginDate, :new.LoginTime,
    Degree, :new.StudentNo, 1);
  commit;
end UpdateFactLevel0;
```

Whenever a new record is inserted into the Lab Activities table in the operational database, an automatic insert will be performed to the Fact Table. Note that since the Degree Code is in another table (e.g. Student Table), we must check the Degree Code from the Student table. A value of 1 is also inserted to the fact measure attribute in the Fact Table.

If we opt for Fact without fact measure, then create view is sufficient, and there is no need for a database trigger.

```
create or replace view ComputerLabFactLevel0 as
select distinct
  LoginDate, LoginTime,
  S.DegreeCode, L.StudentNo
from Student S, LabActivities L
where S.StudentNo = L.StudentNo;
```

In conclusion, dimension tables, which are directly obtained from the operational database tables, can use `create view`, but for the Fact Table, if it has fact measure attributes, the Fact Table must be created using create table but is automatically updated through an activation of the database trigger.

- **PK-FK Constraints**
  In Passive Data Warehousing, as described in all the previous chapters, there is no need to create a Primary Key-Foreign Key (PK-FK) constraint between each dimension key with the fact, simply because once the star schema is created and built, no records will be updated in the star schema (e.g. dimensions nor fact); there will be no insert as the data warehouse is passive and no delete. The PK-FK constraint in the operational databases is needed to maintain data integrity in the database because insert, update and delete of records are frequently performed and maintaining entity integrity and referential integrity is of the utmost importance. But in Passive Data Warehousing, since there is no insert, update or delete of records in the star schema, the need to have a PK-FK constraint is diminished, since there won't be any data integrity issues.

  However, in Active Data Warehousing, where the data warehouse is actively updated when there are changes in the operational database, the PK-FK constraint between dimensions and the fact becomes important to minimise data anomalies due to updates. Therefore, the PK constraint must be enforced in the dimension and the PK-FK constraint in the fact.

  If the dimensions in Level-0 are views from the operational database tables, it is very difficult to assume that the primary keys have been implemented correctly and primary keys cannot be built on top of views. Therefore, it is advisable that dimensions in the Level-0 star schema are created using the `create table` statement, and not through `create view`.

  After each dimension is created using `create table`, a PK constraint should be added. For simplicity, Login Date Dimension and Login Time Dimension are removed from the star schema (because they are one-attribute dimensions), and consequently, LoginDate and LoginTime attributes in the fact become non-dimensional keys. A PK constraint is only added to the Degree Dimension and Student Dimension.

```
alter table DegreeDim
add constraint DegreeDimPK primary key (DegreeCode);

alter table StudentDim
add constraint StudentDimPK primary key (StudentNo);
```

The Fact Table will have two FKs, each referencing these PKs. The Fact Table will also have a composite PK which combines all the four dimension keys.

```
alter table ComputerLabFactLevel0
add constraint DegreeDimFK foreign key (DegreeCode)
references DegreeDim (DegreeCode);

alter table ComputerLabFactLevel0
add constraint StudentDimFK foreign key (StudentNo)
references StudentDim (StudentNo);

alter table ComputerLabFactLevel0
add constraint FactPK
primary key (LoginDate, LoginTime, DegreeCode,
StudentNo);
```

If the fact is created using view (because it is fact without fact measure), it will not be possible to have FK and PK-FK constraints in the fact. In this case, there is no need to implement these constraints because the view is created using a join query with a distinct clause in the select, thereby ensuring that the results are unique.

18.2.1.2   Level-1

In the Level-1 star schema, the Semester Dimension replaces the Login Date Dimension, and the Time Period Dimension replaces the Login Time Dimension. The Semester Dimension contains only two semesters, whereas the Time Period Dimension divides the 24-h day into three time periods: morning (e.g. 6am–11:59am), afternoon (12pm–5:59pm) and night (e.g. 6pm–5:59am).

- **Dimensions**

  Semester and Time Period Dimensions are usually created using the following create table command, whereas the Degree and Student Dimensions may reuse the tables from the Level-0 star schema. Both Semester and Time Period Dimensions must be created manually, as they are not extracted from the operational database tables.

  There is no need to have a database trigger for these two dimensions because the current records are timeless, and no new records will ever be added unless there are rule changes about semesters, which will be covered in the next section. The records in the Time Period Dimension cover the entire 24 h, so there won't be any new records to be inserted in the future.

```
create table SemesterDim
(
  SemesterID             varchar2(10),
  SemesterDescription    varchar2(20),
  StartDate              date,
  EndDate                date,
  constraint SemesterDimPK primary key (SemesterID)
);
```

```
insert into SemesterDim
values ('S1', 'Semester 1',
  to_date('01-JAN', 'DD-MON'),
  to_date('15-JUL', 'DD-MON'));
insert into SemesterDim
values ('S2', 'Semester 2',
  to_date('16-JUL', 'DD-MON'),
  to_date('31-DEC', 'DD-MON'));

create table TimePeriodDim
(
  TimeID             varchar2(10),
  TimeDescription    varchar2(15),
  StarTime           date,
  EndTime            date,
  constraint TimePeriodDimPK primary key (TimeID)
);

insert into TimePeriodDim
values('1', 'Morning',
  to_date('06:00', 'HH24:MI'),
  to_date('11:59', 'HH24:MI'));
insert into TimePeriodDim
values('2', 'Afternoon',
  to_date('12:00', 'HH24:MI'),
  to_date('17:59', 'HH24:MI'));
insert into TimePeriodDim
values('3', 'Night',
  to_date('18:00', 'HH24:MI'),
  to_date('05:59', 'HH24:MI'));
```

- **TempFact**

  A TempFact Table is a temporary table that needs to be created before the final Fact Table is created because at least one of the dimension tables is created manually, not by exporting existing records from the operational database. The TempFact Table is usually created using the following command:

```
create table TempComputerLabFactLevel1 as
select distinct
  LoginTime, LoginDate,
  S.DegreeCode, L.StudentNo
from Student S, LabActivities L
where S.StudentNo = L.StudentNo;
```

  We could also make use of the Fact Table from Level-0 as the source for the TempFact Table on this level.

```
create table TempComputerLabFactLevel1 as
select distinct
  LoginTime, LoginDate,
  DegreeCode, StudentNo
from ComputerLabFactLevel0;
```

  As two new attributes need to be added to the TempFact Table to accommodate SemesterID and TimeID, we cannot use create view to create the TempFact

Table. The following commands add two new attributes to the TempFact Table
and fill in the appropriate values.

```
alter table TempComputerLabFactLevel1
add (SemesterID varchar2(10));

update TempComputerLabFactLevel1
set SemesterID = 'S1'
where to_char(LoginDate, 'MM/DD') >= '01/01'
and to_char(LoginDate, 'MM/DD') <= '07/15';

update TempComputerLabFactLevel1
set SemesterID = 'S2'
where to_char(LoginDate, 'MM/DD') >= '07/16'
and to_char(LoginDate, 'MM/DD') <= '12/31';

alter table TempComputerLabFactLevel1
add (TimeID varchar2(10));

update TempComputerLabFactLevel1
set TimeID = '1'
where to_char(LoginTime, 'HH24:MI') >= '06:00'
and to_char(LoginTime, 'HH24:MI') < '12:00';

update TempComputerLabFactLevel1
set TimeID = '2'
where to_char(LoginTime, 'HH24:MI') >= '12:00'
and to_char(LoginTime, 'HH24:MI') < '18:00';

update TempComputerLabFactLevel1
set TimeID = '3'
where to_char(LoginTime, 'HH24:MI') >= '18:00'
or to_char(LoginTime, 'HH24:MI') < '06:00';
```

Since we need to use create table, we must then create a database trigger
to automatically trigger an insertion into this TempFact Table when there is an
insertion to the Fact Level-0.

```
create or replace trigger UpdateTempFactLevel1
  after insert on ComputerLabFactLevel0
  for each row
  declare
  TempSemesterID        SemesterDim.SemesterID%type;
  TempTimeID            TimePeriodDim.TimeID%type;

begin
  if (to_char(:new.LoginDate, 'MM/DD') >= '01/01')
    and (to_char(:new.LoginDate, 'MM/DD') <= '07/15')
    then
    TempSemesterID := 'S1';
  else
    TempSemesterID := 'S2';
  end if;
```

```
      if (to_char(:new.LoginTime, 'HH24:MI') >= '06:00')
        and (to_char(:new.LoginTime, 'HH24:MI') < '12:00')
        then
        TempTimeID := '1';
      elsif (to_char(:new.LoginTime, 'HH24:MI') >= '12:00')
        and (to_char(:new.LoginTime, 'HH24:MI') < '18:00')
        then
        TempTimeID := '2';
      else
        TempTimeID := '3';
      end if;

      insert into TempComputerLabFactLevel1
      values (:new.LoginTime, :new.LoginDate,
      :new.DegreeCode, :new.StudentNo, TempSemesterID,
      TempTimeID);

      commit;
    end UpdateTempFactLevel1;
```

- **The Fact**

  After the TempFact is created, the final Fact Table for Level-1 can be created.

```
    create table ComputerLabFactLevel1 as
    select
      SemesterID, TimeID, DegreeCode, StudentNo,
      count(StudentNo) as Num_of_Logins
    from TempComputerLabFactLevel1
    group by SemesterID, TimeID, DegreeCode, StudentNo;

    alter table ComputerLabFactLevel1
    add constraint SemesterDimFK foreign key (SemesterID)
    references SemesterDim (SemesterID);

    alter table ComputerLabFactLevel1
    add constraint TimePeriodDimFK foreign key (TimeID)
    references TimePeriodDim (TimeID);

    alter table ComputerLabFactLevel1
    add constraint DegreeDimFK foreign key (DegreeCode)
    references DegreeDim (DegreeCode);

    alter table ComputerLabFactLevel1
    add constraint StudentDimFK foreign key (StudentNo)
    references StudentDim (StudentNo);

    alter table ComputerLabFactLevel1
    add constraint FactPK
    primary key (SemesterID, TimeID, DegreeCode, StudentNo);
```

A database trigger must be created to ensure that for every record inserted into the TempFact Level-1, it will increase the fact measure Number of Logins by one.

```
create or replace trigger UpdateFactLevel1
  after insert on TempComputerLabFactLevel1
  for each row
begin
  update ComputerLabFactLevel1
  set Num_of_Logins = Num_of_Logins + 1
  where SemesterID = :new.SemesterID
    and TimeID = :new.TimeID
    and DegreeCode = :new.DegreeCode
    and StudentNo = :new.StudentNo;
  commit;
end UpdateFactLevel1;
```

18.2.1.3 Level-2

In Level-2, we basically remove the Student Dimension to lower the granularity of the fact measure Number of Logins. All other dimensions are reused. Fact Level-2 reuses Fact Level-1, where the aggregation of the fact measure is increased. There is no need to have a TempFact in Level-2 because we can create the fact in Level-2 by using create view from the fact in Level-1. As a result of this, there won't be any need for a database trigger.

```
create or replace view ComputerLabFactLevel2 as
select SemesterID, TimeID, DegreeCode,
  sum(Num_of_Logins) as Num_of_Logins
from ComputerLabFactLevel1
group by SemesterID, TimeID, DegreeCode;
```

18.2.2   Expiry Date

In Active Data Warehousing, data keeps coming to the data warehouse. As time goes by, the data warehouse collects a good wealth of data, which is beneficial for decision-making. However, the data also spans a long period of time. In some decision-making, decision-makers often want to focus on recent data, as old data (or very old data) in some applications may be meaningless, as the data has already expired. Additionally, if these data are still kept in the data warehouse, the analysis may be clouded by the existence of old data, as data warehousing often uses aggregate functions, which may aggregate all the data in the data warehouse. Therefore, in Active Data Warehousing, it is necessary to remove old data from the data warehouse. The old data should not and will not be removed from the operational databases because the operational databases should not only support day-to-day operations but also act as an archival repository.

In the Lab Activities case study, the star schema uses Semester as a Dimension, without the Year. The fact measure includes all login records. Hence, the measure can be outdated. For example, Number of Logins in Semester 1 for a particular course may include activities, which are simply too old to be considered (e.g. more than 5 years old). This then raises the notion of *Expiry Date* or *Expiry Time*, where the data input to the data warehouse are limited by the time, and as the time goes by, the old data become less important or even irrelevant, and in this case, they should not contribute to the data warehouse anymore. For example, we want to make the data warehouse active for the last 5 years only. Data older than 5 years old will be deleted from the data warehouse. So, this involves removing records, not inserting/updating records.

We are not removing old records from the operational database but removing old records from the Level-0 star schema. However, in the previous section, we have shown that there are two possible versions for Level-0 star schemas: one with fact measure and the other without fact measure. For the one with fact measure, `create table` is used, together with database triggers that trigger an insertion of records to the star schema when there are new records in the transactions (e.g. operational database). For the one without fact measure, we can simply use `create view`, which creates virtual tables from the operational database. These two scenarios will be discussed next.

18.2.2.1   Level-0

For the Level-0 star schema with fact measure, we can create a new Fact Table with a filtering condition on the Login Date (e.g. only to include the last 5 years' data).

```
-- ComputerLabFact (with fact measure)
drop table ComputerLabFactLevel0;
create table ComputerLabFactLevel0 as
select distinct
  LoginDate, LoginTime,
  S.DegreeCode, L.StudentNo,
  count(L.StudentNo) as Num_of_Logins
from Student S, LabActivities L
where S.StudentNo = L.StudentNo
and to_char(LoginDate, 'YYYY') >= '2015'
group by
  LoginDate, LoginTime,
  S.DegreeCode, L.StudentNo;
```

Unfortunately, this has to be done manually (including the FK and PK-FK constraints). Automatic deletion (or creation) can be done in other systems. There is no need to change the database trigger as the Fact Table is rebuilt. The database trigger is only used for the incremental insert of new records, which is not applicable in this case.

For the fact without fact measure, the `create view` command can be used. However, a filtering clause must be used. The filtering clause is used to select which records from the underlying operational database to use in the star schema.

```
-- ComputerLabFact (without fact measure)
create or replace view ComputerLabFactLevel0 as
select distinct
  LoginDate, LoginTime,
  S.DegreeCode, L.StudentNo
from Student S, LabActivities L
where S.StudentNo = L.StudentNo
and to_char(LoginDate, 'YYYY') >= '2015';
```

18.2.2.2   Level-1

The Level-1 star schema includes the creation of TempFact and Fact Tables. As described in the previous section, TempFact for Level-1 has been created using the `create table` command, together with a database trigger for insert (for incremental insert). When trimming old data from the data warehouse, we must execute the `delete` command manually.

```
delete from TempComputerLabFactLevel1
where to_char(LoginDate, 'YYYY') < '2015';
```

The final Fact Table was created using the `create table` command based on the TempFact Table. In other words, when the TempFact changes, we only need to activate a database trigger to trigger a delete to the final Fact Table.

```
create or replace trigger ReduceFactLevel1
  after delete on TempComputerLabFactLevel1
  for each row
begin
  update ComputerLabFactLevel1
  set Num_of_Logins = Num_of_Logins - 1
  where SemesterID = :old.SemesterID
    and TimeID = :old.TimeID
    and DegreeCode = :old.DegreeCode
    and StudentNo = :old.StudentNo;
  commit;
end ReduceFactLevel1;
```

18.2.2.3   Level-2

The Level-2 Fact Table is a virtual table from the fact in Level-1. So, when Level-1 is up-to-date, Level-2 will be up-to-date as well. In other words, in this example, no special maintenance is needed for the Level-2 star schema.

18.2.3 Data Warehouse Rules Changed

One of the important features of Active Data Warehousing is that it always contains up-to-date data. This means there is no time boundary for the lifetime of the data warehouse, except the Expiry Date discussed in the previous section. A consequence of having Active Data Warehousing is that the design that was developed in the beginning when the data warehouse was first built might have changed due to new business rules or new requirements. So, an important question that must be answered when dealing with Active Data Warehousing not only relates to an incremental update of new data but what happens if the rules have changed and what are the implications for the data warehouse.

Using the Lab Activities case study, Semester 1 is from 1 January to 15 July, and Semester 2 is from 16 July to 31 December. Suppose the semester dates change. Instead of Semester 1 being from 1 January to 15 July, it is now from 1 January to 30 June; hence, the two semesters are of equal length. The changes will only be effective from the new year, so this doesn't change the old data that is already in the data warehouse.

18.2.3.1 Level-0

Since the Semester Dimension starts to exist from Level-1 upward, the change of semester dates will not affect the Level-0 star schema as the Level-0 star schema does not have the Semester Dimension. The Level-0 star schema uses the Login Date Dimension when the actual date of the login is recorded.

18.2.3.2 Level-1

Two places that will be affected by the change of the semester dates: Semester Dimension and the TempFact.

- **Semester Dimension**
  For the Semester Dimension, the information about the semester dates is stored in the attributes, Start Date and End Date. These provide additional information about the semester, and they do not affect the accuracy of the fact measure. Dimension attributes, except the key, are considered additional information of the dimension and do not affect the fact in any way. The semester dates are only used when coding the TempFact. So, additional dimension attributes are not critical in terms of the fact.

  When the semester dates change, we can utilise a temporal dimension to store the history of the semester dates. Figure 18.6 shows a temporal dimension for the Semester, called Semester History Dimension. The Start Year and End Year attributes indicate the time period when the record is valid. For the current valid record, the End Year is recorded as either a null, or a maximum year of "9999",

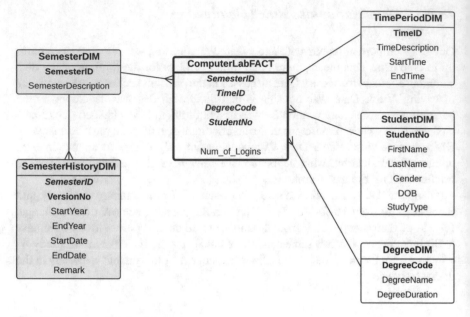

Fig. 18.6 Semester History Dimension

Table 18.1 Semester
Dimension table

SemesterID	Semester description
S1	Semester 1
S2	Semester 2

Table 18.2 Semester History Dimension table

SemesterID	Version no	Start year	End year	Start date	End date	Remark
S1	1	2011	2020	1 January	15 July	
S1	2	2021	null	1 January	30-Jun	New semester dates
S2	1	2011	2020	16 July	31 December	
S2	2	2021	null	1 July	31 December	New semester dates

for example. The Start Date and End Date attributes store the semester dates, whereas the Remark attribute can be used to write any remark for the particular record. Tables 18.1 and 18.2 show an illustration of the Semester Dimension and the History Dimension tables. In this example, the old semester dates were from 2011 to 2020, and the new semester dates start from 2021 onward.

- **TempFact and Fact**
  The rule for the new semester dates will very much affect how the TempFact is maintained or updated. Since TempFact has been created, we only need to change the database trigger. Assuming that the new semester date rule starts in 2021, the only change will be checking the Login Date for SemesterID. The rest remains the same.

```
create or replace trigger UpdateTempFactLevel1
  after insert on ComputerLabFactLevel0
  for each row
  declare
  TempSemesterID        SemesterDim.SemesterID%type;
  TempTimeID            TimePeriodDim.TimeID%type;

begin
  if (to_char(:new.LoginDate, 'YYYY') >= '2021') then
    if (to_char(:new.LoginDate, 'MM/DD') >= '01/01') and
      (to_char(:new.LoginDate, 'MM/DD') <= '06/30') then
      TempSemesterID := 'S1';
    else
      TempSemesterID := 'S2';
    end if;
  else
    if (to_char(:new.LoginDate, 'MM/DD') >= '01/01') and
      (to_char(:new.LoginDate, 'MM/DD') <= '07/15') then
      TempSemesterID := 'S1';
    else
      TempSemesterID := 'S2';
    end if;
  end if;

  if (to_char(:new.LoginTime, 'HH24:MI') >= '06:00') and
    (to_char(:new.LoginTime, 'HH24:MI') < '12:00') then
    TempTimeID := '1';
  elsif
    (to_char(:new.LoginTime, 'HH24:MI') >= '12:00') and
    (to_char(:new.LoginTime, 'HH24:MT') < '18:00') then
    TempTimeID := '2';
  else
    TempTimeID := '3';
  end if;

  insert into TempComputerLabFactLevel1
  values (:new.LoginTime, :new.LoginDate, :new.DegreeCode,
  :new.StudentNo, TempSemesterID, TempTimeID);

  commit;

end UpdateTempFactLevel1;
```

The fact and the database trigger for the fact will not change because it will only affect new records and the semester date rule has been handled in the TempFact.

18.2.3.3   Level-2

The Level-2 star schema is unchanged because it is based on the Level-1 star schema, where the fact in Level-2 is created using the create view command.

The implication is that if Level-1 is updated correctly, the updates will be correctly reflected in Level-2 too. Hence, no special maintenance is needed for the Level-2 star schema.

18.3   Data Warehousing Schema Evolution

Active Data Warehousing is not only about automatic updates. The implication of having an active and dynamic data warehouse is more than just an automatic update when the operational databases are updated. Active Data Warehousing indirectly imposes some degree of interdependency not only between operational databases and data warehouses but also among the star schemas of different granularities within the data warehouse.

As Active Data Warehousing is time-boundless, the data warehousing requirements first used to design the data warehouse might have evolved over time. There might be a need to change the requirements. Consequently, data warehousing schema evolves over time. This is then known as *data warehousing schema evolution*. There are four types of data warehousing schema evolution (Fig. 18.7):

1. Changes to a star schema at one level propagating to upper levels,
2. Changes to a star schema at one level which do not affect the upper levels,
3. Inserting a new star schema into the data warehouse, and
4. Deleting a star schema from a data warehouse.

18.3.1   *Changes Propagating to the Next Levels*

In a data warehouse consisting of star schemas of various levels of granularity, when one star schema changes the structure, the changes may propagate to the upper levels of the star schema. In Active Data Warehousing, star schemas are reactive to changes, so change propagation must be dealt with in the next levels of the star schema. For example, if a Level-1 star schema changes its structure due to new requirements, the changes might be propagated to the Level-2 star schema and so on.

This change propagation only occurs in Active Data Warehousing, not only because of the interconnection between the star schemas in different levels but also due to the reactive nature of Active Data Warehousing, where updates, not only in the data but also in the schema or structure, are propagated immediately to the relevant star schemas.

In the Lab Activities case study, the Semester Dimension does not contain any information about the Year. This means that the analysis is purely based on semester, such as comparing the amount of computer usage between Semester 1 and Semester 2 for a particular degree, but it does not consider the year. This implies that the

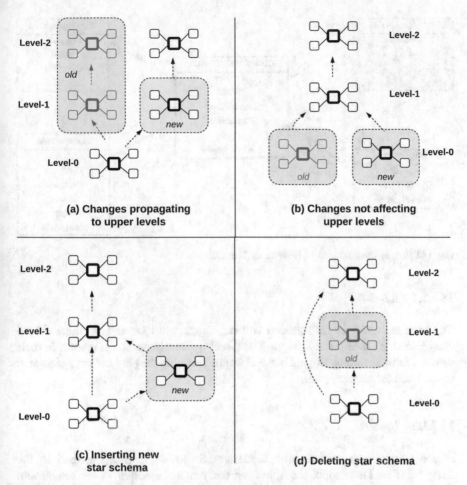

Fig. 18.7 Four cases of data warehousing schema evolution

comparison aggregates the data of all the years according to the semesters. Hence, old data from years ago will still be aggregated with the current data of the two semesters. In other words, the data relating to Semester 1 contains an aggregation of all Semester 1 data, regardless of the year. This kind of analysis will be skewed toward old data, which is why Expiry Date is used to limit the range of data in the data warehouse.

Another way to deal with this situation is by including the Year information into the semester. Therefore, there will be a separate record for each semester/year. In this way, the data analysis can be based not only on semester but also on semester/year, or only year. Adding the Year information into the requirement is an example where the data warehouse schema has evolved. We will see next how the additional requirement to data warehousing might change the schema and how it might propagate to the next levels.

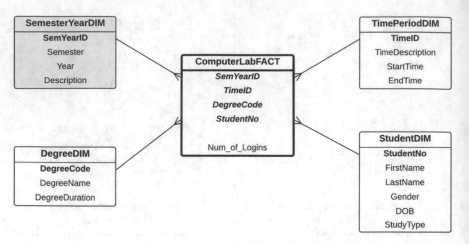

Fig. 18.8 Level-1 star schema with Semester Year Dimension

18.3.1.1 Level-0

The semester and year information will not appear until Level-1, so there are no changes to the Level-0 star schema. The Level-0 star schema stores records in their detailed form, including Login Date and Login Time, but they are not yet aggregated to the semester or year level.

18.3.1.2 Level-1

In the Level-1 star schema, the Semester Dimension is now changed to the Semester Year Dimension, incorporating the year information to the dimension. Other dimensions remain unchanged. Figure 18.8 shows the star schema with the Semester Year Dimension, where the SemYearID attribute is the surrogate key of the dimension. It also includes other attributes, such as Semester, Year and Description.

In the previous version, which included only the Semester Dimension, once the dimension is created and populated (with two records, Semester 1 and Semester 2), this dimension will never change, because any new transaction records can be categorised into either Semester 1 or Semester 2. But now, the dimension includes the Year information. Consequently, when there is a new transaction record which includes the new semester/year, a new record in the Semester Year Dimension must be recorded. In other words, we must now maintain not only the Fact Table of the star schema but also the Semester Year Dimension table.

- **Semester Year Dimension**

  The Semester Year Dimension table must first be created when the data warehouse is built. The Semester Year information will be converted from the Login Date attribute in the operational database.

```
create table SemesterYearDimTemp as
select distinct LoginDate
from LabActivities;

alter table SemesterYearDimTemp
add (
  SemYearID       varchar2(10),
  Semester        varchar2(2),
  Year            varchar2(4),
  Description     varchar2(20)
);

update SemesterYearDimTemp
set Semester = 'S1'
where to_char(LoginDate, 'MM/DD') >= '01/01'
and to_char(LoginDate, 'MM/DD') <= '06/30';

update SemesterYearDimTemp
set Semester = 'S2'
where Semester is null;

update SemesterYearDimTemp
set Year = to_char(LoginDate, 'YYYY');

update SemesterYearDimTemp
set SemYearID = Year || Semester;

create table SemesterYearDim as
select distinct SemYearID, Semester, Year, Description
from SemesterYearDimTemp;
```

After the dimension is created, the dimension must be maintained so that the new transaction records from the operational database will be immediately reflected in this star schema. The following is the process to maintain the Semester Year Dimension table. After converting the Login Date information from the operational database into a suitable format for the Semester Year Dimension table, it will first check if the record exists in the Dimension or not. If the record does not exist, the record is inserted into the Dimension.

```
create or replace trigger AddSemesterYear
  after insert on LabActivities
  for each row
  declare
    TempSemester        SemesterYearDim.Semester%type;
    TempYear            SemesterYearDim.Year%type;
    TempSemesterYearID  SemesterYearDim.SemYearID%type;
    TempDescription     SemesterYearDim.Description%type;
    IsFound number;

begin
  if to_char(:new.LoginDate, 'MM/DD') >= '01/01'
  and to_char(:new.LoginDate, 'MM/DD') <= '06/30' then
    TempSemester := 'S1';
```

```
  else
    TempSemester := 'S2';
  end if;

  TempYear := to_char(:new.LoginDate, 'YYYY');
  TempDescription := null;
  TempSemesterYearID := TempSemester || TempYear;

  select count(*) into IsFound
  from SemesterYearDim
  where SemYearID = TempSemesterYearID;

  if IsFound = 0 then
    insert into SemesterYearDim
    values (TempSemesterYearID, TempSemester, TempYear,
    TempDescription);
    commit;
  end if;

end AddSemesterYear;
```

- **TempFact**

  A TempFact Table can be created in the usual way, but the Fact Table from Level-0 is reused.

  ```
  create table TempComputerLabFactLevel1 as
  select distinct
    LoginTime, LoginDate,
    DegreeCode, StudentNo
  from ComputerLabFactLevel0;
  ```

  Adding the SemYearID attribute into TempComputerLabFactLevel1 table can be tricky as Login Date needs to be converted to SemYearID.

  ```
  alter table TempComputerLabFactLevel1
  add (SemYearID varchar2(10));

  update TempComputerLabFactLevel1
  set SemYearID = to_char(LoginDate, 'YYYY') || 'S1'
  where to_char(LoginDate, 'MM/DD') >= '01/01'
  and to_char(LoginDate, 'MM/DD') <= '06/30';

  update TempComputerLabFactLevel1
  set SemYearID = to_char(LoginDate, 'YYYY') || 'S2'
  where to_char(LoginDate, 'MM/DD') >= '07/01'
  and to_char(LoginDate, 'MM/DD') <= '12/31';
  ```

  Converting the Login Time attribute to TimeID attribute is done as shown in the previous section.

  ```
  alter table TempComputerLabFactLevel1
  add (TimeID number);

  update TempComputerLabFactLevel1
  set TimeID = '1'
  ```

```
where to_char(LoginTime, 'HH24:MI') >= '06:00'
and to_char(LoginTime, 'HH24:MI') < '12:00';

update TempComputerLabFactLevel1
set TimeID = '2'
where to_char(LoginTime, 'HH24:MI') >= '12:00'
and to_char(LoginTime, 'HH24:MI') < '18:00';

update TempComputerLabFactLevel1
set TimeID = '3'
where to_char(LoginTime, 'HH24:MI') >= '18:00'
or to_char(LoginTime, 'HH24:MI') < '06:00';
```

Finally, the database trigger to automatically trigger an insertion to the TempFact is as follows:

```
create or replace trigger UpdateTempFactLevel1
  after insert on ComputerLabFactLevel0
  for each row
  declare
    TempSemYearID
      TempComputerLabFactLevel1.SemYearID%type;
    TempTimeID
      TempComputerLabFactLevel1.TimeID%type;
begin
  if (to_char(:new.LoginDate, 'MM/DD') >= '01/01') and
    (to_char(:new.LoginDate, 'MM/DD') <= '06/30') then
    TempSemYearID := to_char(:new.LoginDate, 'YYYY')
    || 'S1';
  else
    TempSemYearID := to_char(:new.LoginDate, 'YYYY')
    || 'S2';
  end if;

  if (to_char(:new.LoginTime, 'HH24:MI') >= '06:00') and
    (to_char(:new.LoginTime, 'HH24:MI') < '12:00') then
    TempTimeID := '1';
  elsif
    (to_char(:new.LoginTime, 'HH24:MI') >= '12:00') and
    (to_char(:new.LoginTime, 'HH24:MI') < '18:00') then
    TempTimeID := '2';
  else
    TempTimeID := '3';
  end if;

  insert into TempComputerLabFactLevel1
  values (:new.LoginDate, :new.LoginTime,
  :new.DegreeCode, :new.StudentNo,
  TempSemYearID, TempTimeID);

  commit;
end UpdateTempFactLevel1;
```

- **The Fact**

The final Fact Table for Level-1 is created by aggregating the four dimension identifiers and counting the StudentNo. Although the Semester Year Dimension has been created (and maintained using a database trigger) as previously explained, there is no guarantee that the new SemYearID has been inserted into the Semester Year Dimension table. Hence, it would have been easier if the PK-FK constraint between the fact and the dimension is not enforced to guarantee that the fact level will be created successfully. In the end, the Semester Year Dimension table will have all the semester year records anyway; however, we cannot guarantee that the Semester Year Dimension table has been populated with the new semester year record prior to the creation of this Fact Table.

```
create table ComputerLabFactLevel1 as
select
  SemYearID, TimeID, DegreeCode, StudentNo,
  count(StudentNo) as Num_of_Logins
from TempComputerLabFactLevel1
group by SemYearID, TimeID, DegreeCode, StudentNo;
```

Once the Fact Table is created, to maintain this table, a database trigger must be used to ensure that when new records are inserted into the TempFact Table, they will be reflected in the Fact Table, as the fact measure Number of Logins needs to be incremented properly. If it is a new record in the fact (with the same dimension keys), the record will be inserted into the Fact Table. If the record already exists in the Fact Table (with the same dimension keys), the fact measure will simply be incremented by one value.

```
create or replace trigger UpdateFactLevel1
  after insert on TempComputerLabFactLevel1
  for each row
  declare
    IsFound number;
begin
  select count(*) into IsFound
  from ComputerLabFactLevel1
  where  SemYearID = :new.SemYearID
    and TimeID = :new.TimeID
    and DegreeCode = :new.DegreeCode
    and StudentNo = :new.StudentNo;

  if IsFound <> 0 then
    update ComputerLabFactLevel1
    set Num_of_Logins = Num_of_Logins + 1
    where SemYearID = :new.SemYearID
      and TimeID = :new.TimeID
      and DegreeCode = :new.DegreeCode
      and StudentNo = :new.StudentNo;
  else
    insert into ComputerLabFactLevel1
    values (:new.SemYearID, :new.TimeID, :new.DegreeCode,
    :new.StudentNo, 1);
```

```
      end if;
      commit;
   end UpdateFactLevel1;
```

18.3.1.3 Level-2

The changes in the Level-1 star schema, namely, the changes of the Semester Year Dimension, are carried forward to the upper level: Level-2. The changes are reflected at the star schema level, where the Semester Year Dimension is still used (refer to Fig. 18.9 for the Level-2 star schema). The Level-2 star schema basically removes the Student Dimension from Level-1.

From the technical implementation point of view, the Level-2 fact reuses the Level-1 fact, and in Level-1, the fact already takes care of the new Semester Year Dimension. Therefore, the impact to the Level-2 fact is technically minimal, which is shown in the following create view statement.

```
create or replace view ComputerLabFactLevel2 as
select SemYearID, TimeID, DegreeCode,
  sum(Num_of_Logins) as Num_of_Logins
from ComputerLabFactLevel1
group by SemYearID, TimeID, DegreeCode;
```

In short, the propagation of the new semester rules from Level-1 to Level-2 is more at the schema design level rather than at the technical implementation level.

18.3.2 Changes Not Affecting the Next Levels

There are cases where the changes to a star schema on one level will not affect the next levels. Hence, the changes are isolated to that particular star schema.

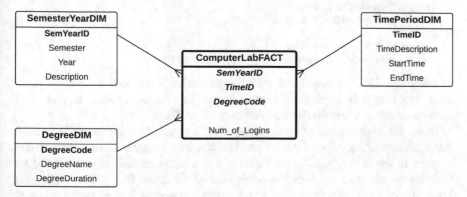

Fig. 18.9 Level-2 star schema with the Semester Year Dimension

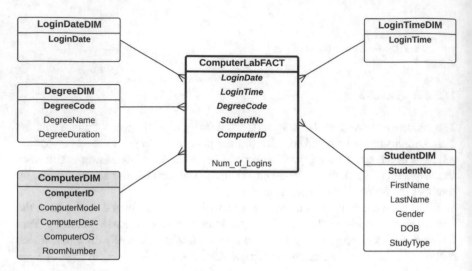

Fig. 18.10 Star schema with Computer Dimension

In the E/R diagram shown in Fig. 18.2, there is a Computer entity which is not used in the star schema. Assuming now that the requirement needs to include the Computer information in the data warehouse, the Computer Dimension needs to be added to the Level-0 star schema, as shown in Fig. 18.10. The other four dimensions are unchanged, and the fact measure also remains the same.

The coding to maintain the new Computer Dimension will be done in the same way as for the other dimensions. For the fact, we will need to get the ComputerID from the Lab Activities table.

```
create table ComputerLabFactLevel0 as
select distinct
  LoginDate, LoginTime,
  S.DegreeCode, L.StudentNo, L.ComputerID,
  count(L.StudentNo) as Num_of_Logins
from Student S, LabActivities L
where S.StudentNo = L.StudentNo
group by
  LoginDate, LoginTime,
  S.DegreeCode, L.StudentNo, L.ComputerID;
```

The FK and PK-FK constraints with the dimension tables need to be done. Also, a database trigger to automatically insert records into the fact needs to be revised to accommodate the ComputerID attribute in the Fact Table.

From the schema point of view, adding the Computer Dimension to the Level-0 star schema will not affect the next levels since the Computer Dimension will not be used in the next levels. In the Level-1 star schema, granularity is reduced by focusing on semester and time period instead of the actual Login Date and Login Time. By reducing the semester and time period granularity, the fact measure is now aggregated. There is no need to include the Computer Dimension in Level-1 as the granularity of the star schema in Level-1 is already reduced.

Adding a dimension to a star schema at one level will not affect the next levels if the dimension is not used in the next levels. Hence, the changes to one level are isolated.

18.3.3   Inserting New Star Schema

The Lab Activities case study has three levels of star schema (refer to Fig. 18.3), where Level-0 contains the raw data with Login Date and Login Time Dimensions, Level-1 is simplified to the Semester Year Dimension and Time Period Dimension (the star schema is previously shown in Fig. 18.8), and Level-2 is at which the Student Dimension is removed so that the fact measure, Number of Logins, could be more aggregated.

The jump in the granularity level from Level-0 to Level-1, that is, from Login Date and Login Time to Semester Year and Time Period (e.g. morning, afternoon, night), might be seen to be wide. It is possible that we would like to insert a new star schema between these levels. Suppose the proposed star schema is at the month and hour granularity level.

We could give this new star schema a Level-1, with the implication that the next levels are shifted upward one level: the old Level-1 becomes the new Level-2, and the old Level-2 becomes the new Level-3. Alternatively, in order to avoid any confusion in the level numbering, we could keep Level-1 and Level-2 as they were, and the new star schema is numbered Sub-Level-1, indicating that it is actually below Level-1. We will use the latter method of numbering: Level-0, Sub-Level-1, Level-1 and Level-2.

The new star schema at the Month and Hour granularity (e.g. Sub-Level-1) is shown in Fig. 18.11.

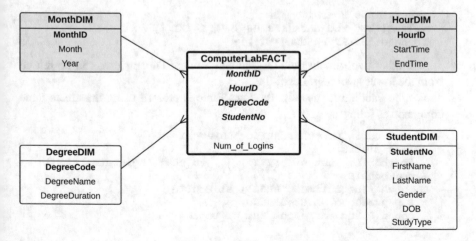

Fig. 18.11 New star schema at the Month and Hour Granularity

- **Creation of the new star schema**

  The creation of this new star schema could be done in the usual way, as described in the previous sections. Let's focus on the two new dimensions first, that is, the Month Dimension and the Hour Dimension. The initial creation of these dimension tables is as follows:

  ```
  create table MonthDim as
  select distinct
    to_char(LoginDate, 'YYYY') || to_char(LoginDate, 'MM')
    as MonthID,
    to_char(LoginDate, 'MON') as Month,
    to_char(LoginDate, 'YYYY') as Year
  from LabActivities;

  alter table MonthDim
  add constraint MonthDimPK primary key (MonthID);

  create table HourDim
  (
    HourID              varchar2(2),
    StarTime            date,
    EndTime             date,
    constraint HourDimPK primary key  (HourID)
  );

  insert into HourDim
  values('00', to_date('00:00', 'HH24:MI'),
  to_date('00:59', 'HH24:MI'));
  insert into HourDim
  values('01', to_date('01:00', 'HH24:MI'),
  to_date('01:59', 'HH24:MI'));
  insert into HourDim
  values('02', to_date('02:00', 'HH24:MI'),
  to_date('02:59', 'HH24:MI'));
  ...
  insert into HourDim
  values('23', to_date('23:00', 'HH24:MI'),
  to_date('23:59', 'HH24:MI'));
  ```

  The other two dimensions, Degree and Student Dimensions, can be reused from the lower level (e.g. Level-0).

  For the Fact Table, usually, the Fact Table is created using the create table command as follows.

  ```
  create table ComputerLabFactSubLevel1 as
  select distinct
    to_char(LoginDate, 'YYYY') || to_char(LoginDate, 'MM')
    as MonthID,
    to_char(LoginTime, 'HH24') as HourID,
    S.DegreeCode, L.StudentNo,
    count(L.StudentNo) as Num_of_Logins
  ```

```
from Student S, LabActivities L
where S.StudentNo = L.StudentNo
group by
  to_char(LoginDate, 'YYYY') || to_char(LoginDate, 'MM'),
  to_char(LoginTime, 'HH24'),
  S.DegreeCode, L.StudentNo;
```

The above method extracts the records directly from the operational database, namely, from tables Student and Lab Activities. Alternatively, we could retrieve the records from the Level-0 Fact Table, using create view, as follows:

```
create or replace view ComputerLabFactSubLevel1 as
select distinct
  to_char(LoginDate, 'YYYY') || to_char(LoginDate, 'MM')
  as MonthID,
  to_char(LoginTime, 'HH24') as HourID,
  DegreeCode, StudentNo,
  sum(StudentNo) as Num_of_Logins
from ComputerLabFactLevel0
group by
  to_char(LoginDate, 'YYYY') || to_char(LoginDate, 'MM'),
  to_char(LoginTime, 'HH24'),
  DegreeCode, StudentNo;
```

There are a couple of differences: (*i*) The create view method which uses the fact Level-0 will be simpler because the view is created and maintained automatically, whereas the create table method needs a separate database trigger to maintain the table for every new record inserted into the operational database. (*ii*) The create view method uses the sum function because it sums the fact measures from the Level-0 Fact Table. The create view method is preferable.

• **Maintaining the star schema**
After the initial creation of this new star schema, we need to implement database triggers so that new transaction records in the operational database can immediately be incorporated into the data warehouse. The Month Dimension needs to be maintained because new transaction records may be for the new month, which is not yet in the data warehouse.

```
create or replace trigger AddMonthYear
  after insert on LabActivities
  for each row
  declare
    TempMonthID    MonthDim.MonthID%type;
    TempYear       MonthDim.Year%type;
    TempMonth      MonthDim.Month%type;
    IsFound number;
begin

  TempMonthID := to_char(:new.LoginDate, 'YYYY') ||
  to_char(:new.LoginDate, 'MM');
```

```
TempYear := to_char(:new.LoginDate, 'YYYY');
TempMonth := to_char(:new.LoginDate, 'MON');

select count(*) into IsFound
from MonthDim
where MonthID = TempMonthID;

if IsFound = 0 then
  insert into MonthDim
  values (TempMonthID, TempMonth, TempYear);
  commit;
end if;
```

```
end AddMonthYear;
```

The Hour Dimension does not need to be maintained manually because the dimension contains the 24-h period when it is created. The Fact Table does not need to be maintained manually either because the Fact Table is a view rather than a table.

- **Impact on the next levels**
  The Level-1 star schema has the Semester Year Dimension and Time Period Dimension. These two dimensions were created directly using the Lab Activities table in the operational database. They do not rely on the lower-level star schemas (e.g. Level-0 and Sub-Level-1). Hence, these dimension tables will not be affected by the new Sub-Level-1 star schema based on the Month and Hour Dimensions. The TempFact was created based on the Fact Table from the Level-0 star schema. Therefore, there won't be any interdependency between the Level-1 star schema and the new Sub-Level-1 star schema. The Fact Table will not be affected either because the Level-2 star schema is based on its TempFact.

  The Level-2 star schema will not be impacted because the dimensions are reused, and the Fact Table is created using the `create view` command based on the Fact Table from the Level-1 star schema.

18.3.4   Deleting Star Schema

Following the scenario in the previous section which has four levels of star schemas, namely, Level-0, the newly inserted Sub-Level-1, Level-1 and Level-2, assume now that we want to delete the Level-1 star schema. This Level-1 star schema has the granularity of Semester Year and Time Period (e.g. morning, afternoon, night) but still maintains the Student Dimension. Note that only the top level, which in this case is Level-2, does not have the Student Dimension, making the aggregation level higher.

There are a couple of implications for the Level-2 star schema when the Level-1 star schema is deleted.

- **Dimensions**

  The Level-2 star schema shares three dimensions with the Level-1 star schema, namely, Semester Year Dimension, Time Period Dimension and Degree Dimension. When deleting the Level-1 star schema, we only remove the Fact Table; the dimension tables are not removed as they are still being used by the Level-2 star schema.

- **Fact**

  The Fact Table for Level-2 is created using the `create view` command based on the Fact Table in Level-1. Since the Level-1 Fact Table is now removed, this will have an impact on the Fact Table for Level-2. Actually, the semantics of the Level-2 and Level-1 star schema in this case study is very similar. The Level-1 star schema has four dimensions, that is, Semester Year, Time Period, Degree and Student Dimensions, whereas the Level-2 star schema has only the first three dimensions, without the Student Dimension. The fact measure is also the same, which is an aggregate value of Number of Logins.

  The creation of the Semester Year Dimension and Time Period Dimension has been discussed at length in the previous section. This was then followed by the creation of the TempFact Table and then the final Fact Table. In the TempFact Table, a temporary Fact Table is created to which two new attributes are added: one attribute to accommodate the Semester Year information and the other attribute for the Time Period information. Once the TempFact is created, the final Fact Table aggregates the records from the TempFact Table based on the four dimension keys: SemYearID, TimeID, Degree Code and StudentNo.

  We could reuse this method to create the new Fact Table for the Level-2 star schema as the Fact Table for the Level-1 star schema is to be removed. The Fact Table could reuse the TempFact, but when it is aggregated, it does not need to include StudentNo as the Student Dimension is not used at this level.

```
create table ComputerLabFactLevelNew2 as
select
  SemYearID, TimeID, DegreeCode,
  count(StudentNo) as Num_of_Logins
from TempComputerLabFactLevel1
group by SemYearID, TimeID, DegreeCode;
```

The database trigger to maintain the Fact Table could also reuse the same database trigger to update Fact Level-1, as previously described.

```
create or replace trigger UpdateFactLevel2
  after insert on TempComputerLabFactLevel1
  for each row
  declare
    IsFound number;
begin
  select count(*) into IsFound
  from ComputerLabFactLevelNew2
  where  SemYearID = :new.SemYearID
    and TimeID = :new.TimeID
    and DegreeCode = :new.DegreeCode;
```

```
    if IsFound <> 0 then
      update ComputerLabFactLevelNew2
      set Num_of_Logins = Num_of_Logins + 1
      where SemYearID = :new.SemYearID
        and TimeID = :new.TimeID
        and DegreeCode = :new.DegreeCode;
  else
      insert into ComputerLabFactLevelNew2
      values (:new.SemYearID, :new.TimeID,
        :new.DegreeCode, 1);
    end if;
    commit;
end UpdateFactLevel2;
```

18.4   Operational Database Evolution

The operational databases which feed into the data warehouse may also evolve over time. This may then impact the data warehouse, as in Active Data Warehousing, updates or changes in the operational databases are reflected in the data warehouse in real time.

As the operational database is changed, the data warehousing requirements may also change. In this section, we are going to see how the changes in the operational database (at various degrees of severity) will change the data warehousing requirements from small changes to the table structure and changes to the entities and relationships to major changes to the operational database.

Some of the changes in the operational database may affect the Level-0 star schema only, while some will propagate to all star schemas in various levels of granularity. In the worst scenario, the entire data warehouse (e.g. all star schemas) will be overhauled, and the old data warehouse is fully decommissioned and replaced with the new one.

18.4.1   Changes in the Table Structure

As the operational or transactional system is used over the time, additional information which was not captured in the past is not being captured, and this additional information may affect the data warehouse. Using the Computer Lab Activities case study, in the original system, the Login Time is recorded but not the Logout Time. Supposing the Logout Time is recorded in the Lab Activities table in the operational database, this information (e.g. Logout Time) or the derived information (e.g. Duration of Login) can be included in the data warehouse. This will impact not only the dimension of the star schema but also the fact measure. So,

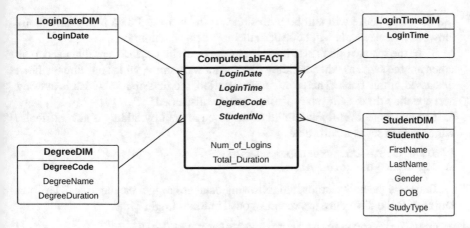

Fig. 18.12   Star schema with Total Duration fact measure

this is an example of how the table structure in the operational database, which in this case is the Lab Activities table structure, is changed by adding new information such as Logout Time.

By having Logout Time information in the operational database, we can measure the duration of the login activity added to the data warehouse. In this case, a new fact measure, Duration (or Total Duration), can be added to the star schema. Figure 18.12 shows a Level-0 star schema, where the new fact measure, Total Duration, has been added to the fact. As a comparison, the original star schema is shown in Fig. 18.3.

Because there is a new fact measure in the Level-0 star schema, the fact measure will likely be carried forward to all upper-level star schemas. This implies that the changes in the table structure in the operational database, in this case the Lab Activities table, affect the Level-0 star schema, which subsequently has an impact on all the upper-level star schemas.

There are two possible options to deal with this. Option 1 is to update the existing star schemas. In this case, the existing Level-0 star schema is updated to include a new fact measure, namely, Total Duration in the Fact Table. All existing records in the Fact Table will have a null value in the Total Duration fact measure because the old records do not have this information. But for every new record added to the Fact Table, Total Duration will be recorded. Technically, two actions need to be taken: Firstly, the Fact Table structure needs to be altered by adding a new column for this new fact measure, and this attribute needs to be filled with a null value in each record. Secondly, to correctly insert the new records into the Fact Table, the database trigger needs to be revised so that the Total Duration information can be inserted correctly for every new record in the Fact Table. This approach needs to be taken in each level of star schema in the data warehouse.

Option 2 is to build a new data warehouse, since the star schema now has new requirements. The old data warehouse, which does not have the new information on Total Duration, will be decommissioned from an Active Data Warehousing. The old data warehouse will still be used since all the historical data is captured there. However, the new data will be kept in the new data warehouse.

Since the second option is to build a new data warehouse, the steps that need to be taken are the same as when building a new data warehouse, which has already been discussed in this book. Therefore, in this section, the first option, which is altering and updating the existing star schemas, will be discussed.

The existing Level-0 Fact Table needs to be altered by adding a new attribute, which is then set as a null value.

```
alter table ComputerLabFactLevel0
add Total_Duration date;
```

To handle the new records, the following database trigger can be used. The Total Duration is the difference between Logout Time and Login Time.

```
create or replace trigger UpdateFactLevelNew0
  after insert on LabActivities
  for each row
  declare
    Degree Student.DegreeCode%type;
begin
  select DegreeCode into Degree
  from Student
  where StudentNo = :new.StudentNo;

  insert into ComputerLabFactLevel0
  values (:new.LoginDate, :new.LoginTime,
    :Degree, :new.StudentNo, 1,
    :new.LogoutTime - :new.LoginTime);
  commit;
end UpdateFactLevelNew0;
```

Changes in the upper levels can be done in a similar way. Figure 18.13 shows an example of the Level-1 star schema where the Total Duration fact measure is added to the fact.

The implementation method to update this star schema is to alter the Fact Table and change the database trigger.

```
alter table ComputerLabFactLevel1
add Total_Duration date;

create or replace trigger UpdateFactLevel1
after insert on TempComputerLabFactLevel1
  for each row
  declare
    IsFound number;
begin
  select count(*) into IsFound
  from ComputerLabFactLevel1
  where  SemYearID = :new.SemYearID
    and TimeID = :new.TimeID
    and DegreeCode = :new.DegreeCode
    and StudentNo = :new.StudentNo;
```

```
  if IsFound <> 0 then
    update ComputerLabFactLevel1
    set Num_of_Logins = Num_of_Logins + 1,
    Total_Duration = Total_Duration + :new.Total_Duration
    where SemYearID = :new.SemYearID
      and TimeID = :new.TimeID
      and DegreeCode = :new.DegreeCode
      and StudentNo = :new.StudentNo;
  else
    insert ComputerLabFactLevel1
    values (:new.SemYearID, :new.TimeID,
      :new.DegreeCode, :new.StudentNo, 1,
      :new.Total_Duration);
  end if;
  commit;
end UpdateFactLevel1;
```

The database trigger checks if the new record inserted into the TempFact is a record with a new Semester Year, which does not yet exist in the Fact Table, and then it will insert the new record with Total Duration, which is the same as the Total Duration in the TempFact. On the other hand, if the record is a record with an existing Semester Year, the trigger will update the fact record and increment the Total Duration by the Duration of the new record in the TempFact.

As the fact is based on the TempFact, the database trigger for the TempFact also needs to be adjusted, as follows. After the TempFact Table is altered, the database trigger inserts the record into the TempFact Table, where Total Duration is the difference between Logout Time and Login Time.

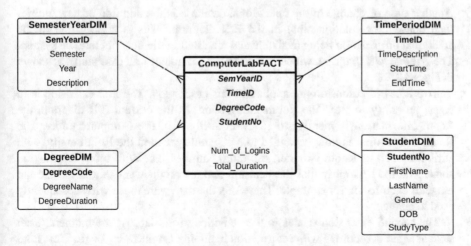

Fig. 18.13 Level-1 star schema with Total Duration fact measure

```
alter table TempComputerLabFactLevel1
add Total_Duration date;

create or replace trigger UpdateTempFactLevel1
  after insert on ComputerLabFactLevel0
  for each row
  declare
    TempSemYearID TempComputerFactLabLevel1.SemYearID%type;
    TempTimeID    TempComputerFactLabLevel1.TimeID%type;
begin
  -- the codes for checking the Semester Year information
  -- based on Login Date are here
  -- for simplicity, the codes are not included here

  -- the codes for checking the Time Period information
  -- based on Login Time are here
  -- for simplicity, the codes are not included here

  insert into TempComputerLabFactLevel1
  values (:new.LoginDate, :new.LoginTime,
  :new.DegreeCode, :new.StudentNo,
  TempSemYearID, TempTimeID,
  :new.LogoutTime - :new.LoginTime);
  commit;
end UpdateTempFactLevel1;
```

18.4.2   Changes in the E/R Schema

Another type of change in the operational database is the addition of new entities (and hence new relationships) in the E/R diagram. This is simply due to the additional information being available and recorded in the new operational system. The original E/R diagram, which is used in the Computer Lab case study, is shown in Fig. 18.2.

Two areas of extension are applied in this case study. The first extension is to create an entity to store the Room information. In the current E/R diagram, the Room information is merely stored as an attribute in the Computer entity. The Room information is now expanded to a new entity, called the Room entity, with attributes such as Room Number, Building, Number, etc. With more information about Rooms, it is clear that the decision-makers might want to analyse the lab usage of each room, for example. Therefore, the data warehouse will subsequently need to change.

The second extension relates to the activities in the lab. At the moment, each student login is recorded with information including Login Date, Login Time (and Logout Time), ComputerID and, of course, StudentNo. With the new audit trial, each login activity will include the software and tools accessed in each login connection as well. This information was not available in the original E/R diagram but will now be included. During one lab attendance, any software accessed by

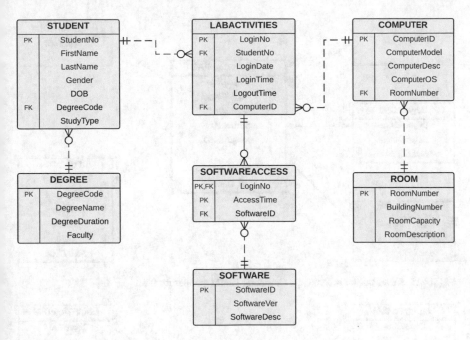

Fig. 18.14 Revised E/R diagram

the student will be recorded. This results in an extension of two entities: Software Access and Software entities.

These two extensions involve the creation of new entities and relationships (as opposed to internal changes within one entity as described in the section above). The new E/R diagram is shown in Fig. 18.14.

The data warehouse is now expanded into two Level-0s. One is at the login granularity and the second is at the software access granularity. Note that for each login, a student may access several software applications. Clearly, these are two levels of granularity which cannot be combined into one star schema.

The Level-0a star schema as shown in Fig. 18.15 focuses on the Computer Room granularity. The fact measures are Number of Logins and Total Duration. Number of Logins basically counts the number of records in the Lab Activity table, as one record in the Lab Activity table is one login activity.

The Level-0b star schema as shown in Fig. 18.16 focuses on the Software Access granularity. The original fact measure Number of Logins is not applicable because Number of Logins is counted per student computer login. In this star schema, one computer login may access several software applications. Therefore, the fact measure must count the Total Access to any software rather than each computer login. The second difference is that there is no Total Duration because there is no information about the duration the student accessed the software at each computer login.

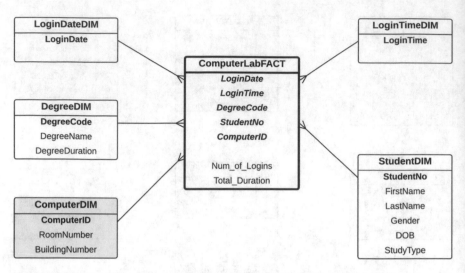

Fig. 18.15   Level-0a star schema at a Computer Room granularity level

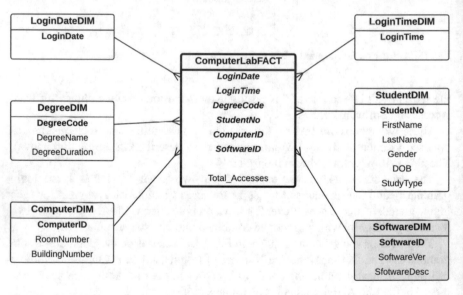

Fig. 18.16   Level-0b star schema at a Software Access granularity level

There are only Login Time and Logout Time but are at the computer login level, not at a specific software level. Therefore, the Total Duration fact measure is not applicable here.

Because the Level-0a star schema is an extension of the original star schema, we can simply reuse the old star schema by adding it to the Computer Dimension. Then, the database trigger needs to also insert ComputerID into the Fact Table.

```
alter table ComputerLabFactLevelNew0a
add ComputerID varchar2(20);

create or replace trigger UpdateFactLevelNew0a
  after insert on LabActivities
  for each row
  declare
    Degree Student.DegreeCode%type;
begin
  select DegreeCode into Degree
  from Student
  where StudentNo = :new.StudentNo;

  insert into ComputerFactLabLevelNew0a
  values (:new.LoginDate, :new.LoginTime,
    Degree, :new.StudentNo, 1,
    :new.LogoutTime - :new.LoginTime,
    :new.ComputerID);
  commit;
end UpdateFactLevelNew0a;
```

The Computer Dimension can be created in the usual way, either using `create table` followed by a database trigger or `create view`. The following illustrates the `create view` method:

```
create or replace view ComputerDim as
select distinct
  C.ComputerID, C.RoomNumber, R.BuildingNumber
from Computer C, Room R
where C.RoomNumber = R.RoomNumber;
```

The Level-0b star schema is a new star schema that needs to be created. The Fact Table basically includes a join operation with the Software Access table, and the count aggregate function counts the join results because the granularity of the fact measure is at each software access level. The Software Access Dimension can be created the same way as the other dimensions.

```
create table ComputerLabFactLevelNew0b as
select distinct
  LoginDate, LoginTime,
  S.DegreeCode, L.StudentNo,
  A.SoftwareID,
  count(*) as Total_Accesses
from Student S, LabActivities L, SoftwareAccess A
where S.StudentNo = L.StudentNo
and L.LoginNo = A.LoginNo
group by
  LoginDate, LoginTime,
  S.DegreeCode, L.StudentNo, A.SoftwareID;
```

The Fact Table needs a database trigger to trigger an automatic update to the Fact Table for every new transaction record added in the operational database. Note that the database trigger is based on an insertion into the Software Access table. For each new record inserted into the Software Access table in the operational database, a new record will be inserted into the Fact Table.

```
create or replace trigger UpdateFactLevelNew0b
  after insert on SoftwareAccess
  for each row
  declare
     StudentNo      LabActivities.StudentNo%type;
     LoginDate      LabActivities.LoginDate%type;
     LoginTime      LabActivities.LoginTime%type;
     ComputerID     LabActivities.ComputerID%type;
     Degree         Student.DegreeCode%type;
begin
  select StudentNo, LoginDate, LoginTime, ComputerID
  into StudentNo, LoginDate, LoginTime, ComputerID
  from LabActivities
  where LoginNo = :new.LoginNo;

  select DegreeCode into Degree
  from Student
  where StudentNo = :StudentNo;

  insert into ComputerLabFactLevelNew0b
  values (LoginDate, LoginTime,
    Degree, StudentNo, ComputerID,
    :new.SoftwareID, 1);

  commit;
end UpdateFactLevelNew0b;
```

Changes to Level-0 star schemas will be propagated to the next levels. The method discussed previously relating to propagating changes to the next levels can be used.

18.4.3   Changes in the Operational Database

All the operational databases, in which the data warehouse is based, may change. The old system is decommissioned and the new system is deployed. Table names may change; table structures may change; the entire design may change. In other words, the old operational database evolves into a new one. These are the most complex changes related to the operational databases, and we need to look at the impact it has on Active Data Warehousing.

The E/R diagram for the operational database that keeps track of the computer lab activities is shown previously in Fig. 18.14, where the Degree entity is included in this schema. Suppose now the system that stores the computer lab activities is changed, and then the Degree entity is no longer there (see Fig. 18.17 for the new E/R diagram for the Computer Lab Activities).

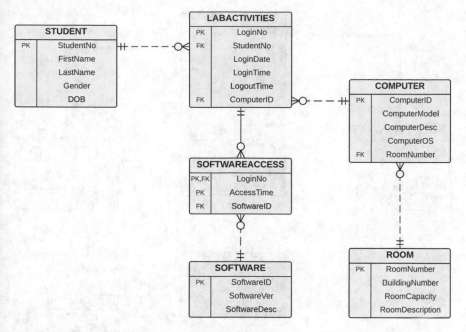

Fig. 18.17 New Computer Lab Activities E/R diagram

There is also a new system that keeps the enrolment records of students, which is shown in the E/R diagram in Fig. 18.18. It keeps track of every individual subject in which each student enrols. It also calculates the WAM (weighted average mark) and GPA (grade point average) of each degree the student has completed at this university. Note that a student may have completed several degrees at this university.

Complexity arises because the data warehouse for the Computer Lab Activities has the Degree Dimension and the Degree entity is no longer in the Computer Lab Activities E/R diagram but now exists in the Student Enrolment E/R diagram. These are two separate operational databases which may run on different systems.

The star schema, as shown in Fig. 18.15, is a Level-0 star schema with five dimensions: Login Date, Login Time, Student, Degree and Computer Dimensions. There are two options to deal with Active Data Warehousing when the operational databases have changed.

- **Option 1: Update existing star schema**

  Option 1 is to update the structure (and data) of the star schema, whenever possible, which in this case is the star schema in Fig. 18.15. Four dimensions, except the Degree Dimensions, are available from the new Computer Lab Activities database. The Degree information needs to be taken from the new Student Enrolment database. Hence, in the database trigger for the fact, it needs to obtain the Degree information from the Student Enrolment system, assuming that the Student Enrolment system is accessible by the database trigger.

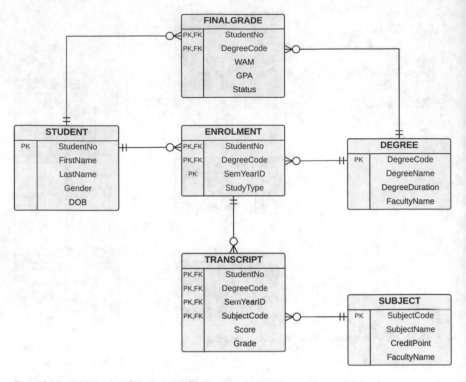

Fig. 18.18 New Student Enrolment E/R diagram

```
create or replace trigger UpdateFactLevelNew0a
  after insert on LabActivities
  for each row
  declare
    Degree
      StudentEnrolmentDatabase.Enrolment.DegreeCode%type;
    TempSem varchar2(2);
    TempSemYearID
      StudentEnrolmentDatabase.Enrolment.SemYearID%type;
begin
  if (to_char(:new.LoginDate, 'MMDD') >= '0101')
  and (to_char(:new.LoginDate, 'MMDD') <= '0630') then
    TempSem = 'S1'
  else
    TempSem = 'S2'
  end if;
  TempSemYearID = to_char(:new.LoginDate, 'YYYY') ||
    TempSem;

  select distinct E.DegreeCode into Degree
  from StudentEnrolmentDatabase.Enrolment E
  where E.StudentNo = :new.StudentNo
  and SemYearID = TempSemYearID;
```

```
    insert into ComputerFactLabLevelNew0a
    values (:new.LoginDate, :new.LoginTime,
      Degree, :new.StudentNo, 1,
      :new.LogoutTime - :new.LoginTime,
      :new.ComputerID);
    commit;
  end UpdateFactLevelNew0a;
```

The Degree Dimension must also be altered so it will obtain the new degree records from the Student Enrolment system, particularly from the Degree table.

- **Option 2: Create new star schema**
  Option 2 is to create new star schemas in all the required granularity levels in the data warehouse. The old star schemas are then decommissioned from the Active Data Warehousing. The old star schemas can still be used as a Passive Data Warehousing to analyse the historical data.

  Because there are two separate operational databases, firstly, we need to create a separate star schema from each operational database, and after that, the two star schemas are combined. Creating two separate star schemas from their respective operational databases is a suitable approach, particularly when the two operational databases are in two different systems, and it may not be possible to access both of them from one system.

  The first star schema is based on the Computer Lab Activities operational database and is shown in Fig. 18.19. This star schema looks very close to the desired final star schema in Fig. 18.15. The only difference is that in the star schema in Fig. 18.19, there is no Degree Dimension. This is simply because the Degree information is no longer available in the operational database of the Computer Lab Activities (refer to Fig. 18.17).

  The second star schema is based on the Student Enrolment operational database as shown in Fig. 18.20. It includes the Degree Dimension, which is

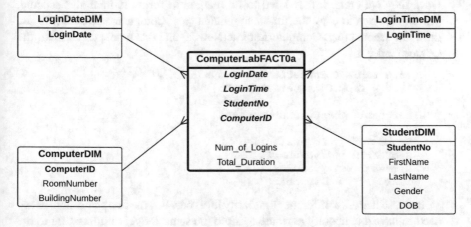

Fig. 18.19 New Computer Lab Activities star schema

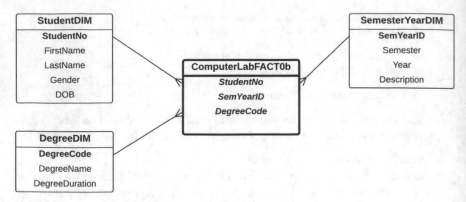

Fig. 18.20 New Student Enrolment star schema

needed in the final star schema, as well as the Student and Semester Year Dimension.

The difference between the first and second star schemas is not only in the Degree Dimension. The granularity seems to be different too as the second star schema uses the Semester Year Dimension, not the Login Date Dimension, as in the first star schema. The second star schema needs to use the Semester Year Dimension because the Student Enrolment system records the semester year information; hence, the date information is irrelevant. Another difference between the first and second star schemas is that in the second star schema, there is no fact measure. The reason for this is that the second star schema will be used as a "lookup table" to obtain the respective Degree information for each student in the first star schema. So, the second star schema is not meant to be an independent star schema.

Both star schemas can be created in the usual way as discussed previously. Once they are created, they need to be merged to form the final star schema. The first step is to copy the first star schema (e.g. ComputerLabFact0a) to the final star schema (e.g. ComputerLabFactNew0a) and add a new column for the Degree Code.

```
create table ComputerLabFactNew0a as
select * from ComputerLabFact0a;

alter table ComputerLabFactNew0a
add (
  DegreeCode varchar2(10),
  SemYearID   varchar2(10),
  Semester    varchar2(2),
  Year        varchar2(4));
```

Because there is a different granularity level between the two star schemas, we need to have a temporary attribute to store the semester year information in the final Fact Table. Then, the SemYearID column is filled with the corresponding semester year information based on the Login Date.

```
update ComputerLabFactNew0a
set Semester = 'S1'
where to_char(LoginDate, 'MM/DD') >= '01/01'
and to_char(LoginDate, 'MM/DD') <= '06/30';

update ComputerLabFactNew0a
set Semester = 'S2'
where Semester is null;

update ComputerLabFactNew0a
set Year = to_char(LoginDate, 'YYYY');

update ComputerLabFactNew0a
set SemYearID = Year || Semester;

alter table ComputerLabFactNew0a
drop (Semester, Year);
```

Finally, the final star schema is updated by filling in the correct Degree Code for each student, where the Semester Year information matches the two star schemas.

```
update ComputerLabFactNew0a
set DegreeCode = ComputerLabFact0b.DegreeCode
where StudentNo = ComputerLabFact0b.StudentNo
and SemYearID = ComputerLabFact0b.SemYearID;

alter table ComputerLabFactNew0a
drop SemYearID;
```

Once the final Fact Table is created, we need a database trigger that will automatically insert records into the final Fact Table when there are new records in the first Fact Table (e.g. ComputerLabFact0a). The trigger needs to convert the Login Date into the equivalent SemYearID, and then based on this SemYearID (as well as the StudentNo), it can search for the correct Degree Code in the second Fact Table (e.g. ComputerLabFact0b). An insertion into the final Fact Table can then be executed.

In short, the second Fact Table (e.g. ComputerLabFact0b) is only used as a lookup table to obtain the Degree Code for each student, based on the Login Date information.

```
create or replace trigger UpdateComputerLabNewFactLevel0a
  after insert on ComputerLabFact0a
  for each row
  declare
    TempSemesterID  SemesterYearDim.Semester%type;
    TempSemYearID   ComputerLabFact0b.SemYearID%type;
    TempDegree      ComputerLabFact0b.DegreeCode%type;
```

```
begin
  if (to_char(:new.LoginDate, 'MM/DD') >= '01/01')
    and (to_char(:new.LoginDate, 'MM/DD') <= '06/30') then
    TempSemesterID := 'S1';
  else
    TempSemesterID := 'S2';
  end if;

  TempSemYearID := to_char(:new.LoginDate, 'YYYY') ||
    TempSemesterID;

  select DegreeCode into TempDegree
  from ComputerLabFact0b
  where StudentNo = :new.StudentNo
  and SemYearID = TempSemYearID;

  insert into ComputerLabFactNew0a
  values (:new.LoginTime, :new.LoginDate,
    TempDegree, :new.StudentNo, :new.ComputerID,
    :new.Num_of_Logins, :new.Total_Duration);
  commit;
end UpdateComputerLabNewFactLevel0a;
```

When there are changes in the Level-0 star schema, these changes may propagate
to the upper levels.

18.5   Summary

Active Data Warehousing is solely about interdependency between the operational
database, which is the source of the data warehouse and the data warehouse itself,
as well as among the star schemas of various levels of granularity in the data
warehouse. This interdependency can be very complex and can be complicated to
maintain.

There are three types of interdependencies studied in this chapter: (i) incremental
updates, which is when new records are inserted into or old records are deleted
from the operational database, they are immediately reflected in the star schema;
(ii) changes in the data warehouse, which is when the data warehouse creates a
new star schema or deletes an existing star schema, as well as changes to the upper
levels of star schemas when the lower-level star schemas change; and (iii) changes
in the operational database, which is when the underlying operational database has
changed and has an impact on the data warehouse.

Since interdependency can be quite complex in many cases, Active Data
Warehousing is used only for applications that really need active data in the data
warehouse. This includes the monitoring system which may rely on real-time data
and not so much on historical data. In this case, an Active Data Warehousing is
appropriate and useful.

18.6 Exercises

18.1 Using the case study discussed in the "Automatic Updates of Data Warehouse" section earlier in this chapter, write the complete SQL commands for Level-0, Level-1 and Level-2 star schemas. Test the system by inserting new records into the operational database, and check that the correct updates have been made in all levels of the star schemas in the data warehouse.

18.2 Using the case study discussed in the "Changes Propagating to the Next Levels" section in this chapter, write the complete SQL commands for Level-0, Level-1 and Level-2 star schemas, where the Semester Year Dimension is added to the Level-1 star schema. Test the system by adding new transaction records with new Semester Year information, which is not in the previous transaction records, and check if the Semester Year Dimension is correctly implemented, as well as in the fact of each level of the star schemas.

18.3 Using the case study discussed in Sect. 18.4.1, write the complete SQL commands for Level-0 and Level-1 star schemas, where the Total Duration fact measure is added to the star schemas. Test the system by adding new transaction records with Logout Time information, and check if the Total Duration fact measure is calculated correctly in both levels of the star schemas.

18.7 Further Readings

Active Data Warehousing is about schema evolution, that is, how schema evolution affects the data warehouse. The schema evolution can either involve the databases or the data warehouses. [1] discusses the effects of evolution on the database design. Other database design books, such as [2–5] and [6], described the concept of evolution in database design.

Active Data Warehousing uses triggers and views to maintain the effect of changes. Further details on these SQL commands can be found in various SQL books, including [7–10].

Many of the well-known database textbooks explain the use of database triggers and views using SQL [11–24].

References

1. S.W. Ambler, P.J. Sadalage, *Refactoring Databases: Evolutionary Database Design*. Addison-Wesley Signature Series (Fowler) (Pearson Education, London, 2006)
2. S. Bagui, R. Earp, Database design using entity-relationship diagrams, in *Foundations of Database Design* (CRC Press, New York, 2003)

3. M.J. Hernandez, *Database Design for Mere Mortals: A Hands-On Guide to Relational Database Design*. For Mere Mortals (Pearson Education, London, 2013)
4. N.S. Umanath, R.W. Scamell, *Data Modeling and Database Design* (Cengage Learning, Boston, 2014)
5. T.J. Teorey, S.S. Lightstone, T. Nadeau, H.V. Jagadish, Database modeling and design: logical design, in *The Morgan Kaufmann Series in Data Management Systems* (Elsevier, Amsterdam, 2011)
6. G. Simsion, G. Witt, Data modeling essentials, in *The Morgan Kaufmann Series in Data Management Systems* (Elsevier, Amsterdam, 2004)
7. J. Melton, *Understanding the New SQL: A Complete Guide*, 2nd edn., vol. I (Morgan Kaufmann, Burlington, 2000)
8. C.J. Date, *SQL and Relational Theory—How to Write Accurate SQL Code*, 2nd edn. Theory in practice (O'Reilly, Newton, 2012)
9. A. Beaulieu, *Learning SQL: Master SQL Fundamentals* (O'Reilly Media, Newton, 2009)
10. M.J. Donahoo, G.D. Speegle, *SQL: Practical Guide for Developers*. The Practical Guides (Elsevier Science, Amsterdam, 2010)
11. C. Coronel, S. Morris, *Database Systems: Design, Implementation, and Management* (Cengage Learning, Boston, 2018)
12. T. Connolly, C. Begg, *Database Systems: A Practical Approach to Design, Implementation, and Management* (Pearson Education, London, 2015)
13. J.A. Hoffer, F.R. McFadden, M.B. Prescott, *Modern Database Management* (Prentice Hall, Englewood Cliffs, 2002)
14. A. Silberschatz, H.F. Korth, S. Sudarshan, *Database System Concepts*, 7th edn. (McGraw-Hill Book, New York, 2020)
15. R. Ramakrishnan, J. Gehrke, *Database Management Systems*, 3rd edn. (McGraw-Hill, New York, 2003)
16. J.D. Ullman, J. Widom, *A First Course in Database Systems*, 2nd edn. (Prentice Hall, Englewood Cliffs, 2002)
17. H. Garcia-Molina, J.D. Ullman, J. Widom, *Database Systems: The Complete Book* (Pearson Education, London, 2011)
18. P.E. O'Neil, E.J. O'Neil, *Database: Principles, Programming, and Performance*, 2nd edn. (Morgan Kaufmann, Los Altos, 2000)
19. R. Elmasri, S.B. Navathe, *Fundamentals of Database Systems*, 3rd edn. (Addison-Wesley-Longman, Reading, 2000)
20. C.J. Date, *An Introduction to Database Systems*, 7th edn. (Addison-Wesley-Longman, Reading, 2000)
21. R.T. Watson, *Data Management—Databases and Organizations* 5th edn. (Wiley, London, 2006)
22. P. Beynon-Davies, *Database Systems* (Springer, Berlin, 2004)
23. M. Kifer, A.J. Bernstein, P.M. Lewis, *Database Systems: An Application-Oriented Approach* (Pearson/Addison-Wesley, Reading, 2006)
24. W. Lemahieu, S. vanden Broucke, *Principles of Database Management: The Practical Guide to Storing, Managing and Analyzing Big and Small Data* (Cambridge University, Cambridge, 2018)

Part VI
OLAP, Business Intelligence, and Data Analytics

Chapter 19
Online Analytical Processing (OLAP)

The illustration depicted in Fig. 19.1 shows a typical journey of data from an operational database, data warehousing and OLAP to data analytics. All of the previous chapters focus on the left-hand side, the transformation process from the operational database to the data warehouse, where we discuss in length the step-by-step transformation process of various cases, using extensive case studies. In this chapter, we are going to focus on the middle part, that is, the *OLAP* part (or *Online Analytical Processing*) and *Business Intelligence* reporting.

An OLAP is an SQL query that retrieves data from the data warehouse. OLAP queries focus on data that would generally be used by management in the decision-making process. Hence, data that shows ranking, trends, sub-totals, grand totals, etc. would be of great interest to management. OLAP queries retrieve data from a data warehouse, but the retrieved results are not in an appropriate format for a presentation to management. These data are considered "raw". The retrieved data need to be properly formatted and presented to management using various reporting tools, which usually include graphs. These reports and graphs are called Business Intelligence. Therefore, the role of OLAP is to retrieve the necessary data from the data warehouse and to present this "raw" data to Business Intelligence for processing for presentation.

19.1 Sales Data Warehousing

A Sales data warehouse will be used throughout this chapter to illustrate how OLAP queries can retrieve data about sales. Figure 19.2 shows a layperson's view of a Sales data warehouse using a cube. It is clear that the cube has three dimensions, Time, Product and Location; each cell contains a Total Sales (Sales$) figure for a particular time, product and location.

The Sales star schema is shown in Fig. 19.3. Each dimension contains several attributes. The Time Dimension contains TimeID, Month and Year attributes,

© The Author(s), under exclusive license to Springer Nature Switzerland AG 2021
D. Taniar, W. Rahayu, *Data Warehousing and Analytics*, Data-Centric Systems
and Applications, https://doi.org/10.1007/978-3-030-81979-8_19

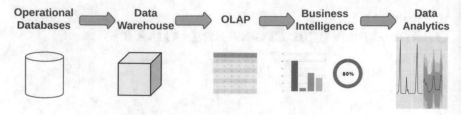

Fig. 19.1 A Data Journey: from operational to analytics

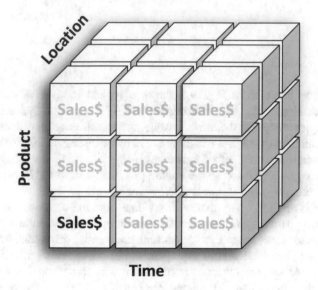

Fig. 19.2 A Sales cube

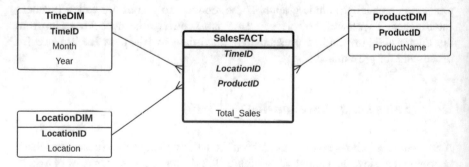

Fig. 19.3 A Sales star schema

where the TimeID attribute is the key identifier. The Product Dimension contains ProductID as the key identifier and the ProductName attribute. The Location Dimension contains LocationID and Location attributes. The Fact Table contains all three key identifiers from the dimensions as well as Total Sales as the fact measure.

Table 19.1 Time dimension

TimeID	Month	Year
201801	Jan	2018
201802	Feb	2018
201803	Mar	2018
201804	Apr	2018
201805	May	2018
201806	Jun	2018
...	...	...
201907	Jul	2019
201908	Aug	2019
201909	Sep	2019
201910	Oct	2019
201911	Nov	2019
201912	Dec	2019

Table 19.2 Location dimension

LocationID	Location
MEL	Melbourne
SYD	Sydney
ADL	Adelaide
PER	Perth

Table 19.3 Product dimension

ProductID	ProductName
C01	Clothing
S02	Shoes
A01	Accessories
C02	Cosmetics
K01	Kids and Baby

The following are the contents of the three dimension tables. For simplicity, the Sales Data Warehouse only contains data for a period of 2 years. Hence, the Time Dimension table, as shown in Table 19.1, contains 24 months. The Location Dimension table, as shown in Table 19.2, contains only four locations in Australia. The Product Dimension table, as shown in Table 19.3, contains five different product categories.

19.2 Basic Aggregate Functions

OLAP queries usually concentrate around aggregation and aggregate functions. The basic aggregate functions, which are common to SQL, are count, sum, avg, max and min, as well as the use of group by clause when using an aggregate function. These aggregate functions, as well as group by, have been used extensively in the previous chapters. However, for completeness of the discussions on OLAP, these basic aggregate functions are described again.

19.2.1 count *Function*

The count function counts the number of records in a table. Generally, there are three forms of count:

1. count(*), which counts the number of records in a table
2. count(attribute), which counts the number of records (excluding null values) in that specified attribute
3. count(distinct attribute), which counts the number of unique records of the specified attribute

The following are the examples of the count function, and the respective query results are shown in Tables 19.4, 19.5, 19.6, and 19.7. Notice that there are 24 records in the Time Dimension table (see Table 19.4) but only 2 years in Table 19.5 as it uses distinct. Table 19.6 uses distinct and hence the results are 12 months, but Table 19.7 shows there are 24 months in the Time Dimension table. In this case, there is no difference between count(*) and count(attribute) as there are no null values in the Time Dimension table (e.g. both Tables 19.4 and 19.7 show 24 records).

```
select count(*) as Number_of_Records
from TimeDim;

select count(distinct Year) as Number_of_Years
from TimeDim;

select count(distinct Month) as Number_of_Months
from TimeDim;

select count(Month) as Number_of_Months
from TimeDim;
```

Table 19.4 count(*)

Number of records
24

Table 19.5 count(distinct Year)

Number of years
2

Table 19.6 count(distinct Month)

Number of months
12

Table 19.7 count(Month)

Number of months
24

Table 19.8 `sum(Total_Sales)`

Total Sales
2828318

19.2.2   *sum Function*

Given an attribute, the sum function calculates the total of the values in the specified attribute. Hence, the attribute required by the sum function must be a numerical attribute.

The following SQL command calculates the Total Sales in Melbourne in 2019. The result is shown in Table 19.8.

```
select sum(Total_Sales) as Total_Sales
from SalesFact S, TimeDim T
where S.TimeID = T.TimeID
and LocationID = 'MEL'
and Year = 2019;
```

19.2.3   *avg, max and min Functions*

The avg, max and min functions calculate the average, maximum and minimum value of the given attribute of a table. In an OLAP context, the attribute is a numerical attribute. The following retrieves the average, maximum and minimum Total Sales in Melbourne in 2019 from the Sales Fact Table.

```
select avg(Total_Sales) as Average_Sales
from SalesFact S, TimeDim T
where S.TimeID = T.TimeID
and S.LocationID = 'MEL'
and T.Year = 2019;

select max(Total_Sales) as Maximum_Sales
from SalesFact S, TimeDim T
where S.TimeID = T.TimeID
and S.LocationID = 'MEL'
and T.Year = 2019;

select min(Total_Sales) as Minimum_Sales
from SalesFact S, TimeDim T
where S.TimeID = T.TimeID
and S.LocationID = 'MEL'
and T.Year = 2019;
```

The result of each of the above queries is one figure, which is average, maximum and minimum Total Sales, respectively. Tables 19.9, 19.10, and19.11 show the query results. These results show that Total Sales in Melbourne in 2019 range from $25,496 to $73,439, with an average of $47,138.63.

Table 19.9 Average sales

Average sales
47138.63

Table 19.10 Maximum sales

Maximum sales
73439

Table 19.11 Minimum sales

Minimum sales
25496

Table 19.12 Total sales by product

ProductName	Total sales
Clothing	715423
Cosmetics	633632
Shoes	645881
Accessories	355433
Kids and Baby	477949

19.2.4 *group by Clause*

Basic aggregate functions are often used in conjunction with group by clause in SQL. The group by clause is used to break down the aggregate value into several categories specified by the group by clause.

The following query retrieves the sum of Total Sales in Melbourne in 2019. The return value as shown previously in Table 19.8 is a single value which represents the sum of Total Sales.

```
select sum(Total_Sales) as Total_Sales
from SalesFact S, TimeDim T
where S.TimeID = T.TimeID
and LocationID = 'MEL'
and Year = 2019;
```

If we want to see the breakdown of this Total Sales by each Product, the group by clause can do it. The results are shown in Table 19.12.

```
select ProductName, sum(Total_Sales) as Total_Sales
from SalesFact S, TimeDim T, ProductDim P
where S.TimeID = T.TimeID
and S.ProductID = P.ProductID
and LocationID = 'MEL'
and Year = 2019
group by ProductName
order by ProductName;
```

19.3 Cube and Rollup

Basic OLAP queries usually consist of new group by capabilities, namely, group by cube and group by rollup. The details are discussed in the following sections.

19.3.1 Cube

In OLAP, it is typical to produce a matrix visual or table which crosses two pieces of categorical information and contains sub-totals. Table 19.13 shows the Total Sales, together with sub-totals and the grand total of two Products (e.g. Clothing and Shoes) in two locations (e.g. Melbourne and Perth).

There are nine figures for Total Sales in Table 19.13. If we only use basic aggregate functions and group by clause, we need four separate SQL statements to retrieve all of these nine values. The first SQL retrieves each individual Total Sales of Clothing and Shoes in both locations, Melbourne and Perth, which is a total of four values. The second SQL retrieves the sub-totals of Clothing and Shoes in both locations (e.g. one sub-total for Clothing and another for Shoes). The third SQL retrieves the sub-totals of Melbourne and Perth for both products (e.g. one sub-total for Melbourne and another for Perth). Finally, the last SQL retrieves the grand total of Total Sales of both products in both locations. The group by clause is used in the first three SQL commands, but not the last one. The results are shown in Tables 19.14, 19.15, 19.16, and 19.17.

Table 19.13 Total sales by product

	Melbourne	Perth	
Clothing	$1,477,348	$861,268	$2,338,616
Shoes	$1,311,316	$897,153	$2,208,469
	$2,788,664	$1,758,421	$4,547,085

Table 19.14 Product-location total sales

ProductName	Location	Total sales
Clothing	Melbourne	1477348
Clothing	Perth	861268
Shoes	Melbourne	1311316
Shoes	Perth	897153

Table 19.15 Product sub-totals

ProductName	Total sales
Clothing	2338616
Shoes	2208469

Table 19.16 Location
sub-totals

Location	Total sales
Melbourne	2788664
Perth	1758421

Table 19.17 Grand total

Total sales
4547085

```
select ProductName, Location,
  sum(Total_Sales) as Total_Sales
from SalesFact S, ProductDim P, LocationDim D
where S.ProductID = P.ProductID
and S.LocationID = D.LocationID
and S.LocationID in ('MEL', 'PER')
and ProductName  in ('Clothing', 'Shoes')
group by ProductName, Location;

select ProductName, sum(Total_Sales) as Total_Sales
from SalesFact S, ProductDim P, LocationDim D
where S.ProductID = P.ProductID
and S.LocationID = D.LocationID
and S.LocationID in ('MEL', 'PER')
and ProductName  in ('Clothing', 'Shoes')
group by ProductName;

select Location, sum(Total_Sales) as Total_Sales
from SalesFact S, ProductDim P, LocationDim D
where S.ProductID = P.ProductID
and S.LocationID = D.LocationID
and S.LocationID in ('MEL', 'PER')
and ProductName  in ('Clothing', 'Shoes')
group by Location;

select sum(Total_Sales) as Total_Sales
from SalesFact S, ProductDim P, LocationDim D
where S.ProductID = P.ProductID
and S.LocationID = D.LocationID
and S.LocationID in ('MEL', 'PER')
and ProductName  in ('Clothing', 'Shoes');
```

To avoid using four SQL commands, the group by cube clause can be used.
The SQL is almost the same as the previous query expressed in group by, but
we now use group by cube. The results are shown in Table 19.18, and all the
nine Total Sales figures are there, although the formatting of the query result is not
presented as professionally as required. Nonetheless, the results are all there.

```
select ProductName, Location, sum(Total_Sales) as Total_Sales
from SalesFact S, ProductDim P, LocationDim D
where S.ProductID = P.ProductID
and S.LocationID = D.LocationID
and S.LocationID in ('MEL', 'PER')
and ProductName  in ('Clothing', 'Shoes')
```

```
group by cube (ProductName, Location)
order by ProductName, Location;
```

Rows 1–3 in Table 19.18 (e.g. Clothing (null)) indicate the sub-total for Clothing, whereas rows 4–6 (e.g. Shoes (null)) are the sub-total for Shoes. Hence, the sub-total for each Product is there. Rows 7 and 8 (e.g. (null) Melbourne and (null) Perth) show the sub-totals for Melbourne and Perth, respectively. Notice from these rows, the (null) in the query result indicates sub-total, not the (null) value of the column. Finally, the last row, which is (null) (null), indicates the grand total of Total Sales from both cities and products.

So, in short, the group by cube clause can be used to produce a matrix visual or table.

19.3.2   Rollup

group by rollup is quite similar to group by cube. If the above query is expressed using group by rollup as follows, the results are shown in Table 19.19. The sub-totals for Clothing and Shoes are still there, but not the sub-totals for the two cities. So the rollup results are a subset of those of cube. The rollup results indicate that if there is a (null) in the categorical column (e.g. the DimensionID attribute), there cannot be a not (null) value in the right-side column. Hence, (null) Melbourne and (null) Perth will be excluded from the rollup results.

Table 19.18 Product-location cube

	ProductName	Location	Total sales
1	Clothing	Melbourne	1477348
2	Clothing	Perth	861268
3	Clothing	(null)	2338616
4	Shoes	Melbourne	1311316
5	Shoes	Perth	897153
6	Shoes	(null)	2208469
7	(null)	Melbourne	2788664
8	(null)	Perth	1758421
9	(null)	(null)	4547085

Table 19.19 Product-Location rollup

	ProductName	Location	Total sales
1	Clothing	Melbourne	1477348
2	Clothing	Perth	861268
3	Clothing	(null)	2338616
4	Shoes	Melbourne	1311316
5	Shoes	Perth	897153
6	Shoes	(null)	2208469
7	(null)	(null)	4547085

Table 19.20 Location-Product rollup

	Location	ProductName	Total sales
1	Melbourne	Clothing	1477348
2	Melbourne	Shoes	1311316
3	Melbourne	(null)	2788664
4	Perth	Clothing	861268
5	Perth	Shoes	897153
6	Perth	(null)	1758421
7	(null)	(null)	4547085

On the other hand, Clothing (null) and Shoes (null), as well as (null) (null), are still in the rollup results (e.g. rows 3, 6 and 7). Therefore, the order of the attributes (or columns) in the group by rollup is important.

```
select ProductName, Location, sum(Total_Sales) as Total_Sales
from SalesFact S, ProductDim P, LocationDim D
where S.ProductID = P.ProductID
and S.LocationID = D.LocationID
and S.LocationID in ('MEL', 'PER')
and ProductName  in ('Clothing', 'Shoes')
group by rollup (ProductName, Location)
order by ProductName, Location;
```

If we change the order of the attributes in the group by rollup to (Location, ProductName) as in the SQL below, the results will be different (see Table 19.20).

```
select Location, ProductName, sum(Total_Sales) as Total_Sales
from SalesFact S, ProductDim P, LocationDim D
where S.ProductID = P.ProductID
and S.LocationID = D.LocationID
and S.LocationID in ('MEL', 'PER')
and ProductName  in ('Clothing', 'Shoes')
group by rollup (Location, ProductName)
order by Location, ProductName;
```

In contrast, the order of the attributes in group by cube does not matter because group by cube produces all combinations between the attributes.

19.3.3 Rollup vs. Cube

The difference between rollup and cube will become clearer if there are more than two attributes in the group by. The following SQL uses three attributes (e.g. ProductName, Location and TimeID) in the cube. The results are shown in Table 19.21. For simplicity, the query focuses on January data only (e.g. 201801 and 201901). The results show all possible sub-totals are included: individual Product, Location, Time and combinations of these.

Table 19.21 Product-Location-Time cube

	ProductName	Location	TimeID	Total sales
1	Clothing	Melbourne	201801	60325
2	Clothing	Melbourne	201901	64823
3	Clothing	Melbourne	(null)	125148
4	Clothing	Perth	201801	37279
5	Clothing	Perth	201901	41479
6	Clothing	Perth	(null)	78758
7	Clothing	(null)	201801	97604
8	Clothing	(null)	201901	106302
9	Clothing	(null)	(null)	203906
10	Shoes	Melbourne	201801	61119
11	Shoes	Melbourne	201901	63968
12	Shoes	Melbourne	(null)	125087
13	Shoes	Perth	201801	44556
14	Shoes	Perth	201901	27112
15	Shoes	Perth	(null)	71668
16	Shoes	(null)	201801	105675
17	Shoes	(null)	201901	91080
18	Shoes	(null)	(null)	196755
19	(null)	Melbourne	201801	121444
20	(null)	Melbourne	201901	128791
21	(null)	Melbourne	(null)	250235
22	(null)	Perth	201801	81835
23	(null)	Perth	201901	68591
24	(null)	Perth	(null)	150426
25	(null)	(null)	201801	203279
26	(null)	(null)	201901	197382
27	(null)	(null)	(null)	400661

```
select
  ProductName, Location, TimeID,
  sum(Total_Sales) as Total_Sales
from SalesFact S, ProductDim P, LocationDim D
where S.ProductID = P.ProductID
and S.LocationID = D.LocationID
and S.LocationID in ('MEL', 'PER')
and ProductName  in ('Clothing', 'Shoes')
and TimeID in ('201801', '201901')
group by cube (ProductName, Location, TimeID)
order by ProductName, Location, TimeID;
```

If we use group by rollup instead, as in the following SQL query, the results are shown in Table 19.22. It is clear that (null) followed by non (null) in the categorical columns will be excluded in the rollup results. Hence, the rollup results are a subset of the cube results. Also be aware that the order of the attributes

Table 19.22
Product-Location-Time rollup

	ProductName	Location	TimeID	Total sales
1	Clothing	Melbourne	201801	60325
2	Clothing	Melbourne	201901	64823
3	Clothing	Melbourne	(null)	125148
4	Clothing	Perth	201801	37279
5	Clothing	Perth	201901	41479
6	Clothing	Perth	(null)	78758
7	Clothing	(null)	(null)	203906
8	Shoes	Melbourne	201801	61119
9	Shoes	Melbourne	201901	63968
10	Shoes	Melbourne	(null)	125087
11	Shoes	Perth	201801	44556
12	Shoes	Perth	201901	27112
13	Shoes	Perth	(null)	71668
14	Shoes	(null)	(null)	196755
15	(null)	(null)	(null)	400661

in the group by rollup is important as a different ordering of attributes will produce different results.

```
select
  ProductName, Location, TimeID,
  sum(Total_Sales) as Total_Sales
from SalesFact S, ProductDim P, LocationDim D
where S.ProductID = P.ProductID
and S.LocationID = D.LocationID
and S.LocationID in ('MEL', 'PER')
and ProductName in ('Clothing', 'Shoes')
and TimeID in ('201801', '201901')
group by rollup (ProductName, Location, TimeID)
order by ProductName, Location, TimeID;
```

19.3.4   Partial Cube and Partial Rollup

Partial cube or partial rollup means that one or more attributes are taken out from the cube and rollup. For example, if the full cube is group by cube(A,B,C) where *A*, *B* and *C* are attributes, an example of a partial cube is attribute *A*, which is taken out from the cube but is still within the group by group by A cube(B,C). We need to understand what it means by a full cube cube(A,B,C) in the first place. cube(A,B,C) means that each of the attributes *A*, *B* and *C* (and their combinations) will have a sub-total in the query results; the attribute will have a (null) entry in the respective column. When an attribute (e.g. attribute *A*) is taken out from the cube, as in group by A cube(B,C), it means that attribute *A* will not have a sub-total anymore, as it is not in the cube any longer.

Table 19.23
Product-Location-Time
partial cube

	ProductName	Location	TimeID	Total sales
1	Clothing	Melbourne	201801	60325
2	Clothing	Melbourne	201901	64823
3	Clothing	Melbourne	(null)	125148
4	Clothing	Perth	201801	37279
5	Clothing	Perth	201901	41479
6	Clothing	Perth	(null)	78758
7	Clothing	(null)	201801	97604
8	Clothing	(null)	201901	106302
9	Clothing	(null)	(null)	203906
10	Shoes	Melbourne	201801	61119
11	Shoes	Melbourne	201901	63968
12	Shoes	Melbourne	(null)	125087
13	Shoes	Perth	201801	44556
14	Shoes	Perth	201901	27112
15	Shoes	Perth	(null)	71668
16	Shoes	(null)	201801	105675
17	Shoes	(null)	201901	91080
18	Shoes	(null)	(null)	196755

Let's compare group by cube (A, B, C) with group by A cube (B, C).
Using the Product-Location-Time cube in the previous section, the SQL for the full
cube is as follows. The results are shown in Table 19.21 with 27 rows in the query
results.

```
select
  ProductName, Location, TimeID,
  sum(Total_Sales) as Total_Sales
from SalesFact S, ProductDim P, LocationDim D
where S.ProductID - P.ProductID
and S.LocationID = D.LocationID
and S.LocationID in ('MEL', 'PER')
and ProductName  in ('Clothing', 'Shoes')
and TimeID in ('201801', '201901')
group by cube (ProductName, Location, TimeID)
order by ProductName, Location, TimeID;
```

The next OLAP is partial cube, where ProductName is taken out from the cube,
as in the following SQL command. Looking at the full cube results shown previously
in Table 19.21, it is clear that rows 19–27 (which are the last nine rows from the full
cube results) will be excluded from the partial cube results because of (null) in the
ProductName column (see Table 19.23).

Table 19.24
Product-Location-Time
partial rollup

	ProductName	Location	TimeID	Total sales
1	Clothing	Melbourne	201801	60325
2	Clothing	Melbourne	201901	64823
3	Clothing	Melbourne	(null)	125148
4	Clothing	Perth	201801	37279
5	Clothing	Perth	201901	41479
6	Clothing	Perth	(null)	78758
7	Clothing	(null)	(null)	203906
8	Shoes	Melbourne	201801	61119
9	Shoes	Melbourne	201901	63968
10	Shoes	Melbourne	(null)	125087
11	Shoes	Perth	201801	44556
12	Shoes	Perth	201901	27112
13	Shoes	Perth	(null)	71668
14	Shoes	(null)	(null)	196755

```
select
  ProductName, Location, TimeID,
  sum(Total_Sales) as Total_Sales
from SalesFact S, ProductDim P, LocationDim D
where S.ProductID = P.ProductID
and S.LocationID = D.LocationID
and S.LocationID in ('MEL', 'PER')
and ProductName  in ('Clothing', 'Shoes')
and TimeID in ('201801', '201901')
group by ProductName, cube (Location, TimeID)
order by ProductName, Location, TimeID;
```

If we change the above partial cube to partial rollup, as in the following SQL command, rows 7–8 and 16–17 from the partial cube results in Table 19.23 will be excluded from the partial rollup results (see Table 19.24 which contains only 14 rows now).

```
select
  ProductName, Location, TimeID,
  sum(Total_Sales) as Total_Sales
from SalesFact S, ProductDim P, LocationDim D
where S.ProductID = P.ProductID
and S.LocationID = D.LocationID
and S.LocationID in ('MEL', 'PER')
and ProductName  in ('Clothing', 'Shoes')
and TimeID in ('201801', '201901')
group by ProductName, rollup (Location, TimeID)
order by ProductName, Location, TimeID;
```

Table 19.25 Product-Location cube with `grouping`

	ProductName	Location	Total sales	Product	Location
1	Clothing	Melbourne	1477348	0	0
2	Clothing	Perth	861268	0	0
3	Clothing	(null)	2338616	0	1
4	Shoes	Melbourne	1311316	0	0
5	Shoes	Perth	897153	0	0
6	Shoes	(null)	2208469	0	1
7	(null)	Melbourne	2788664	1	0
8	(null)	Perth	1758421	1	0
9	(null)	(null)	4547085	1	1

19.3.5 *grouping and decode Functions*

The cube and rollup queries use the (null) entry to indicate that it is about the sub-total. While in OLAP we do not normally consider the presentation of the query results (because the presentation will be handled by Business Intelligence—see the last section in this chapter), sometimes we would like to change the wording of (null) to something more meaningful. To achieve this, the `grouping` and `decode` functions can be used.

The `grouping` function is a binary function that will produce 0 or 1. Let's look at the following cube example where the `grouping` function is used. The results are shown in Table 19.25. Note the last two columns show the `grouping` function results, where 1 indicates that the corresponding column has a (null) value.

```
select
  ProductName, D.Location, sum(Total_Sales) as Total_Sales,
  grouping(ProductName) as Product,
  grouping(D.Location) as Location
from SalesFact S, ProductDim P, LocationDim D
where S.ProductID = P.ProductID
and S.LocationID = D.LocationID
and S.LocationID in ('MEL', 'PER')
and ProductName  in ('Clothing', 'Shoes')
group by cube (ProductName, D.Location)
order by ProductName, D.Location;
```

Now we can incorporate the `grouping` function into the `decode` function. The `decode` function is like an `if then else` statement in a programming language. It takes four parameters. If the first parameter matches with the second parameter, then the result is the third parameter; else the result is the fourth parameter.

The following is the SQL command to use both `decode` and `grouping` functions. The results are shown in Table 19.26. Note the (null) entries in ProductName and Location columns are now reworded more appropriately (e.g. All Products or All Locations).

```
select
  decode(grouping(ProductName),1,'All Products',ProductName)
    as Product_Name,
  decode(grouping(Location),1,'All Locations',Location)
    as Location,
  sum(Total_Sales) as Total_Sales
from SalesFact S, ProductDim P, LocationDim D
where S.ProductID = P.ProductID
and S.LocationID = D.LocationID
and LocationID in ('MEL', 'PER')
and ProductName  in ('Clothing', 'Shoes')
group by cube (ProductName, Location)
order by ProductName, Location;
```

19.4   Ranking

In addition to cube and rollup, ranking is one of the most common operations in Business Intelligence reporting. Naturally, management decision-making is often based on reports that show the ranking of certain aspects of the business.

19.4.1   Rank

OLAP ranking queries can be expressed using the `rank() over` function. The following example shows a ranking report based on the Total Sales of each Product. The data is limited to 2019 data of sales in Melbourne. The results are shown in Table 19.27.

Table 19.26
Product-Location cube with decode

	ProductName	Location	Total sales
1	Clothing	Melbourne	1477348
2	Clothing	Perth	861268
3	Clothing	All Locations	2338616
4	Shoes	Melbourne	1311316
5	Shoes	Perth	897153
6	Shoes	All Locations	2208469
7	All Products	Melbourne	2788664
8	All Products	Perth	1758421
9	All Products	All Locations	4547085

Table 19.27   Product sales ranking

	ProductName	Total sales	Sales rank	Custom rank
1	Clothing	715423	5	1
2	Shoes	645881	4	2
3	Cosmetics	633632	3	3
4	Kids and Baby	477949	2	4
5	Accessories	355433	1	5

```
select
  ProductName,
  sum(Total_Sales) as Total_Sales,
  rank() over(order by sum(Total_Sales)) as Sales_Rank,
  rank() over(order by sum(Total_Sales) desc) as Custom_Rank
from SalesFact S, ProductDim P, TimeDim T
where S.ProductID = P.ProductID
and S.TimeID = T.TimeID
and S.LocationID = 'MEL'
and Year = 2019
group by ProductName
order by Custom_Rank;
```

The rank() over function uses order by Total Sales, which sorts
the records based on the Total Sales. The first rank() over sorts the Total Sales
from the smallest to the largest value; hence, the ranking is from 5 to 1, where rank
5 has the largest Total Sales figure, and rank 1 is the lowest. However, in business
reporting, the top figure should be ranked 1. In order to do this, the order by
clause must use desc so that the sorting is from the largest to the smallest.

In case there is a tie between two Total Sales (e.g. just assume that Shoes and
Cosmetics have exactly the same Total Sales), there are two ways to present the
rankings: one is using the rank() over function, and the other is using the
dense_rank() over function. To contrast these two ranking functions, let's
look at these examples.

The following is the normal ranking method using the rank() over function.
The results are shown in Table 19.28. In this example, we assume the Total Sales of
Shoes and Cosmetics are the same; both are ranked 2. Since they are tied, there is
no product ranked 3. The next ranking is rank 4.

```
select
  ProductName,
  sum(Total_Sales) as Total_Sales,
  rank() over(order by sum(Total_Sales) desc) as Rank
from SalesFact S, ProductDim P, TimeDim T
where S.ProductID = P.ProductID
and S.TimeID = T.TimeID
and S.LocationID = 'MEL'
and Year = 2019
group by ProductName
order by Rank;
```

Table 19.28 Product Sales ranking with (normal) rank

	ProductName	Total sales	Rank
1	Clothing	715423	1
2	Shoes	645881	2
3	Cosmetics	645881	2
4	Kids and Baby	477949	4
5	Accessories	355433	5

Table 19.29 Product Sales ranking with dense rank

	ProductName	Total sales	Dense rank
1	Clothing	715423	1
2	Shoes	645881	2
3	Cosmetics	645881	2
4	Kids and Baby	477949	3
5	Accessories	355433	4

Now it is the same query but uses dense_rank() over instead. The results in Table 19.29 show that no rank is skipped, even though there is a tie. So, in this case, after two rank 2, the next one is rank 3, so it is a dense rank.

```
select
  ProductName,
  sum(Total_Sales) as Total_Sales,
  dense_rank() over(order by sum(Total_Sales) desc)
    as Dense_Rank
from SalesFact S, ProductDim P, TimeDim T
where S.ProductID = P.ProductID
and S.TimeID = T.TimeID
and S.LocationID = 'MEL'
and Year = 2019
group by ProductName
order by Dense_Rank;
```

If there is no tie, there will be no difference between rank() over and dense_rank() over functions.

There is another function, row_number() over, which is not meant for ranking but shows a similar behaviour to ranking. The row_number() over function actually gives a row number based on the order by clause inside row_number() over function. The following is the same SQL command as above but uses the row_number() over function.

```
select
  ProductName,
  sum(Total_Sales) as Total_Sales,
  row_number() over(order by sum(Total_Sales) desc)
    as Row_Number
from SalesFact S, ProductDim P, TimeDim T
where S.ProductID = P.ProductID
and S.TimeID = T.TimeID
and S.LocationID = 'MEL'
and Year = 2019
group by ProductName
order by Row_Number;
```

Again assuming that Shoes and Cosmetics have exactly the same Total Sales, Table 19.30 shows the result of the row_number() over query. Note that in case of a tie, the row number will keep the increment. There is no particular ordering method for the records that are tied, which in this case are rows 2 and 3. Row

Table 19.30 Product Sales
row number

	ProductName	Total Sales	Row Number
1	Clothing	715423	1
2	Shoes	645881	2
3	Cosmetics	645881	3
4	Kids and Baby	477949	4
5	Accessories	355433	5

Table 19.31 Top-2 Products
with highest total sales

	ProductName	Total sales	Product rank
1	Clothing	715423	1
2	Shoes	645881	2

numbers are given to the query results based on the `order by` clause in the `row_number() over`.

19.4.2   Top-N and Top-Percent Ranking

When the ranking list is longer, the management often likes to focus on a few top ranks only. This is called *Top-N* ranking, where N is a number in the ranking list. The following SQL command retrieves the Top-2 Products based on their Total Sales. The results are shown in Table 19.31, showing Clothing and Shoes are the two top Products with the highest Total Sales. The data is limited to only Year 2019 in Melbourne.

Looking at the inner query, the inner query retrieves the ranking of all Products based on the Total Sales. The outer query filters these results by choosing only the Top-2 records.

```
select *
from
  (select
    ProductName,
    sum(Total_Sales) as Total_Sales,
    rank() over(order by sum(Total_Sales) desc)
      as Product_Rank
  from SalesFact S, TimeDim T, LocationDim L, ProductDim P
  where S.TimeID = T.TimeID
  and S.LocationID = L.LocationID
  and S.ProductID = P.ProductID
  and Year = 2019
  group by ProductName)
where Product_Rank <= 2;
```

Another way to choose the top few ranked items is by specifying the top rank percentage rather than specifying an integer number to indicate the top N records in the complete ranking list. The `percent_rank() over` function can be used.

Table 19.32 Percent rank of
all products

	ProductName	Total sales	Percent Rank
1	Clothing	715423	1
2	Shoes	645881	0.75
3	Cosmetics	633632	0.5
4	Kids and Baby	477949	0.25
5	Accessories	355433	0

This function calculates the ranking in terms of the percentage. The following is the SQL query that shows the percentage ranking of all Products.

```
select ProductName, sum(Total_Sales) as Total_Sales,
  percent_rank() over (order by sum(Total_Sales))
    as Percent_Rank
from SalesFact S, TimeDim T, ProductDim P
where S.TimeID = T.TimeID
and S.ProductID = P.ProductID
and Year = 2019
and LocationID = 'MEL'
group by ProductName
order by Percent_Rank desc;
```

The results are shown in Table 19.32. The Percent Rank ranges from 0 (the bottom rank) to 1 (the top rank). Because our dataset only has five Products, the Percent Rank increments by 0.25 (or 25%) per Product. If we want to select the Top-2 Products, it could be tricky to use the percent_rank function. However, if we would like to retrieve the top 75% of the Product, then we can specify a condition in the outer query, such as the following:

```
select *
from (
  select ProductName, sum(Total_Sales) as Total_Sales,
    percent_rank() over (order by sum(Total_Sales))
      as Percent_Rank
  from SalesFact S, TimeDim T, ProductDim P
  where S.TimeID = T.TimeID
  and S.ProductID = P.ProductID
  and Year = 2019
  and LocationID = 'MEL'
  group by ProductName
  order by Percent_Rank)
where Percent_Rank >= 0.75;
```

19.4.3   Partition

In the previous sections, ranking is applied to one set of data, which is the Total Sales of each Product, and then the ranking method is to rank each Product based on its Total Sales. However, in some cases, we may not want to compare the Total Sales

Table 19.33 Product Sales ranking partition by Product

	ProductName	TimeID	Total sales	Rank by product
1	Clothing	201902	224253	1
2	Clothing	201901	207673	2
3	Clothing	201903	170238	3
4	Shoes	201902	182786	1
5	Shoes	201903	177432	2
6	Shoes	201901	167820	3

between Products, but we would like to rank each product internally, for example, based on the Month. So, each Product has its own internal ranking, that is, to rank the Month's Total Sales. In this case, we need a *partition*.

partition by is a clause within the rank() over function, which partitions the dataset into several partitions, and each partition will have its own internal ranking. The following shows an example of an internal ranking of Clothing and Shoes based on the monthly Total Sales for the period of January to March 2019. The results as shown in Table 19.33 show that Clothing has its own internal ranking based on monthly Total Sales and so does Shoes.

```
select
  ProductName, TimeID,
  sum(Total_Sales) as Total_Sales,
  rank() over(partition by ProductName
    order by sum(Total_Sales) desc) as Rank_by_Product
from SalesFact S, ProductDim P
where S.ProductID = P.ProductID
and TimeID in ('201901', '201902', '201903')
and P.ProductName in ('Clothing', 'Shoes')
group by ProductName, TimeID
order by ProductName, Rank_by_Product;
```

The results combine the two partitions (e.g. Clothing partition and Shoes partition) into one table in the query results because SQL queries produce the results in a single table format. If we do not carefully sort the results according to the partitioned attribute, we might mix the two partitions in the query results, which may confuse the readers. Nevertheless, the job of OLAP queries is to retrieve the required data. Business Intelligence Reporting tools will be used for presentation purposes. So, Table 19.33 might be presented as "two" cards or sheets, where the first sheet is for the Clothing and the second for the Shoes (refer to an illustration in Fig. 19.4).

A partition can be applied to multiple attributes. The following SQL command partitions the data based on Product as well as Time. It means Product has its own internal ranking based on Total Sales, and so does Time. The results are shown in Table 19.34. When multiple partitions are used in one query, the results can be confusing because both partitions are jumbled into one table. It will be easier to imagine the illustration in Fig. 19.5. The picture shows that the Product partition has "two" sheets, one for Clothing and the other for Shoes, each with its own ranking.

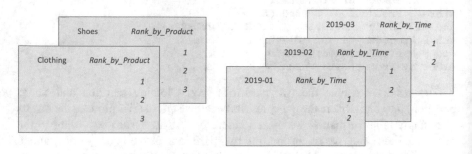

Shoes   Rank_by_Product

Clothing   Rank_by_Product

1

2

1

3

2

3

Fig. 19.4   Ranking partition by Product

Table 19.34   Product Sales ranking partition by Product and Time

	ProductName	TimeID	Total sales	Rank by product	Rank by time
1	Clothing	201902	224253	1	1
2	Clothing	201901	207673	2	1
3	Clothing	201903	170238	3	2
4	Shoes	201902	182786	1	2
5	Shoes	201903	177432	2	1
6	Shoes	201901	167820	3	2

Fig. 19.5   Ranking partition by Product and Time

It also shows that the Time partition has "three" sheets because there are 3 months
(e.g. January to March), again each with their own ranking based on the Total Sales.

```
select
  ProductName, TimeID,
  sum(Total_Sales) as Total_Sales,
  rank() over(partition by ProductName
    order by sum(Total_Sales) desc) as Rank_by_Product,
  rank() over(partition by TimeID
    order by sum(Total_Sales) desc) as Rank_by_TimeID
from SalesFact S, ProductDim P
```

```
where S.ProductID = P.ProductID
and TimeID in ('201901', '201902', '201903')
and P.ProductName in ('Clothing', 'Shoes')
group by ProductName, TimeID
order by ProductName;
```

Inspecting Table 19.33, the ranking based on Product is probably easy to see, as the Clothing rankings are in one group and the Shoes rankings are in the other. But checking the ranking for the Time can be confusing because of the absence of visual grouping. Row 1 for Clothing in 201902 is ranked 1. We need to find which other Product in 201902 is ranked 2. In this case, it is Shoes, which is on row 4. So, row 1 is on the same group with row 4, which is the ranking for 201902 in which Clothing has a higher Total Sales than Shoes. Based on this method, we can see that row 2 is in the same group as row 6, that is, 201901 ranking in which again Clothing has more Total Sales than Shoes. But for 201903 (see row 3 and row 5), Shoes is higher than Clothing.

19.5 Cumulative and Moving Aggregate

Another report that is common in business is cumulative and moving aggregates. Cumulative aggregate gets the cumulative sum, whereas moving aggregate is commonly used to get the moving average over a certain window size period. The former is usually used as an indicator of performance target, whereas the latter is often used to smooth out some outlier figures in a certain period of time.

19.5.1 Cumulative Aggregate

Cumulative aggregate uses the rows unbounded preceding clause in the sum(sum() over) function. The following is the SQL command to calculate the Cumulative Total Sales every month in 2019 in the Location of MEL.

Particularly, notice two things in this query: one is the use of sum (sum (Total_Sales)) over, and the other is the rows unbounded preceding. The first is to get the sum of sum (Total_Sales). Note that sum (Total_Sales) calculates the sum of Total Sales per group by, whereas the first sum is to get the Cumulative Total Sales. The second important thing is rows unbounded preceding. The unbounded indicates that the cumulative total sales starts from the beginning or the first record of the query result. The query results are shown in Table 19.35. The cumulative column is highlighted.

Table 19.35 Cumulative sales

	LocationID	TimeID	Total Sales	**Cumulative**
1	MEL	201901	243,080	243,080
2	MEL	201902	259,353	502,433
3	MEL	201903	246,368	748,801
4	MEL	201904	237,193	986,714
5	MEL	201905	234,513	1,221,227
6	MEL	201906	235,630	1,456,857
7	MEL	201907	251,330	1,708,187
8	MEL	201908	239,575	1,947,762
9	MEL	201909	208,165	2,115,927
10	MEL	201910	228,710	2,384,637
11	MEL	201911	207,694	2,592,331
12	MEL	201912	235,987	2,828,318

```
select LocationID, S.TimeID,
  to_char(sum(Total_Sales), '999,999,999') as Total_Sales,
  to_char(sum(sum(Total_Sales)) over
    (order by LocationID, S.TimeID
    rows unbounded preceding), '999,999,999') as Cumulative
from SalesFact S, TimeDim T
where S.TimeID = T.TimeID
and Year = 2019
and LocationID in ('MEL')
group by LocationID, S.TimeID;
```

Cumulative aggregate can incorporate partition, which is often used where the dataset is partitioned into several groups, and each group has its own cumulative aggregate. Table 19.35 contains data from one Location (e.g. MEL). Suppose we need one cumulative report of two Locations; in this case, we need to partition the data based on Location. The following SQL command calculates the Cumulative Total Sales of two Locations (e.g. MEL and PER), individually. Notice the partition by LocationID clause in the sum(sum (Total_Sales)) function. The results are shown in Table 19.36. Notice that the Cumulative Total Sales resets when a new Location (e.g. PER) started.

```
select LocationID, TimeID as Time,
  to_char(sum(Total_Sales), '999,999,999') as Total_Sales,
  to_char(sum(sum(Total_Sales)) over
    (partition by LocationID
    order by LocationID, TimeID
    rows unbounded preceding), '999,999,999') as Cumulative
from SalesFact S, TimeDim T
where S.TimeID = T.TimeID
and Year = 2019
and LocationID in ('MEL', 'PER')
group by Location, TimeID;
```

Table 19.36 Cumulative Sales partitioned by Location

	LocationID	TimeID	Total Sales	Cumulative
1	MEL	201901	243,080	243,080
2	MEL	201902	259,353	502,433
3	MEL	201903	246,368	748,801
4	MEL	201904	237,193	986,714
5	MEL	201905	234,513	1,221,227
6	MEL	201906	235,630	1,456,857
7	MEL	201907	251,330	1,708,187
8	MEL	201908	239,575	1,947,762
9	MEL	201909	208,165	2,115,927
10	MEL	201910	228,710	2,384,637
11	MEL	201911	207,694	2,592,331
12	MEL	201912	235,987	2,828,318
13	PER	201901	150,456	150,456
14	PER	201902	147,412	297,868
15	PER	201903	135,137	433,005
16	PER	201904	136,877	569,882
17	PER	201905	151,685	721,567
18	PER	201906	136,491	858,058
19	PER	201907	130,951	989,009
20	PER	201908	172,590	1,161,599
21	PER	201909	156,086	1,317,685
22	PER	201910	167,189	1,484,874
23	PER	201911	176,590	1,661,464
24	PER	201912	141,242	1,802,706

19.5.2 Moving Aggregate

Syntactically, the difference between cumulative aggregate and moving aggregate is very small. In cumulative aggregate, it uses rows unbounded preceding, whereby the unbounded states that the starting of the cumulative is from the first row of the query results. In moving aggregate, the starting will not be from the beginning but from how many records behind the current record in the query results. Hence, in moving aggregate, we use rows n proceedings, where n is an integer number which indicates how many rows it is behind the current row.

The following SQL command gives an example of Moving 3-Month Average Total Sales. In this case, $n = 2$ because the average calculation started from 2 months before the current month, and hence it becomes the 3-Month Average. The only exception is the average of the first 2 months, which are the first month Total Sales (in row 1) and the average of 2 months' Total Sales (in row 2). After these, the Average will be based on 3 months (see Table 19.37 for the complete results).

Also notice in the SQL command that avg(sum (Total_Sales)) is used instead of sum(sum (Total_Sales)). This is because we wanted to

Table 19.37 Moving
3-Month Average Sales

	LocationID	TimeID	Total Sales	Avg 3 Months
1	MEL	201901	243,080	243,080
2	MEL	201902	259,353	251,217
3	MEL	201903	246,368	249,600
4	MEL	201904	237,913	247,878
5	MEL	201905	234,513	239,598
6	MEL	201906	235,630	236,019
7	MEL	201907	251,330	240,491
8	MEL	201908	239,575	242,178
9	MEL	201909	208,165	233,023
10	MEL	201910	228,710	225,483
11	MEL	201911	207,694	214,856
12	MEL	201912	235,987	224,130

calculate the average of 3 months Total Sales; hence, the avg is used. The (sum
(Total_Sales) is still used because it calculates the sum of Total Sales for each
group by.

```
select LocationID, S.TimeID,
  to_char(sum(Total_Sales), '999,999,999') as Total_Sales,
  to_char(avg(sum(Total_Sales)) over
    (order by LocationID, S.TimeID
    rows 2 preceding), '999,999,999') as Avg_3_Months
from SalesFact S, TimeDim T
where S.TimeID = T.TimeID
and Year = 2019
and LocationID in ('MEL')
group by LocationID, S.TimeID;
```

19.6   Business Intelligence Reporting

The data retrieved by OLAP queries, as shown in the previous sections, are rather
raw, in terms of their presentation. For example, the cube and rollup results which
contain sub-group totals may not be presented in a clear and intuitive way for readers
to see. OLAP query results containing multiple partitions can be confusing as the
data from multiple partitions are combined into one query result report. In short,
OLAP queries are not meant for presentation; rather their main job is only to retrieve
the required data from the data warehouse. Once the requested data is retrieved, it
will be passed on to the *Business Intelligence* (BI) reporting tool to ensure the data
is presented professionally for management. This section demonstrates some of the
reports or graphs that may be produced by any BI tool, by taking OLAP query results
as the input.

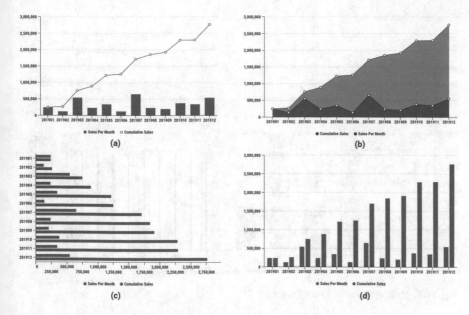

Fig. 19.6   Cumulative Sales in Melbourne

19.6.1   *Cumulative and Moving Aggregate*

An OLAP cumulative query, which is expressed in the following SQL, may produce Total Sales and Cumulative Sales presented in a graph for a better presentation (rather than in a table, as in Table 19.35). Figure 19.6 shows four different ways to present a cumulative report. Figure 19.6a shows the Monthly Total Sales in a bar graph and the Cumulative Total Sales in a line graph. Looking at this graph, we can see clearly the increase of the Cumulative Sales month by month, and hence the cumulative becomes intuitive. Figure 19.6b shows a different style for presenting a cumulative result, that is, using the line area graph for both Total Sales and its cumulative result. The other two graphs show both Total Sales and the cumulative result in bar graphs (e.g. vertical and horizontal styles). These are four of many possible ways to present a cumulative report. Nevertheless, the cumulative report presents one of these ways, which will be much clearer than in table format.

```
select TimeID, sum(Total_Sales) as Total_Sales,
  sum(sum(Total_Sales)) over (order by TimeID
    rows unbounded preceding) as Cumulative
from SalesFact S, TimeDim T
where S.TimeID = T.TimeID
and Year = 2019
and LocationID in ('MEL')
group by TimeID;
```

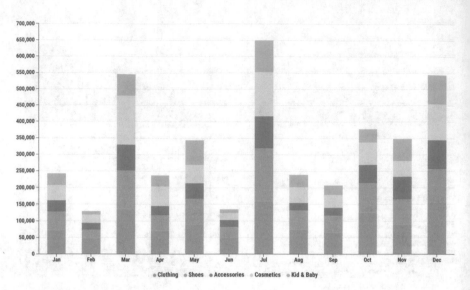

Fig. 19.7 Drilling down into each Product

If you need to drill down into each Product in each month, you can invoke the following OLAP query using simple group by, and the results are presented using a stacked bar graph in Fig. 19.7.

```
select S.TimeID, ProductName, sum(Total_Sales) as Total_Sales
from SalesFact S, TimeDim T, ProductDim P
where S.TimeID = T.TimeID
and S.ProductID = P.ProductID
and Year = 2019
and LocationID in ('MEL')
group by S.TimeID, ProductName;
```

Sometimes, it is necessary to compare the cumulative results of two Locations, for example. Figure 19.8 shows the cumulative result of two Locations, Melbourne and Sydney. We can either invoke two cumulative queries, one for Melbourne and the other for Sydney, and then the graph will combine both results. Alternatively, combining both cumulative queries into one cumulative OLAP query will produce a combined table. Either way, the graph will simply take the results from the OLAP query and present them in a graph, as in Fig. 19.8. Having the cumulative results of two Locations is a good way to compare the performance of two Locations and to see how one Location's cumulative results catches up with the other toward the end of the year.

The last graph in this subsection is a moving aggregate. Figure 19.9 shows the Moving 3-Month Average. Notice the ups in March and July are balanced out through the rather low Total Sales in the respective previous months. The data for

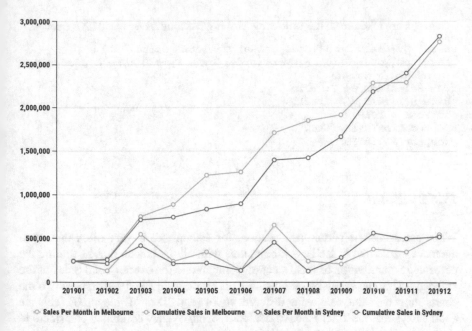

Fig. 19.8 Cumulative Sales of two Locations: Melbourne and Sydney

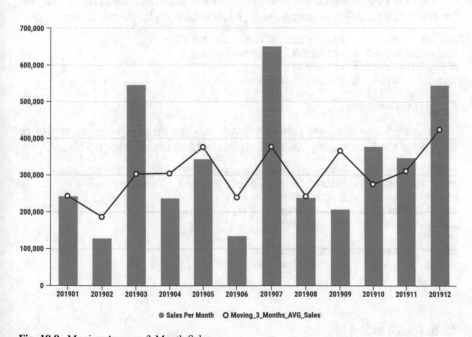

Fig. 19.9 Moving Average 3-Month Sales

this graph are taken from the following Moving 3-Month Average OLAP query:

```
select S.TimeID, sum(Total_Sales) as Total_Sales,
  avg(sum(Total_Sales)) over (order by S.TimeID rows 2 preceding)
     as Avg_3_Months
from SalesFact S, TimeDim T
where S.TimeID = T.TimeID
and Year = 2019
and LocationID in ('MEL')
group by S.TimeID;
```

19.6.2   Ratio

When assessing one particular Location, for example, it is very often that man-agement would like to see how it compares with all the others, that is, finding the percentage contribution of this Location compared to the overall Total Sales of the company. This can easily be seen using a pie chart, as in Fig. 19.10. These two pie charts show the same data using different styles (e.g. 2D or 3D pie chart). This pie chart shows that the Total Sales of Melbourne and Sydney combined contribute to almost two-thirds of all Total Sales of the company.

The following query retrieves the Total Sales of each Location. The BI tool will be able to calculate the percentage of each Location and plot the pie charts.

```
select Location, sum(Total_Sales) as Total_Sales
from SalesFact S, TimeDim T, LocationDim L
where S.TimeID = T.TimeID
and S.LocationID = L.LocationID
and Year = 2019
group by Location;
```

From each Location, we can drill down into the Product level, as shown in Fig. 19.11. The OLAP query simply adds Product into the `group by` clause.

```
select Location, ProductName, sum(Total_Sales) as Total_Sales
from SalesFact S, TimeDim T, LocationDim L, ProductDim P
where S.TimeID = T.TimeID
and S.LocationID = L.LocationID
and S.ProductID = P.ProductID
and Year = 2019
group by Location, ProductName
order by Location, ProductName;
```

19.6.3   Ranking

Ranking is one of the most important tools in decision-making, that is, finding the top performing Products or Locations. In OLAP, the `rank()` `over` function can

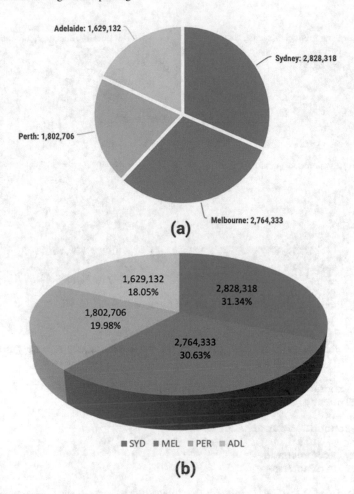

Fig. 19.10 Total Sales ratio in each Location

be used to calculate the ranking. Alternatively, we can simply use order by to sort the items in the report. Figure 19.12 shows four different ways of ranking the Products. We can use the normal vertical bar chart as in Fig. 19.12a or a horizontal bar chart as in Fig. 19.12b. Figure 19.12c is rather unusual but may still be used. The pie chart in Fig. 19.12d may not clearly show the ranking, especially when in this case, the first two Products performed approximately equally.

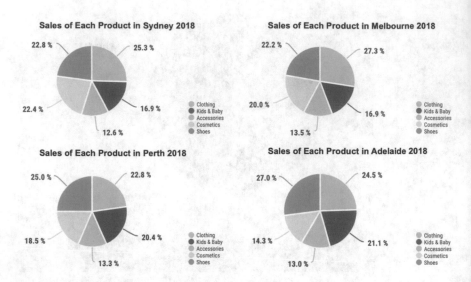

Fig. 19.11 Product Sales performance in various Locations

The OLAP query to retrieve the data for these charts is a simple `group by` query.

```
select ProductName, sum(Total_Sales) as Total_Sales
from SalesFact S, TimeDim T, ProductDim P
where S.TimeID = T.TimeID
and S.ProductID = P.ProductID
and Year = 2019
group by ProductName
order by ProductName;
```

Another useful ranking is to rank the Locations to see which Locations perform well. This is shown in Fig. 19.13. This chart uses a different style, that is, a half-pie chart. It can be intuitive, depending on the data itself. In this case, the top Location is on the most left, going down to the most right. The query simply uses a similar `group by` clause but grouping and ordering based on Location rather than Product.

```
select Location, sum(Total_Sales) as Total_Sales
from SalesFact S, TimeDim T, LocationDim L
where S.TimeID = T.TimeID
and S.LocationID = L.LocationID
and Year = 2019
group by Location
order by Location;
```

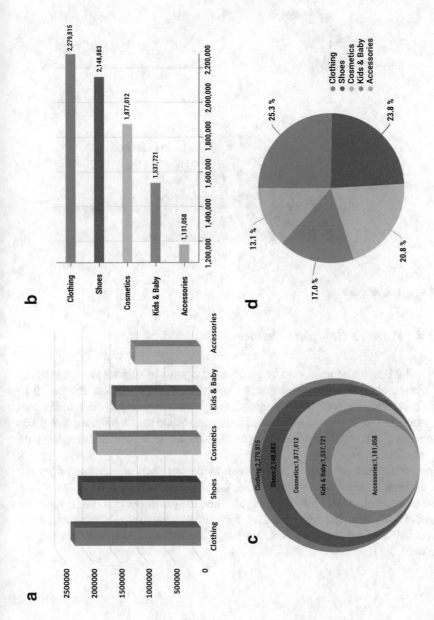

Fig. 19.12 Product ranking

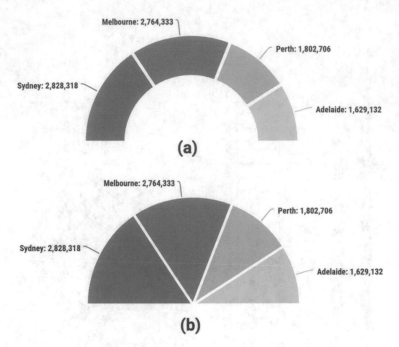

Fig. 19.13 Location ranking

19.6.4 A More Complete Report

A typical BI report combines many graphs and figures in various presentations to give management a more complete view of the business operation. Figure 19.14 shows the performance of various products (e.g. Clothing, Shoes, etc.), in the past year, as well as their respective cumulative figures. Some total figures are also given in the report. This report requires multiple OLAP queries to provide the report with adequate data.

The report (and all other reports shown in the previous subsections) does not have to have its parameters (e.g. which Location, or which Product, or which Year) statically embedded into the query. Rather, the parameters should be specified dynamically during runtime by the user, for example, the user may want to specify a certain Year of data for the report, etc. So, it is a dynamic BI report rather than a static one.

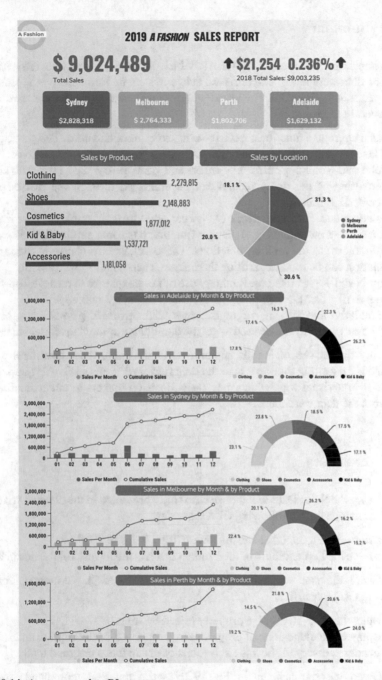

Fig. 19.14 A more complete BI report

19.7   Summary

This chapter mainly focuses on OLAP, which is the SQL query to retrieve data from the data warehouse. The retrieved data will then be formatted by the Business Intelligence Reporting tool. The OLAP queries discussed in this chapter are divided into several categories:

(a) Basic aggregate functions: count, sum, avg, max and min. The group by clause is often used in conjunction with these basic aggregate functions.
(b) Cube and Rollup: group by cube and group by rollup. The simple formatting of the query results can be enhanced through the decode and grouping functions.
(c) Ranking and Partition: rank() over and dense_rank() over functions. The row_number() over function has some similarities (as well as differences) to the ranking functions. The partition clause in the ranking function can be used to partition the dataset, each with its own ranking.
(d) Top-N and Top-Percentage Ranking: use of nested queries to retrieve Top-N and percent_rank function to retrieve Top-Percentage rankings.
(e) Cumulative and Moving Aggregate: row unbounded preceding or row n proceeding can be used to get the cumulative or moving aggregate values.

Finally, the last section in this chapter gives an illustration as to how the Business Intelligence reporting tool may take the data retrieved from the OLAP queries and present it in a visual way, which may assist decision-makers in understanding the data from the data warehouse.

19.8   Exercises

19.1 Using the Sales Data Warehouse case study presented in the chapter, write the SQL commands for the following OLAP queries:

(a) Retrieve the Product that has the lowest Total Sales in 2019.
(b) Retrieve the Percent Rank of the Product that has the lowest Total Sales in 2019.

19.2 Given the star schema in Fig. 19.15, write the SQL commands for the following OLAP queries:

(a) Display the Top-10 average property prices by suburb.
(b) Display the average price of properties by property-type description and suburb. It is not necessary to show the sub-totals or group totals or grand total.

19.3 Given the star schema in Fig. 19.16, write the SQL commands for the following OLAP queries. The Sales Method Dimension table lists all kinds of sales methods, such as In-Store, Online, Phone Order, etc.

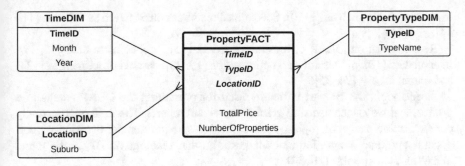

Fig. 19.15 Property Sales star schema

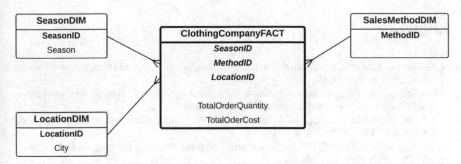

Fig. 19.16 Clothing Company star schema

(a) Perform a Cumulative Sum of Total Order Cost of all Online orders (Instruction: use all dimensions). Note that for each City, there should be a separate Cumulative Sum.

(b) Perform another Cumulative Sum of Total Order Cost but partitioned based on the SalesMethodID: one partition for Online orders, one partition for In-Store orders and one partition for Phone Order orders. Hints: It must also be partitioned based on Location (or Suburb). Hence, for example, Online orders for Melbourne will have one set of cumulative Total Order Cost, whereas Online orders for Sydney will have a separate set of cumulative Total Order Cost.

(c) Show the total order costs of each source order and rank them.

(d) Display the source order that generates the highest total order cost.

19.9 Further Readings

SQL is the basis for the OLAP implementation. OLAP queries usually use various OLAP functions, such as `cube`, `rollup`, etc., as well as ranking functions and aggregate functions (e.g. moving aggregates). Most database textbooks cover SQL in various depth, such as [1–10]. Specialized resources on SQL are as follows: [11–14].

Recent work on OLAP covers missing data [15], probabilistic data [16, 17], interval data [18], processing and optimization [19, 20], parallel processing [21–23] and spatial OLAP [24, 25].

In our book, the BI part is mainly used for presenting the OLAP results in a professional way, such as using graphs, charts and reports. The field of BI actually covers a wide range of topics, such as business requirements, decision-support systems and data integration and analytics. Further readings on BI can be found in the following books: [26–32].

References

1. C. Coronel, S. Morris, *Database Systems: Design, Implementation, and Management* (Cengage Learning, New York, 2018)
2. T. Connolly, C. Begg, *Database Systems: A Practical Approach to Design, Implementation, and Management* (Pearson Education, New York, 2015)
3. J.A. Hoffer, F.R. McFadden, M.B. Prescott, *Modern Database Management* (Prentice Hall, Englewood Cliffs, 2002)
4. A. Silberschatz, H.F. Korth, S. Sudarshan, *Database System Concepts, Seventh Edition* (McGraw-Hill, New York, 2020)
5. R. Ramakrishnan, J. Gehrke, *Database Management Systems*, 3rd edn. (McGraw-Hill, New York, 2003)
6. J.D. Ullman, J. Widom, *A First Course in Database Systems*, 2nd edn. (Prentice Hall, Englewood Cliffs, 2002)
7. H. Garcia-Molina, J.D. Ullman, J. Widom, *Database Systems: The Complete Book* (Pearson Education, New York, 2011)
8. P.E. O'Neil, E.J. O'Neil, *Database: Principles, Programming, and Performance*, 2nd edn. (Morgan Kaufmann, Los Altos, 2000)
9. R. Elmasri, S.B. Navathe, *Fundamentals of Database Systems*, 3rd edn. (Addison-Wesley-Longman, Reading, 2000)
10. C.J. Date, *An Introduction to Database Systems*, 7th edn. (Addison-Wesley-Longman, Reading, 2000)
11. J. Melton, *Understanding the New SQL: A Complete Guide, Second Edition*, vol I (Morgan Kaufmann, Los Altos, 2000)
12. C.J. Date, *SQL and Relational Theory—How to Write Accurate SQL Code*, 2nd edn. Theory in practice (O'Reilly, New York, 2012)
13. A. Beaulieu, *Learning SQL: Master SQL Fundamentals* (O'Reilly Media, New York, 2009)
14. M.J. Donahoo, G.D. Speegle, *SQL: Practical Guide for Developers*. The Practical Guides (Elsevier Science, Amsterdam, 2010)
15. M.B. Kraiem, K. Khrouf, J. Feki, F. Ravat, O. Teste, New OLAP operators for missing data, in *Actes des 13èmes journées francophones sur les Entrepôts de Données et l'Analyse en Ligne, Business Intelligence and Big Data, EDA 2017, Lyon, France, 3-5 mai 2017*, ed. by O. Boussaïd, F. Bentayeb, J. Darmont. RNTI, vol. B-13, pp. 53–66 (Éditions RNTI, 2017)
16. X. Xie, X. Hao, T.B. Pedersen, P. Jin, J. Chen, OLAP over probabilistic data cubes I: aggregating, materializing, and querying, in *Proceedings of the 32nd IEEE International Conference on Data Engineering, ICDE 2016, Helsinki, Finland, May 16-20, 2016* (IEEE Computer Society, New York, 2016), pp. 799–810

17. X. Xie, K. Zou, X. Hao, T.B. Pedersen, P. Jin, W. Yang, OLAP over probabilistic data cubes II: parallel materialization and extended aggregates. IEEE Trans. Knowl. Data Eng. **32**(10), 1966–1981 (2020)
18. C. Koncilia, T. Morzy, R. Wrembel, J. Eder, Interval OLAP: analyzing interval data, in *Data Warehousing and Knowledge Discovery—Proceedings of the 16th International Conference, DaWaK 2014, Munich, Germany, September 2-4, 2014*, ed. by L. Bellatreche, M.K. Mohania. Lecture Notes in Computer Science, vol. 8646 (Springer, Berlin, 2014), pp. 233–244
19. K. Zeng, S. Agarwal, I. Stoica, iolap: Managing uncertainty for efficient incremental OLAP, in *Proceedings of the 2016 International Conference on Management of Data, SIGMOD Conference 2016, San Francisco, CA, USA, June 26–July 01, 2016*, ed. F. Özcan, G. Koutrika, S. Madden (ACM, New York, 2016), pp. 1347–1361
20. M. Golfarelli, S. Graziani, S. Rizzi, Shrink: An OLAP operation for balancing precision and size of pivot tables. Data Knowl. Eng. **93**, 19–41 (2014)
21. A.A.B. Lima, C. Furtado, P. Valduriez, M. Mattoso, Parallel OLAP query processing in database clusters with data replication. Distrib. Parallel Databases **25**(1–2), 97–123 (2009)
22. M.W.M. Ribeiro, A.A.B. Lima, D. de Oliveira, OLAP parallel query processing in clouds with c-pargres. Concurr. Comput. Pract. Exp. **32**(7), e5590 (2020)
23. D. Taniar, C.H.C. Leung, W. Rahayu, S. Goel, *High Performance Parallel Database Processing and Grid Databases* (Wiley, New York, 2008)
24. S. Bimonte, M. Bertolotto, J. Gensel, O. Boussaid, Spatial OLAP and map generalization: Model and algebra. Int. J. Data Warehous. Min. **8**(1), 24–51 (2012)
25. S. Bimonte, O. Boucelma, O. Machabert, S. Sellami, A new spatial OLAP approach for the analysis of volunteered geographic information. Comput. Environ. Urban Syst. **48**, 111–123 (2014)
26. D. Loshin, Business intelligence: the Savvy Manager's guide, in *The Morgan Kaufmann Series on Business Intelligence* (Elsevier, Amsterdam, 2012)
27. B. Brijs, *Business Analysis for Business Intelligence* (CRC Press, New York, 2016)
28. R. Sherman, *Business Intelligence Guidebook: From Data Integration to Analytics* (Elsevier, Amsterdam, 2014)
29. L.T. Moss, S. Atre, Business intelligence roadmap: the complete project lifecycle for decision-support applications, in *Addison-Wesley Information Technology Series* (Addison-Wesley, Reading, 2003)
30. C. Vercellis, *Business Intelligence: Data Mining and Optimization for Decision Making* (Wiley, New York, 2011)
31. L. Bulusu, *Open Source Data Warehousing and Business Intelligence* (CRC Press, New York, 2012)
32. W. Grossmann, S. Rinderle-Ma, Fundamentals of business intelligence, in *Data-Centric Systems and Applications* (Springer, Berlin, 2015)

Chapter 20
Pre- and Post-Data Warehousing

Pre- and post-data warehousing refers to the activities before and after a data warehouse is created. Figure 20.1 shows the context of pre- and post-data warehousing.

Pre-Data Warehousing is the transformation process from operational databases to data warehousing. This has been the focus of the previous chapters, which cover a systematic approach in transforming operational databases to star schemas. Before operational databases are ready to be transformed to a data warehouse, much preparation must be done. This is known as data preparation, in particular data cleaning. In the data cleaning process, identifying dirty data is a crucial step. In this chapter, the pre-data warehousing section will focus on exploring dirty data.

Post-Data Warehousing covers the activities after a data warehouse is created. Chapter 19 focuses on Online Analytical Processing (OLAP), which is a querying mechanism to retrieve data from star schemas. The data which have been retrieved by OLAP can later be plotted for better visualisation and presentation using Business Intelligence (BI) tools. This will lead to data analytics. However, before data analytics can be undertaken, it is important to understand what kind of data we have in the data warehouse, which is critical before further data analysis is undertaken. This is also known as data exploration because the data in the data warehouse is explored. The aim is to understand the data in the data warehouse so that more effective and targeted data analytics can be performed. Hence, in this chapter, the Post-Data Warehousing section will focus on exploring data warehouses.

20.1   Pre-Data Warehousing: Exploring Dirty Data

Data cleaning is to clean data that is dirty (e.g. incorrect, inconsistent, etc). The crucial step in data cleaning is to identify data incorrectness, inconsistency, etc. In

© The Author(s), under exclusive license to Springer Nature Switzerland AG 2021   541
D. Taniar, W. Rahayu, *Data Warehousing and Analytics*, Data-Centric Systems and Applications, https://doi.org/10.1007/978-3-030-81979-8_20

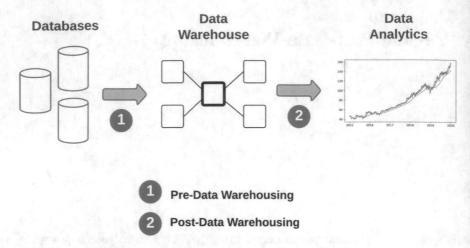

Fig. 20.1 Pre- and Post-Data Warehousing

particular, the focus is on *duplication* problems, *relationship* problems, *inconsistent and incorrect values* and *null value* problems in the operational databases.

20.1.1   Duplication Problems

Duplication problems may occur at three different levels: (i) between records, (ii) between attributes and (iii) between tables.

20.1.1.1   Data Duplication Between Records

Primary keys in relational databases are used to prevent record redundancies in tables. This is very important to prevent data anomalies. Hence technically, record redundancies do not exist in relational tables if the tables have primary keys. However, it is well known that data transformation from operational databases to data warehouses is often a long journey. This is often due to different systems where the operational databases are implemented, different geographical locations, different data governance, ownership, access control and many other reasons. Often, operational databases, which become input to the data transformation process, must go through long processes, such as exporting the database from one system to another, changing the data structure in order to adjust to the new system and environment, compressing the database for more efficient data transportation and so on. The list can easily become very long. Consequently, after this long journey, the database, the metadata or even the data itself may be corrupted to some degree.

It is not surprising that the operational database at the end of this journey may be different from the original operational database.

In some instances, the notion of the primary key is lost. Consequently, record redundancies may exist. Therefore, we can no longer assume that the operational databases are cleaned and consistent according to relational theory. Firstly, record redundancy may exist. Therefore, before we start with the systematic data transformation process to create a star schema, we need to check if a record redundancy exists in the operational database. Record duplication due to a missing primary key can be checked using the `having count(*) > 1` condition in the SQL statement.

The following SQL statement lists all records having duplicate records:

```
select <<PK attribute>>, count(*)
from <<operational database table>>
group by <<PK attribute>>
having count(*) > 1;
```

To clean this table, the `distinct` clause can be used to remove duplicates, and then the clean records are copied to a new table.

```
create table <<new table>> as
select distinct *
from <<old table>>;
```

In some cases, a table may have duplicate records but with different PK values. For example, two records may have different CustomerIDs, but actually they are records for the same customer. This may have happened intentionally or unintentionally. A mistake may occur unintentionally due to data key-in errors (as well as missing PK constraints during the transformation process), or it can be intentional because different systems give a different PK to the same customer. This cannot be detected using the above SQL command because the PKs are different. Detecting redundant records with different PK values can be tricky. One option is to `group by` all non-PK attributes. But if the number of non-PK attributes is large, this could be cumbersome.

```
select <<all non-PK attributes>>, count(*)
from <<operational database table>>
group by <<all non-PK attributes>>
having count(*) > 1;
```

If we are certain that these are duplicate records and hence must be removed, removing them can also be tricky because the PKs are different. One simple solution is to create a new table but only import the non-PK attributes. Then, the new table is appended with a SurrogateKey.

```
create table <<new table>> as
select distinct <<list of all non-PK attributes>>
from <<old table>>;

alter table <<new table>>
add (SurrogateKey number(6));
```

```
create sequence SurrogateKey_seq_ID
start with 1
increment by 1
maxvalue 99999999
minvalue 1
nocycle;

update <<new table>>
set SurrogateKey = SurrogateKey_seq_ID.nextval;
```

Another possible scenario relating to PKs and duplicate records is exactly the opposite to the above, that is, different records sharing the same PK values. This may happen because these different records were imported from different systems, and for some reason, there is no PK integrity across different systems. Hence, when they are combined into a single table, these different records may end up with the same PK values. If we assume that they are totally different records, the PKs must then be made different. Hence, new PK values must be assigned.

These duplicate records may be checked manually, which can be time-consuming, especially if there are a lot of duplicate records. If the data warehouse does not use this PK attribute from this operational database, there is no need to clean these PKs because they will not affect the creation of the star schema and will not affect the fact measures of the Fact Tables. In other words, not all errors found in the operational database must be corrected. The rule of thumb is if the errors will not affect the data warehouse, we can safely ignore them. However, if the errors in the operational database must be correct, then we can append the table with a new surrogate key.

20.1.1.2   Data Duplication Between Attributes

This may occur after merging two tables into a single table. Suppose the merged table has two attributes with different names but the same meaning. For example, in the Patient Emergency case study, one attribute is called heart_rate_apache, and the other is heartrate. These two attributes may come from different tables but should have the same meaning. To check if the values of these two attributes are the same, the following SQL command can be used:

```
select count(*)
from Emergency_Patient
where heart_rate_apache <> heartrate;
```

It will return a number of records where the two attributes have different values. If there is none, the two attributes can be collapsed into one, using the following alter table command.

```
alter table Emergency_Patient
drop column heartrate;
```

20.1.1.3   Duplication Between Tables

The above data duplication scenarios are related to a single table. Duplications may also be related to multiple tables, especially relating to metadata. For example, there is a table called Patient, and there is another table called Emergency_ Patient. If these were to be merged into one table, technically, the `union` operator in SQL requires both tables to be union-compatible. If one table has more attributes than the other, the `union` operator will not work. Firstly, we need to check the number of attributes of both tables.

```
select count(*)
from All_Tab_Columns
where Table_Name = 'Patient';

describe table Patient;

select count(*)
from All_Tab_Columns
where Table_Name = 'Emergency_Patient';

describe table Emergency_Patient;
```

If both tables have many identical attributes as well as some distinct attributes, and if both tables are to be merged, both tables must be union-compatible. Hence, each table needs to be appended with the attributes that the table doesn't have but which exist in the other table. Then, a `union` operation can be performed to merge the two tables. After merging the two tables, duplicate records can be handled as discussed above.

20.1.2   Relationship Problems

Primary Keys and Foreign Keys are used in relational databases to enforce referential integrity. By enforcing referential integrity, data anomalies can be prevented. However, during the transformation process, the original operational databases may be changed due to unknown reasons, as mentioned in the previous section. As a result, PK-FK constraints may unintentionally be removed.

Firstly, we can check if the FK value of one table matches the PK value of the other table, confirming referential integrity. Obviously, this is unnecessary if we are certain that the PK-FK constraint actually exists. But if it does not, the following SQL can be used to check for invalid FK values:

```
select *
from <<table 1>>
where <<FK>> not in
  select <<PK>>
  from <<table 2>>;
```

If there are records in Table 20.1 where the FK values do not match the PK of Table 20.2, resolving this issue might be tricky. The easiest way is to make these FK values null.

```
update <<table 1>>
set <<FK>> = null
where <<FK>> not in
  select <<PK>>
  from <<table 2>>;
```

Another scenario for the PK-FK issue is where the FK value in Table 20.1 is null, but it should not be. This referential integrity constraint should have been prevented if the PK-FK relationship is defined as a compulsory relationship, meaning that the FK should not be null. Finding null FKs is easy. Once they are found, assigning the correct PK value can be tricky and time-consuming, as the update must be done one by one.

The third possible issue causing PK-FK referential integrity constraint violation is due to different coding or formatting. For example, the FK uses MEL, whereas the PK uses MELB (for Melbourne). Technically, they do not match and hence, referential integrity is violated. However, from the business rule perspective, they should have been considered as a match. Resolving this issue is the same as those of the null FK problem, where the FK needs to be updated to match the PK of the referencing table.

20.1.3 Inconsistent Values

Inconsistent values are one of the most common problems in databases. Inconsistent value problems may appear at two levels: (i) record level and (ii) between attributes.

20.1.3.1 Inconsistent Values at a Record Level

One common inconsistent value problem is that the unit of measurement used in one attribute is not uniform throughout all records. For example, the height attribute of a record uses centimetres (e.g. 175 cm), whereas another record uses meters (e.g. 1.75 m). Detecting the use of different units of measurement in a column could be done by specifying the range of values in the search. For example:

```
select count(*)
from Patient
where height > 100;

select count(*)
from Patient
where height < 2.5;
```

Table 20.1 Dirty Data Matrix

	Attribute level	Between records	Between attributes	Between tables
1. Duplicate	**A** N/A (duplicates must be between records or between attributes)	**B** Duplicate PK; duplicate records but different PK; duplicate PK but different records	**C** Two attributes with identical values	**D** Two tables to be merged have common attributes
2. Relationship	**E** N/A (relationship must be between records (vertical), between attributes (horizontal) or between tables	**F** see cells **J**, **N** and **R**	**G** see cells **K** and **S**	**H** FK values do not exist in PK; FK is null, but should not be; FK=PK but different format
3. Inconsistent values	**I** N/A (inconsistency exists only between two records, two attributes or two tables)	**J** Different units; different coding/standard; different precision	**K** Two attributes must be the same, but they are not the same; two attributes must not be the same but two attributes have conflicting values	**L** see cell **H**
4. Incorrect Values	**M** Incorrect spelling; incorrect range; wrong entry; incorrect data type	**N** Incorrect values in time-series tables	**O** see cell **K**	**P** see cell **H**
5. Null Values	**Q** Null values replaced by zeroes or empty strings	**R** Missing values or missing records in time-series tables	**S** Two attributes, but only one attribute should have a value	**T** Valid FK but the value is incorrect

Table 20.2 Dirty Data Matrix: Simplified

	Attribute level	Between records	Between attributes	Between tables
1. Duplicate	A N/A (duplicates must be between records or between attributes)	B Duplicate PK; duplicate records but different PK; duplicate PK but different records	C Two attributes with identical values	D Two tables to be merged have common attributes
3. Inconsistent Values	I N/A (inconsistency exists only between two records, two attributes or two tables)	J Different units; different coding/standard; different precision	K Two attributes must be the same, but they are not the same; two attributes must not be the same but the same; two attributes have conflicting values	L FK values do not exist in PK; FK=PK but different format
4. Incorrect Values	M Incorrect spelling; incorrect range; wrong entry; incorrect data type	N Incorrect values in time-series tables	O Correct values independently but becomes incorrect collectively	P Valid FK but the value is incorrect
5. Null Values	Q Null values replaced by zeroes or empty strings	R Missing values or missing records in time-series tables	S Two attributes, but only one attribute should have a value	T FK is null, but should not be

The first one checks if the records use centimetres (assuming the height of a patient must be over 100 cm), whereas the second one checks if the records use metres (assuming that the maximum height is less than 2.5 m). To convert one unit of measurement to another (e.g. from meters to centimetres), we need to update the records.

```
update Patient
set height = height * 100
where height < 2.5;
```

Another common inconsistent value problem is due to different codings or standards. For example, using a Flight Trip table as an example, the airport attributes (e.g. source and destination of a trip) record the cities to be visited on a trip. Some trips to/from Jakarta may use the IATA standard and the Jakarta airport code, which is CGK. Other records may use the abbreviation JKT to indicate Jakarta. Hence, this creates data inconsistencies among records. Detecting data inconsistency due to different codings or standards can be tricky and time-consuming, as domain experts need to go through the records to see if there are obvious conflicting codings or standards being used. Other examples of data inconsistency due to different codings or standards are MEL vs. MELB (for Melbourne) or NSW vs. N.S.W (for New South Wales), etc. There are endless possibilities of data inconsistency due to different codings and standards.

Another data inconsistency problem is due to the different precision of numerical attributes. For example, some records may use six digits of precision after the decimal point, whereas other records may be rounded to the nearest integer. This can easily happen in many applications which involve various measurements, especially scientific applications, medical applications, etc.

20.1.3.2   Inconsistent Values Between Attributes

When merging two tables into a single table, it is possible that there are two attributes with the same meaning in the merged table, but the values are different. For example, in an Emergency Patient table, there are two attributes related to maximum heart rate: one attribute called h1_heartrate_max and the other attribute d1_heartrate_max. The values of these two attributes are different. Further investigation must be undertaken to determine whether the two attributes have different meanings and use, and subsequently, the values are different.

As another example, in the Employee table, the meaning of two attributes is the same (e.g. salary and gross salary), but the values are different, so this must be further investigated as the two attributes should have the same values.

The opposite of the above is where two similar attributes are required to have different values. For example, in a Flight table with Pilot and Co-Pilot attributes, the Co-Pilot attributes must be either null (in the case of a small plane that does not require a Co-Pilot) or different from the Pilot attributes (since the Pilot cannot be the Co-Pilot for the same flight).

Another problem of data inconsistency between attributes is where there are two different attributes but the values of these attributes are in conflict with to each other. For example, one attribute is d1_heartrate_max and the other is d1_heartrate_min, but the minimum attribute has a higher value than the maximum attribute in the same records. This is a data conflict.

20.1.4   Incorrect Values

Incorrect values may occur in three levels: (i) an attribute level, (ii) between record level (iii) between table level.

20.1.4.1   Incorrect Value Problem at an Attribute Level

The most common incorrect value problem is due to incorrect spelling. This may range from a simple misspelling to the use of extremely unusual characters in the attribute value.

Another problem relating to incorrect values is because the value of an attribute falls outside the perceived correct data range. For example, a negative value for height and an extremely high human body temperature (e.g. 100 degree Celsius) are considered incorrect values.

The most difficult incorrect value problem to detect (and to correct) is when the value of an attribute is simply wrong, for example, if an employee's residence is in Melbourne but Sydney is recorded as his city of residency. The value Sydney is not misspelt nor out of range; it is simply wrong because it should not be Sydney, it should be Melbourne.

Another common mistake at an attribute level is an incorrect data type, for example, if an attribute of numerical values is declared as a string data type. This attribute will not be able to use numerical functions to perform common numerical calculations as the data is stored as a string rather than as a number.

20.1.4.2   Incorrect Value Problem Between Records

Suppose the records are ordered in a certain order, and one attribute should have ascending values. If the attribute value of a record is less than the previous record (e.g. violating the ascending order rule), then this attribute value is said to have an incorrect value problem.

20.1.4.3   Incorrect Value Problem Between Tables

In the aforementioned Relationship Problems section, the problem is due to the missing PK-FK referential integrity between tables. However, even when PK-FK referential integrity is maintained between two tables, an incorrect value in the FK may still occur. In this case, the FK value of a record is correctly pointing to an existing PK value of another table, but it is pointing to the wrong PK value. For example, in the Employee-Department relationship, the Department No (which is the PK of the Department table) is an FK in the Employee table. An employee is recorded as working in the Accounting Department. The Accounting Department exists in the Department table; hence, there is no violation in the referential integrity. However, that particular employee does not work in the Accounting Department but in the Engineering Department. So in this case, it is an incorrect FK value—not a referential integrity error.

20.1.5   Null Value Problems

The null value problems exist at three levels: (i) an attribute level, (ii) between records and (iii) between attributes.

20.1.5.1   Null Value Problems at an Attribute Level

In the operational database, we could have a not null constraint imposed on a table. However, due to the transformation process, where the table is being transported and transformed from one system to another, in many instances, table constraints, such as a not null constraint, are lost. Therefore, it might be possible that an attribute has null values when it shouldn't have null values. The treatment of null values, which should not be null, may depend on the business rule.

Another common problem related to null values is that many times a zero (0) is recorded instead of null, if it is a numerical attribute. For a string attribute, it is common to store an empty string "", instead of a null. The problem will become harder when null and zero are both used in a numerical attribute or when the null and empty string are used together in a string attribute. This could easily create confusion as to whether to consider all zero (0) values as null values or as zero (0) values.

20.1.5.2   Null Value Problems Between Records

Null values may exist in between records, especially in a time-series (or sensor-produced) table. In a time-series (or a sensor-produced) table, the records are ordered by the timestamp. An example of a time-series table is a forex (foreign exchange) table recording the currency exchange rate from USD (US dollars) to AUD (Australian dollars). In this table, there are two attributes: date and the exchange rate. If the table is sorted by date, there might be some missing records if the exchange rates on those dates have not been recorded in the table. This does not mean that there were no exchange rates for those missing dates, but the exchanged values were not recorded, perhaps due to some technical difficulties or errors during the recording phase. However, we know that there is an exchange rate every day.

To fill-in the missing exchange rates, some interpolation techniques may be used. For example, using the forex example, the missing exchange rate data of 1 day may be estimated using an average function between the exchange rate of the previous day and the next day. There are other more sophisticated interpolation techniques used for different circumstances depending on the application.

20.1.5.3   Null Value Problems Between Attributes

Null value problems may exist in the context of correlating multiple attributes. For example, there are two attributes: discount 1 and discount 2. Assuming that we can only get one kind of discount; if discount 1 has a value, discount 2 should be null and vice versa.

20.1.6   Summary

Table 20.1 summarises the different data errors discussed in the previous sections.

A lot of these problems are due to the relationships; they very rarely occur in isolation. There are actually three different levels of relationships: (*i*) between records, (*ii*) between attributes and (*iii*) between tables. These three columns in Table 20.1 show data mistakes due to relationships. For example, duplicate values in PK (cell **B**) show that the PK attribute value of one record is identical to that of another record, so it is a relationship between or among records. The inconsistency value in two attributes (cell **K**) shows the correctness of one attribute value depends on the relationship to another attribute (e.g. a min heart rate attribute must not have a value larger than a max heart rate attribute), so it is a relationship or a correlation between records. An incorrect value problem between tables is shown as an incorrect FK value (cell **P**), so it is a relationship between tables through the PK-FK relationship.

Problems not related to relationships are shown only in cells **M** and **Q**—incorrect values or null problems at an attribute level. Even the Duplicate, Relationship and Inconsistent Value problems do not exist at an attribute level (e.g. cells **A**, **E** and **I** are N/A). Hence, the Relationship row could be absorbed into the Inconsistent and Null Problem rows; consequently, the Data Error Matrix can be simplified (see Table 20.2 for the Simplified Data Error Matrix). Note also cell **O** (Incorrect Values between Attributes) is similar to cell **K** (Inconsistent Values between Attributes).

For example, there are two attributes: Title and Qualification. An entry in the title is "Mr" with Qualification of a "PhD". Both entries are correct, independently. But when considering both attributes, the Title should be "Dr". So, this is a kind of inconsistency or incorrectness.

20.2   Post-Data Warehousing: Exploring the Extended Fact Table

Once a data warehouse is created, before any intensive data analytics is done, it is crucial to understand what kind of data we have in the data warehouse. This is *Post-Data Warehousing*. This data exploration is not to find dirty data, as in the Pre-Data Warehousing; rather it is carried out to understand more about the data we have in the data warehouse. Without much understanding of the data in the data warehouse, data analytics would be difficult. Therefore, understanding or exploring the data in the data warehouse is necessary before undertaking more comprehensive and thorough analysis.

20.2.1   Extended Fact Table

This first chapter in this book shows a flow diagram from an operational database to a data warehouse, to OLAP and Business Intelligence and to data analytics. This flow depicts the use of data at various levels in the organisation. Data in the operational database is used for transaction management to support daily business operations. Data in the data warehouse or star schema is used at various levels of granularity through OLAP and Business Intelligence to assist the decision-making process. Finally, data analytics discovers trends, patterns and knowledge from the data in the data warehouse. In other words, the final goal of data warehousing is to assist decision-makers and managers to undertake data analytics in a more efficient and targeted manner, as the data has been cleaned, transformed, integrated and stored in a structured model, namely, star schemas. So, the main question of data analytics for data warehousing is how do star schema data support data analytics.

Figure 20.2 shows an illustration of a typical star schema with a number of dimensions, each with a dimension key and the fact with potentially a number of fact measures. The fact measures are the central focus of a star schema, which can be viewed from one or multiple dimensions. Once the data is stored in the Fact Table and in the associated dimension tables, querying data, including data exploration, is straightforward. However, apart from having the fact measures, the Fact Table only contains the dimension keys. Further details of each dimension are only available in the respective dimensions. Therefore, to query a star schema, a join needs to be performed between the Fact Table and the necessary dimension tables. This is known as a *star join*.

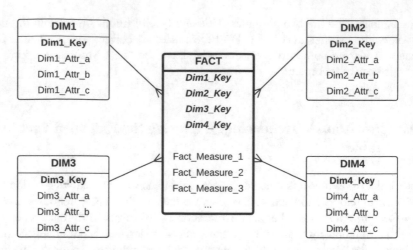

Fig. 20.2 Star schema

Table 20.3 An Extended Fact Table

Dim1 Key	Dim1 Attr-a	Dim1 Attr-b	Dim1 Attr-c	Dim2 Key	Dim2 Attr-a	Dim3 Key	Fact Measure-1	Fact Measure-2	Fact Measure-3

A sample result of a star join query is shown in Table 20.3. This is also called an *Extended Fact Table*, as the Fact Table is extended to include some dimension attributes. These dimension attributes provide additional information to the respective dimension key, which is normally a surrogate key. Hence, having more information about the dimension through the attributes of that dimension provides users with more meaningful results. This Extended Fact Table then becomes the main source for data analysis.

The Extended Fact Table, as illustrated in Table 20.3, includes three additional attributes from Dimension-1 and two additional attributes from Dimension-2 (the additional columns from the dimensions are highlighted). Dimension keys are usually categorical attributes, whereas other dimension attributes could be categorical or non-categorical attributes. Fact measures are usually non-categorical attributes (e.g. numerical values). Using an Emergency Patient example, Dimension-1 could be the Patient Dimension, where the Dimension Key is the PatientID, and the three attributes could be sex (categorical), ethnicity (categorical) and age (numerical). Dimension-2 could be the Hospital Dimension, where the key is HospitalID and the additional attribute is Hospital Type. The fact measures are various clinical measurements of the patient, such as blood glucose, and various blood/urine test results. Note that in the Extended Fact Table, not all dimension keys nor all attributes from each dimension may be included. In this example, only three additional attributes from Dimension-1 and one additional attribute from Dimension-2 are included. A total of three dimension keys and three fact measures from the Fact Table are included in this Extended Fact Table.

In general, the Extended Fact Table is obtained from the Level-0 star schema, where the fact measures are raw and non-aggregated values. However, in some circumstances, it may be better to use higher-level star schemas for data analysis, where the fact measures are already in an aggregated form. The main importance is not the granularity level of the Extended Fact Table; rather the dimension keys as well as some additional attributes from the dimension (and, of course, the fact measures) are included. This kind of file structure is commonly used as an input to data analysis in many data science projects. However, there are also many data mining techniques and algorithms which do not assume that the input data is in this structure. Therefore, the main distinction between the traditional data mining approach and the data warehousing analytic approach is the underlying data structure, where the star schema is already structured in such a way that is ready for data analysis, whereas in traditional data mining, each technique and algorithm may require the data to be structured in a different way.

20.2.2 A Typical Data Science Project

In a typical data science project, usually you are given a csv file to work with. It is common that the csv file contains hundreds of columns and hundreds of thousands of rows. The first task is to understand what the data (or the file) is about. The topic of the project is generally outside our area of expertise; therefore, it would be beneficial to work with domain experts on this data science project.

In this section, we are going to use a typical Emergency Patient data as a case study. A snapshot of the csv file is shown in Fig. 20.3. The file is huge, with lots of numbers in hundreds of columns and hundreds of thousands of rows. If we zoom in to the file, we can see something like that in Fig. 20.4. We could perhaps see there is glucose in the heading of columns EC and ED, as well as hemoglobin in the heading of columns EG and EH. We could expect to see many column headings related to the blood test results (and other measurements) of the patients.

Depending on the technicality, this csv file could be imported into a table in a Relational DBMS. The structure of the table, which can be displayed using the SQL command describe Emergency_Patient, is shown in Table 20.4. In the original csv file, the last column is column HI. This is equivalent to 217 attributes, which is verified by the following SQL command:

```
select count(*)
from user_tab_columns
where table_name='EMERGENCY_PATIENT'
```

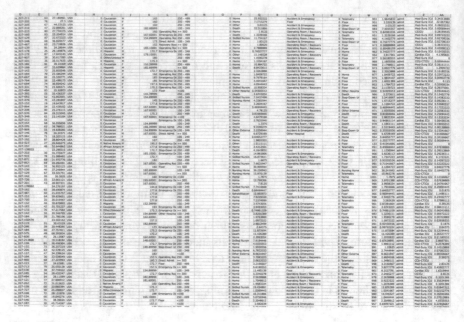

Fig. 20.3 A snapshot of the csv file

EA	EB	EC	ED	EE	EF	EG	EH	EI	EJ
d1_creatinine	d1_creatinine	d1_glucose_m	d1_glucose_m	d1_hco3_max	d1_hco3_min	d1_hemaglobi	d1_hemaglobi	d1_hematocrit	d1_hematocrit
1.5833522	1.5720425			21	21	12	9.3999996	34	28
0.95001131	0.85953403	162.16216	108.10811	26	23	11.6	10.1	36	30
1.9113322	1.391088	201.8018	122.52252	19.4	16.9	9.8999996	9.1000004	30	27
0.7125085	0.5767926	129.72974	93.693695	26	21	15.5	13.7	45	41
1.0065596	0.81429541	90.090088	68.468468	23	18	11.1	10.1	35	30
1.3345397	1.0857272	156.75676	142.34235	20	17	15.9	8.3999996	47	25
0.99524993	0.7690568	225.22522	97.297295	26.5	22	14.5	13.9	42	42
0.9047727	0.6559602	144.14415	91.891891	20.6	16.6	12.9	12	37	36
1.2666818	1.0178692	190.991		28	23	14.7	10.1	42	29
0.87084371	0.80298573	154.95496	106.3063	25.799999	23.9	12.6	11.7	36	34
0.8482244				22		14.4		43	
0.74643743	0.56548291	151.35135	93.693695	23.5	18	15.4	14	46	42
1.1875142	0.8482244	257.65765	154.95496	27.6	23.4	12.5	12.2	37	36
3.0196788	2.2053833	154.95496	124.32433	23.200001	21.200001	16.4	12.7	49	39
1.6172812	1.1309658	154.95496	82.882881	20.700001	10.4	12.5	11.8	37	36
10.845963	3.8679032	163.96396	84.684685	16.299999	6.8000002	17.799999	14.2	52	41
1.5607328	1.3458494	118.91892	75.675674	26	22	10.2	8.1999998	32	25
1.0178692	0.9047727	111.71171		25	24	15.2	14.8	45	44
0.75774711	0.66726983	97.297295		28	24	16.700001	14.1	50	45
1.0857272	0.92739201	225.22522	144.14415	25.299999	24.1	10.9	10.1	32	30
0.95001131	0.78036642	136.93694	104.5045	24.299999	22.1	12	11.2	40	35
0.78036642	0.73512781			29	27	13.5	12.4	40	37
0.74643743	0.72381812	140.54054	108.10811	27	25	12.6	12.3	37	35
0.9839403	0.89346302	398.19821	207.20721	28	22.4	7.8000002	6.4000001	19	19
0.95001131	0.31667045	106.3063	99.099098	30	27.299999	15.5	14	46	42
1.1422705	1.1083466	180.18018	180.18018	21.1	21.1	16	16	46	47
0.95001131	0.83691472	113.51351	97.297295	26.200001	25.6	9.8999996	8.5	29	25
0.66726983	0.5202443	189.18919	140.54054	15.6	13	14.4	5.4000001	42	16
3.8565936	1.4815652	434.23422	115.31532	20.4	18.9	9.5	2	32	28
1.2327528	0.99524993	151.35135	113.51351	24	21	13.4	12.4	40	38
0.42976701	0.41845736			22	21	9.1000004	9	28	28

Fig. 20.4 A zoom in to the csv file

Table 20.4 Emergency Patient Table structure

Attributes
data_source, encounter_id, hospital_id, patient_id, age, bmi, country, elective_surgery, ethnicity, gender, height, hospital_admit_source, hospital_bed_size, hospital_bed_size_numeric, hospital_death, hospital_disch_location, hospital_los_days, hospital_type, icu_admit_source, icu_admit_type, icu_death, icu_disch_location, icu_id, icu_los_days, icu_stay_type, icu_type, pre_icu_los_days, pregnant, readmission_status, smoking_status, teaching_hospital, weight,
albumin_apache, arf_apache, bilirubin_apache, bun_apache, creatinine_apache, diagnosis_apache, fio2_apache, gcs_eyes_apache, gcs_motor_apache, gcs_unable_apache, gcs_verbal_apache, glucose_apache, heart_rate_apache, hematocrit_apache, intubated_apache, map_apache, paco2_apache, paco2_for_ph_apache, paco2_apache, ph_apache, resprate_apache, sodium_apache, temp_apache, urineoutput_apache, ventilated_apache, wbc_apache,
d1_diasbp_invasive_max, d1_diasbp_invasive_min, d1_diasbp_max, d1_diasbp_min, d1_diasbp_noninvasive_max, d1_diasbp_noninvasive_min, d1_heartrate_max, d1_heartrate_min, d1_mbp_invasive_max, d1_mbp_invasive_min, d1_mbp_max, d1_mbp_min, d1_mbp_noninvasive_max, d1_mbp_noninvasive_min, d1_padias_invasive_max, d1_padias_invasive_min, d1_pamean_invasive_max, d1_pamean_invasive_min, d1_pasys_invasive_max, d1_pasys_invasive_min, d1_resprate_max, d1_resprate_min, d1_spo2_max, d1_spo2_min, d1_sysbp_invasive_max, d1_sysbp_invasive_min, d1_sysbp_max, d1_sysbp_min, d1_sysbp_noninvasive_max, d1_sysbp_noninvasive_min, d1_temp_max, d1_temp_min, h1_diasbp_invasive_max, h1_diasbp_invasive_min, h1_diasbp_max, h1_diasbp_min, h1_diasbp_noninvasive_max, h1_diasbp_noninvasive_min, h1_heartrate_max, h1_heartrate_min, h1_mbp_invasive_max, h1_mbp_invasive_min, h1_mbp_max, h1_mbp_min, h1_mbp_noninvasive_max, h1_mbp_noninvasive_min, h1_padias_invasive_max, h1_padias_invasive_min, h1_pamean_invasive_max, h1_pamean_invasive_min, h1_pasys_invasive_max, h1_pasys_invasive_min, h1_resprate_max, h1_resprate_min, h1_spo2_max, h1_spo2_min, h1_sysbp_invasive_max, h1_sysbp_invasive_min, h1_sysbp_max, h1_sysbp_min, h1_sysbp_noninvasive_max, h1_sysbp_noninvasive_min, h1_temp_max, h1_temp_min,
d1_albumin_max, d1_albumin_min, d1_bilirubin_max, d1_bilirubin_min, d1_bun_max, d1_bun_min, d1_calcium_max, d1_calcium_min, d1_creatinine_max, d1_creatinine_min, d1_glucose_max, d1_glucose_min, d1_hco3_max, d1_hco3_min, d1_hemaglobin_max, d1_hemaglobin_min, d1_hematocrit_max, d1_hematocrit_min, d1_inr_max, d1_inr_min, d1_lactate_max, d1_lactate_min, d1_platelets_max, d1_platelets_min, d1_potassium_max, d1_potassium_min, d1_sodium_max, d1_sodium_min, d1_wbc_max, d1_wbc_min, h1_albumin_max, h1_albumin_min, h1_bilirubin_max, h1_bilirubin_min, h1_bun_max, h1_bun_min, h1_calcium_max, h1_calcium_min, h1_creatinine_max, h1_creatinine_min, h1_glucose_max, h1_glucose_min, h1_hco3_max, h1_hco3_min, h1_hemaglobin_max, h1_hemaglobin_min, h1_hematocrit_max, h1_hematocrit_min, h1_inr_max, h1_inr_min, h1_lactate_max, h1_lactate_min, h1_platelets_max, h1_platelets_min, h1_potassium_max, h1_potassium_min, h1_sodium_max, h1_sodium_min, h1_wbc_max, h1_wbc_min,
d1_arterial_pco2_max, d1_arterial_pco2_min, d1_arterial_ph_max, d1_arterial_ph_min, d1_arterial_po2_max, d1_arterial_po2_min, d1_pao2fio2ratio_max, d1_pao2fio2ratio_min, h1_arterial_pco2_max, h1_arterial_pco2_min, h1_arterial_ph_max, h1_arterial_ph_min, h1_arterial_po2_max, h1_arterial_po2_min, h1_pao2fio2ratio_max, h1_pao2fio2ratio_min, apache_3j_diagnosis, apache_3j_hospital_death_prob, apache_3j_score, apache_4a_hospital_death_prob, apache_4a_icu_death_prob, apsiii, apache_2_diagnosis, apache_post_operative, aids, cirrhosis, diabetes_mellitus, hepatic_failure, immunosuppression, leukemia, lymphoma, solid_tumor_with_metastasis, apache_3j_bodysystem, apache_2_bodysystem, prediction

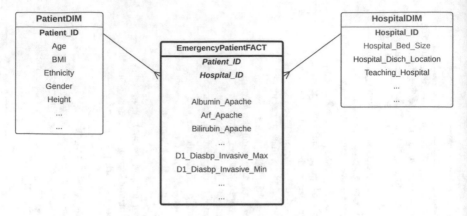

Fig. 20.5 Emergency Patient star schema

Looking at the table structure as shown in Table 20.4, the attributes can be grouped into three: basic information about the patient (e.g. patient_id, age, gender, height, bmi, etc.), information about the hospital (e.g. hospital_id, hospital_type, teaching_hospital, etc.) and hundreds of measurements of the patients (e.g. heartrate_apache, temp_apache, urineoutput_apache, etc.).

This table looks like a Fact Table. From a data warehousing point of view, it is a *Level-0 Fact Table*. The fact measures are the hundreds of measurements of the patients and two identifiers of the dimensions: one is patient_id (or encounter_id, which is a visit_id and can be regarded as a surrogate key), and the other is hospital_id (which is the key of the hospital). The table is actually a *Lower-Than-3NF Fact Table* because the table includes information about the patient and the hospital. It is actually a join between the Fact Table and the two dimension tables: Patient and Hospital Dimensions. Hence, the star schema is something like that shown in Fig. 20.5.

If we only consider the identifiers of Patient and Hospital Dimensions and all of the fact measures, it is exactly the same as the Fact Table of the Emergency Patient star schema shown in Fig. 20.5. However, in data science projects, because we also need information about each dimension, therefore, the csv table also includes some important information from each of the dimension: Patient and Hospital Dimensions. In other words, although the csv file is technically not a pure Fact Table, it can be considered an *Extended Fact Table* with information from the dimension tables. So, what we are going to do is explore this Extended Fact Table.

Since the csv file has a data source column, which is the first attribute in Table 20.4, the csv file possibly comes from multiple star schemas, where each Fact Table focuses on one data source, such as an Australian data source, American data source, Brazilian data source, etc. Each data source represents the data to which it belongs, and each data source is an independent star schema. So, imagine that there are, for example, ten data sources; the csv file combines these ten Fact Tables. In short, the csv file is an Extended Fact Table merged from multiple star schemas.

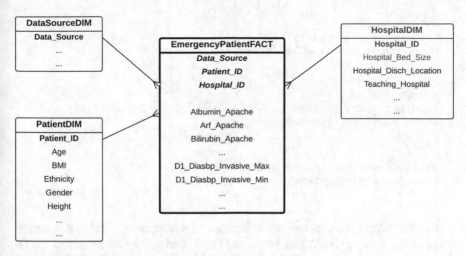

Fig. 20.6 Emergency Patient star schema with Data Source

The combined Fact Table then has three dimensions: Data Source, Patient and Hospital Dimensions. This star schema is shown in Fig. 20.6. The csv file is a join between these four tables to include some attributes from the dimension tables into the csv file for analysis.

Once we understand the structure of the table from a data warehousing perspective, the next step is to explore the data in the Extended Fact Table. During data exploration, it would be useful to work with domain experts. The domain experts may shed some light on the information obtained from the data exploration, which will enable a deeper understanding of the data. There are three types of data exploration to be discussed: (i) explore single attributes, (ii) search records and (iii) explore multiple attributes.

20.2.3   Explore Individual Attributes

20.2.3.1   Basic Statistics

A basic exploration is to count the number of records in the table. To explore one particular attribute, we need to know if there are null values in the attribute. In this example, we explore an attribute called `heart_rate_apache`. This basic counting confirms that the total of null and not null values in attribute `heart_rate_apache` matches the total number of records in the table.

```
select count(*)
from Emergency_Patient;

482425
```

```
select count(*)
from Emergency_Patient
where heart_rate_apache is not null;
```

418028

```
select count(*)
from Emergency_Patient
where heart_rate_apache is null;
```

64397

```
select count(heart_rate_apache)
from Emergency_Patient;
```

418028

Further data exploration can be undertaken on this attribute. Note that a normal heart rate is between 60 and 100. We would like to find out say the minimum, maximum and median heart rate of patients in this table. The min function in the following SQL shows that the minimum heart rate is zero (0), which is most likely not the real minimum heart rate. Remember that in the Exploring Dirty Data section, often there are mistakes when null values are replaced by zeroes (0). The next SQL shows that there are 13 records (from over 400,000 records) with zeroes (0) for heart rate.

The next investigation is to find the maximum heart rate, which is 300 beats per minute, and then the median, which is 98 beats per minute. However, the median is based on the entire dataset. It should exclude the records with zeroes (0) for heart rate. Even when we include a condition to exclude the zero (0) heart rate records, the median remains the same, which is 98. The next question is how many records are there with the median value, and the answer is 7758 records. That's why even when the 13 records with zero (0) for heart rate are excluded from the dataset, the median remains unchanged, simply because there are a lot of records that are the median value.

Finally, we use the avg function to calculate the average (or mean) of the heart rate, which is 94.3321812, slightly lower than the median of 98. The median value is the value of the patient that is exactly in the middle of the dataset. So, when the average value is lower than the median value, this means that the data is skewed to the low side, with a long tail of low heart rate pulling the average (mean) down more than the median.

```
select min(heart_rate_apache)
from Emergency_Patient;
```

0

```
select count(heart_rate_apache)
from Emergency_Patient
where heart_rate_apache = 0;
```

13

```
select max(heart_rate_apache)
from Emergency_Patient;
```

```
300

select median(heart_rate_apache)
from Emergency_Patient;

98

select median(heart_rate_apache)
from Emergency_Patient
where heart_rate_apache > 0;

98

select count(heart_rate_apache)
from Emergency_Patient
where heart_rate_apache = 98;

7758

select avg(heart_rate_apache)
from Emergency_Patient;

94.3321812
```

20.2.3.2   Count Distribution: Histogram

After understanding the basic statistics, such as min, max, median and mean, the next step is to understand the distribution of the data. A normal distribution, sometimes called the *bell curve*, is a distribution that occurs naturally in many situations. A bell curve is shown in Fig. 20.7. The graph shows that the majority of the data falls into the middle category. The middle line is the average. This means the data count that is equivalent to the mean value should be higher than the other

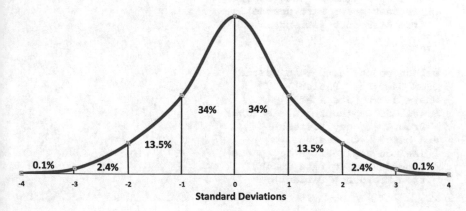

Fig. 20.7 Bell curve

categories. The data count will gradually decrease as it moves away from the mean value.

The bell curve uses *standard deviation*, which is a measure of how spread out the numbers are. The mean to plus/minus 1 standard deviation is the areas to the right and to the left of the mean. These two regions contain most of the data as about 68% of the entire dataset falls into this region. The next region is from plus/minus 1 standard deviation (from the mean) to plus/minus 2 standard deviation. Roughly 95% of the dataset falls into these four regions. This means that in a normal distribution, 95% of the data must be between minus 2 standard deviation (from the mean) to plus 2 standard deviation (from the mean). Only a small minority of the data is outside this range.

Now we would like to examine if the heart_rate_apache attribute fits into the bell curve. Firstly, we need to get the mean and standard deviation. The mean has already been calculated as 94.3321812. The standard deviation can be calculated using the stddev function in SQL.

```
select stddev(heart_rate_apache)
from Emergency_Patient;

30.1489292
```

The standard deviation is 30.1489292, and the mean is 94.3321812. This means the mean to plus/minus 1 standard deviation is 94.3321812 ± 30.1489292. Using the following SQL command, we can count the number of records where the heart_rate_apache attribute is in between these values:

```
select count(heart_rate_apache)
from Emergency_Patient
where heart_rate_apache >=
  (select avg(heart_rate_apache)
  from Emergency_Patient)
and heart_rate_apache <
  (select avg(heart_rate_apache) +
    stddev(heart_rate_apache)
  from Emergency_Patient);

175995

select count(heart_rate_apache)
from Emergency_Patient
where heart_rate_apache <
  (select avg(heart_rate_apache)
  from Emergency_Patient)
and heart_rate_apache >=
  (select avg(heart_rate_apache) -
    stddev(heart_rate_apache)
  from Emergency_Patient);

77612
```

Note that the total number of records, as reported in the previous section, is 418028. Based on the above SQL commands, the percentage of records that fall into the category between the mean and plus 1 standard deviation is 175995/418028 = 42.1%. The left side of the mean, which is between minus 1 standard deviation and the mean, is 77612/418028 = 18.6%.

Using the same methodology, we can find the next areas, which are from minus 2 standard deviation to minus 1 standard deviation (to the left of the mean) and from plus 1 standard deviation to plus 2 standard deviation (to the right of the mean). The SQL commands are very similar to the previous ones, as follows:

```
select count(heart_rate_apache)
from Emergency_Patient
where heart_rate_apache >=
  (select avg(heart_rate_apache) +
    stddev(heart_rate_apache)
  from Emergency_Patient)
and heart_rate_apache <
  (select avg(heart_rate_apache) +
    stddev(heart_rate_apache)*2
  from Emergency_Patient);

51734

select count(heart_rate_apache)
from Emergency_Patient
where heart_rate_apache <
  (select avg(heart_rate_apache) -
    stddev(heart_rate_apache)
  from Emergency_Patient)
and heart_rate_apache >=
  (select avg(heart_rate_apache) -
    stddev(heart_rate_apache)*2
  from Emergency_Patient);

100918
```

Dividing by the total number of records, the percentages are 24.1% (to the left of the mean) and 12.4% (to the right of the mean). We can use the same calculations to obtain the rest. If we plot a bar chart overlaying with the ideal bell curve, the result is shown in Fig. 20.8.

Notice that the data distribution of the heart_rate_apache attribute does not follow the exact normal distribution. There is a jump from minus 2 to minus 1 standard deviation. Also, the total percentage on the right of the mean is around 56.7% compared to only 43.3% on the left of the mean. So there are a lot more records that fall into the right side of the mean, which is why the median is bigger than the mean (e.g. 98 > 94.3321812), as previously calculated.

By showing the data distribution using the standard deviation, we can see how the data is largely distributed, and perhaps this will give some understanding of the data of this particular attribute.

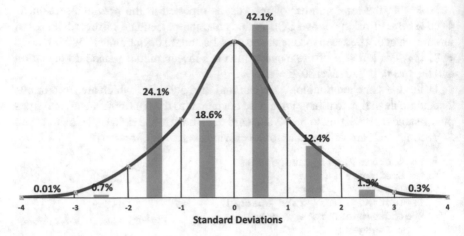

Fig. 20.8 Histogram of heart_rate_apache attribute

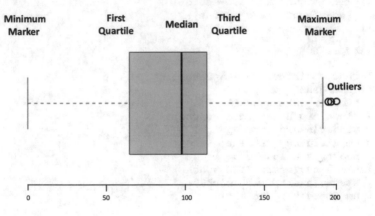

Fig. 20.9 Boxplot

20.2.3.3   Value Distribution: Boxplots

Another way to show data distribution of an attribute is by using *boxplots*. Boxplots show the distribution of the value, not necessarily the count, similar to the histograms. Boxplots, like histograms, are a graphical depiction of numerical data through their quartiles. They use median as the middle quartile, which means there is one quartile below the median and another quartile above the median. The median is called the second quartile (Q2), and the other quartiles are Q1 and Q3. Below Q1, there is a minimum (or low) marker, and above Q3, there is a maximum (or high) marker. Figure 20.9 shows a boxplot visualisation. The *x*-axis represents the range of the values of the data. A boxplot is a very simple but effective way to visualise outliers. Any values below the minimum marker are outliers, and so are

Fig. 20.10 Boxplot and bell curve

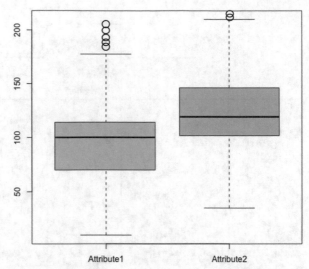

the values above the maximum marker. Hence, you can think about the minimum and maximum markers as the boundaries of the data distribution.

When we plot a graph, usually the y-axis represents the value of the data. Figure 20.10 shows the boxplots of two attributes, and the y-axis represents the values of the data. A boxplot is like plotting the entire dataset (of an attribute) in a vertical style so that we can "see" the entire dataset of an attribute at once.

To plot a boxplot, we first need to know the median (which is the second quartile, which also means the middle, or 50%), and then we need to find Q1 (which is the first quartile, or top 75%) and Q3 (which is the third quartile, or top 25%). The next step is to find the minimum (or low) and maximum (or high) markers.

To determine these markers, we need to know the *InterQuartile Range* or IQR. IQR is the length between Q1 and Q3 ($IQR = Q3 - Q1$). Based on IQR, we can then determine the minimum and maximum markers. The minimum marker is $1.5 \times IQR$ below Q1, and the maximum marker is $1.5 \times IQR$ above Q3. Figure 20.11 shows the relationship between a boxplot and the bell curve. The length between Q1 and Q3 is 50%, which is slightly smaller than the range between minus 1 and plus 1 standard deviation. The figure also shows that any values below minus 2.698 standard deviation or above plus 2.698 standard deviation are considered outliers.

To get the boxplot of the `heart_rate_apache` attribute, we need to get the median, Q1 and Q3. The SQL command to get these values is as follows:

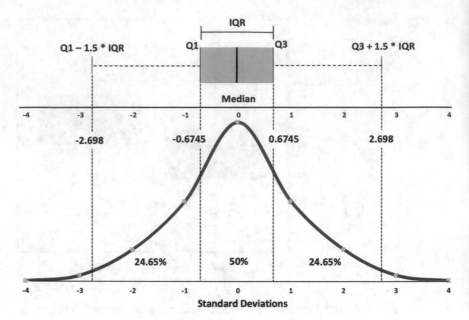

Fig. 20.11 Boxplots of two variables

```
select
  max(Q3) as Q3,
  max(Median) as Median,
  max(Q1) as Q1
from (
  select
    heart_rate_apache,
    percentile_cont(0.25) within group
      (order by heart_rate_apache) over() as Q1,
    median(heart_rate_apache) over() as Median,
    percentile_cont(0.75) within group
      (order by heart_rate_apache) over() as Q3
  from Emergency_Patient);
```

The values obtained are Median=98, Q1=65 and Q3=114. From Q3 and Q1, we can get IQR, which is 49. Then, the minimum and maximum markers are $1.5 \times IQR$ away from the respective quartiles. Hence, the maximum marker = 187.5, whereas the minimum marker = −8.5.

The following SQL command shows that there are 893 outlier values which are above the maximum marker, with a maximum value of 300. The minimum value is 0.

```
select count(heart_rate_apache)
from Emergency_Patient
where heart_rate_apache > (114+73.5);

893

select max(heart_rate_apache)
from Emergency_Patient;
```

```
300

select min(heart_rate_apache)
from Emergency_Patient;

0

select count(heart_rate_apache)
from Emergency_Patient
where heart_rate_apache < (65-73.5);

0
```

The boxplot for the heart_rate_apache attribute is shown in Fig. 20.12. The boxplot shows a box bounded by Q1 and Q3. The maximum and minimum markers are also shown. The outliers above the maximum marker are quite dense in the beginning and become sparse toward the end, with a maximum value of 300.

20.2.4   Search Records

Once we have some idea about the data distribution of an attribute as demonstrated in the previous section, the next step is to dig into some records to understand more about the attribute. The boxplot of the heart_rate_apache attribute shows that there are 893 outlier records which have values higher than the maximum marker of 187.5 (=114+73.5). Also, from the SQL commands, we know there were 13 records with 0 heart rate and there were 64397 records with null values in the attribute. So, these are some pointers on how to investigate this attribute further.

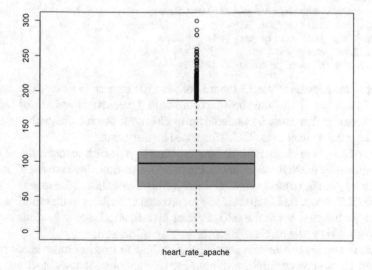

Fig. 20.12 Boxplot of the heart_rate_apache attribute

Table 20.5   Snapshot of outlier heart rate apache

Hospital death	Hospital Los Days	ICU death	ICU Los Days	Smoking status	Heart rate apache
0	14	0	2.5732	Current smoker	191
0	20	0	19.850201	Never smoked	194
0	12	0	5.1097002	Current smoker	280
1	1	1	1.0652	Unknown	205
0	8	0	4.9119	Unknown	200
0	49	0	26.733801	Current smoker	204
0	19	0	16.5408	Unknown	190
0	10	0	1.9793	Current smoker	193
1	1	1	.99690002	Unknown	189
1	7	1	2.9331	Current smoker	190
0	5	0	1.0165	Current smoker	196
0	0	0	2		190
1	0	1	6		188

The following SQL command retrieves the outlier records. Since there are hundreds of attributes in the Emergency Patient table, we only choose a few attributes, such as hospital deaths, hospital LoS (length of stay) days, ICU deaths, ICU LoS days, smoking status and heart rate apache. In this investigation, we are trying to understand the characteristics of the outlier heart rate apache records. A snapshot of the data retrieved by the SQL is shown in Table 20.5. Since there are 893 records, it is perhaps difficult to browse through each record, but at least we have a snapshot of the results, which shows death, LoS and smoking status.

```
select
  hospital_death, hospital_los_days,
  icu_death, icu_los_days,
  smoking_status, heart_rate_apache
from Emergency_Patient
where heart_rate_apache > 187.5;
```

Next is to investigate the 13 records of zero (0) `heart_rate_apache`. The results detailed in Table 20.6 show a record with a negative Hospital Los Days, so further investigation must be undertaken to check all the records with a negative LoS in hospital as well as in ICU. This could be dirty data.

The last step is to check null the `heart_rate_apache` records. Based on the results shown by the SQL command in the previous section, there are 64397 records. It is certainly not possible to view all the records. A snapshot of the results is shown in Table 20.7. From this snapshot, one record that is striking is that there was one death in the hospital not in the ICU. Further investigation should be done to check the records where `hospital_death <> icu_death`.

These examples show some possibilities on how to find out more about records based on the some conditions explored in the previous sections, such as outlier values, etc.

Table 20.6 Zero heart rate apache

Hospital Death	Hospital Los Days	ICU Death	ICU Los Days	Heart Rate Apache
1	2.967361	1	.01805554	0
1	.9826389	1	.69791669	0
1	2.3472221	1	.4375	0
0	0	0	1	0
0	0	0	1	0
1	0	1	1	0
1	1	0	8	0
0	1	0	18	0
1	2	1	1	0
1	0	1	3	0
0	−8	0	7	0
0	0	0	2	0
0	1	0	278	0

Table 20.7 Snapshot of null heart rate apache

Hospital Death	Hospital Los Days	ICU Death	ICU Los Days	Heart Rate Apache
0	27	0	1	
0	4	0	2	
0	6	0	3	
0	7	0	4	
0	9	0	3	
1	187	0	19	
0	10	0	6	
0	32	0	15	
0	24	0	5	
0	6	0	3	
0	22	0	2	
1	3	1	3	
1	21	1	8	
0	13	0	5	
0	11	0	2	
0	86	0	3	
0	14	0	9	

20.2.5 Explore Multiple Attributes

The previous sections explored the attributes individually. However, we can also explore multiple attributes and try to find some correlations. Let's explore another attribute, called the `resprate_apache` attribute. The normal respiratory rate or (`resprate_apache`) is between 12 and 20.

First, let's examine the number of entries in this attribute. Based on the following SQL commands, there are 416561 not null entries and 65864 null entries in this attribute.

```
select count(resprate_apache)
from Emergency_Patient;

416561

select count(*)
from Emergency_Patient
where resprate_apache is null;

65864
```

The next step is to plot a boxplot. The SQL command below calculates the quartiles and median. The IQR is calculated as follows: $1.5 \times (29 - 10) = 28.5$. The number of outliers above the maximum marker is 3318 records. Figure 20.13 shows the boxplot for the two attributes: heart_rate_apache and resprate_apache.

```
select
  max(Q3) as Q3,
  max(Median) as Median,
  max(Q1) as Q1
from (
  select
```

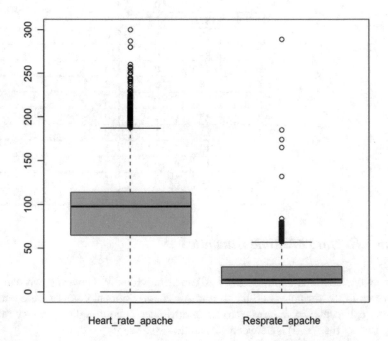

Fig. 20.13 Boxplots of the heart_rate_apache and resprate_apache attributes

```
    resprate_apache,
    percentile_cont(0.25) within group
      (order by resprate_apache) over() as Q1,
    median(resprate_apache) over() as Median,
    percentile_cont(0.75) within group
      (order by resprate_apache) over() as Q3
  from Emergency_Patient);
```

```
  Q3      Median     Q1
  29       14        10
```

```
select count(resprate_apache)
from Emergency_Patient
where resprate_apache > (29+28.5);
```

```
3318
```

The maximum makers for `resprate_apache` is (29+28.5) and for `heart_rate_ apache` is (114+73.5). The following SQL command retrieves the number of records that fall into the outlier category of both attributes. There are 29 records. By looking at these 29 records, perhaps nothing is conclusive.

```
select count(*)
from Emergency_Patient
where resprate_apache > (29+28.5)
and heart_rate_apache > (114+73.5);
```

```
29
```

Further investigation to find the relationship between these two attributes could be undertaken. For example, we could look at the range of the first and the third quartile of `resprate_apache`, which is between 10 and 29, and find out the median of `heart_rate_apache` and compare this with the median of `heart_rate_apache` for the normal range of `resprate_apache`, as defined by the medical authorities, which is between 12 and 20.

```
select median(heart_rate_apache)
from Emergency_Patient
where resprate_apache > 10
and resprate_apache < 29;
```

```
95
```

```
select median(heart_rate_apache)
from Emergency_Patient
where resprate_apache > 12
and resprate_apache < 20;
```

```
92
```

Q3 of `resprate_apache` is 29. We would like to find out the median of `heart_rate_apache` when `resprate_apache` is increased from 29 to 40 and later to 60 and compare the results. The following SQL commands show that there are not many implications on `heart_rate_apache` when `resprate_apache` is increased. Hence, further analysis must be undertaken.

```
select median(heart_rate_apache)
from Emergency_Patient
where resprate_apache > 29
and resprate_apache < 40;

110

select median(heart_rate_apache)
from Emergency_Patient
where resprate_apache > 41
and resprate_apache < 60;

112

select median(heart_rate_apache)
from Emergency_Patient
where resprate_apache > 29
and resprate_apache < 60;

110
```

This section gives a short illustration on the complexity of data exploration of big data. The correlation between attributes must be further investigated, preferably working with domain experts to find more meaningful knowledge about the data. Since there are hundreds of attributes with measurements, choosing the right attributes to look at is an important guess.

20.3   Summary

This chapter gives the context of data warehousing, in terms of the pre- and post-activities. Understanding pre- and post-data warehousing is important in the whole context of data science and data analytics.

1. Pre-Data Warehousing
   Finding dirty data before creating a data warehouse is of the utmost importance. As the famous quote says, "Garbage In Garbage Out", so finding dirty data and cleaning them is crucial as the quality of the data warehouse depends very much on the quality of the operational databases, which are the raw materials of the data warehouse.
   Most of the dirty data is due to relationships between records, between attributes and between tables. Isolated dirty data not associated with relationships might also occur. Finding dirty data in many instances is not straightforward and requires some detective work. When errors are found, they need to be corrected. Finding dirty data is one thing; correcting them is another.
2. Post-Data Warehousing
   The relationship between data warehousing and data science/analytics is also important. Similar to pre-data warehousing, where understanding the data in the operational databases is one key aspect, in post-data warehousing, understanding

the data in the data warehouse is another key aspect and is an important factor in the success of data analytics. Hence, a thorough exploration of the data in data warehousing plays an important role in data analytics.

20.4 Exercises

20.1 The following are the tables in the operational database of the university student enrolment system:

1. **Subject** (<u>UnitCode</u>, UnitTitle, UnitOverview, CreditPoints, UnitLocation, Unit-Mode)
2. **Subject_Offering** (<u>UnitCode</u>, UnitTitle, <u>UnitLocation</u>, UnitMode, <u>Semester Year</u>, DayOrNight, ChiefExaminer, ExamLocation)
3. **Course** (<u>CourseCode</u>, CourseName, CourseLocation, CourseMode)
4. **Student** (<u>StudentID</u>, Surname, Title, GivenNames, Address1, Address2, Address3, Address4, Address5, Address6, Address7, Address8, Address9, Address10, Address11, EmailAddress, VisaCategory)
5. **Student_Course** (<u>StudentID, CourseCode, StartSemesterYear</u>, EnrolmentStatus, LastSemesterYear, Completed, WAM)
6. **Student_WES** (<u>StudentID, UnitCode, SemesterYear</u>, UnitEnrolmentStatus, Mark, Grade)
7. **Enrolment** (<u>StudentID</u>, Surname, Title, GivenNames, Mark, Grade, Teach-Period, <u>UnitCode</u>, UnitLocation, UnitMode, DayOrEvening, <u>SemesterYear</u>, StudentStatus, DiscontinuedDate, ExamLocation, CourseCode, CourseLocation, CourseMode, CourseAttendedType, CourseStatus, CorrespondenceCategory, Address1, Address2, Address3, Address4, Address5, Address6, Address7, Address8, Address9, Address10, Address11, EmailAddress, VisaCategory, UnitTitle)
8. **Moodle_Activities** (<u>StudentID, UnitCode</u>, DateTime, URL)

Before creating a data warehouse from this operational database, we need to explore these tables and see if there are any errors, inconsistencies, etc. Your tasks are to explore the tables in the operational database and write the SQL commands to find the data errors in the tables.

(a) Find students who enrolled in an incorrect Course Code.
(b) Find incorrect Grades in the Enrolment table.
(c) Find out if the Mark in the Enrolment table use different precision.
(d) Find out if there are duplicate records in the Moodle Activities table.
(e) Find out if there is any mismatch between Unit Title and Unit Code in the Enrolment table.
(f) Find students who have passed a unit but have re-enrolled in the same unit again. Check the Student WES table.

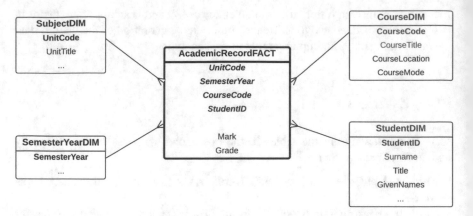

Fig. 20.14   Academic Record star schema

(g) Check if the Course Location in the Course table is the same as in the Enrolment table.
(h) Check if there are records where Address2 is not null but Address1 is null.
(i) Check if the Visa Category uses uniform coding or standards.
(j) Check if there are any null Email Addresses.

20.2   Figure 20.14 shows an Academic Record star schema based on the operational database in Question 1. This star schema is built for one faculty only. So all the Subjects and Courses are from one faculty. This data warehouse is used to analyse the academic performance of students and subject fail rates and grade distributions in that faculty.

The star schema contains four dimensions: Subject, SemesterYear, Course and Student. The fact measures of the fact are Mark and Grade. Your tasks are to explore this data warehouse. Write the SQL commands to answer the following queries:

(a) Find the grade distribution of a subject in a semester. This information is useful to see student performance in the subject.
(b) Find the Top-5 subjects with the highest failure rates in a semester. This information is useful for the faculty to find out which subjects have a high failure rate so an investigation can be undertaken to determine the reasons for this.
(c) Find the Top-5 subjects with the highest HD and D rates (combined) in a semester. (Notes: HD or High Distinction is the highest grade, equivalent to an A Grade in some systems, whereas D or Distinction is similar to a B Grade.) This information is useful for the faculty to find out which subjects appear to be "easy", as most students in these subjects performed well.
(d) Find students who achieved the highest mark in each subject in a semester. The top student in each subject will receive a commendation letter from the faculty.

(e) Find the mean and standard deviation of the marks in a subject in a semester. This information is needed to build a histogram of marks for the subject. Also find the number of students who fall within minus/plus 1 standard deviation.

(f) Find the median, Q1 and Q3, and calculate IQR to find the maximum and minimum markers. This information is needed to build a boxplot.

20.5 Further Readings

The main topic of this chapter is data exploration in pre and post data warehousing, including data preparation, data cleaning in pre-data warehousing and basic statistical analysis in post-data warehousing. There are excellent references on data analytics which cover data preprocessing, such as [1] and [2]. Data cleaning is an important topic in itself and is widely discussed in the following books: [3–6].

SQL has been used extensively in pre- and post-data warehousing. A good reference for SQL for data cleaning, wrangling and analytics is [7]. General and more thorough discussions on SQL itself can be found in the following books: [8–11].

Basic statistical analysis for exploring and analysing data from the data warehouse is usually covered in business statistics textbooks, such as [12–14]. There are plenty of textbooks on data mining which also cover data preparation, data cleaning and data preprocessing, as well as data exploration. Further references on the data mining and data science books are as follows: [15–20].

References

1. J. Moreira, A. Carvalho, T. Horvath, *A General Introduction to Data Analytics* (Wiley, London, 2018)
2. T.A. Runkler, *Data Analytics: Models and Algorithms for Intelligent Data Analysis* (Springer, Berlin, 2012)
3. J.W. Osborne, *Best Practices in Data Cleaning: A Complete Guide to Everything You Need to Do Before and After Collecting Your Data* (SAGE Publications, New York, 2013)
4. I.F. Ilyas, X. Chu, *Data Cleaning* (Association for Computing Machinery and Morgan and Claypool Publishers, California, 2019)
5. T. Dasu, T. Johnson, Exploratory data mining and data cleaning, in *Wiley Series in Probability and Statistics* (Wiley, London, 2003)
6. V. Ganti, A. Das Sarma, *Data Cleaning: A Practical Perspective.* Synthesis Lectures on Data Management (Morgan and Claypool Publishers, California, 2013)
7. A. Badia, SQL for data science: data cleaning, wrangling and analytics with relational databases, in *Data-Centric Systems and Applications* (Springer, Berlin, 2020)
8. J. Melton, *Understanding the New SQL: A Complete Guide*, 2nd edn., vol. I (Morgan Kaufmann, Burlington, 2000)
9. C.J. Date, *SQL and Relational Theory—How to Write Accurate SQL Code*, 2nd edn. Theory in practice (O'Reilly, Newton, 2012)
10. A. Beaulieu, *Learning SQL: Master SQL Fundamentals* (O'Reilly Media, Newton, 2009)

11. M.J. Donahoo, G.D. Speegle, *SQL: Practical Guide for Developers*. The Practical Guides (Elsevier, Amsterdam, 2010)
12. D.M. McEvoy, *A Guide to Business Statistics* (Wiley, London, 2018)
13. N. Bajpai, *Business Statistics* (Pearson, London, 2009)
14. M. Berenson, D. Levine, K.A. Szabat, T.C. Krehbiel, *Basic Business Statistics: Concepts and Applications* (Pearson Higher Education AU, London, 2012)
15. I.H. Witten, E. Frank, M.A. Hall, C.J. Pal, Data mining: practical machine learning tools and techniques, in *The Morgan Kaufmann Series in Data Management Systems* (Elsevier, Amsterdam, 2016)
16. J. Han, J. Pei, M. Kamber, Data mining: concepts and techniques, in *The Morgan Kaufmann Series in Data Management Systems* (Elsevier, Amsterdam, 2011)
17. N. Ye, Data mining: theories, algorithms, and examples, in *Human Factors and Ergonomics* (CRC Press, New York, 2013)
18. D.J. Hand, H. Mannila, P. Smyth, Principles of Data Mining, in *A Bradford Book* (Bradford Book, New York, 2001)
19. K.J. Cios, W. Pedrycz, R.W. Swiniarski, L.A. Kurgan, *Data Mining: A Knowledge Discovery Approach* (Springer, New York, 2007)
20. V. Kotu, B. Deshpande, *Data Science: Concepts and Practice* (Elsevier, Amsterdam, 2018)

Chapter 21
Data Analytics for Data Warehousing

Post-Data Warehousing discussed in Chap. 20 focuses on *data exploration* of the Extended Fact Table, which is the preliminary step in data analytics. This chapter extends it to *data analytics*. The main focus of data analytics for data warehousing is the analysis of data which has already been structured in a data warehouse schema, with a particular focus on fact measures, supported by the dimensions. Hence, the main question is how the data in the data warehouse can be used for efficient data analytics.

This chapter firstly discusses the main differences between traditional data mining and data analytics for data warehousing. Although many of the traditional data mining techniques may be adapted to data warehousing, it is important to understand how these techniques use the data that is already specifically structured in data warehousing.

This chapter describes data analytics for data warehousing and how it differs from traditional data mining, as the input data for data analysis is structurally different as data analytics for data warehousing is to analyse the data in the star schema. Therefore, the main difference between traditional data analytics and data warehousing analytics lays in the data requirements.

Three main data analytics techniques are explored in this chapter: (*i*) statistical analysis using regression, (*ii*) clustering analysis and (*iii*) classification using regression trees.

© The Author(s), under exclusive license to Springer Nature Switzerland AG 2021   577
D. Taniar, W. Rahayu, *Data Warehousing and Analytics*, Data-Centric Systems
and Applications, https://doi.org/10.1007/978-3-030-81979-8_21

21.1   Traditional Data Mining Techniques vs. Data Analytics for Data Warehousing

21.1.1   Traditional Data Mining Techniques

Traditional data mining is a collection of techniques to discovery patterns, correlations and knowledge on the given input data. Data mining techniques use intelligent methods applied to the input data to extract data patterns. There are various data mining techniques, including but not limited to association rules, sequential patterns, classification and clustering.

Data mining tasks can be classified into two categories: (*i*) descriptive data mining and (*ii*) predictive data mining. *Descriptive data mining* describes the dataset and presents interesting general properties of the data. This somehow summarises the data in terms of its properties and its correlation with others. For example, within a set of data, some data have common similarities with the members in that group, and hence the data is grouped into one cluster. Another example is that when certain data exists in a transaction, another type of data would follow.

Predictive data mining builds a prediction model which makes inferences from the available set of data and attempts to predict the behaviour of new datasets. For example, for a class or category, a set of rules is inferred from the available dataset, and when new data arrives, the rules can be applied to this new data to determine to which class or category it belongs. Prediction is made possible because the model consisting of a set of rules is able to predict the behaviour of new information.

Many data mining techniques have been proposed in the academic world, and some of which have been used widely by practitioners in industry. However, the aim of this section is to highlight the differences between data mining techniques (now called the *traditional data mining*) and data analytics specifically for data warehousing.

As mentioned earlier, data analytics for data warehousing uses data from star schemas—a star schema has a rigid structure consisting of facts and dimensions. Traditional data mining, in contrast, does not have this constraint as each data mining algorithm assumes a certain data structure for the algorithm. To emphasise this matter in traditional data mining, we will describe association rules and decision trees.

Association rules discover association relationships or correlations among a set of items. Association analysis is widely used in transaction data analysis, such as a market basket. A typical example of an association rule in a market basket analysis is finding the rule (cereal → milk), indicating that if cereal is bought in a purchase transaction, it is a likely chance that milk will also appear in the same transaction. Association rule mining is one of the most widely used data mining techniques. Since its introduction in the early 1990s through the Apriori algorithm, association rule mining has received huge attention across various research communities. The association rule mining methods aim to discover rules based on the correlation between different attributes/items found in the dataset. To discover such rules,

Table 21.1 Transaction dataset for association rule mining

TransactionID	Items purchased
1	Bread, Jam, Milk
2	Bread, Cereal, Juice, Milk
3	Bread, Cereal, Jam, Juice
4	Bread, Cereal, Juice
5	Coffee, Milk, Oat

Table 21.2 Association rules

Association rules	Confidence
Bread→Cereal	75%
Bread→Juice	75%
Cereal→Bread	100%
Cereal→Juice	100%
Jam→Bread	100%
Juice→Bread	100%
Juice→Cereal	100%
Bread→(Cereal,Juice)	75%
Cereal→(Bread,Juice)	100%
Juice→(Bread,Cereal)	100%
(Bread,Cereal)→Juice	100%
(Bread,Juice)→Cereal	100%
(Cereal,Juice)→Bread	100%

association rule mining algorithms at first capture a set of significant correlations present in a given dataset and then deduce meaningful relationships from these correlations. Since the discovery of such rules is a computationally intensive task, many association rule mining algorithms have been proposed.

Table 21.1 shows an example of a market basket transaction dataset containing five transactions. Each transaction contains a number of items being purchased within a transaction.

Based on the predefined support and confidence thresholds that the user sets, the association rule mining algorithm takes the dataset in Table 21.1 and produces the association rules shown in Table 21.2.

An association rule is expressed in $A \rightarrow B$, indicating that the majority of transactions in the dataset that contain A must also contain B. The confidence level for an association rule is the percentage of transactions satisfying the $A \rightarrow B$ rule. Take the first rule (e.g. Bread→Cereal) as an example. 75% confidence means that three-quarter of transactions that contain Bread also contain Cereal.

Association rules are not solely confined to $A \rightarrow B$, where both items A and B are single items; they can also be a multiple itemset, containing more than one single item. For example, Bread→(Cereal,Juice) and (Bread,Cereal)→Juice contain multiple itemsets. In (Bread,Cereal)→Juice, all the transactions that contain Bread and Cereal also contain Juice; hence, the confidence is 100%.

Table 21.3 Normalised transaction dataset for association rule mining

TransactionID	Product
1	Bread
1	Jam
1	Milk
2	Bread
2	Cereal
2	Juice
2	Milk
3	Bread
3	Cereal
3	Jam
3	Juice
4	Bread
4	Cereal
4	Juice
5	Coffee
5	Milk
5	Oat

From a data warehousing point of view, there are three main issues with the traditional association rules:

1. **Unnormalised data**

   Items Purchased (or List of Products). The second column is an aggregate list of products for each transaction. This is a multivalued attribute. This data structure can easily be normalised by making the second column an atomic attribute. The new structure still contains two columns, namely, TransactionID and Product, but each cell is atomic. A normalised version of the input dataset is shown in Table 21.3. However, this is not a star schema structure. The data seems to be a dimension, but it is not a Product Dimension, as a Product Dimension is usually a list of all available products. The market basket is not a Product Dimension as it has the notion of transactions, where items or products are bought in a transaction.

2. **No numerical fact measure**

   The algorithm does not consider any numerical fact measure. In this market basket example, the number of units for each product being purchased is not considered at all. The association rules are only about product correlation. On the other hand, data warehousing (and star schemas) is about fact measures, which are numerical values; and these are not used in the association rule algorithm.

3. **One dimension**

   The original association rule algorithm considers only one dimension, which is the Product. It does not consider other dimensions, such as Location or Branch, Loyalty Cards, etc. Recent works have focused on multi-dimensional association rules; however association rules generally focus on categorical data, whereas the focus of data warehousing is numerical fact measures.

Another most common data mining technique is classification. *Classification* is the process of assigning new instances (or objects) to predefined categories or classes. There are many different techniques for classification. A *decision tree* is one of the most popular classification techniques. In a decision tree, the objective is to create a set of rules that can be used to differentiate one target class from another. The target class is labeled with categorical values. The data used to build a decision tree, often called the *training dataset*, contains attributes and the corresponding target class. The classifier algorithm then builds a classification model, in this case, a decision tree, based on the training dataset. Once a decision tree model is built, any incoming data (known as the *testing dataset*) use the decision tree to determine the target class for the incoming record.

A training dataset for a decision tree usually consists of several attributes and one class/category attribute. The class/category attribute is used to identify the classes/categories of the records. This class/category attribute is also called the target class attribute. To simplify, the class/category attribute is called the target class, whereas the others are simply known as attributes. The attributes can be categorical or continuous, whereas the target class is a categorical attribute.

Figure 21.4 shows an example of the training dataset for a decision tree. The attributes in this training dataset are categorical attributes, where each attribute has a number of distinct countable values. For example, the possible values for Weather are Fine, Shower and Thunderstorm, whereas the possible values for Temperature are Hot, Mild and Cool. The target class is Yes or No, indicating whether or not to walk, depending on the weather conditions.

A decision tree is a tree consisting of nodes and directed arcs. There is one root node which then branches to lower-level nodes. The lower-level nodes can be non-leaf nodes or leaf nodes. A non-leaf node corresponds to a question or a test on an attribute, whereas an arc or a branch represents the outcome of the test (e.g. Weather = Fine). Hence, a decision tree is literally a flowchart containing if conditions. However, the branches from a non-leaf node can have as many branches as possible according to the possible values of the question at that node. For example, for the Weather leaf node, the branches are the possible outcomes of the Weather, such as Fine, Shower and Thunderstorm.

A leaf node contains a final decision or target class for a branch from the root node to the leaf node. A node is assigned as a leaf node when all the records at the node belong to the same class or the majority of records belong to a class.

Figure 21.1 shows an example of a decision tree structure from the data shown in Table 21.4. Each non-leaf node asks a question stated by the node label, whereas the answers are indicated by the branches that come out from the node. For example, the root node Weather asks if it is Fine, Shower or Thunderstorm. The next question will depend on the answer to the current question. For example, if the Weather is Fine, then the next question is about the Temperature: is it Hot, Mild or Cool?

The leaf node shows the target class or category, which in this case is a binary class: Yes or No, indicating to walk or not to walk, depending on the values from the other attributes (e.g. Weather, Temperature, Time and Day). Using this decision tree, following the left-most branches (e.g. the Weather is Fine, the Temperature

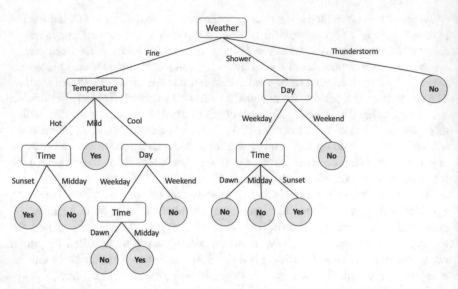

Fig. 21.1 A decision tree

Table 21.4 Decision training dataset

No	Weather	Temperature	Time Time	Day	Walk (Target Class)
1	Fine	Hot	Sunset	Weekday	Yes
2	Fine	Mild	Sunset	Weekend	Yes
3	Shower	Mild	Midday	Weekday	No
4	Thunderstorm	Mild	Sunset	Weekend	No
5	Fine	Mild	Midday	Weekday	Yes
6	Shower	Hot	Sunset	Weekday	Yes
7	Fine	Cool	Dawn	Weekend	No
8	Shower	Mild	Dawn	Weekday	No
9	Thunderstorm	Hot	Midday	Weekday	No
10	Thunderstorm	Cool	Dawn	Weekend	No
11	Fine	Hot	Midday	Weekday	No
12	Thunderstorm	Cool	Midday	Weekday	No
13	Fine	Cool	Midday	Weekday	Yes
14	Shower	Hot	Dawn	Wcckend	No
15	Fine	Cool	Dawn	Weekday	No

is Hot, and the Time is Sunset), the decision is to walk, regardless of the Day. On the right-most branch, if the Weather is Thunderstorm, the answer is definitely a No, regardless of all other criteria. Moreover, some criteria might not be used in a decision tree at all.

The shape of the decision tree depends on which attribute is picked first at the root node, which attributes are picked next, etc. For the same training dataset, it is

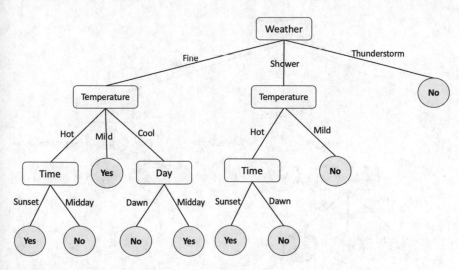

Fig. 21.2 Decision tree: version 2

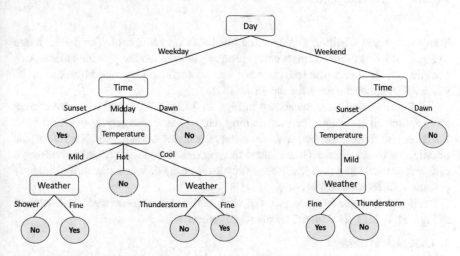

Fig. 21.3 Decision tree: version 3

most likely that there are several possible equivalent decision trees that can look very different, but all of which correctly summarise the decision rule of the training dataset and can be used to correctly classify the training dataset. For example, Figs. 21.2, 21.3, and 21.4 are three other possible decision trees from the same training dataset shown in Table 21.4. Figure 21.2 chooses the order of the attributes in the table when constructing the decision tree; use the Weather attribute first, followed by Temperature and Time. In this case, there is no need to use the Day attribute, because all branches from this decision tree have landed into the target class. Figure 21.3 uses a different order of the attributes: firstly Day and then Time,

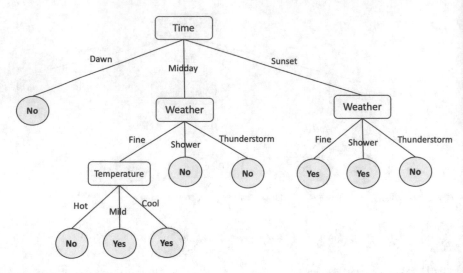

Fig. 21.4 Decision tree: version 4

Temperature and Weather. The decision tree looks very different. The decision tree in Fig. 21.4 looks more compact and is probably better than the previous three. Also note that the final decision tree does not use the attribute Day, and all rules can be generated without the need for the attribute Day.

All four decision trees, as shown in Figs. 21.1, 21.2, 21.3, and 21.4, are correct and capture all the rules for this training dataset. However, some decision trees are taller and more complex, whereas others are leaner and simpler. Therefore, the question as to which attribute is used to represent a node is critical. Choosing a different attribute as a non-leaf node at any level of the decision tree will ultimately produce a different tree structure.

From the data warehousing point of view, traditional decision trees as illustrated in Figs. 21.1, 21.2, 21.3, and 21.4 also face issues:

1. **Categorical data**
   A decision tree uses categorical data, whereas a star schema (and extended Fact Table) focuses on fact measures, which are numerical values. In star schemas, only dimension keys are categorical values.
2. **Target class**
   In a decision tree, the target class is a categorical value. In the context of data warehousing and star schema, it is unclear if the target class was intended to be a fact measure; however, a fact measure needs to be a numerical value, not a categorical value.

All of these issues are related to the training data structure used by the decision tree algorithm. In data warehousing, the data is already structured in a certain format, namely, star schema, and this is not easily adopted by the aforementioned decision tree algorithm. The prediction value (or the target class) should be numerical values, and the attributes used in the decision tree must also use other fact measures, which are also numerical values. Classification algorithms for data warehousing must consider these factors.

21.1.2 Data Analytics Requirements in Data Warehousing

Data analytics in data warehousing must use the data that is already structured and stored in data warehouses. The data structure is a star schema consisting of fact measures and dimension keys. Since the fact measures are the central part of a data warehouse, data analysis in data warehousing is applied to the fact measures, and because the fact measures are numerical values, data analysis in data warehousing is data analysis of numerical values.

This chapter focuses on three data analysis techniques suitable for a data warehousing context. The first is regression or prediction. In time-series data, regression can be used to predict the possible values of future data by analysing their past trends. It contains two variables, the timestamp and the value. After observing the value over a period of time, the regression model can be used to estimate future values. If the input dataset is a star schema which consists of fact measures and dimension keys, in a time-series model, one of the dimensions that will be used in the regression is a Time Dimension (or Timestamp) and one fact measure that contains the values being observed. Figure 21.5 illustrates a star schema containing three fact measures and four dimension keys, one being Timestamp. A regression uses Timestamp and one of the fact measures (e.g. fact measure 1). By observing all the values in fact measure 1 against the Timestamp, the regression model predicts the future values of fact measure 1. The arrows in Fig. 21.5 illustrate the timeline and the values of fact measure 1 against the timeline. As can be seen from this

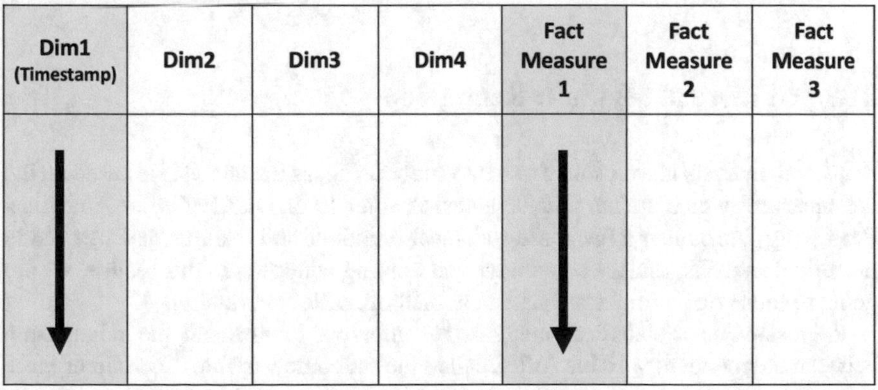

Dim1 (Timestamp)	Dim2	Dim3	Dim4	Fact Measure 1	Fact Measure 2	Fact Measure 3
↓				↓		

Fig. 21.5 Fact Table and time-series regression

Dim1	Dim2	Dim3	Dim4	Fact Measure 1	Fact Measure 2	Fact Measure 3

Fig. 21.6 Data analysis of fact measures

illustration, the input dataset is a Fact Table, and the analysis takes this Fact Table and produces a regression model.

The second and third data analytics for data warehousing (namely, clustering and classification) are related to the analysis of fact measures. Figure 21.6 shows a typical Fact Table with a number of fact measures and dimension keys. The clustering of the Fact Table records is based on their fact measure values. Classification, including the target class, also uses fact measures. The focus is clearly on numerical values stored in the fact measures of the Fact Table. Compared with the decision tree described above, the aforementioned decision tree uses categorical data, whereas the decision tree for data warehousing uses fact measure numerical values, including the target class.

In summary, data analytics for data warehousing focuses on fact measure numerical values. The data analysis techniques to be discussed are (i) regression, (ii) clustering and (iii) classification.

21.2   Statistical Method: Regression

Statistical analysis is well suited to data warehousing, as the analysis is based on the fact measures which are numerical values. Chapter 19 on OLAP (Online Analytical Processing) introduces a few basic statistical functions and features that are used in numerical analysis, such as cumulative and running windows. In this section, we are going to focus on an important statistical method, called *regression*.

Regression is a statistical method that attempts to estimate the relationship between a dependent variable (often called the outcome variable) and one or more independent variables (often called predictors, covariates or features). For example, the outcome variable is Total Sales and the predictor is the Time. Since Fact Tables, especially fact measures, are often time-based (Time Dimension), a regression model can be used to predict future values of the observed fact measure. The regression model tries to fit the model into the dataset, and this model can then be used to predict incoming data.

21.2.1   Simple Linear Regression

Simple linear regression is the process of finding the equation of a line that is the best fit for a series of data. To find an equation of a straight line, we only need to calculate two values, namely, (i) slope and (ii) intercept.

Slope is often called the "gradient", which indicates how steep the line is. In terms of a mathematical equation, slope is a replication factor of the x-axis giving the value of the y-axis, which is shown as follows:

$$y = slope \times x \tag{21.1}$$

Figure 21.7 illustrates three different slopes: *slope*=1 produces a perfect diagonal line, *slope*=2 doubles the value of the y-axis for every x-axis value, and *slope*=0.5 gives half of the value of x-axis to y-axis. The higher the slope value, the more vertical the line will be.

Intercept indicates the height of the line. It also indicates the intercepting point with the y-axis. The three examples in Fig. 21.7 intercept with the 0 point in y-axis, and hence, the intercept is 0. Figure 21.8 shows one line starting at 0 and the other starting at 2; the latter indicates that the intercept is equal to 2. The following

Fig. 21.7 Three different slope values

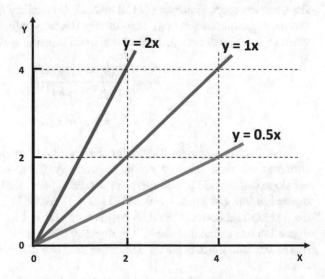

Fig. 21.8 Intercepts

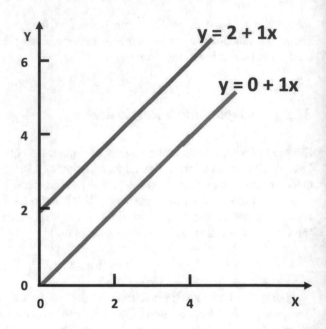

mathematical equation covers both intercept and slope:

$$y = intercept + (slope \times x) \tag{21.2}$$

A simple linear regression method is to construct a straight line to represent the data. Given the dataset in x and y, the regression model needs to calculate the slope and intercept. Equations (21.3) and (21.4) calculate the slope (denoted by b_1) and intercept (denoted by b_0), respectively (Note: $\bar{x}$ refers to the average x value, whereas x_i refers to each individual x value. The same applies for y).

$$b_1 = \frac{\sum (x_i - \bar{x})(y_i - \bar{y})}{\sum (x_i - \bar{x})^2} \tag{21.3}$$

$$b_0 = \bar{y} - b_1 \bar{x} \tag{21.4}$$

Table 21.5 is a sample dataset for the regression model. The Time and Value attributes are the x-axis and y-axis, respectively. Using Eqs. (21.3) and (21.4), we get slope = 8.39285714 and intercept = 6.28571429. Hence, using Eq. (21.2), the regression line becomes $y = 6.28571429 + (8.39285714 \times x)$. Figure 21.9 shows the results. The dots are the data points from Table 21.5, whereas the line is the regression line calculated above. The regression line shows the trend of the data and it tries to fit into the data points. This regression line can then be used as a prediction model.

The regression model can also be computed using SQL commands. Firstly, we need to calculate the averages (e.g. $\bar{x}$ and $\bar{y}$). Then, the slope and intercept can be

Table 21.5 A sample dataset
for regression

Time	Value
1	11
2	27
3	34
4	38
5	45
6	61
7	63

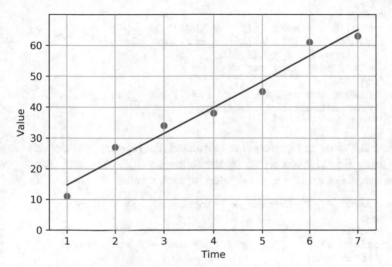

Fig. 21.9 Linear regression represented as a line

calculated using Eqs. (21.3) and (21.4). Finally, the regression model is calculated using Eq. (21.2).

The averages can simply be calculated using the `avg` function, which is as follows:

```
select avg(x) as x_bar, avg(y) as y_bar
from dataset;
```

Next is to calculate the slope, which needs to make use of $\bar{x}$ and $\bar{y}$. The following is the SQL command to calculate the slope, based on Eq. (21.3). This SQL command basically "attaches" $\bar{x}$ and $\bar{y}$ to each record of x and y, where the "attachment" is carried out using a Cartesian product between the original dataset and the $\bar{x}$ and $\bar{y}$. Attaching $\bar{x}$ and $\bar{y}$ is necessary because for each x and y, we need to subtract them from $\bar{x}$ and $\bar{y}$.

```
select
  sum((x - x_bar) * (y - y_bar)) /
  sum((x - x_bar) * (x - x_bar)) as slope
from (
```

```
     select avg(x) as x_bar, avg(y) as y_bar
     from dataset) av, dataset;
```

Once the slope is calculated, we can then calculate the intercept, using Eq. (21.4), which needs to use the slope, as well as $\bar{x}$ and $\bar{y}$. The following SQL command combines the slope and intercept calculation in one SQL. The most inner subquery is to calculate $\bar{x}$ and $\bar{y}$, whereas the next inner query is to calculate the slope.

```
select slope, y_bar_max - (slope * x_bar_max) as intercept
from (
  select
    sum((x - x_bar) * (y - y_bar)) /
    sum((x - x_bar) * (x - x_bar)) as slope,
    max(x_bar) as x_bar_max,
    max(y_bar) as y_bar_max
  from (
    select avg(x) as x_bar, avg(y) as y_bar
    from dataset) av, dataset
);
```

Once the slope and intercept are calculated, the regression model can be calculated using Eq. (21.2), as shown by the following SQL command. The regression model calculates the new *y_pred* value for each *x* value.

```
select x, y, (intercept + (slope * x)) as y_pred
from
  dataset,
  (select slope, y_bar_max - (slope * x_bar_max) as
    intercept
  from (
    select
      sum((x - x_bar) * (y - y_bar)) /
      sum((x - x_bar) * (x - x_bar)) as slope,
      max(x_bar) as x_bar_max,
      max(y_bar) as y_bar_max
    from (
      select avg(x) as x_bar, avg(y) as y_bar
      from dataset) av, dataset
  ))
order by x;
```

The above `select` results can be stored in a table, using the `create table` statement.

```
create table linear_regression as
select x, y, (intercept + (slope * x)) as y_pred
from
  dataset,
  (select slope, y_bar_max - (slope * x_bar_max) as
    intercept
  from (
    select
      sum((x - x_bar) * (y - y_bar)) /
```

```
            sum((x - x_bar) * (x - x_bar)) as slope,
          max(x_bar) as x_bar_max,
          max(y_bar) as y_bar_max
       from (
          select avg(x) as x_bar, avg(y) as y_bar
          from dataset) av, dataset
    ))
  order by x;
```

The regression model calculated using the above SQL commands contains discrete values of the regression. For each x value, there are y and y_pred values; the former is from the original dataset, whereas the latter is the value of the regression model. Figure 21.10 shows the regression model represented as discrete values instead of a line. The y_pred value is shown as a data point label.

A simple linear regression model can be well understood if it is applied to a real dataset. Figure 21.11 shows a typical star schema that stores information about the population of cities. This star schema has various dimensions, including Year, City, Country of Birth, Language, Gender and Religion. Potentially, more dimensions will be included. The fact is Population, which is the number of people.

For the regression model, only two attributes are used, which are Year and Population. The Year is from the Year Dimension and the Population is from the fact. In this example, we focus on two cities in Australia, Melbourne and Sydney, and the population data is based on the last 70 years of data from 1950 to 2020. Figure 21.12 shows the regression model for population in both cities. Note the population trend increases in both cities. From the graphs, we see that there was a significant population increase in both cities around 1970. The increase persisted in Sydney until around the year 2000, whereas in Melbourne, the increase was only lasted until around the year 1990. Both cities are now increasing in population, but

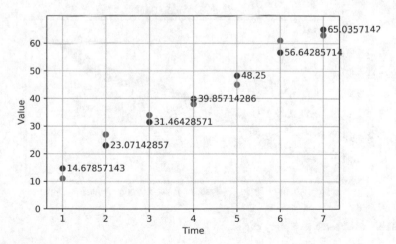

Fig. 21.10 Linear regression represented as discrete values

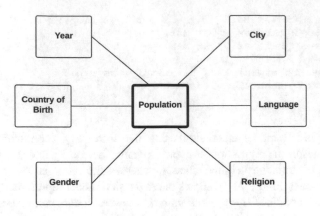

Fig. 21.11 Population star schema

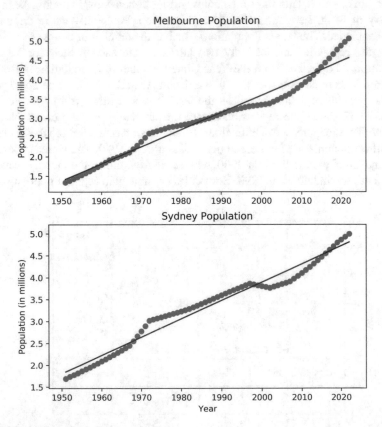

Fig. 21.12 Melbourne vs. Sydney population

Melbourne seems to be increasing at a faster rate than Sydney. In summary, a simple linear regression model can be used to show the trend of the data.

21.2.2   Polynomial Regression

Although simple linear regression can be used to show the global trend of the data, it requires the relation between dependent and independent variables (e.g. y-axis and x-axis) to be linear. This means that the data globally shows a linear behaviour, whether it is going up or down, or simply static. The regression model is then represented as a straight line.

What if the data distribution is more complex and may not be linear? Can the linear models be used to fit non-linear data? How can we generate a curve that best captures the data that may not be linear? *Polynomial regression* might be the answer to these questions.

The equation for polynomial regression is shown in Eq. (21.5). The degree is indicated by n. When $n=1$, it becomes simple linear regression as shown previously in Eq. (21.2), which is $y = b_0 + (b_1 \times x)$. The calculation of b_2, b_3, ..., b_n is out of the scope of this book. There are plenty of tools and libraries that plot polynomial regression by specifying the degree of the regression (e.g. the value of n in Eq. (21.5)).

$$y = b_0 + (b_1 \times x) + (b_2 \times x^2) + (b_3 \times x^3) + ... + (b_n \times x^n) \qquad (21.5)$$

Figure 21.13 shows a number of polynomial regressions of various degrees. Since the dataset shows a steep jump early in the series (e.g. x-axis = 5), the entire dataset

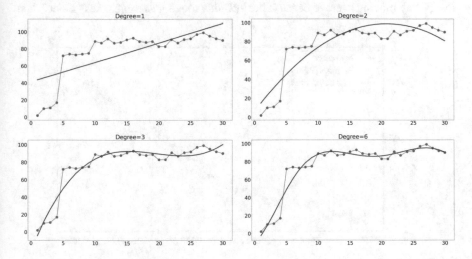

Fig. 21.13 Polynomial regression

does not show a simple linear correlation between the two axes. Therefore, simple linear regression (e.g. degree=1) may not fit in well with this dataset. When the degree is increased to 2, the curve fits in better with the dataset, as it shows a general trend when data jumps in the beginning of the series. Higher degrees, such as degree=3 and 6, fit the regression curve with more detail to the dataset. However, a higher degree raises the complexity and may also overfit the model to the data. Nevertheless, having an option to have a high degree of regression allows the model to better fit into the data which does not show linear behaviour.

21.2.3   Rolling Windows vs. Regression

The polynomial regression examples from the previous section seem to indicate that the polynomial regression model is able to fit the non-linear data better, as the curve of the polynomial regression is closer to the original data. The polynomial regression curves seem to smooth out the data points.

In Chap. 19 on OLAP, moving windows or rolling windows can be used to smooth out data spikes. It will be interesting to compare moving windows (rolling windows) and polynomial regression, especially in the context of smoothing out data points.

Figure 21.14 shows Microsoft stock from 2015 to 2020. The graph also shows a 30-day rolling average (30-day moving average) and a 120-day rolling average (120-day moving average). It is clear that the 30-day moving average is able to smooth the original data; the line that represents the 30-day moving average is also very close to the original data. The 120-day moving average is coarser, as the average takes the data of 120 days, but it is smoother than the 30-day moving average. The line for the 120-day moving average is generally lower than the original data and the 30-day moving average because the 120-day moving average takes "older" data

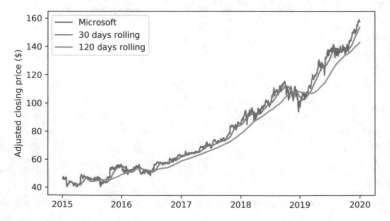

Fig. 21.14 Rolling windows model

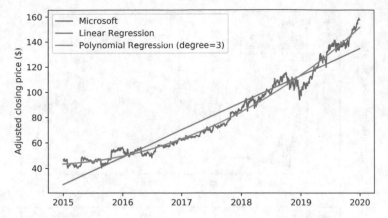

Fig. 21.15 Regression model

(e.g. up to 4 months old), and since older data is lower than newer data, the 120-day moving average is lower than the original data and the 30-day moving window. Nevertheless, both moving window methods are able to smooth the original data points.

Figure 21.15 shows the regression model of the same data as the previous figure. The linear regression model shows a straight line that represents the increasing trend of the Microsoft stock data. This means that the stock increased steadily over the last 5 years (e.g.2015–2020). The polynomial regression (in this example, degree=3) does not smooth out the data regardless of the increase of the degree in the polynomial regression model, like the moving average or rolling windows. This is because this particular dataset has a linear behaviour and, consequently, using polynomial regression will not give much insight into the data. If smoothing is required, then rolling windows will be more appropriate, and calculating moving windows will be much faster than polynomial regression.

Figure 21.16 illustrates an example where polynomial regression would be suitable. This is also stock data, but in this example, it is an airline's stock data, namely, AirAsia. The first figure shows the original data, together with 30-day rolling windows and 120-day rolling windows. Note that both rolling window methods are able to smooth out the original data. Notably, the 120-day rolling window is much smoother than the 30-day rolling window.

In contrast, the second figure shows the regression models. The linear regression shows a straight but slightly increasing line, whereas the polynomial regression model (in this case, degree=3) shows the upward and downward trend of this particular stock.

Comparing the rolling windows and the polynomial regression for this particular example, clearly the polynomial regression model shows the trend of this stock much more clearly. For the linear regression model, the main aim of linear regression is to show the global trend in the long run, without paying much attention to the local details and deviations. On the other hand, rolling mean or rolling

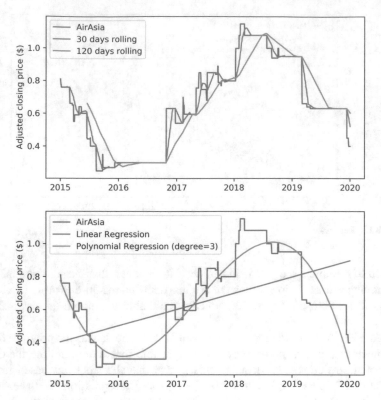

Fig. 21.16 Rolling windows vs. regression

window is used to smooth out the graph due to local fluctuations. The main aim of polynomial regression is not to smooth out the data like rolling mean but to show the global trend of the data but in a strict sense as in linear regression; hence, polynomial regression is suitable for non-linear data.

21.2.4   Non-Time-Series Regression

Regression analysis is primarily used for two conceptually distinct purposes. First, regression analysis is widely used for prediction and forecasting, where its use has substantial overlap with the field of machine learning. Second, in some situations, regression analysis can be used to infer causal relationships between the independent and dependent variables (e.g. y-axis and x-axis).

Regression is for predictive analysis, but prediction does not necessarily use time-series data. Both independent and dependent variables can be non-temporal,

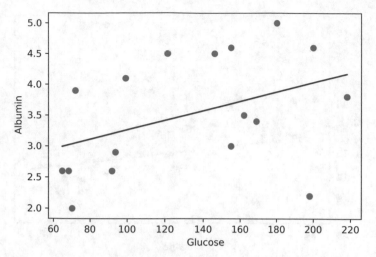

Fig. 21.17 Linear regression: Glucose-Albumin

and hence, a casual relationship between the independent and dependent variables is not a temporal-based relationship.

If prediction is not temporal, how would prediction be used in non-time-series regression? A regression model is built using a training dataset. Once the regression model is built, the testing data can then be used to check the accuracy of the prediction by calculating the residual values between the testing data and the regression model; the lower the residual value, the higher the prediction accuracy. Hence, it is not predicting the future temporal data but evaluating the incoming data to see if they match accurately with the regression curve.

Figure 21.17 takes a small sample of data from the Emergency Patient Extended Fact Table from Chap. 20 on Pre-and Post-Data Warehousing. Only two fact measures, namely, Glucose and Albumin, are taken for this example. The linear regression model is developed. The regression line shows an increasing trend of Glucose-Albumin. To test this regression model, future records consisting of these two fact measures will be compared against the regression line, and the prediction accuracy will be calculated.

Another example is shown in Fig. 21.18. This consists of two other fact measures, namely, Age and WBC (White Blood Cell). The regression line shows a decreasing trend.

Both of these examples illustrate the regression model applied to non-time-series data, where the regression line will be used to predict the future data. Ideally, the future data will arrive near and around the line, that is, when the prediction has a high accuracy. When the incoming data is a distance from the prediction line, the accuracy of the prediction is considered to be low.

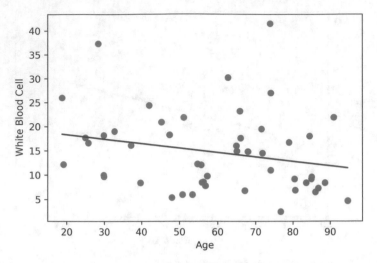

Fig. 21.18   Linear regression: Age-WBC (White Blood Cell)

21.3   Clustering Analysis

Clustering is a data mining technique to find groups or clusters in data. Members or objects within a cluster are considered to be closer or similar. This also implies that an object is closer to another object within the same cluster than to an object in a different cluster. The group does not have a category label that tags the cluster with prior identifiers. Hence, clustering is unsupervised learning, where the target label of training data is unknown. The clustering algorithm tries to form groups or clusters from the data characteristics, not based on cluster labels.

One example of clustering analysis is to cluster customers according to their buying behaviours. The clustering process forms groups of customers, where customers within a cluster (or a group) have similar buying habits. It does not matter what the cluster or the group is. The most important thing is the membership of the cluster. After obtaining the clusters and understanding the members of each cluster, we may identify each cluster with a label. But this is not the point of clustering. Clustering deals with cluster formation using the characteristics (or attributes) of its potential members.

It is easier to understand the clustering results if the clusters are visually presented. Figure 21.19 illustrates the clusters. In this example, each record is presented by two attributes, Age and Salary. From this figure, it is clear that the records are grouped into three clusters.

One key measure of clustering is the *similarity measure* between objects. The similarity is expressed by a similarity function, which is a metric to measure how similar two objects are. The opposite of a similarity function is a distance function,

Fig. 21.19 Clustering

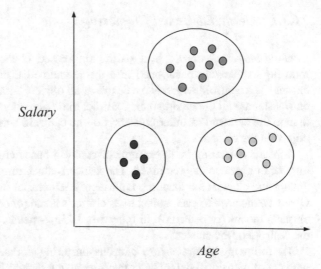

which is used to measure the distance between two data objects. The further the distance, the greater is the difference between the two data objects. Therefore, the distance function is exactly the opposite of the similarity function, although both of them may be used for the same purpose: to measure two data objects in terms of their suitability for a cluster. Data objects within one cluster should be as similar as possible, compared with data objects from a different cluster. Therefore, the aim of a clustering algorithm is to ensure that the intra-cluster similarity is high and the inter-cluster similarity is low.

One of the most common distance measures is the *Euclidean distance*. Equation (21.6) calculates the distance between two data points x_i and x_j using Euclidean distance. The two data points x_i and x_j have h dimensions. The distance between the two data points in each dimension is calculated by subtraction. Euclidean distance is good for attributes that are continuous, such as numerical fact measures. For other types of data, other problem-specific measures may be used.

$$dist(x_i, x_j) = \sqrt{\sum_{k=1}^{h}(x_{ik} - x_{jk})^2} \qquad (21.6)$$

There are many types of clustering methods, but two of them are widely used. They are (i) centroid-based clustering and (ii) density-based clustering. In the centroid-based clustering, the number of desired clusters is predefined, and hence, the clustering process is to process the objects and to divide the objects into the predefined number of clusters. Density-based clustering, in contrast, does not require the number of clusters to be predefined as the clustering process will determine the ideal number of clusters.

21.3.1   Centroid-Based Clustering

Centroid-based clustering is a partitioning-based clustering where objects are mutually exclusively partitioned into the predefined k number of clusters. Each cluster has a centroid which is the centre of all objects within the cluster. Objects in one cluster are all drawn into their centroid; these objects are closer to their centroid than to the centroid of other clusters. So, the centroid acts like the gravity pull of objects in the cluster.

k-means clustering is an example of centroid-based clustering. In k-means, the number of cluster k is predefined. The k-means clustering algorithm processes the dataset and clusters the data into k clusters. Each cluster has a centroid (or means), where the distance of each object to its cluster's mean (or centroid) is closer to other cluster's means (or centroid). In other words, in k-means, the mean of a cluster is the centroid of the cluster.

The following illustrates how k-means clustering works. For simplicity, let's start with one-dimensional data. Using the Emergency Patient ICU data, the attribute to be used is Glucose. Hence, the data is a fact measure. So, clustering in the data warehousing uses fact measures. In this case, only one fact measure is used. Because only one fact measure is used, the data is said to be one-dimensional data. This is not to be confused with dimensions in the star schema.

Actually, two attributes from the Fact Table are used: one is the fact measure, which is Glucose in this example, and the second is the identifier of the object itself. So, each Glucose value to be processed has an ID, which is the recordID or the patientID. Since we are only concerned with the fact measure, as the clustering is to cluster the Glucose value, the data are still known as one-dimensional data and not two-dimensional data. If the data points are plotted as a one-dimensional graph, the axis label is the fact measure label (e.g. Glucose), and the axis value is the fact measure value, which is a numerical value.

Given the following dataset D (e.g. Glucose), which consists of 17 patient records:

D = {162, 93, 68, 154, 121, 198, 99, 180, 169, 64, 72, 154, 218, 145, 70, 91, 200}

Suppose the predefined k=3. The initial step is to choose three data points from the existing dataset D to become the initial centroids (or means) of the three clusters. Suppose we choose the first, the middle and the last data points from D, which are 162, 169, and 200; hence, $m_1 = 162$, $m_2 = 169$ and $m_3 = 200$.

The k-means algorithm consists of a number of repetitions. In each repetition, the new centroids or means are calculated. Because of the new means, the composition of the previous clusters might change. When they change, the new means will need to be recalculated, and the same process repeats. The algorithm terminates when there is no change to the cluster compositions. The final cluster compositions become the final outcome of the algorithm.

Following up the example above, the k-means process is as follows:

- Step 1:
  - The initial means are $m_1 = 162$, $m_2 = 169$ and $m_3 = 200$.
  - The mid-point between m_1 and m_2 is 165.5. This means any data points less than 165.5 belong to Cluster 1, which is {64, 68, 70, 72, 91, 93, 99, 121, 145, 154, 154, 162}. Cluster 1 has 12 data points with a total of 1293.
  - The mid-point between m_2 and m_3 is 184.5. Hence, Cluster 2 has data between 165.5 and 184.5. Cluster 2 has two data points only, which are {169, 180}, with a total of 349.
  - Cluster 3 has three data points above 184.5, which is {198, 200, 218}, with a total of 616.

- Step 2:
  - The new means of the three clusters from Step 1 are calculated, and they are $m_1 = 107.75$, $m_2 = 174.5$ and $m_3 = 205.33$. Because of the new means, the cluster compositions changed.
  - The mid-point between m_1 and m_2 is now 141.13. Hence, Cluster 1 has all data points below this mid-point. Cluster 1 now has eight data points = {64, 68, 70, 72, 91, 93, 99, 121} with a total of 678.
  - The mid-point between m_2 and m_3 is now 189.92. Cluster 2 now ranges from 107.75 to 189.92. Six data points fall into this category. Cluster 2 now has {145, 154, 154, 162, 169, 180} with a total of 964.
  - Cluster 3 still has three data points which are above 189.92, which are {198, 200, 218}. The members of Cluster 3 are unchanged although the centroid of this cluster has changed slightly from 200 to 205.33.

- Step 3:
  - The means of the three clusters are $m_1 = 84.75$, $m_2 = 160.67$ and $m_3 = 205.33$. Note that m_3 did not change, because the members of Cluster 3 have not changed.
  - The mid-point between m_1 and m_2 is 122.71. Cluster 1 has eight data points, which are unchanged from Step 2. The members of Cluster 1 are {64, 68, 70, 72, 91, 93, 99, 121}.
  - The mid-point between m_2 and m_3 is 183. Cluster 2 is also unchanged with six data points {145, 154, 154, 162, 169, 180}.
  - Cluster 3 has not changed since the last iteration.

Since the members of each cluster are not changed in Step 3, the clustering process terminates with three clusters: Cluster 1 = {64, 68, 70, 72, 91, 93, 99, 121}, Cluster 2 = {145, 154, 154, 162, 169, 180}, and Cluster 3 = {198, 200, 218}. Figure 21.20 shows a graphical representation of the data points and their cluster compositions.

Clustering two-dimensional (or higher-dimensional) data is slightly more complex than that of one-dimensional data, although the concept of using a centroid in

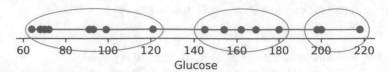

Fig. 21.20 Clustering of one-dimensional data

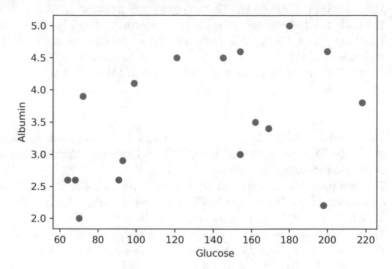

Fig. 21.21 Two-dimensional data

each cluster is the same. The following illustrates the clustering of two-dimensional data. Figure 21.21 shows two-dimensional data consisting of Glucose (x-axis) and Albumin (y-axis). These are two fact measures from the Emergency Patient ICU case study. Again, it is called two-dimensional data because of the x-axis and y-axis, and is not to be confused with two dimensions in the star schema. These attributes, Glucose and Albumin, are fact measures, and therefore, the clustering process is to cluster the star schema records based on the values of the two fact measures.

The main tool of centroid-based clustering is the similarity measure (or the distance measure). The main question is how is the similarity measure used in two-dimensional data. Because the unit measurement of the two axes is different, at least in this case, how the data is plotted in a two-dimensional map may produce a different mental map of what the clusters should look like. For example, the map in Fig. 21.21 can be stretched in two different directions. If we stretched it horizontally, it will look something like that in Fig. 21.22. As a result, the data looks like a one-dimensional data, because the fluctuation in the y-axis is quite negligible. This means the distance in the context of the y-axis is very small, which makes it unimportant.

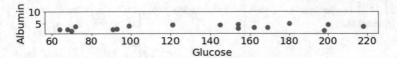

Fig. 21.22 Two-dimensional data stretched horizontally

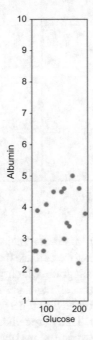

Fig. 21.23 Two-dimensional data stretched vertically

On the other hand, if the map is stretched vertically, similar to that in Fig. 21.23, the distance in the x-axis becomes smaller, and the distance in the y-axis becomes larger. This will also produce different clusters.

Therefore, in two-dimensional data (as well as in higher-dimensional data), it is important that the distance in each axis is normalised so that the distance unit of all dimensions is uniform. After the Glucose and Albumin values are normalised (e.g. normalised into values ranging from 0 to 10), the normalised data is shown in Fig. 21.24. Once the data is normalised, the clustering process begins.

As in the clustering of one-dimensional data, the initial step is to choose the initial centroids from the dataset. Once the initial centroids are chosen, we then allocate each data point from the dataset into a cluster.

In the one-dimensional clustering example, we use the mid-point between the means of two adjacent clusters. Once the mid-point is identified, it will be easy to assign each data point to the intended cluster, as the mid-points will serve as the boundary of the cluster. So using this method, the boundaries for each cluster are defined by the mid-points between two means.

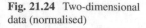

Fig. 21.24 Two-dimensional
data (normalised)

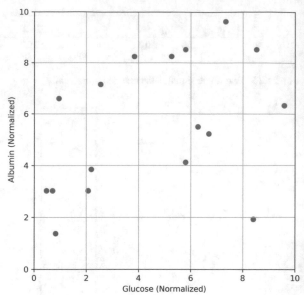

In two-dimensional clustering, a similar method can be used. To identify the mid-points in two-dimensional data, we can use the concept of the *Voronoi diagram* or VD. A Voronoi diagram is a diagram which partitions the map into mutually exclusive areas or cells, where each cell has a *generator point*. The border between two adjacent VD cells is created by having a perpendicular line right in the middle between the two generator points. This perpendicular line will divide the area into two areas, where the distance between the mid-point and the two generator points will be equal, as the mid-point is exactly in the middle of the two generator points.

Figure 21.25 shows the VD cells with the initial centroids. In this case, we assume that the predefined number of clusters is $k=3$. Note the initial centroids are chosen from points in the dataset. Once the VD cells are constructed by partitioning the map based on the centroids using the perpendicular bisector, the membership of each cluster (or VD cell) becomes very clear.

Once the initial clusters are formed, we need to recalculate the mean or the centroid of each cluster (or VD cell). Calculating the mean or centroid of two-dimensional data is simply performed by averaging the x-axis and y-axis of the data points in each cluster. Once the new centroids are calculated, the VD cells will be adjusted. Note that from this point onward, the centroid is very unlikely to be an existing data point in the dataset; rather it is likely to be an imaginary point.

Figure 21.26 shows that the new centroids (e.g. CA_1, CB_1 and CC_1) have moved away from the initial centroids. Note that the borders of the VD cells have also shifted, resulting in an object movement from Cluster 2 to Cluster 3 (the object is denoted by a circular dot with white in the middle).

The process continues by calculating the new means of the three VD cells. Because Cluster 1 does not lose an object to nor receive an object from other

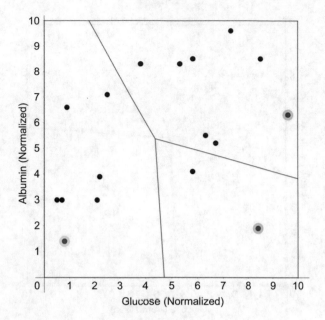

Fig. 21.25 Initial clusters using Voronoi diagram

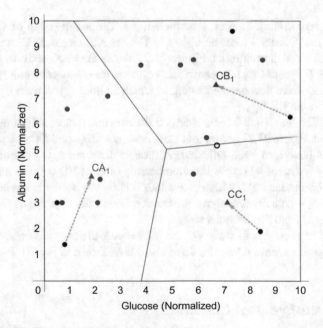

Fig. 21.26 Step-1 Clustering process using Voronoi diagram

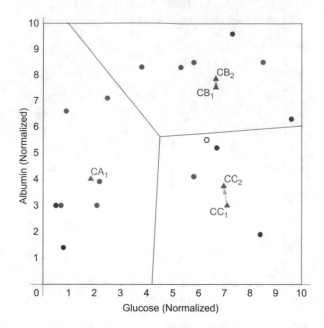

Fig. 21.27 Step-2 Clustering process using the Voronoi diagram

clusters, the centroid of Cluster 1 is unchanged. The membership of Cluster 1 is unchanged too. Cluster 2 loses an object to Cluster 3. As a result, Clusters 2 and 3 need to recalculate the centroids. Figure 21.27 shows the new centroids of Clusters 2 and 3 (e.g. CB_2 and CC_2). Because these clusters have new centroids, the borders of the VD cells have now been adjusted, in which another object from Cluster 2 has moved to Cluster 3.

Because Clusters 2 and 3 have changed their memberships, new centroids must be calculated. Figure 21.28 shows that CB_2 becomes CB_3, and CC_2 becomes CC_3. The borders have also been adjusted. Although there are new centroids, and of course new borders of VD cells, the membership of each VD cell has not changed. The object on the far right, which looks like it is on the border between Cluster 2 and Cluster 3, is actually just above the border; hence, the object is in Cluster 2, so it is unchanged from the previous step.

Because the members of each VD are unchanged, the centroids will not change either. The process terminates with three clusters as shown in Fig. 21.29.

21.3.2   Density-Based Clustering

Density-based clustering is also a partitioning-based clustering, but the way the objects are partitioned is not based on the centroid of the cluster. Density-based clustering creates a chain of neighbourhood objects, where one is close to another

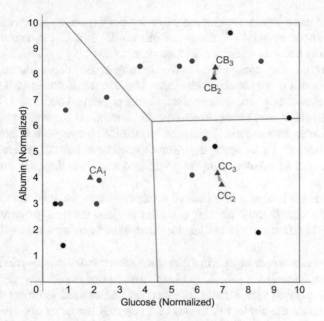

Fig. 21.28 Step-3 Clustering process using the Voronoi diagram

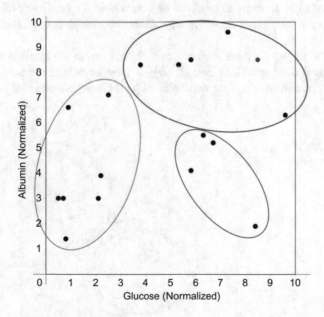

Fig. 21.29 k-means final clustering results

object(s) and these objects are close to another object(s) and so on, creating a chain of neighbourhood objects. The feature of this cluster is that it is dense due to the tight proximity between one object and another.

DBSCAN is the most well-known density-based clustering method. In DBSCAN, there are five important elements: *MaxDist* (maximum distance), *MinPts* (minimum number of points), core points, border points and outliers.

MaxDist, which is a user-defined criteria, determines the maximum distance allowed between two objects. Note that in DBSCAN, two objects are clustered together if they are not far apart; the distance must be at most MaxDist. If the two objects have a distance more than the predefined MaxDist, they will not be in the same cluster.

MinPts, which is also a user-defined criteria, defines the minimum number of objects that a cluster must have. If a cluster has objects in a group which have satisfied the MaxDist criteria but has less than MinPts objects, this will not form a cluster.

Suppose we have nine objects in a two-dimensional space as shown in Fig. 21.30. The objects are labeled A, B, ..., I. MaxDist from one object, say object B, can be drawn as a circle with a radius of MaxDist and centred at object B. If there are objects inside the circle, this means that object B has other objects within the proximity specified by MaxDist. In the case of object B, it has another three objects in the vicinity, namely, objects A, C and D, as these three objects are inside the circle centred at B. In order to label the objects within the proximity of B, an arch is drawn from B to each of these three neighbourhood objects (e.g. B-A, B-C and B-D).

Looking at object C, there are only two objects inside the circle centred at C; hence, we have C-B and C-D. Not all objects have objects in the neighbourhood specified by the MaxDist. For example, object I has no other objects in the proximity.

Fig. 21.30 DBSCAN
clustering concepts

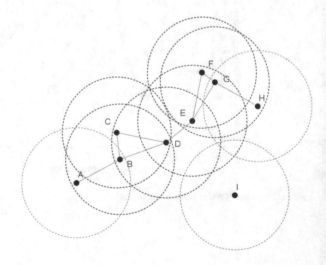

Apart from MaxDist and MinPts, there are three types of objects or points, namely, *core points*, *border points* and *outlier points*. *Core points* are points that satisfy the MinPts criteria, whereas *border points* are points that do not satisfy the MinPts criteria but are connected to one of the core points. *Outlier points* are points that do not have any objects within the proximity.

In Fig. 21.30, we assume that the user has specified MinPts=3. Hence, all objects, except objects *A*, *H* and *I*, are core points, as they have satisfied the criteria MinPts=3. Objects *A* and *H* do not satisfy the MinPts=3 constraint but are connected to one of the core points. Therefore, they are border points. Object *I* is clearly an outlier point as it is not connected to a core or border point.

A cluster consists of core points and border points. So in this case, all objects, except object *I*, are within one cluster.

If MinPts=4, the cluster composition will still be unchanged. The only difference is that objects C and F will become border points instead of core points. But they will still be part of the cluster because a cluster consists of core points and all of their border points.

Suppose object *I* has a close neighbour, say object *J*. Both objects *I* and *J* will be outliers, because it does not satisfy the MinPts criteria (e.g. MinPts=3). So even though objects *I* and *J* satisfy MaxDist, since they do not satisfy MinPts, they cannot form a cluster. Unless MinPts is reduced to 2, then *I* and *J* will form a cluster.

Apart from the way DBSCAN clusters objects, there are two other differences between density-based clustering and centroid-based clustering. In centroid-based clustering, like k-means, the desired number of clusters k is predefined. On the other hand, DBSCAN does not require the number of clusters to be predetermined. The other difference is that DBSCAN may identify outlier objects as they are not part of any clusters. In contrast, centroid-based partitioning will use all objects when creating the clusters; hence, no objects will be left out.

In summary, in centroid-based clustering, users must specify k, whereas in density-based clustering, users need to specify MaxDist and MinPts.

Using the same example as the centroid-based clustering (e.g. the Glucose-Albumin data), Fig. 21.31 shows the results of DBSCAN clustering. In this case, MinPts=3. Three clusters are created with one object as an outlier. It can be seen clearly that density-based clustering forms clusters or neighbourhoods where one object is close to another object, which, in turn, is close to another object, etc., potentially creating a long chain of objects in the neighbourhood.

21.4 Classification Using Regression Trees

A decision tree is one of the most widely used classification method in data mining. Traditionally, a decision tree is usually applied to categorical data, as discussed in the early part of this chapter. For example, the decision trees shown in Figs. 21.1, 21.2, 21.3, and 21.4 are typical decision trees on categorical data. The decision

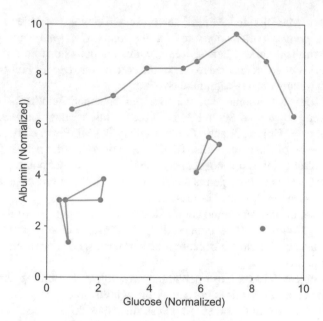

Fig. 21.31 Glucose-Albumin clustering using DBSCAN

depicted on each node in the tree evaluates the data against categorical criteria, such as Weather = Sunny, Temperature = Hot, etc. The target class of this decision tree is also categorical data, which is to walk or not to walk.

Data warehousing, in the form of an Extended Fact Table, focuses on numerical values as stored in the fact measures of the Fact Table. Therefore, it is expected that the classification for data warehousing must be based on numerical values (or continuous values) rather than categorical data. The target class of the classification for data warehousing is also a numerical measure. Table 21.6 shows a snapshot of an Emergency Patient Extended Table, a case study that has been discussed in this chapter and Chap. 20. The first column is PatientID; for simplicity, it is coded alphabetically. There are many fact measures in this Extended Fact Table, but the focus is on Glucose, Albumin as well as Mortality Prediction. Note that in this case, the classification uses the fact measures, namely, Glucose and Albumin, and the target class is Mortality Prediction. There are a lot more fact measures in this case study, but for simplicity, only Glucose and Albumin are used. There are other attributes related to the dimensions as well, but in this case, only PatientID is used as an identifier of the record.

The classification technique for numerical measures that is going to be used in this chapter is the *regression tree*. A regression tree has some similarities to the traditional decision tree, with one exception, namely, the regression tree works with continuous values, which serves the purpose of classification for data warehousing.

Classification is a predictive data mining technique, which means that firstly, the classification model is built using the training dataset. Once the model is built, the

Table 21.6 Emergency Patient Extended Fact Table

Patient ID	...	Glucose	Albumin	...	...	Mortality Prediction
A		162.2	3.5			0.573189504
B		93.7	2.9			0.22
C		68.5	2.6			0.217082562
D		155.0	3.0			0.534815242
E		121.0	4.5			0.475139465
F		198.0	2.2			0.969279952
G		99.1	4.1			0.552492172
H		180.2	5.0			0.752011091
I		169.0	3.4			0.799263517
J		64.9	2.6			0.383771494
K		72.1	3.9			0.45
L		155.0	4.6			0.862259153
M		218.0	3.8			0.99
N		146.0	4.5			0.813710339
P		91.9	2.6			0.496873792
Q		200.0	4.6			0.745266898
R		70.3	2.0			0.132516482

target class of the incoming data can be predicted by using the classification model. This is also the case with the regression tree. The first task is to build a regression tree using a training dataset. In this chapter, the training dataset to be used is the data in Table 21.6. The next subsections will go through, step by step, how the regression tree is constructed using this dataset.

21.4.1 Selecting the Root Node

The first task in building a decision tree, as well as a regression tree, is to choose the root node (for the tree). There are two aspects in selecting a root node: (1) select an attribute (in this case, it is either Glucose or Albumin) as the root node, and (2) determine a condition to each branch of the root node. Determining the condition can be tricky because the attribute values are continous values (or numerical values), which is different from the traditional decision trees, as such shown in Fig. 21, whereby the nodes contain categorical values. The root node is the Weather attribute, and the branches are the three categorical values of the Weather attribute: Fine, Shower and Thunderstorm. These branches act as a condition: if the Weather is Fine, then it will go to the left sub-tree; if the Weather is Shower, it will go to the middle sub-tree; and if the Weather is Thunderstorm, it will go to the right sub-tree.

However, the Glucose and Albumin attributes are continuous values. The criteria for each branch needs to be decided, such as if Glucose <100, it will go to the left sub-tree, or else (Glucose ≥ 100), it will go to the right sub-tree. The same applies to Albumin, if Albumin is chosen as the root node. So the main question for selecting the root node is which attribute and what conditions.

Note that the target class for this Emergency Patient case study, as noted by the training dataset in Table 21.6, is attribute Mortality Prediction, which is a numerical value. Ideally, when the dataset is partitioned by the root node, let say Glucose > 100, the partition should contain data that is cohesive in terms of its target class (e.g. the Mortality Prediction attribute value). So, the tree may say that if Glucose < 100, then, for example, the Mortality Prediction is 0.7, as all the data in the training dataset with Glucose <100 has a Mortality Prediction of 0.7. This is an ideal case, where the partition has a cohesive target value. If the partition has large differences in their target value, then the chosen partitioning attribute and conditions are said to be suboptimal, because the partition needs to be re-partitioned, as the current partition is not conclusive enough. In other words, a partitioning attribute and its conditions must be chosen carefully so that the partitions will be optimal, whereby the data in each partition is as cohesive as possible.

Let's assume that Glucose is chosen as the attribute for the root node. If we say the condition is Glucose <100 for the left partition, the question is why not Glucose <101, for example. Note that Glucose is a numerical attribute, so technically, any number can be used as a condition. But let's focus on the training dataset in Table 21.6. There are 17 records in this training dataset, 2 of which have an identical Glucose value, which are Patients D and L with Glucose = 155. Since the job of the root node is to partition the dataset into two partitions, the condition for the partitioning is severely limited. To partition 17 records (where 2 of them are identical) into 2 partitions (based on the Glucose value), there are only 15 possibilities. The first possibility is that the left partition gets one record, and the right partition the rest. The second possibility is that the left partition gets two records, and the right partition 15, and so on. The last possibility is that the left partition gets 16 records, and the right partition only 1.

One possible partitioning is that the left partition gets seven records and the right partition gets ten records. This is shown in Fig. 21.32. The x-axis is Glucose and the y-axis is Mortality Prediction. Each object is labelled with the PatientID. In this case, the left partition gets Patients J, C, K, R, P, B and G, and the right partition the rest. The partitioning condition for the left partition is Glucose $< x$ value (the condition for the right partition is exactly the opposite). Patient G has Glucose 99.1, and Patient E 121. If we want the smallest seven records to go to the left partition and the rest to go to the right partition, then the condition can be Glucose $< x$, where x can be any Glucose value between 99.1 and 121, exclusive. The partitioning results will be the same, that is, for Glucose <100 or Glucose <110, as long as the condition is any value between Patients G and E. Therefore, the best option is to choose an x value right between Patients G and E, which is 110.05. Hence, the condition becomes as follows: if Glucose <110.05, it goes to the left partition; else, it goes to the right partition.

Using this method, with 17 records, 2 of which are identical, there are only 15 ways of partitioning. The first option is to choose a middle point between Patients

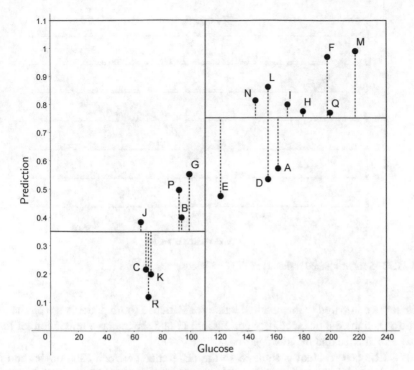

Fig. 21.32 An example of partitioning for Glucose

J and C, which is Glucose 66.7 (the left partition only has Patient J, and the right partition the rest). The second option is to choose a middle point between Patients C and R, which is Glucose 69.4, and so on.

The next question is which of the 15 possible ways to partition the dataset based on Glucose is the best partitioning method. As mentioned above, the best partitioning method will produce partitions containing data (in each partition) that is as cohesive as possible, in terms of their target class. To check the cohesiveness of the partitioning method, we use the *sum of squared residuals* or SSR. Firstly, we need to understand what *residual* means.

A residual value of an object is the difference between this object and the average value of all objects in the same partition. Using the example shown in Fig. 21.32, where the split is based on the middle point between Patients G and E (e.g. Glucose <110.05), the average Mortality Prediction value for the left partition is 0.350391, which is roughly in the middle of 3 and 4 on the *y*-axis, whereas the average for the right partition is 0.751494. Note that the average value is denoted by a horizontal line. The horizontal line on the left of the vertical partitioning line is the average for the left partition, whereas the horizontal line on the right of the vertical partitioning line is the average for the right partition.

The residual of each object is denoted by the dotted line between the object and the average value of the partition. For example, the residual value for Patient G is

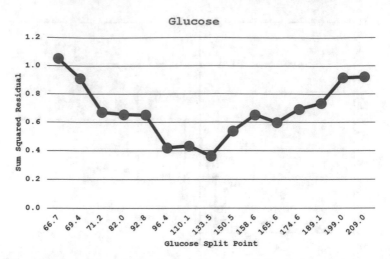

Fig. 21.33 Sum of squared residuals (SSR) for Glucose

quite large compared to the residual value for Patient J (both patients are on the left partition), since the distance between Patient G and the average line is longer than between Patient J and the average line.

In SSR, each residual is squared so the difference between each object and its average is always a positive value. The sum of squared residuals is the sum of all squared residuals of all objects (in both partitions, with respect to their own average). Given the *left* and *right* partitions with n and m number of objects, respectively, $left=\{r_1, r_2, ..., r_n\}$ and $right=\{s_1, s_2, ..., s_m\}$, the SSR formula is given in Eq. (21.7).

$$SSR = \sum_{i=1}^{n}(r_i - \bar{r})^2 + \sum_{j=1}^{m}(s_j - \bar{s})^2 \qquad (21.7)$$

To determine the best partitioning for Glucose, we need to calculate the SSR for all possible partitioning, of which there are 15. Figure 21.33 shows the SSR value for each partitioning. The lowest SSR indicates the best partitioning method, which, in this case, is the middle point between Patients E and N, which is 133.5. The SSR is equal to **0.3622**. Therefore, if Glucose is chosen as the root node, the condition for the branches would be Glucose <133.5 and Glucose ≥ 133.5.

However, there is another possibility for the root node, that is, Albumin. Albumin has 17 records too, but Patients C, J and P have identical Albumin (2.6); Patients E and N also have identical Albumin (4.5); and Patients L and Q have identical Albumin (4.6). Hence, Albumin only has 12 partitioning options.

The SSR of each partitioning is shown in the graph in Fig. 21.34. It shows that Albumin = 3.2, which is the middle point between Patients D and I (see Fig. 21.35), has the lowest SSR (**0.7748**).

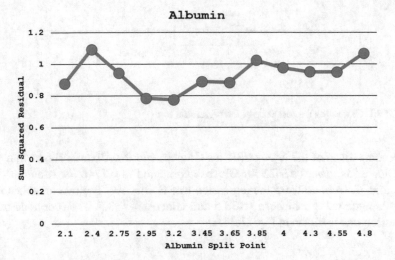

Fig. 21.34 Sum of squared residuals (SSR) for Albumin

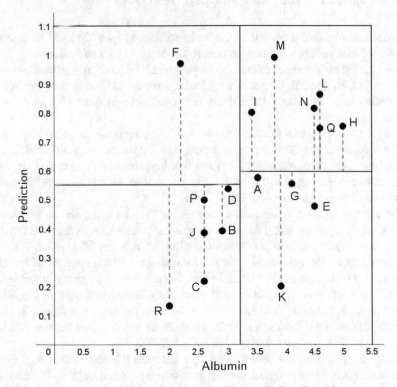

Fig. 21.35 A partitioning for Albumin (middle point between Patients D and I)

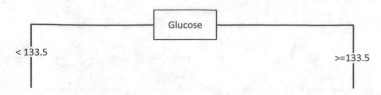

Fig. 21.36 Glucose as the root node of the regression tree

A comparison of the lowest SSR for Glucose and for Albumin shows that SSR for Glucose is lower (**0.3622** for Glucose compared to **0.7748** for Albumin), and therefore, the root node for the regression tree is Glucose. The two branches from the root node will be Glucose <133.5 and Glucose ≥133.5. The root node of the regression tree is shown in Fig. 21.36.

21.4.2   Level 1: Processing the Left Sub-Tree

The root node basically partitions the training dataset (see Table 21.6) into two partitions based on the condition, which is Glucose <133.5 and Glucose ≥133.5. Figure 21.37 shows the split of the training dataset. The left partition contains eight records, J, C, R, K, P, B, G and E, all of which are on the left side of the vertical partitioning line (Glucose = 133.5), whereas the right partitions contain the other nine records.

The next task is to focus on the left sub-tree. A recursive approach is required to build a regression tree. This means once the training dataset is partitioned into the left and the right partitions, each partition will be considered a new dataset. Now, we focus on the left sub-tree. Therefore, the new dataset for this sub-tree is the eight records previously mentioned.

The processing itself is identical to processing the root node, that is, to calculate the SSR of all possible split points and choose the lowest SSR as the next node for the regression tree. So firstly, let's evaluate Glucose as a candidate node. Note that we only focus on the new dataset for the left sub-tree, which consists of the eight records, of which none are identical; hence, there will only be seven possible split points. The SSR of each possible split point for Glucose is shown in Fig. 21.38. From the graph, it is clear that the lowest SSR among the seven possible split points is the third from the left, which is 71.2, which is the middle point between Patients R and K. The SSR value for this split point is **0.0983**.

Figure 21.39 shows the partitioning line between Patients R and K. The left horizontal line is the average of the left partition with three objects, Patients J, C and R, whereas the right horizontal line is the average of the right partition with five objects, Patients K, P, B, G and E. Note that we are focusing on the eight objects for the left sub-tree. The other nine objects are not considered as they are on the right sub-tree. Hence, these nine objects are shown as grey dots in the figure to illustrate

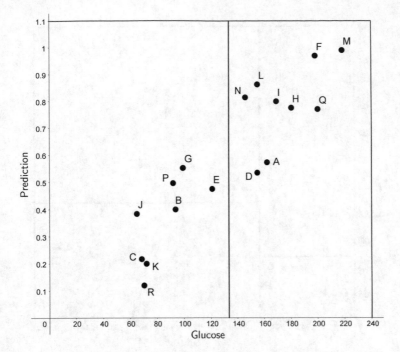

Fig. 21.37 Glucose as the first partitioning node for the regression tree

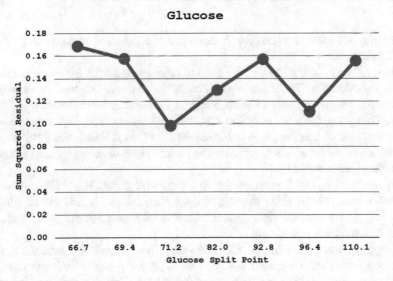

Fig. 21.38 SSR of Glucose

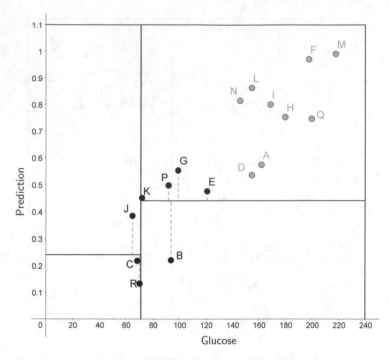

Fig. 21.39 Glucose as a candidate partitioning node for the left sub-tree

where they are located, even though they are not considered in the calculation for the left sub-tree.

Now, let's consider Albumin. We need to calculate the SSR for all possible split points for Albumin. There are eight records for the left sub-tree, as previously mentioned. However, three of them, Patients C, J and P, have the same Albumin value. Therefore, there are only five possible split points. The SSR value for each split point is shown in the graph in Fig. 21.40. The lowest SSR is the third from the left, which is Albumin = 3.4, which is the middle point between Patients B and K. The SSR value is **0.0923**.

Figure 21.41 shows the partitioning line between Patients B and K for Albumin. The left horizontal line is the average of the left objects, R, P, J, C and B, which is equal to 0.29, whereas the right horizontal line is the average of the right objects, K, G and E, which is equal to 0.49.

A comparison of the lowest SSR between Glucose and Albumin for the left sub-tree shows they are quite close. But the SSR for Albumin is lower than that for Glucose (e.g. **0.0923** for Albumin and **0.0983** for Glucose). Hence, Albumin is chosen as the node for the left sub-tree.

Figure 21.41 shows that the Albumin partitions into left and right, with the left having five objects and the right only having three objects. As the process is recursive, the next question is when will the process terminate and what are the

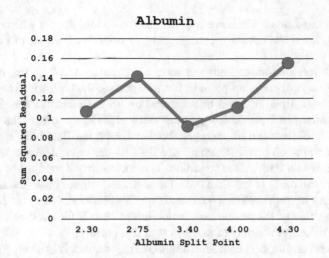

Fig. 21.40 SSR of Albumin

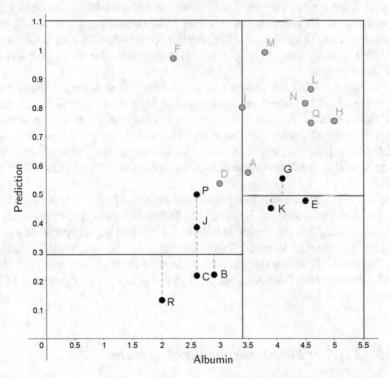

Fig. 21.41 Albumin as a candidate partitioning node for the left sub-tree

termination conditions. There are two termination conditions: (i) when the objects within a partition are cohesive enough and (ii) when the number of objects in a partition is very small.

The first termination condition is easy to understand. When the objects are cohesive enough, there is no point in further splitting the partition. The degree of cohesiveness of a partition can be measured by SSR. Remember that SSR is the sum of squared residuals. The bigger the value of SSR, the bigger the difference between each object with its average. This is why the smallest SSR is preferable. The first termination condition could use SSR as the criteria, for example, SSR <0.05 is used as a termination condition. This means that if the SSR of a partition (say the left partition) is less than 0.05, the partition will not be processed. Hence, the regression tree will produce a left node for this partition, where the left node is the average value of the target class, which in this case is the average value of the Mortality Prediction attribute of the objects of that partition.

The second termination condition also makes sense, that is, when the number of objects is reasonably small, there is no point in further splitting the partition. The threshold is set by the user and it is usually around 10% of the training dataset. This means that if a partition has less than 10% records, the partition will not be processed further. The process for this partition terminates with a leaf node, and the leaf node will have a target class, which is the average value of the objects of this partition.

Returning to the left sub-tree, in which Albumin is used as the partitioning node, as shown in Fig. 21.41, there are five records in the left partition and three records in the right partition. Assuming that the second termination condition is that the number of records must be less than or equal to three, then the right partition which consists of three object, Patients K, G and E, will terminate. The target class is the average Mortality Prediction value of these three objects, which in this case is equal to **0.493**.

The left partition contains more than three objects. Hence, it doesn't satisfy the second termination condition. However, the SSR of this partition is very small (e.g. **0.087**). Assuming that the threshold is SSR <0.09, then the left partition will also terminate. The average Mortality Prediction value for the left partition is equal to **0.290**. Figure 21.42 shows the regression tree with the root node and the left sub-tree. The two branches from Albumin on the left sub-tree are all terminated with respective leaf nodes.

21.4.3   Level 1: Processing the Right Sub-Tree

Processing the right sub-tree is similar to the left sub-tree. Firstly, calculate the SSR value for each of the possible split points from both attributes, Glucose and Albumin. The dataset is confined to the right partition dataset. In this case study, the right partition dataset contains nine records (see the right partition in Fig. 21.37). For Glucose, of the nine records, two are identical, Patients D and L, with Glucose

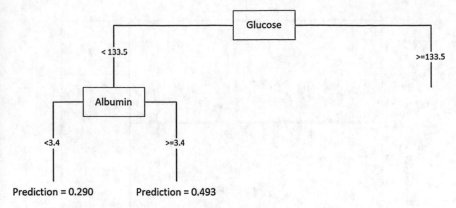

Fig. 21.42   Regression tree with root node and left sub-tree

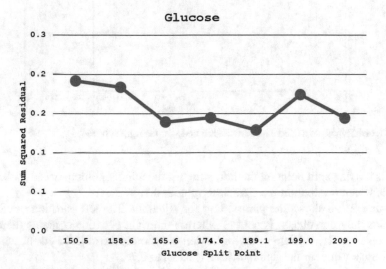

Fig. 21.43   SSR of Glucose

155. Therefore, there are only seven possible split points. The SSR value of each split point is shown in Fig. 21.43. Of the seven SSR values, the lowest is **0.129**, which is the SSR of the middle point between Patients H and F. The split point is Glucose 189.

The left and right partitions created by the split point between Patients H and F for Glucose are shown in Fig. 21.44. The left partition contains six objects (e.g. N, L, I, H, D and A), whereas the right partition contains only three objects (e.g. F, Q and M).

For Albumin, of the nine objects, two are identical, which are Patients L and Q with Albumin 4.6. Therefore, there are only seven possible split points for Albumin. The SSR value for each of the seven split points is shown in Fig. 21.45. The lowest

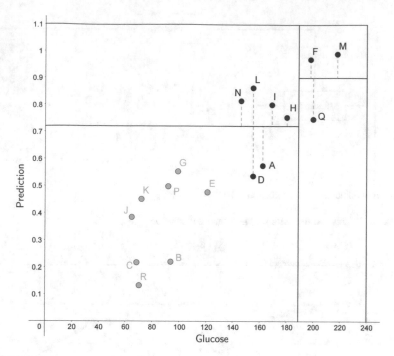

Fig. 21.44 Glucose as a candidate partitioning node for the right sub-tree

SSR is the first split point on the left, which is the middle point between Patients F and D, which is Albumin 2.6. The SSR value is **0.1537**.

Figure 21.46 shows the partitioning for Albumin. The left partition obviously contains only one object, Patient F, whereas the right partition contains the other eight objects. The average value for the left partition is the object F itself, whereas the average value for the right partition is equal to 0.76.

Comparing the SSR between Glucose and Albumin, Glucose has **0.129**, whereas Albumin has **0.1537**. Therefore, the split node for the right sub-tree is Glucose, where the condition is Glucose <189.

Since Glucose is the partitioning node, we need to check if the two branches from Glucose (less than 189 to the left and greater or equal to 189 to the right) satisfy any of the two terminating conditions. The left partition (where Glucose <189) has more than three objects, and the SSR for the left partition is 0.901; neither satisfy the predefined terminating conditions, which are maximum three objects with SSR <0.9. Therefore, the left partition needs to be processed further.

The right partition (where Glucose ≥189) consists of three objects only (refer to Fig. 21.44). Hence, the right partition will terminate. The target class for this branch is the average Mortality Prediction value of the three objects, which is **0.902**.

The regression tree so far is shown in Fig. 21.47.

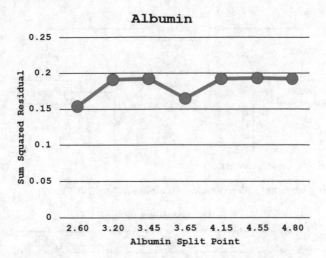

Fig. 21.45 SSR of Albumin

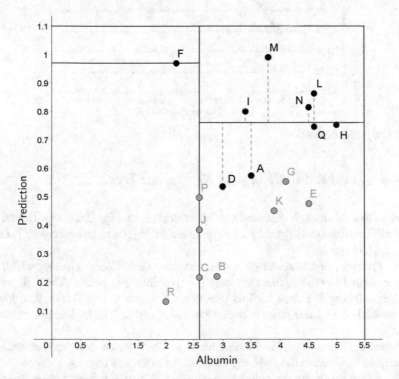

Fig. 21.46 Albumin as a candidate partitioning node for the right sub-tree

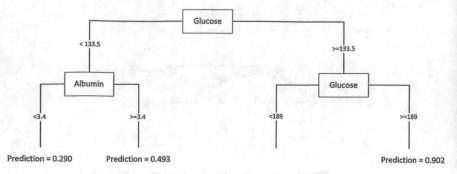

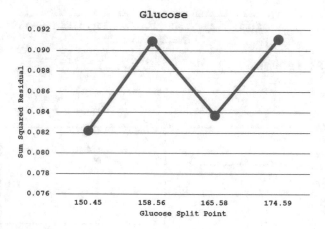

Fig. 21.47 Regression tree with root node, left sub-tree and right sub-tree

Fig. 21.48 SSR of Glucose

21.4.4 Level 2: Finalising the Regression Tree

There is one sub-branch that needs to be processed, that is, Glucose <189 on the right sub-tree. Figure 21.44 shows that there are six objects to be processed: Patients N, L, I, H, D and A.

For Glucose, of the six objects, two have the same Glucose value, which are Patients D and L. Hence, there are only four possible split points. The SSR values of all possible split points for this partition are shown in Fig. 21.48. The lowest SSR **0.082** is the middle point between Patients N and D. The middle point value is 150.45.

Figure 21.49 shows the partitioning based on Glucose. The left partition has only one object, N, whereas the right partition has the other five objects.

For Albumin, all the six objects have distinct Albumin values; hence, there are five possible split points. The SSR values for Albumin are shown in Fig. 21.50. The lowest SSR, which is the middle point between Patients A and N (4.0), is **0.047**.

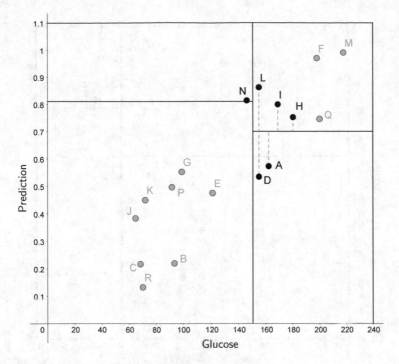

Fig. 21.49 Glucose as a candidate partitioning node on Level 2

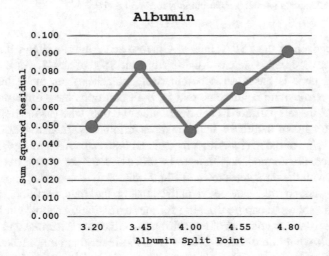

Fig. 21.50 SSR of Albumin

Figure 21.51 shows the partitioning based on Albumin. Both partitions only have three objects left.

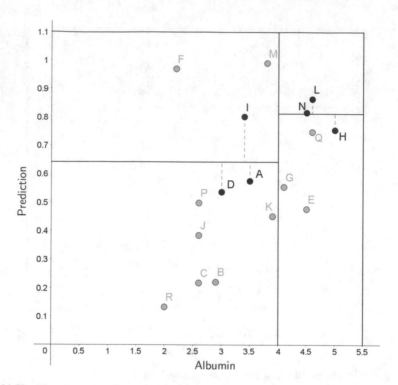

Fig. 21.51 Albumin as a candidate partitioning node on Level 2

A comparison of the SSR values of Glucose and Albumin shows that Glucose has an SSR of **0.082**, whereas Albumin has an SSR of **0.047**. As a result, the partitioning node is based on Albumin. Both branches from the Albumin node satisfy the terminating conditions, that is, both partitions have three objects, and furthermore, the left partition has an SSR value of 0.065, whereas the right partition has an SSR value of 0.006. So, both branches terminate with an average Mortality Prediction value of 0.636 (for Albumin <4) and 0.809 (for Albumin ≥4).

Since there are no more partitions to be processed, the regression tree process is terminated with the final tree shown in Fig. 21.52.

This regression tree can be visualised using the map partitioning method. Figure 21.53 shows a map partitioning for the regression tree shown in Fig. 21.52. Because the dataset contains three attributes, Glucose, Albumin and Mortality Prediction, it would be difficult to draw a three-dimensional map. Instead, a two-dimensional map is used. On this map, the x-axis is Glucose and the y-axis is Albumin. The dots on the map are the values corresponding to the x-axis and y-axis. In order to fit the third attribute, namely, the Mortality Prediction attribute, on to the map, we use a colour code to show the severity of the Mortality Prediction value, from 0 to 1 from low to high. Hence, each object on the map is colour-coded accord-

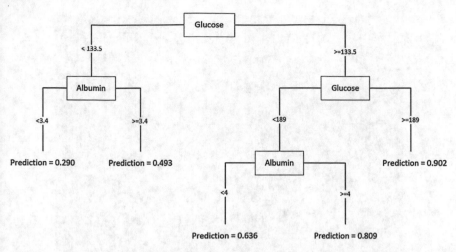

Fig. 21.52 The complete regression tree

ing to the Mortality Prediction value. The darker the colour, the higher the Mortality Prediction value. In this way, we fit three attributes into a two-dimensional map.

The vertical lines and horizontal lines are the partitioning nodes in the regression tree. The root node is the middle vertical line, where Glucose is somewhere between 130 and 140 (e.g. it is actually 133.5, according to the regression tree). The left side of this vertical line indicates the left sub-tree of the regression tree, whereas the right side of this vertical line is the right sub-tree.

On the left sub-tree, the area is horizontally divided by Albumin at around 3.5 (e.g. 3.4 to be precise, according to the regression tree). This horizontal line divides the area into two areas, above and below the horizontal line. The top area contains three objects, and the average Mortality Prediction value is 0.493. The bottom area contains five objects with an average value of 0.290. These two areas indicate the leaf nodes of the left sub-tree.

On the right side of the vertical line in the middle indicates the objects where Glucose $\geq$ 133.5, which is the right sub-tree. This area is further divided. The second vertical line at Glucose 189 divides the right area into two areas: the left side of the vertical line and the right side of the vertical line. The right side of the vertical line where Glucose $\geq$ 133.5 is a leaf node as it contains only three objects, and the leaf node value is 0.902, which is the average Mortality Prediction value.

The left side of the vertical line where Glucose $\geq$ 133.5 is further divided into two areas, partitioned by Albumin horizontal line at 4. The top area is a leaf node with three objects and an average value of 0.809, and the bottom area is another leaf node also with three objects and an average value of 0.636.

In conclusion, a regression tree can be seen as a grid partitioning method, using vertical and horizontal lines to partition the area into two. Further splits are done until all areas are finalised with an average prediction value.

As classification is a predictive data mining technique, once a regression tree is built using a training dataset, incoming data (from the testing dataset) is matched

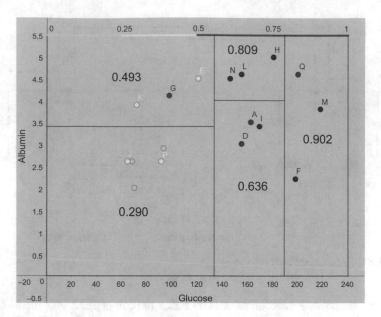

Fig. 21.53 Glucose-Albumin complete partitioning

against the tree, and the prediction value of the incoming data and the predicted value given by the regression tree are compared. For example, looking at the map partitioning in Fig. 21.53, if the incoming data has Glucose <133.5 and Albumin <3.4, then the expected prediction value is 0.290, as shown in the bottom left-hand grid.

A comparison of the regression tree and decision tree shows there are a number of differences:

- the regression tree is a binary tree, where each node has two branches. The decision tree is an n-ary tree, where each node may have more than two branches.
- a regression tree works with numerical attributes, whereas a decision tree works with categorical attributes. The target class of a regression tree is also a numerical attribute.
- an attribute that has been used as a node in the upper level of the tree can be reused in the lower level of the sub-tree but with different conditions for the branches. In a decision tree, once an attribute is used in the upper level of the tree, the attribute cannot be used in the sub-tree. Because a regression tree is a binary tree, n-ary splits are formed using multiple binary splits, which is why the same attribute can be used in many different nodes. On the other hand, because multiple branches are allowed for a node in a decision tree, once the attribute is used, the same attribute cannot be used in the sub-tree.

21.5   Data Warehousing: The Middle Man

Data analytics[1] is often considered in isolation. In many cases, the data is already prepared in one big file or table. Data analytics uses sophisticated methods, and this has attracted many researchers. The problems that data analytics algorithms need to solve are generally interesting, useful and genuine, and the results produced are effective as they solve real problems.

In the data lifecycle as shown in Fig. 1.6 of Chap. 1 and repeated in different illustrations in Fig. 20.1 of Chap. 20 as well as in Fig. 21.54, data analytics is the top-end activity in the data lifecycle. The features of data analytics are (i) the input data is already well prepared, (ii) the algorithms are sophisticated and complex, and (iii) the problems to solve are attractive and the results are expected to be directly useful.

The attractiveness of the problems, the sophistication of the solutions and the usefulness of the results are certainly the significant strengths of work on data analytics. However, the input data is often too simplistic, or at least the assumption that the data is already readily prepared for data analytics often neglects the fact that preparing such an input data is in many cases, if not all, actually the major work in the data lifecycle. The input data for data analytics often exists in the form of an Extended Fact Table, as explained in Chap. 20. The pipeline from the operational databases that keep the transactions and raw data to the input data for data analytics is very long; it often occupies as much as 80% (or sometimes even more) of the entire lifecycle. Therefore, we need to put much effort to this preparation and transformation work in order to value the work and the results produced by data analytics algorithms.

Having the correct input data for the data analytics algorithms, or in fact for any algorithms and processes, is critical, as the famous quote "garbage in garbage out" had said. Even when the original data is correct, but when it is presented inaccurately to a data analytics algorithm, it may consequently produce incorrect reasoning. The Bookshop case study discussed in Chap. 6 gave an illustration of how the "correct" Book Price in the Book Dimension may lead to an incorrect correlation with the fact measure: Number of Books Sold. The causal relationship between the TimeID, Book Price and Number of Books Sold might lead to a wrong conclusion to the Number of Books Sold (see Table 6.3). Although technically, the Book Price in the Book Dimension is correct, and the fact table is also correct, the combined information may lead to an incorrect conclusion if the data is not carefully transformed. This shows the importance of orchestrating the input data for data analytics.

The transformation process from operational databases to the data that is ready to be used by data analytics engines has a long process, and paying careful attention to this long chain of processes is equally, if not more, important than the data analytics

---

[1] In this book, we use the term "data analytics" broadly that also covers data mining and machine learning.

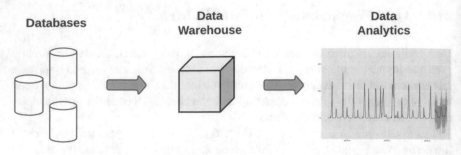

Fig. 21.54   Data warehouse as the middle man

engine itself. The operational databases that feed the data into this process usually come from multiple data sources;[2] not all of them are from the same system or environment; some might not even be a database system. The transformation from the operational databases to data warehouses covers many operations and steps, such as data cleaning, aggregation, summarisation, extraction, integration and filtering, just to name a few. It is not surprising that these whole processes occupy most of the time in the data processing and analytics lifecycle. Even when the transformation process is executed smoothly, the input data for the data analytics engine may still pose some issues, as illustrated in the Bookshop case study in Chap. 6.

Therefore, avoiding to acknowledge the long and hard processes required to prepare input data for the data analytics engine is similar to not correctly select suitable materials when building a product, even though we know that good products are built from good materials. Using a building construction domain as an illustration, in the past, constructing a tall office building needed to mix the concrete or semen on the building site by the builder themselves. Nowadays, readily made walls have been prepared separately and only need to be assembled at the building construction site. Using this latest technique, construction is more modular, the speed is much faster, and more mass production becomes possible. The role of data warehousing in data analytics is quite similar; the transformation process is more systematic and modular.

It is pretty common for data scientists to bypass data warehousing, that is, creating the input data for data analytics directly from the operational databases by applying data cleaning, transformation, integration, etc. A data warehouse is actually a placeholder; the processes to transform operational databases to data warehouses are systematic, and if data warehousing is used correctly, it can help data analytics. In the transformation process to data warehousing, the process to focus on what to focus (e.g. the fact measures) is identified. In data warehousing, the concept of granularity is deeply engraved. Other elements are also considered, including multi-fact, multi-input, etc. In other words, data warehousing prepares the data from the operational databases for better use in the data analysis stage.

---

[2] Chapter 13 introduces the concept of multi-input operational databases in data warehousing

Therefore, understanding data warehousing is invaluable to data analytics. Data warehousing is not about something in the past (e.g. historical data), but it acts as a "middle man" to the data analytics; it not only makes the data analytics process smoother but also enhances the capabilities of data analytics. The middle man is also capable of extending the data warehousing to deal with multi-model data, including semistructured and non-structured data. Therefore, data warehousing plays a vital role in the entire data processing lifecycle.

Data analytics, on the other hand, can also contribute to the data sources, namely, operational databases and data warehouses. The results and models produced by the data analytics engine can provide useful feedback to the operational databases and, subsequently, to the data warehouses. This is the role of a data analytics engine that has not been explored widely, despite its potential.

Data analytics has a wide range of tools that makes the analysis processes efficient and fast. There are plenty of available libraries, such as those in Pandas, Python, etc. Due to this, the use of SQL in data analytics is somewhat limited. However, using SQL in data analytics is not the primary purpose of data warehousing. The purpose of data warehousing is to convert raw materials into readily used materials as the building block in a data analytics engine. The use of SQL in this chapter is merely to illustrate that SQL can be used as a data analytics language, considering the wealth of the relational database engine, such as query optimisation, indexing, data storage, etc., which might give a cutting-edge benefit to the data analytics engine.

Since data analytics technology is already so advanced, there are a number of roles that data analytics can play in the entire data processing lifecycle, apart from the main task of data analytics, which is analysing the data itself. Through the results and models produced by the data analytics engine, people understand more about the relationship in the data, whether it be between attributes or between records. This is invaluable knowledge to the operational databases. It is widely acknowledged that in industry, even some basic relational theory concepts, such as Primary Key-Foreign Key relationship, are not adhered in practice. Data duplications are very common, primarily due to inexpensive data storage. Normalisation is mostly neglected. Hence, data consistency is a common issue. Feeding back the knowledge about the data from the data analytics engine to the operational databases is invaluable. The clustering model obtained by the data analytics engine can help the data designer to arrange the data more efficiently. The knowledge on relationships between records and attributes may enhance metadata management in the operational databases. Other data analytics models and results may improve data quality, data structures, indexing, etc., if feedback adequately to the operational databases. Consequently, data warehousing will also improve. Figure 21.55 shows the loop feedback cycle in the data processing.

Label 1 covers the basic transformation steps from operational databases to data warehousing. Label 2 shows the strength of data warehousing, whereby the granularity concept is put into practice. Label 3 covers BI dashboards, which often deal with highly aggregated data. Label 4 prepares the Extended Fact Table, which feeds into the data analytics engine (Label 5). Label 6 shows that both results from

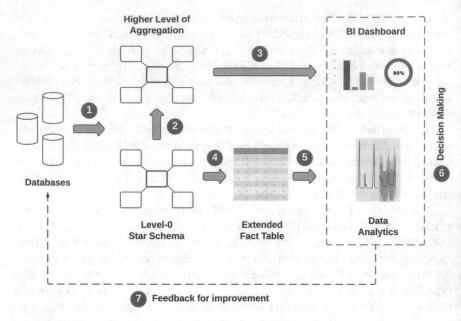

Fig. 21.55 Data warehousing and analytics: the complete picture

dashboards and data analytics engine help management in their decision-making process. And finally, Label 7 shows how data analytics engine results and models help improve the operational databases through a feedback process.

Data analytics (including data mining and machine learning) is only a small part in the data processing lifecycle. But data analytics, data mining and machine learning can play an essential role beyond data analysis for decision-making. Its loop feedback to the data sources (e.g. databases and data warehouses) is also very important. On the other hand, data warehousing plays an important middle role to support data analytics and is a good mediator for the operational databases.

Data warehousing is truly the fuel for the data analytics engine and hence the subtitle of the book "Fueling for Data Engine". We hope you enjoy reading the entire book as much as we enjoy writing this book.

21.6 Summary

Data analytics for data warehousing focuses on data in the star schema. This means that the focus on data analytics is primarily on fact measures, which are numerical values. Therefore, data analytics techniques for data warehousing analyse numbers, not categorical data. Because of this nature, statistical analysis is well-suited to data warehousing.

This chapter introduces three data analytics techniques suitable for data warehousing: (i) regression, (ii) cluster analysis and (iii) classification. Regression is a traditional statistical technique that can be used for prediction. Once a regression model is built using a training dataset, incoming data (from the testing dataset) can be evaluated against the model to check its accuracy.

Two cluster analyses are presented in this chapter, namely, centroid-based clustering and density-based clustering, both of which are partitioning-based clustering methods, where the data is partitioned into a number of clusters. Centroid-based clustering uses the centroid of each cluster to pull its members to the cluster. As a result, each cluster has members around its centroid. A Voronoi diagram can be used to easily form centroid-based clusters. Density-based clustering uses the concept of neighbours and neighbourhood. As a result, a cluster may form a long chain of neighbours, where one neighbour is related to another neighbour within the cluster.

A regression tree is used to illustrate the classification for data warehousing. It partitions the data into grids, and the grids describe the rules of the regression tree. So, there are similarities between regression tree grid-based partitioning and partitioning-based clustering, such as centroid-based or density-based. However, regression tree grid-based partitioning uses a different similarity measure, as the similarity among objects in each grid uses the value of the target class. In contrast, cluster analysis does not have a target class.

21.7 Exercises

21.1 Figures 21.17 and 21.18 show the simple linear regression models for Glucose-Albumin and Age-WBC (White Blood Cell).

The Glucose-Albumin data consists of 17 data points, as previously shown in Table 21.6, whereas the Age-WBC data consists of 49 data points, in the format of (Age, WBC), which are as follows:

{(74, 27), (56.3, 8.6), (55.9, 8.5), (80.4, 9.2), (56.8, 7.8), (29.8, 18.2), (19.1, 12.2), (32.7, 19), (28.3, 37.3), (73.8, 41.5), (62.7, 30.2), (25.7, 16.6), (66, 17.6), (64.9, 16), (37.2, 16.1), (94.6, 4.7), (39.7, 8.4), (29.9, 9.7), (48, 5.4), (42, 24.4), (71.5, 19.5), (47.4, 18.3), (24.9, 17.7), (50.8, 6), (45.3, 20.9), (76.7, 2.5), (55.7, 12.1), (86.8, 7.3), (83.5, 8.4), (84.4, 18), (18.7, 26), (85, 9.6), (57.3, 9.8), (80.6, 6.9), (65.8, 23.2), (67.9, 14.8), (84.9, 9.2), (67.1, 6.8), (51.1., 21.9), (88.6, 8.4), (29.8, 10), (65, 14.9), (74, 11), (78.9, 16.7), (53.4, 6), (86, 6.5), (71.8, 14.5), (54.7, 12.3), (90.9, 21.9)}.

Using the SQL commands, calculate the intercept and slope for both datasets. To check your solutions, the intercept and slope for the Glucose-Albumin are 2.4959 and 0.0077, respectively. The intercept and slope for the Age-WBC are 20.1455 and 0.0918, respectively.

21.2 Given dataset D with 20 data points, $D = \{5, 19, 25, 21, 4, 1, 17, 23, 8, 7, 6, 10, 2, 20, 14, 11, 27, 9, 3, 16\}$ and $k = 3$, run through step by step the k-means

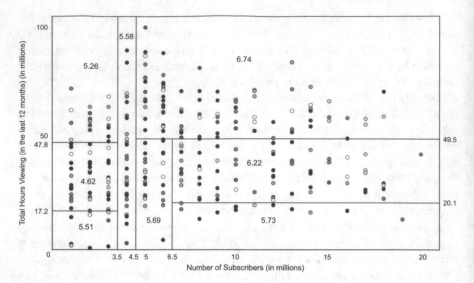

Fig. 21.56 YouTube Influencers

algorithm to produce the desired clusters. Assume that the initial centroids chosen from D are $m_1 = 8, m_2 = 7$ and $m_3 = 6$, which are the three data points in the middle of dataset D.

21.3 Figure 21.56 shows a map partitioning of YouTube Influencers. The x-axis is the number of subscribers, whereas the y-axis is the total hours of viewing in the last 12 months. Each dot on the map represents a YouTuber, and his/her income is represented by the gradient colour of the dot, where the darker the colour the higher the income of that particular YouTuber. The number inside each grid (e.g. 5.26 in the top left-hand corner) represents the income of the YouTubers in that category. The income is written in an e^x format, where 5.26 in this example is the x value. The income of $e^{5.26}$ is roughly equivalent to \$192,000.

Draw the regression tree based on this map partitioning.

21.8   Further Readings

Data mining is mature a discipline which has produced a wealth of techniques and methods. There are excellent textbooks on data mining for further reading in this area: [1–7]. The parallel processing of data mining techniques is described in here: [8].

Some basic business statistics methods, such as regression, etc., can be found in many textbooks on business statistics. Some of them are listed here: [9–11].

Data analytics is often taught as a general course on data analysis but often covers data mining methods. Further readings in data analytics are as follows: [12, 13].

References

1. I.H. Witten, E. Frank, M.A. Hall, C.J. Pal, *Data mining: practical machine learning tools and techniques*, in *The Morgan Kaufmann Series in Data Management Systems* (Elsevier, Amsterdam, 2016)
2. J. Han, J. Pei, M. Kamber, Data mining: concepts and techniques, in *The Morgan Kaufmann Series in Data Management Systems* (Elsevier, Amsterdam, 2011)
3. N. Ye, Data mining: theories, algorithms, and examples, in *Human Factors and Ergonomics* (CRC Press, New York, 2013)
4. X. Wu, V. Kumar, The top ten algorithms in data mining, in *Chapman & Hall/CRC Data Mining and Knowledge Discovery Series* (CRC Press, New York, 2009)
5. D.J. Hand, H. Mannila, P. Smyth, Principles of data mining, in *A Bradford book* (Bradford Book, New York, 2001)
6. M. Bramer, *Principles of Data Mining*, 4th edn. Undergraduate Topics in Computer Science (Springer, Berlin, 2020)
7. K.J. Cios, W. Pedrycz, R.W. Swiniarski, L.A. Kurgan, *Data Mining: A Knowledge Discovery Approach* (Springer, New York, 2007)
8. D. Taniar, C.H.C. Leung, W. Rahayu, S. Goel, *High Performance Parallel Database Processing and Grid Databases* (Wiley, New York, 2008)
9. D.M. McEvoy, *A Guide to Business Statistics* (Wiley, New York, 2018)
10. N. Bajpai, *Business Statistics* (Pearson, London, 2009)
11. M. Berenson, D. Levine, K.A. Szabat, T.C. Krehbiel, *Basic Business Statistics: Concepts and Applications* (Pearson Higher Education AU, London, 2012)
12. J. Moreira, A. Carvalho, T. Horvath, *A General Introduction to Data Analytics* (Wiley, New York, 2018)
13. T.A. Runkler, *Data Analytics: Models and Algorithms for Intelligent Data Analysis.* (Springer, Bücher, 2012)

Printed in the United States
by Baker & Taylor Publisher Services